LINEAR ALGEBRA
Gateway to Mathematics

Robert Messer

Albion College

HarperCollins*CollegePublishers*

Sponsoring Editor: George Duda
Project Management: Publication Services
Cover Designer: Zena Scarpulla
Cover/Photo: 1990 Rudi V. Starrex/Photonica
Production Administrator: Randee Wire
Printer and Binder: R.R. Donnelley & Sons Company
Cover Printer: R.R. Donnelley & Sons Company

Linear Algebra: Gateway to Mathematics

Library of Congress Cataloging-in-Publication Data
Messer, Robert.
    Linear algebra : gateway to mathematics / Robert Messer.
        p.   cm.
    Includes index.
    ISBN 0-06-501728-5
    1. Algebras, Linear.   I. Title.
QA184.M47   1993
512'.5–dc20   93-21043
                                                              CIP

9 8 7 6 5 4 3 2 1

To Pat and Mark

Preface

Linear algebra is a central course in many undergraduate mathematics programs. It prepares students for the demands of physical chemistry, relativity, quantum mechanics, and mathematical models in biology, economics, and other quantitative areas of study. At the same time, it must serve as a transition from the mechanical manipulations of algebra and calculus to more theoretical upper-level mathematics courses.

Expectations

Students should enter the course with the equivalent of two or three semesters of college-level mathematics (typically calculus with perhaps an introduction to differential equations and some work with vectors in the plane and three-dimensional Euclidean space). They will leave with an understanding of the basic results of linear algebra and an appreciation of the beauty and utility of mathematics. They will also be fortified with the mathematical maturity required for subsequent courses in abstract algebra, real analysis, and elementary topology. Students who have additional background in dealing with the mechanical operations of vectors and matrices will benefit from seeing this material placed in a more general context.

Objectives

This textbook is designed to resolve the conflict between the abstractions of linear algebra and the needs and abilities of the students who may have dealt only briefly with the theoretical aspects of previous mathematics courses.

Students will repeatedly encounter discussions of the advantages of dealing with the general theory. The text points out how the similarities in numerous examples can be understood and clarified under a unifying concept, and how a general theorem reduces the need for computation. At the same time, a careful attempt is made to distinguish between the abstract notions and the particular examples to which they apply. Applications of mathematics are integrated into the text to reinforce the theoretical material and to illustrate its usefulness.

Another theme of this book, and perhaps the most important one, is the recognition that many students will at first feel uncomfortable or at least unfamiliar with the theoretical nature inherent in many of the topics of linear algebra. There are numerous discussions of the logical structure of proofs and the need to translate terminology into notation, and suggestions about efficient ways to discover a proof. Frequent connections are made with familiar properties of algebra, geometry, and calculus. Although complex numbers make a brief appearance as required for the discussion of the roots of the characteristic polynomial of a matrix, real numbers are consistently used as scalars. This strengthens the ties between linear algebra and geometry and sets the stage for the introduction of other fields of scalars in an advanced course in linear algebra.

This book combines the many simple and beautiful results of elementary linear algebra with some powerful computational techniques to demonstrate that theoretical mathematics need not be difficult, mysterious, or useless.

Organization

The first chapter introduces the fundamental concepts of vector spaces. Mathematical notation and proof techniques are carefully discussed in the course of deriving the basic algebraic properties of vector spaces. This work is immediately tied to three families of concrete examples. With this background, students can appreciate the matrix reduction technique discussed in the second chapter as a powerful tool rather than as the main focal point of the course.

Chapter 3 continues the study of vector spaces, culminating in results about the dimension of a vector space. Chapter 4 then introduces the structures of inner products and norms to quantify some of the geometric concepts discussed previously.

Chapter 5 deals with matrix multiplication and inverses. An introduction to Markov chains provides a fascinating opportunity to examine some nontrivial applications of these results. Chapter 6 relates this material to the general concept of linear functions between vector spaces.

Chapter 7 begins with a careful presentation of mathematical induction. An inductive definition leads to the straightforward although somewhat computational presentation of the theory of determinants. Enough guidance is given so that the student will not become lost or begin to feel that everything is as complicated as the most intricate detail. The final chapter, on eigenvalues and eigenvectors, ties together many of the major topics of the course.

Pedagogical Features

Several distinctive features assist students as they study for the course, work on assignments, and review for examinations. Instructors will also find these features helpful in preparing lectures and assignments.

Mathematical strategy sessions. These discussions will nurture students in their understanding of definitions, use of proof techniques, and familiarity with mathematical notation.

Crossroads. Students will appreciate these indications of how topics fit together, where ideas will be used in futures sections, and how a concept unifies material from previous courses. Suggestions are also given for related readings and explorations.

Quick examples. These are concise examples and typical problems making direct use of the material in the text. Students can expect to find examples similar to these on homework assignments and examinations.

Exercises. Each section contains a variety of exercises. Some are routine drill problems, but the emphasis is on honest problems (at a range of levels of difficulty) that illustrate the concepts presented in the section or their development in relation to other topics. Occasional open-ended problems encourage students to develop their mathematical creativity. Exercises often set the stage for topics appearing in later sections. Pay particular attention to exercises whose numbers are boxed; they are referred to at some other point in the text.

Answers to approximately one-third of the exercises appear at the end of the book. Students can check their work against the answers provided for this representative sample of the exercises. Complete solutions are presented for typical exercises that demonstrate important concepts or are particularly instructive.

Projects.   Each chapter contains suggested projects for student investigation. Some projects are in the form of a guided tour with interesting side issues for students to think about. Others are mere sketches with suggested references and an invitation to explore the resources available in the mathematics section of the library. These projects may be used for independent investigation or for a group of students to work on cooperatively. Students who want to focus on a portion of one of these projects will nevertheless see the larger context for their work. These sections may also be used as reading assignments to make students aware of the role of linear algebra in mathematics and its application to other disciplines. This may plant the seeds for more ambitious investigations in a senior seminar or undergraduate research project.

References for related material.   Additional references are provided for students interested in pursuing topics related to the course or previewing more advanced topics.

Chapter summaries.   A concise overview of the accomplishments of each chapter will reinforce the theme of placing details in a larger context. A checklist of concepts reorganizes the material from the chapter in the categories of computation, theory, and application. This will aid students in studying for exams without converting the course into a list of terms and techniques to be memorized.

Chapter review exercises.   Additional sets of problems recall the highlights of each chapter. These exercises are again at a variety of levels. They often bring together topics from more than one section. This encourages students to see the development of the material from a broad perspective rather than filing each section in a separate mental compartment.

Supplements.   LINALG is a tool kit which has a "pull down" menu format and context-sensitive help. LINALG is a linear algebra program for use by instructors for classroom demonstrating as well as by students for exploring and solving problems. It is available in IBM format and can be obtained by writing to the Department of Mathematics at the University of Arizona.

Our **Instructor's Manual** will be available to adopters of this text. It can be obtained by contacting your local HarperCollins College Publishers representative. It will contain answers to exercises not found at the back of the book, suggestions or ideas to bring out in lectures, classroom activities, sample exams, and a discussion of the use of software packages in teaching linear algebra from this text.

Acknowledgments

This book began as an experiment in on-site publishing. The original version was written, edited, typeset, and tested at Albion College. It has undergone numerous revisions in response to helpful comments from many people. George Duda, editor at Harper-Collins, has contributed wisdom and guidance in adding magic to transform the original manuscript into a distinctive textbook. It was a pleasure to work with Shirley Cicero,

Carol Zombo, James Harris, and Shelley Clubb. David Mason did an amazing job of removing stylistic and grammatical irregularities from the manuscript. The reviewers, listed below, made valuable suggestions for exercises, examples, clarifications, and other improvements.

John F. Cavalier	West Virginia Institute of Technology
Stuart Goff	Keene State College
Sydney Graham	Michigan Technological University
Benny P. Lo	Ohlone College
Roger Lutz	Western New Mexico University
Andy Magid	University of Oklahoma
Ana Mantilla	University of Arizona
Donald Passman	University of Wisconsin-Madison
Patrick Stewart	Dalhousie University

Colleagues Paul Anderson, Ruth Farro, John Fink, Raymond Greenwell, William Higgins, Lisa Holden, Norman Loomer, and Eric Wilson checked over the solutions to the exercises. The following students, professors, and other readers have made suggestions for improvements in preliminary editions of this book. Their help is gratefully acknowledged.

Albion College: Geoffrey Armstrong, Thomas Basso, Charmian Bultema, Josh Cassada, Sascha Chin, Matthew Chittle, Sue Colton, Douglas Copley, Kenneth Davis, Michelle Diener, Erik Eid, Paul Enns, Penny Eveningred, Joe Gibson, Jafar Hasan, John Hill, Diane Hines, David Holden, Jacob Hooker, Michelle Hribar, Michael Juchno, Michael Kidder, Phillip Koppers, David Kutcipal, Adam LaPratt, Kevin Lepard, Martin Ludington, Vincent Magnotta, Steven Malinak, John Mark Meldrim, Pat Messer, Joseph Meyers, Renée Miller, Debra Neumeyer, Martha O'Kennon, David Oliver, Andrea Ondracek, Amanda Parke, Gregory Parker, Michael Parr, Sarah Paukstis, John Peternel, Polly Reeder, Christopher Richardson, Karl Schwartz, Bryan Sladek, Wendy Stretch, Ronald Targosz, Alex Tashian, Linda Valachovic, Greg Wallender, Robert Wells, Aaron Werbling, Charles Yun

Indiana State University: Steven Edwards

Ripon College: Norman Loomer

Spring Arbor College: Charles Carey, Katie Marema, Benjamin Rohrer

Wittenberg University: William Higgins, Eric Wilson

Your comments about any aspect of this book are welcome. The names will be acknowledged of those whose comments or suggestions are incorporated in future editions. You may contact the author at the address below.

Robert Messer
Department of Mathematics
Albion College
Albion, Michigan 49224
e-mail: RAM@ALBION.BITNET

Contents

Chapter 4: Inner Product Spaces

Chapter 5: Matrices

Chapter 6: Linearity

Chapter 7: Determinants

Chapter 8: Eigenvalues and Eigenvectors

Vector Spaces

Welcome to the study of linear algebra. Vector spaces are the central concept of this course. So after you have had a brief chance to brush up on the basics of sets and logic, the definition of vector space will be presented in the second section. By looking at vector spaces from a mathematical point of view, you will be able to appreciate the common features of three important families of vector spaces: Euclidean spaces, spaces of matrices, and spaces of functions. This approach will reveal the applications of vector spaces to geometry, tabulation of data, and calculus as a unified subject rather than a bag of mathematical tricks.

1.1 Sets and Logic

In any encounter with mathematics, you are likely to hear about sets. Think back to your algebra classes in high school. You searched for solution sets to various types of equations, and you considered systems, or sets, of equations. Sets appeared everywhere in geometry: sets of points in the plane forming circles, lines, triangles, and quadrilaterals in many varieties (you would need a set just to contain all the terminology), as well as spheres, cylinders, cones, cubes and other polyhedra in three-dimensional space.

Whether your calculus course was a pump or a filter, you undoubtedly looked at various sets associated with a function (domain, range, image, zeros), sets of points defining a partition of an interval for a Riemann sum, and the interval of convergence (a set of points on the number line) of a power series. The graph of a function is nothing more than the set of pairs of x-coordinates and y-coordinates that satisfy the defining relation of the function. You developed techniques for locating interesting points (critical points and points of inflection) on such sets. You found ways to calculate the area of a set bounded by graphs of a variety of common functions.

Sets are also of fundamental importance in linear algebra. They form the underlying fabric of vector spaces, the primary object of our study. Furthermore, the notation and terminology of sets will give us a language to describe the ideas and results of linear algebra.

The notion of a set is fundamental to modern mathematics in another sense. It is the basic idea from which other mathematical terms are defined. Thus, we cannot define the concept of set in terms of more elementary concepts. One way out of this embarrassing situation is to give a system of axioms for set theory and hope that the axioms can be judiciously chosen to capture all the essential properties our intuition would like sets to have. This is comparable to the approach we will take with vector spaces. As interesting as this might be, axiomatic set theory would require a separate course covering material that sheds relatively little light on linear algebra. An alternative approach is to find a few synonyms for the word *set* and to make an agreement among friends as to the use of this concept and some associated notation.

A **set**, then, is to be thought of as a collection (or a family, aggregate, or ensemble) of elements (or objects, or points). We will usually be dealing with sets whose elements are mathematical objects, such as numbers, geometric points, or vectors (whatever they are), rather than physical objects such as people, trees, or photons (whatever they are).

We will denote the fact that some object x is an **element** (or a member) of a set S by writing $x \in S$. For example, if E denotes the set of positive even integers, then $2 \in E$.

If a set contains only a few elements, it may be convenient to describe the set by listing its elements, for example:

$$S = \{2, 3, 5, 7\},$$

or, for a more arbitrary set:

$$X = \{x_1, x_2, \ldots, x_n\}.$$

A set with a large number (or even an infinite number) of elements can best be described by giving some rule to determine which objects are members and which are not. This can be done informally, as the set E was described earlier, or by listing enough elements to establish an obvious pattern:

$$E = \{2, 4, 6, 8, \ldots\}.$$

Another common and very useful notation is to give a rule that determines whether any particular element of some other, known set is to be an element of the set being defined. For example, if $\mathbb{N}$ denotes the set of positive integers, we can write

$$E = \{x \in \mathbb{N} \mid x \text{ is divisible by } 2\}.$$

Mathematical Strategy Session   Mathematical gadgets are frequently described in terms of their relation to other objects or by the operations we can perform on them. This operational approach is very precise in specifying exactly how the gadget works, but it often leaves us wondering if we would recognize one in a dark alley. Let us consider the situation with sets.

A set is described in relation to the elements it contains. Given a set and an element, we can decide whether or not the element is in the set. This operational view of the set is the extent to which we can know the set. Although this may appear vague, it actually clears up several questions about sets.

First of all, the order in which elements are listed in a set is irrelevant to the set. Thus, $\{2, 3, 5, 7\}$ and $\{3, 7, 2, 5\}$ contain the four prime numbers less than 10 (and no other elements); hence, they are the same set. Similarly, if an element is repeated in the list, its containment in the set is the same as if it appears only once. Thus, $\{2, 2, 3, 3, 3, 5, 5, 5, 5, 5, 7, 7, 7, 7, 7, 7, 7\}$ is merely an extravagant way to denote the same set of four primes.

Suppose S and T are two sets of similar kinds of objects (for example, sets of real numbers or sets of points in a plane). Often we want to combine the elements of S with the elements of T to form a new set known as the **union** of S and T. The union of S and T is denoted $S \cup T$. It can be defined in terms of the logical connective *or* as follows:

$$S \cup T = \{x \mid x \in S \text{ or } x \in T\}.$$

Recall that mathematicians use the word *or* in the inclusive sense. Thus, $S \cup T$ consists of the elements that are in S or in T, including those in both S and T.

In a similar fashion we can form a new set composed of the elements common to S and T. This set, the **intersection** of S and T, is denoted $S \cap T$. It can be defined in terms of the logical connective *and* as follows:

$$S \cap T = \{x \mid x \in S \text{ and } x \in T\}.$$

Along with the union and intersection operations on sets, there is another relation among sets that is important in mathematics. If all the elements of one set S turn out to be elements of another set T, then S is a **subset** of T. We write $S \subseteq T$ to denote this containment relation.

Mathematical Strategy Session The intuitive idea of set containment is simple: one set is part of another set. For example, the interval $[2, 3]$ on the number line is a subset of the interval $[0, 4]$. In order to deal with set containment in situations where intuition fails, it is useful to analyze this concept as a logical implication. To say that two sets S and T satisfy the containment relation $S \subseteq T$ means

$$\text{if } x \in S, \text{ then } x \in T.$$

That is, we are given the hypothesis that an element x is in the set S, and from this assumption we must derive the conclusion that the same element x is in the set T. We can write this implication more concisely with the symbols

$$x \in S \quad \Longrightarrow \quad x \in T,$$

where the arrow denotes logical implication.

Crossroads   The same logical process of deriving consequences from stated assumptions is involved whether you are analyzing a novel, interpreting a legal principle, or applying a law of physics. In mathematics, furthermore, the concept of logical implication is almost a way of life. Before we go any further, let us examine this concept a little more closely.

An implication relates two statements. We refer to one statement as the assumption, the hypothesis, or the antecedent of the implication. It is the given condition we can use in a proof with no further justification. The other statement is called the result, the conclusion, or the consequent. It is the condition we are trying to derive.

Be careful not to confuse the conclusion of an implication with the entire implication (involving the hypothesis as well as the conclusion). It is possible for the implication to be true even if the conclusion is false. Here is a table that gives the truth value of an implication $P \Longrightarrow Q$ in terms of the four combinations of truth values for the hypothesis P and the conclusion Q:

P	Q	$P \Longrightarrow Q$
False	False	True
False	True	True
True	False	False
True	True	True

That is, an implication is true unless it permits a false conclusion to follow from a true hypothesis.

You may be surprised that a false hypothesis is allowed in a true implication. Moreover, if the hypothesis is always false, the first two rows of the truth table show that the implication is automatically true. Actually, the conventions in the truth table for implication are useful in mathematics and quite natural in everyday language. In algebra, for example, we want the implication

$$x > 2 \quad \Longrightarrow \quad x^2 > 4$$

to be true for any real number x. In particular, we must agree that the implication is true in the following three situations:

Value of x	*Hypothesis* ($x > 2$)	*Conclusion* ($x^2 > 4$)
-3	False	True
-1	False	False
3	True	True

Similarly, in everyday conversation we accept the statement "If it is raining, then the streets are wet" without checking the current weather conditions.

If you are curious about alternative truth tables for implication, you may wish to read the article "Material Implication Revisited" in the March 1989 issue of *American Mathematical Monthly*. The author, Joseph Fulda, discusses the problems that arise with alternative views of implication.

Mathematical Strategy Session   Here is how you can use logical implication to give a detailed proof of set containment $[2, 3] \subseteq [0, 4]$. We start by assuming a real number x is in the interval $[2, 3]$. We write what this means in terms of inequalities and use properties of inequalities to string together a sequence of implications. Our goal is to reach the desired conclusion that x is in the interval $[0, 4]$. The result of all this will constitute a proof that the first statement $x \in [2, 3]$ implies the last statement $x \in [0, 4]$; in other words, that $[2, 3] \subseteq [0, 4]$. This is more complicated to describe than it is to do:

$$x \in [2, 3] \implies 2 \leq x \leq 3$$
$$\implies 0 \leq 2 \text{ and } 2 \leq x \text{ and } x \leq 3 \text{ and } 3 \leq 4$$
$$\implies 0 \leq x \text{ and } x \leq 4$$
$$\implies 0 \leq x \leq 4$$
$$\implies x \in [0, 4].$$

Several sets are so important that mathematicians use special symbols to denote them. You have already encountered the set of positive integers $\mathbb{N}$ in an example earlier in this section. The special symbol $\mathbb{R}$ denotes the set of real numbers (numbers that can be expressed in ordinary decimal notation: positive, negative, zero, rational, or irrational). Originally, the set of natural numbers and the set of real numbers were denoted by symbols in boldface type. Because boldface chalk is not readily available, professors often wrote these special symbols on the chalkboard with a double stroke. This convention has made its way into print.

One particularly simple set appears with annoying regularity. This is the empty set, denoted $\emptyset$. It is the set with no elements; that is, $\emptyset = \{\ \}$.

Quick Example   *Suppose S is any set. Prove that* $\emptyset \subseteq S$

The containment $\emptyset \subseteq S$ translates into the logical implication

$$x \in \emptyset \implies x \in S.$$

Because the empty set has no elements, $x \in \emptyset$ is always false, no matter what the element x is. The truth table for $P \implies Q$ shows that an implication with a false hypothesis is always true. The truth of this implication is what we need to prove $\emptyset \subseteq S$. ∎

The operations of union and intersection on sets and the relations of equality and containment among sets satisfy many laws of a rather algebraic nature. For example, for two sets S and T the law

$$S \cap T = T \cap S$$

is reminiscent of the commutative laws of addition and multiplication of real numbers. We will also encounter a commutative law for addition of vectors. All these commutative laws are similar in form even though the objects under consideration and the operations used to combine the objects differ in the various contexts.

Mathematical Strategy Session   Often in mathematics you can abstract enlightening analogies from constructions that superficially appear quite different. Look for such underlying similarities whenever you explore new topics in mathematics. This will reassure you that the new material is not totally alien. More important, they will guide you in asking and answering questions in the new situation.

Let us examine two standard techniques for proving that two sets are equal. The first technique involves manipulating the expression on one side of the equation into the form on the other side. In particular, we will start with one side of the equation and string together a chain of expressions each of which is equal to the previous expression. The goal is to end the chain with the expression on the other side of the equation. You probably recall using this method to verify identities in your trigonometry course. Indeed, it is a very general technique that you should consider anytime you need to prove that two expressions denote the same quantity.

Quick Example   *Prove that $S \cap T = T \cap S$ for sets S and T.*

Begin by reformulating the expression on the left side of the equality in terms of the logical connective *and* and repeatedly rewriting the set $S \cap T$ so as to transform it into $T \cap S$:

$$S \cap T = \{x \mid x \in S \text{ and } x \in T\}$$
$$= \{x \mid x \in T \text{ and } x \in S\}$$
$$= T \cap S.$$

The key here is that the truth of the conjunction of the condition $x \in S$ with the condition $x \in T$ does not depend on which is listed first.   ∎

Another technique for proving equality is more specific to the context of sets. It involves showing that every element in one set is an element of the second and, conversely, that every element of the second set is an element of the first. Because a set is completely described by the elements it contains, these two steps will establish that the sets are identical. When you encounter difficulties in forging a chain of equalities to connect the right and left sides of an equation involving sets, you should try this proof technique. It has the advantage of breaking the proof up into two simpler parts.

Quick Example *Prove that $S \cap \emptyset = \emptyset$ for any set S.*

Earlier in this section we proved that the empty set is a subset of any set. In particular, $\emptyset \subseteq S \cap \emptyset$. Let us create a string of implications to prove the reverse containment, $S \cap \emptyset \subseteq \emptyset$.

$$x \in S \cap \emptyset \implies x \in S \text{ and } x \in \emptyset$$
$$\implies x \in \emptyset.$$

The key here is that if two conditions hold (here we have the condition $x \in S$ connected with the word *and* to the condition $x \in \emptyset$), then in particular the second condition holds.

Exercises 1.1

1. Suppose S and T are sets. Prove the following properties.

   a. $S \cap T \subseteq S$

   b. $S \cap T \subseteq T$

   c. $S \subseteq S \cup T$

   d. $T \subseteq S \cup T$

2. Suppose R, S, and T are sets. Prove the following properties.

   a. $S \cup T = T \cup S$

   b. $(R \cup S) \cup T = R \cup (S \cup T)$

   c. $S \cup \emptyset = S$

3. Investigate what happens when the identities in Exercise 2 are rewritten with the intersection symbol in place of the union symbol.

4. In Exercises 2 and 3 you may have noticed the nice analogy between the operations of union and intersection on sets and the operations of addition and multiplication on numbers.

   a. What role does the empty set play in this analogy?

   b. Try to reinterpret some basic algebraic formulas (such as $a \cdot (b + c) = a \cdot b + a \cdot c$ and $(a + b)^2 = a^2 + 2ab + b^2$) in terms of set-theoretical identities.

   c. Which of these formulas are true for sets? Can you prove them or provide counterexamples?

5. There is an analogy between the containment relation $\subseteq$ in set theory and the order relation $\leq$ among real numbers. You are undoubtedly familiar with order properties that hold for all real numbers x, y, and z, such as:

   $$x \leq x.$$
   If $x \leq y$ and $y \leq x$, then $x = y.$
   If $x \leq y$ and $y \leq z$, then $x \leq z.$
   Either $x \leq y$ or $y \leq x.$

a. Replace x, y, and z by sets, and replace $\leq$ by $\subseteq$. Which statements remain true? Provide a proof or a counterexample for each.

b. What happens with properties such as the following implication?

$$\text{if } x \leq y, \text{ then } x + z \leq y + z.$$

c. Investigate other statements about sets that are derived by analogy from other basic facts about real numbers.

6. Why are analogies important in mathematics?

7. **a.** Consider a set, such as $\{0\}$, with one element. (Be careful not to confuse $\{0\}$ with the empty set $\varnothing = \{\}$, which has zero elements.) Show that there are two subsets of this set.

b. How many subsets does a two-element set, such as $\{0, 1\}$, have?

c. How many subsets does a three-element set have?

d. Suppose you knew how many subsets a set with k elements has. Let s denote this number of subsets. Now enlarge the set by including a new element. In terms of s, how many subsets does this new set (with $k + 1$ elements) have?

e. Based on your results, find a simple formula for s as a function of k. Does your formula work for the empty set (where $k = 0$)?

8. Is it possible to add two sets? For sets of real numbers, we might define addition of sets in terms of addition of the elements in the sets. Let us introduce the following meaning to the symbol $\oplus$ for adding a set A of real numbers to another set B of real numbers:

$$A \oplus B = \{a + b \mid a \in A \text{ and } b \in B\}.$$

a. List the elements in the set $\{1, 2, 3\} \oplus \{5, 10\}$.

b. List the elements in the set $\{1, 2, 3\} \oplus \{5, 6\}$.

c. List the elements in the set $\{1, 2, 3\} \oplus \varnothing$.

d. If a set A contains m real numbers and a set B contains n real numbers, can you predict the number of elements in $A \oplus B$? If you run into difficulties, can you determine the minimum and maximum numbers of elements possible in $A \oplus B$?

e. Does the commutative law hold for this new addition? That is, does $A \oplus B = B \oplus A$?

f. Reformulate other laws of real-number addition in terms of this new addition of sets. Which of your formulas are true? Can you prove them or provide counterexamples?

g. What about laws that combine set addition with union and intersection? For example, does $(A \cup B) \oplus C = (A \oplus C) \cup (B \oplus C)$?

h. Is there any hope of extending the other operations of arithmetic to sets of real numbers? What about algebra? Limits? Power series?

1.2 Basic Definitions

In elementary physics courses a vector is often defined to be a quantity that has magnitude and direction. Primary examples are such quantities as velocity, force, and momentum. The essential properties of these vectors are that they can be meaningfully added together and multiplied by real numbers, called scalars in this context. For example, two forces acting on an object yield a resultant force; also, a force on an object can be doubled in magnitude without changing its direction.

Another common interpretation is to represent a vector as an arrow or a directed line segment in the plane. Again, there is a rule for adding arrows and a rule for multiplying them by real numbers. We will consider these rules in Section 1.9.

The goal of this chapter is to develop a notion of vectors that encompasses not only these physical and geometric interpretations, but also the many mathematical applications that arise in advanced calculus and differential equations, for example. In each of these interpretations, it is a mistake to consider a single vector in isolation. Of fundamental importance, rather, is the relation among all the vectors under consideration, and especially the algebraic operations of adding a pair of vectors and multiplying a vector by a real number.

We begin, then, with a discussion about a set, or space, of vectors. The vectors are thus elements of such a vector space. We require that there be some way of adding two vectors to produce another vector and some way of multiplying a vector by a real number to produce another vector. Just as it is far too early to specify the actual objects that can form vector spaces, it is far too early to specify exactly what is meant here by addition and multiplication. However, based on experience and tradition, it is customary to consider only those notions of addition and multiplication that obey the eight rules, or axioms, listed in the following definition.

As you read through these axioms, consider their application when you interpret vectors as physical quantities (velocity, force, or momentum) and as geometric objects (directed line segments in the plane or in three-dimensional space). Also notice that when real numbers are used for vectors as well as for scalars, these axioms state familiar laws of addition and multiplication of ordinary arithmetic.

Definition 1.1

A **vector space** is a set V with operations of addition and scalar multiplication. The elements of V are called **vectors.** The operation of **addition** combines any two vectors $\mathbf{v} \in V$ and $\mathbf{w} \in V$ to produce another vector in V denoted $\mathbf{v} + \mathbf{w}$. The operation of **scalar multiplication** combines any real number $r \in \mathbb{R}$ and any vector $\mathbf{v} \in V$ to produce another vector in V denoted $r\mathbf{v}$. A real number used in this operation is called a **scalar.** These operations must satisfy the following eight axioms for all $\mathbf{v}, \mathbf{w}, \mathbf{x} \in V$ and $r, s \in \mathbb{R}$.

1. $\mathbf{v} + \mathbf{w} = \mathbf{w} + \mathbf{v}$ (commutative law of addition)
2. $(\mathbf{v} + \mathbf{w}) + \mathbf{x} = \mathbf{v} + (\mathbf{w} + \mathbf{x})$ (associative law of addition)

3. There is a vector in V,   (additive identity law)
   denoted **0**, such that
   $\mathbf{v} + \mathbf{0} = \mathbf{v}$.

4. For each vector **v** in V there   (additive inverse law)
   is a vector in V, denoted $-\mathbf{v}$,
   such that $\mathbf{v} + (-\mathbf{v}) = \mathbf{0}$.

5. $r(\mathbf{v} + \mathbf{w}) = r\mathbf{v} + r\mathbf{w}$ (distributive law)

6. $(r + s)\mathbf{v} = r\mathbf{v} + s\mathbf{v}$ (distributive law)

7. $r(s\mathbf{v}) = (rs)\mathbf{v}$ (associative law of multiplication)

8. $1\mathbf{v} = \mathbf{v}$ (scalar identity law)

Crossroads When we speak of a vector space based on this definition, we literally do not know what we are talking about. The point is that we are talking about a lot of possible things at the same time. You are invited to look ahead to see some of the examples of vector spaces that we will be dealing with. But when you return to this section, be very careful not to attach any significance beyond what is stated in the definition to the $+$ sign used between vectors, to the *additive identity* vector **0** mentioned in Axiom 3, or to the *additive inverse* $-\mathbf{v}$ of a vector **v** mentioned in Axiom 4. On the other hand, the $+$ sign between the real numbers in Axiom 6 denotes ordinary addition of real numbers; likewise, ordinary multiplication of real numbers is called for on the right side of the equation in Axiom 7; and the ordinary unit 1 of the real number system appears in Axiom 8.

Students sometimes feel uncomfortable when introduced to a new level of abstraction. This feeling is understandable. However, in a short time the axioms will seem like old friends, and vectors will be as familiar as constants and variables in ordinary algebra. You will soon be dealing with vector quantities in much the same way you automatically solve an equation $8x + 3 = 15$ without having to think of a concrete interpretation for each symbol. This facility will free your mind for problem solving at a higher level.

In the next few sections you will have ample opportunity to become familiar with the eight axioms of a vector space. Right now, let us concentrate on the requirement that the operations of addition and scalar multiplication always yield another element of the vector space. We say that a vector space is **closed** under the operations of addition and scalar multiplication. This is an appropriate term for indicating that we can never get out of the set by performing the specified operation.

You are undoubtedly aware of the closure of familiar number systems (natural numbers, integers, rational numbers, real numbers) under the operations of ordinary addition and multiplication. A perennial source of difficulty arises because the real numbers are not closed under division (a single exception, not being able to divide by zero, is the spoiler).

To develop your familiarity with the idea of closure, several exercises ask you to check various subsets of real numbers for closure under familiar operations of arithmetic. If you think the set is closed, you need to start with two arbitrary elements in the set. Introduce some notation for these two elements and spell out what it means for them to be in the set. From these hypotheses, show that the operation applied to the two elements yields another element in the set. On the other hand, if you think the set is not closed, you should present an explicit counterexample. List two elements in the set and show that when you apply the operation, the result is no longer in the set.

Quick Example *Determine whether the sets* [0, 1] *and* {0, 1, 2} *are closed under ordinary multiplication of real numbers.*

Let x and y denote arbitrary elements of [0, 1]. That means $0 \le x \le 1$ and $0 \le y \le 1$. Multiply the three expressions in $0 \le x \le 1$ by the positive number y to obtain $0 \le xy \le y$. Since we also know that $y \le 1$, we can conclude that $0 \le xy \le 1$. In other words, $xy \in [0, 1]$. Thus, [0, 1] is closed under multiplication.

The set {0, 1, 2} is not closed under multiplication. As an explicit counterexample, multiply 2, an element of the set, with itself to obtain 4, which is not in the set. ■

Exercises 1.2

1. Give examples of quantities that have direction and magnitude. Which of these can be combined in some form of addition and scalar multiplication? In each case, how would you determine whether the operations satisfy the vector space axioms?

2. Give some rules of ordinary real number arithmetic that do not have counterparts among the vector space axioms. Will any of these rules hold when interpreted in the context of a vector space?

3. Locate a text on abstract algebra and look up the axioms for the structure known as a group. How do these axioms compare with the axioms for a vector space?

4. Talk to your friends about their views of a number such as 15. Do they think of it in completely abstract terms for its role in arithmetic and algebra, do they envision three rows of five unspecified objects, or do they think in concrete terms of 15 pencils or other specific objects? Does the size of the number influence the responses? Do math students tend to think about numbers differently than students in other fields of study? How aware are people of the level of abstraction with which they view numbers? Did they reach this level of abstraction with a sudden flash of insight or by working with numbers over a period of time?

5. New symbolism should be handled with care. The vector space axioms involve two kinds of addition as well as two kinds of multiplication.

   a. Rewrite the eight axioms using $\oplus$ and $\odot$ for the vector space operations to distinguish them from the addition (denoted by $+$) and multiplication (denoted by juxtaposition) of real numbers.

    b. In an expression written without this distinction, explain how the context will always eliminate any confusion as to which addition and which multiplication are intended.

6.   Suppose r is a real number and $\mathbf{v}$ is a vector in a vector space.

    a. Explain how the use of the minus sign in the expression $(-r)\mathbf{v}$ differs from its use in the expression $r(-\mathbf{v})$.

    b. Why is $-r\mathbf{v}$ an ambiguous expression?

7.   Prove that the following subsets of $\mathbb{R}$ are closed under ordinary addition.

    a. $[5, \infty)$

    b. $\{3, 6, 9, 12, \dots \}$

    c. $\{0\}$

    d. $\varnothing$

8.   Which of the following subsets of $\mathbb{R}$ are closed under ordinary multiplication? In each case, prove that the set is closed or provide an explicit counterexample.

    a. $[5, \infty)$

    b. $[0, 1)$

    c. $(-1, 0)$

    d. $\{-1, 0, 1\}$

    e. $\{1, 2, 4, 8, 16, \dots \}$

9.   **a.** What is the smallest subset of $\mathbb{R}$ that contains $\frac{1}{2}$ and is closed under addition?

    b. What is the smallest subset of $\mathbb{R}$ that contains $\frac{1}{2}$ and is closed under multiplication?

1.3 Properties of Vector Spaces

If you look ahead a few sections, you will see an abundance of interesting and useful examples of vector spaces. Keep in mind that all vector spaces have certain features that arise as logical consequences of the eight axioms. In this section and the next we will look at some of these consequences. In any example some of these results will be obvious. One advantage of proving these results in the context of the general vector space is that we avoid having to worry about them every time we want to consider a new example of a vector space. These results follow from the eight axioms purely by logic, so whenever the eight axioms are satisfied these further results will automatically hold. By the end of this chapter, the number of general results magnified by the quantity of examples will amply demonstrate the efficiency of our approach to the study of vector spaces.

    Beyond this practical consideration, the axiomatic method is of fundamental importance for its power of unification. By studying a variety of examples in the context of a general theory, we are able to see underlying patterns in situations that superficially appear to have little in common. Only by such conceptual unification can hundreds of years of mathematical discoveries be successfully organized and comprehended in a course lasting only a few months.

Mathematical Strategy Session A **theorem** is a statement that has a proof. The statement of a theorem often takes the form of one condition (the **hypothesis**) implying some other condition (the **conclusion**). The purpose of the **proof** is to show that the conclusion, no matter how obviously true (or obviously false) it may seem, follows logically from the hypothesis under consideration. Theorems previously proved may, of course, be used in subsequent proofs.

In the following theorem, the first sentence establishes the general context of the theorem. The assumption that V is a vector space gives a set with operations that satisfy the eight axioms. The second statement is an implication that offers the additional hypothesis that $\mathbf{v}$ is an arbitrary element of V. With all this going for us, we need to show that the equality in the conclusion follows as a logical consequence.

Theorem 1.2

Suppose V is a vector space. If $\mathbf{v}$ is any element of V, then $\mathbf{0} + \mathbf{v} = \mathbf{v}$.

Proof This theorem is a slight modification of the additive identity axiom. Notice that the commutative law of addition allows us to interchange the order in which the vectors are added together. The plan, then, is to start with the vector $\mathbf{0} + \mathbf{v}$ and write different expressions for it as allowed by the axioms until we see that it is equal to $\mathbf{v}$.

$$\mathbf{0} + \mathbf{v} = \mathbf{v} + \mathbf{0}$$ *by the commutative law of addition, with $\mathbf{0}$ taking the place of the vector $\mathbf{v}$ in the axiom and $\mathbf{v}$ here taking the place of $\mathbf{w}$ in the axiom;*

$$= \mathbf{v}$$ *by the additive identity axiom.* ∎

Mathematical Strategy Session Often the proof of an equality is accomplished most easily and quickly by the technique illustrated in the proof of Theorem 1.2. Start with the more complicated side of the equality, make some modifications to produce a series of expressions, each equal to the preceding one (and hence to the original one), and end with the other side of the equality. The string of equalities will prove that the first expression equals the last.

The format of the following theorem is very common in mathematics. It is known as a *uniqueness* theorem because of the conclusion that only one object satisfies the stated condition. This theorem claims that the additive inverse of any vector is unique. See Exercise 5 at the end of this section for a companion result about the uniqueness of the additive identity.

Theorem 1.3

Suppose V is a vector space. For any $\mathbf{v} \in V$ there is only one vector $-\mathbf{v}$ in V that satisfies the additive inverse condition of Axiom 4.

Proof   Suppose we suspected that a vector $\mathbf{v}$ in a vector space had another additive inverse $\sim\mathbf{v}$ in addition to the regular one $-\mathbf{v}$. That is, $\mathbf{v} + (\sim\mathbf{v}) = \mathbf{0}$ as well as $\mathbf{v} + (-\mathbf{v}) = \mathbf{0}$. We want to show that $\sim\mathbf{v} = -\mathbf{v}$. Well, if we were dealing with real numbers, we could just subtract $\mathbf{v}$ from both sides of the equation $\mathbf{v} + (\sim\mathbf{v}) = \mathbf{0}$ and immediately obtain the conclusion. Let us try the same idea here, but with each step justified by one of the axioms.

We are given that

$$\mathbf{v} + (\sim\mathbf{v}) = \mathbf{0}.$$

Add $-\mathbf{v}$ to the left of both sides:

$$-\mathbf{v} + (\mathbf{v} + (\sim\mathbf{v})) = -\mathbf{v} + \mathbf{0}.$$

Use the commutative law of addition to regroup the terms on the left side, and use the additive identity axiom to simplify the right side to obtain

$$(-\mathbf{v} + \mathbf{v}) + (\sim\mathbf{v}) = -\mathbf{v}.$$

Then

$$
\begin{aligned}
(\mathbf{v} + (-\mathbf{v})) + (\sim\mathbf{v}) &= -\mathbf{v} && \textit{by the commutative law of addition;} \\
\mathbf{0} + (\sim\mathbf{v}) &= -\mathbf{v} && \textit{by the additive inverse axiom;} \\
\sim\mathbf{v} &= -\mathbf{v} && \textit{by Theorem 1.2.} \quad\blacksquare
\end{aligned}
$$

♟ Mathematical Strategy Session   The preceding proof of Theorem 1.3 illustrates another useful technique for proving equalities. Start with an equality known to be true, perform the same operation to both sides of the equation, modify the expression on either side of the equation, and in the end deduce the desired equality. Actually, now that we see the key idea of the proof (adding $-\mathbf{v}$ to both sides of the equation $\mathbf{v} + (\sim\mathbf{v}) = \mathbf{0}$), we can tidy things up and prove that $\sim\mathbf{v} = -\mathbf{v}$ by a straightforward string of equalities:

$$
\begin{aligned}
\sim\mathbf{v} &= \mathbf{0} + (\sim\mathbf{v}) \\
&= (\mathbf{v} + (-\mathbf{v})) + (\sim\mathbf{v})
\end{aligned}
$$

$$= (-\mathbf{v} + \mathbf{v}) + (\sim\mathbf{v})$$
$$= -\mathbf{v} + (\mathbf{v} + (\sim\mathbf{v}))$$
$$= -\mathbf{v} + \mathbf{0}$$
$$= -\mathbf{v}.$$

You should be able to supply the reason for each of these steps. Also, you should notice that everything that happened in this proof happened in the other proof as well.

You may wonder what justification there is for doing the same thing to both sides of an equality. After all, big troubles would arise if we were allowed to erase the exponents from both sides of the equality $(-2)^2 = 2^2$. The answer to this problem involves the meaning of equality and the principle of substitution. This principle says that when one of two equal quantities is substituted for the other in any expression, the resulting expression will be equal to the original one. For example, instead of adding $-\mathbf{v}$ to both sides of the equation $\mathbf{v} + (\sim\mathbf{v}) = \mathbf{0}$, we can think of substituting $\mathbf{0}$ for $\mathbf{v} + (\sim\mathbf{v})$ in the expression $-\mathbf{v} + (\mathbf{v} + (\sim\mathbf{v}))$ to obtain the equation $-\mathbf{v} + (\mathbf{v} + (\sim\mathbf{v})) = -\mathbf{v} + \mathbf{0}$. This principle can also be applied to the equation $(-2)^2 = 2^2$. We would use the square root function to try to erase the exponents. So the principle of substitution says that $\sqrt{2^2}$ is equal to $\sqrt{(-2)^2}$. Fortunately, however, $\sqrt{(-2)^2}$ is equal to $+2$ and not -2.

The next theorem gives a relation between the real number 0 and the zero vector $\mathbf{0}$ (the additive identity that is guaranteed to exist by Axiom 3). Two proofs are given. Be sure to notice the similarities.

Theorem 1.4

Suppose V is a vector space. If $\mathbf{v}$ is any element of V, then $0\mathbf{v} = \mathbf{0}$.

Proof
$$0\mathbf{v} = (0 + 0)\mathbf{v} = 0\mathbf{v} + 0\mathbf{v};$$
$$0\mathbf{v} + (-(0\mathbf{v})) = (0\mathbf{v} + 0\mathbf{v}) + (-(0\mathbf{v}));$$
$$\mathbf{0} = 0\mathbf{v} + (0\mathbf{v} + (-(0\mathbf{v})));$$
$$\mathbf{0} = 0\mathbf{v} + \mathbf{0};$$
$$\mathbf{0} = 0\mathbf{v}.$$

Alternative proof
$$0\mathbf{v} = 0\mathbf{v} + \mathbf{0}$$
$$= 0\mathbf{v} + (0\mathbf{v} + (-(0\mathbf{v})))$$
$$= (0\mathbf{v} + 0\mathbf{v}) + (-(0\mathbf{v}))$$

$$= (0 + 0)\mathbf{v} + (-0\mathbf{v}))$$
$$= 0\mathbf{v} + (-(0\mathbf{v}))$$
$$= \mathbf{0}. \quad \blacksquare$$

The final theorem in this section lists some useful consequences of the vector space axioms. In the following exercises you are asked to prove these statements. Keep track of which parts you have proved; you can use them in later proofs. Of course, you must refrain from using any of these statements prior to proving them.

Theorem 1.5

Suppose $\mathbf{v}$ is any element of a vector space V, and $r, s \in \mathbb{R}$. The following results hold.

    a. $-\mathbf{v} + \mathbf{v} = \mathbf{0}$

    b. $r\mathbf{0} = \mathbf{0}$

    c. If $r\mathbf{v} = \mathbf{0}$, then $r = 0$ or $\mathbf{v} = \mathbf{0}$.

    d. $(-1)\mathbf{v} = -\mathbf{v}$

    e. If $-\mathbf{v} = \mathbf{v}$, then $\mathbf{v} = \mathbf{0}$.

    f. $-(-\mathbf{v}) = \mathbf{v}$

    g. $(-r)\mathbf{v} = -(r\mathbf{v})$

    h. $r(-\mathbf{v}) = -(r\mathbf{v})$

    i. If $\mathbf{v} \neq \mathbf{0}$ and $r\mathbf{v} = s\mathbf{v}$, then $r = s$.

Exercises 1.3

In this set of exercises, $\mathbf{v}$, $\mathbf{w}$, $\mathbf{x}$, and $\mathbf{y}$ represent arbitrary vectors in a vector space V, and r and s represent arbitrary real numbers. Your proofs should show how the results can be derived from the axioms and previously proved results. Provide justification for any step that is not an obvious application of one of the axioms. Use enough parentheses to avoid ambiguous expressions.

  1.   Prove Theorem 1.5, part a: $-\mathbf{v} + \mathbf{v} = \mathbf{0}$.

  2.   Prove that $\mathbf{v} + \mathbf{v} = 2\mathbf{v}$.

  3.   Prove that $\frac{1}{2}\mathbf{v} + \frac{1}{2}\mathbf{v} = \mathbf{v}$.

  4.   **a.** Prove that $\mathbf{0} + \mathbf{0} = \mathbf{0}$.

        b. Prove that if the vector $\mathbf{v}$ satisfies $\mathbf{v} + \mathbf{v} = \mathbf{v}$, then $\mathbf{v} = \mathbf{0}$.

  5.   Prove that there is only one vector that satisfies the condition of the additive identity in Axiom 3.

  6.   Prove Theorem 1.5, part b: $r\mathbf{0} = \mathbf{0}$. In words, multiplying any real number r times the additive identity vector $\mathbf{0}$ yields the additive identity vector $\mathbf{0}$.

7. Prove Theorem 1.5, part c: if $r\mathbf{v} = \mathbf{0}$, then $r = 0$ or $\mathbf{v} = \mathbf{0}$. You may find it easier to prove the logically equivalent statement: if $r \neq 0$ and $r\mathbf{v} = \mathbf{0}$, then $\mathbf{v} = \mathbf{0}$.

8. Prove Theorem 1.5, part d: $(-1)\mathbf{v} = -\mathbf{v}$. You may want to use the strategy employed in one of the proofs of Theorem 1.3. Alternatively, just show that $\mathbf{v} + ((-1)\mathbf{v}) = \mathbf{0}$ and use Theorem 1.3 itself.

9. **a.** Prove that $-\mathbf{0} = \mathbf{0}$.

   b. Prove Theorem 1.5, part e: if $-\mathbf{v} = \mathbf{v}$, then $\mathbf{v} = \mathbf{0}$.

10. Prove Theorem 1.5, part f: $-(-\mathbf{v}) = \mathbf{v}$. Give a proof that uses Theorem 1.3 and another proof that uses Theorem 1.5, part d, and Axioms 7 and 8.

11. **a.** Prove Theorem 1.5, part g: $(-r)\mathbf{v} = -(r\mathbf{v})$.

    b. Prove Theorem 1.5, part h: $r(-\mathbf{v}) = -(r\mathbf{v})$.

12. Prove Theorem 1.5, part i: if $\mathbf{v} \neq \mathbf{0}$ and $r\mathbf{v} = s\mathbf{v}$, then $r = s$.

13. Prove that if a vector space has more than one element, then it has an infinite number of elements.

14. Prove that any two of the following expressions are equal:

$$\mathbf{v} + (\mathbf{w} + (\mathbf{x} + \mathbf{y}))$$
$$\mathbf{v} + ((\mathbf{w} + \mathbf{x}) + \mathbf{y})$$
$$(\mathbf{v} + \mathbf{w}) + (\mathbf{x} + \mathbf{y})$$
$$(\mathbf{v} + (\mathbf{w} + \mathbf{x})) + \mathbf{y}$$
$$((\mathbf{v} + \mathbf{w}) + \mathbf{x}) + \mathbf{y}$$

The point is that even though addition of vectors always combines exactly two vectors to produce a third, it doesn't really matter which pairs are added together first. Consequently, it is customary to eliminate the parentheses when dealing with sums of vectors. The vector represented by any one of the five expressions above will be denoted $\mathbf{v} + \mathbf{w} + \mathbf{x} + \mathbf{y}$. To show why this sloppiness is legitimate when an arbitrary number of vectors is to be summed would require an elaborate system for keeping track of parentheses. If you are familiar with proofs by induction, you may want to regard this as a challenge. In any case, we want to be free from writing all these parentheses. From now on we will rely on our happy experiences with real numbers where a similar problem is ignored.

15. **a.** Write down the ways of grouping the sum of five vectors. (Keep the vectors in the same order.)

    b. How many ways can the sum of six vectors be grouped?

    c. Can you discover a pattern that would enable you to determine how many ways the sum of n vectors can be grouped?

16. **a.** Suppose $a, b, c, d \in \mathbb{R}$. Prove that

$$(a\mathbf{v} + b\mathbf{w}) + (c\mathbf{v} + d\mathbf{w}) = (a + c)\mathbf{v} + (b + d)\mathbf{w}.$$

    b. State a generalization for sums with an arbitrary number of terms.

1.4 Subtraction and Cancellation

In the system of real numbers, any element $y \in \mathbb{R}$ has an additive inverse $-y$. It is then possible to define the operation of subtraction in terms of addition of this additive inverse. That is, we can define $x - y$ to be $x + (-y)$. This allows us to develop the properties of subtraction in terms of known results about addition and additive inverses.

This is a natural approach to take with vector spaces, where the definition furnishes an operation of addition and each vector is required to have an additive inverse. Axiom 4 guarantees that each vector $\mathbf{v}$ has an additive inverse $-\mathbf{v}$. Furthermore, Theorem 1.3 states that each vector has a unique additive inverse. Let us then adopt the following definition of subtraction.

Definition 1.6

Let V be a vector space. The operation of **subtraction** combines two vectors $\mathbf{v} \in V$ and $\mathbf{w} \in V$ to produce a vector denoted $\mathbf{v} - \mathbf{w}$ and defined by the formula

$$\mathbf{v} - \mathbf{w} = \mathbf{v} + (-\mathbf{w}).$$

Crossroads   Although the definition of vector space mentions only the operations of addition and scalar multiplication, we now see that the third operation of subtraction is implicit in that definition. Our job is to satisfy ourselves that this is a reasonable definition for a concept called subtraction. In fact, the properties of this new operation are very reminiscent of the properties of subtraction of real numbers. Keep in mind that all these properties are inherent in each of the examples of vector spaces in the upcoming sections, as well as in any other system that satisfies the definition of vector space.

Theorem 1.7

Suppose $\mathbf{v}$, $\mathbf{w}$, and $\mathbf{x}$ are elements of a vector space V and $r, s \in \mathbb{R}$. The following identities hold.

    a. $\mathbf{v} - \mathbf{0} = \mathbf{v}$

    b. $\mathbf{0} - \mathbf{v} = -\mathbf{v}$

    c. $\mathbf{v} - \mathbf{w} = -\mathbf{w} + \mathbf{v}$

    d. $-\mathbf{v} - \mathbf{w} = -\mathbf{w} - \mathbf{v}$

    e. $-(\mathbf{v} + \mathbf{w}) = -\mathbf{v} - \mathbf{w}$

    f. $\mathbf{v} - (-\mathbf{w}) = \mathbf{v} + \mathbf{w}$

    g. $(\mathbf{v} - \mathbf{w}) + \mathbf{w} = \mathbf{v}$

h. $(\mathbf{v} + \mathbf{w}) - \mathbf{w} = \mathbf{v}$

i. $\mathbf{v} - (\mathbf{w} + \mathbf{x}) = (\mathbf{v} - \mathbf{w}) - \mathbf{x}$

j. $\mathbf{v} - (\mathbf{w} - \mathbf{x}) = (\mathbf{v} - \mathbf{w}) + \mathbf{x}$

k. $\mathbf{v} - (-\mathbf{w} + \mathbf{x}) = (\mathbf{v} + \mathbf{w}) - \mathbf{x}$

l. $\mathbf{v} - (-\mathbf{w} - \mathbf{x}) = (\mathbf{v} + \mathbf{w}) + \mathbf{x}$

m. $r(\mathbf{v} - \mathbf{w}) = r\mathbf{v} - (r\mathbf{w})$

n. $(r - s)\mathbf{v} = r\mathbf{v} - (s\mathbf{v})$

Here are proofs of two of these identities. Once you catch on to the techniques of using the definition of subtraction along with the appropriate axioms and previous results, you should have no trouble proving the rest of these identities.

Quick Example   *Prove Theorem 1.7, part a:* $\mathbf{v} - \mathbf{0} = \mathbf{v}$.

$$\mathbf{v} - \mathbf{0} = \mathbf{v} + (-\mathbf{0}) \qquad \textit{definition of subtraction;}$$
$$= \mathbf{v} + \mathbf{0} \qquad \textit{Section 1.3, Exercise 9;}$$
$$= \mathbf{v} \qquad \textit{additive identity axiom.} \quad \blacksquare$$

Quick Example   *Prove Theorem 1.7, part e:* $-(\mathbf{v} + \mathbf{w}) = -\mathbf{v} - \mathbf{w}$.

$$-(\mathbf{v} + \mathbf{w}) = (-1)(\mathbf{v} + \mathbf{w}) \qquad \textit{Theorem 1.5, part d;}$$
$$= (-1)\mathbf{v} + (-1)\mathbf{w} \qquad \textit{distributive law of Axiom 5;}$$
$$= -\mathbf{v} + (-\mathbf{w}) \qquad \textit{Theorem 1.5, part d;}$$
$$= -\mathbf{v} - \mathbf{w} \qquad \textit{definition of subtraction.} \quad \blacksquare$$

The following theorem shows that the familiar cancellation property of real numbers also holds in the context of vector spaces.

Theorem 1.8   Cancellation Theorem

Suppose $\mathbf{v}$, $\mathbf{w}$, and $\mathbf{x}$ are vectors in a vector space V. If $\mathbf{v} + \mathbf{x} = \mathbf{w} + \mathbf{x}$, then $\mathbf{v} = \mathbf{w}$.

The proof of this theorem involves starting with the equation $\mathbf{v} + \mathbf{x} = \mathbf{w} + \mathbf{x}$ as a hypothesis and deriving the conclusion $\mathbf{v} = \mathbf{w}$. The key idea should be clear: we want to subtract $\mathbf{x}$ (or, equivalently, add $-\mathbf{x}$) on both sides of the equation to achieve the cancellation. An easy way to show the steps in such a proof is to write a sequence of statements, each of which implies the succeeding one. We will use the symbol $\Longrightarrow$ to denote that the statement on the left (or immediately above) implies the statement on the right. From the transitive property of logical implication, we can conclude that the first statement in the sequence implies the last.

Proof $\mathbf{v} + \mathbf{x} = \mathbf{w} + \mathbf{x}$ $\Longrightarrow$ $(\mathbf{v} + \mathbf{x}) + (-\mathbf{x}) = (\mathbf{w} + \mathbf{x}) + (-\mathbf{x})$

$\Longrightarrow$ $\mathbf{v} + (\mathbf{x} + (-\mathbf{x})) = \mathbf{w} + (\mathbf{x} + (-\mathbf{x}))$

$\Longrightarrow$ $\mathbf{v} + \mathbf{0} = \mathbf{w} + \mathbf{0}$

$\Longrightarrow$ $\mathbf{v} = \mathbf{w}$

If you examine this proof, you will notice that after the first step, the same vector reappears on both sides of the equation, although in various forms. This suggests an alternative proof of the cancellation theorem that gains in simplicity of presentation what it loses in obscurity of the key idea that motivated the original proof.

Alternative proof $\mathbf{v} = \mathbf{v} + \mathbf{0}$

$= \mathbf{v} + (\mathbf{x} + (-\mathbf{x}))$

$= (\mathbf{v} + \mathbf{x}) + (-\mathbf{x})$

$= (\mathbf{w} + \mathbf{x}) + (-\mathbf{x})$

$= \mathbf{w} + (\mathbf{x} + (-\mathbf{x}))$

$= \mathbf{w} + \mathbf{0}$

$= \mathbf{w}$ ■

After working the following exercises, you should be convinced that the laws of addition, subtraction, and scalar multiplication in a vector space are very similar to the corresponding laws of ordinary arithmetic. From this point on, we will be more tolerant of certain ambiguities in writing expressions. For example, the sum of vectors $\mathbf{v}_1$, $\mathbf{v}_2, \ldots, \mathbf{v}_n$ will be denoted $\mathbf{v}_1 + \mathbf{v}_2 + \cdots + \mathbf{v}_n$, with no parentheses to indicate grouping. The point is, of course, that all groupings give expressions that represent the same vector (see Exercise 14 of Section 1.3). Similarly, we will write $-r\mathbf{v}$ for either $-(r\mathbf{v})$ or $(-r)\mathbf{v}$ since, again, these expressions denote equal quantities (see Theorem 1.5, parts g and h).

Exercises 1.4

In Exercises 1 to 12, give step-by-step proofs of the identities stated in Theorem 1.7. Follow the format of the proofs given for part a and part e. If you are careful to take only one step at a time, it should be obvious what result you are using; hence, you will not need to write down the justification of each step.

1. Prove Theorem 1.7, part b: $\mathbf{0} - \mathbf{v} = -\mathbf{v}$.

2. Prove Theorem 1.7, part c: $\mathbf{v} - \mathbf{w} = -\mathbf{w} + \mathbf{v}$.

3. Prove Theorem 1.7, part d: $-\mathbf{v} - \mathbf{w} = -\mathbf{w} - \mathbf{v}$.

4. Prove Theorem 1.7, part f: $\mathbf{v} - (-\mathbf{w}) = \mathbf{v} + \mathbf{w}$.

5. Prove Theorem 1.7, part g: $(\mathbf{v} - \mathbf{w}) + \mathbf{w} = \mathbf{v}$.

6. Prove Theorem 1.7, part h: $(\mathbf{v} + \mathbf{w}) - \mathbf{w} = \mathbf{v}$.

7. Prove Theorem 1.7, part i: $\mathbf{v} - (\mathbf{w} + \mathbf{x}) = (\mathbf{v} - \mathbf{w}) - \mathbf{x}$.

8. Prove Theorem 1.7, part j: $\mathbf{v} - (\mathbf{w} - \mathbf{x}) = (\mathbf{v} - \mathbf{w}) + \mathbf{x}$.

9. Prove Theorem 1.7, part k: $\mathbf{v} - (-\mathbf{w} + \mathbf{x}) = (\mathbf{v} + \mathbf{w}) - \mathbf{x}$.
10. Prove Theorem 1.7, part l: $\mathbf{v} - (-\mathbf{w} - \mathbf{x}) = (\mathbf{v} + \mathbf{w}) + \mathbf{x}$.
11. Prove Theorem 1.7, part m: $r(\mathbf{v} - \mathbf{w}) = r\mathbf{v} - (r\mathbf{w})$.
12. Prove Theorem 1.7, part n: $(r - s)\mathbf{v} = r\mathbf{v} - (s\mathbf{v})$.
13. Suppose $\mathbf{v}$ and $\mathbf{w}$ are vectors in a vector space with $\mathbf{v} \neq \mathbf{0}$. Suppose r and s are real numbers. Prove that if $r\mathbf{v} + \mathbf{w} = s\mathbf{v} + \mathbf{w}$, then $r = s$.
14. Suppose $\mathbf{v}$, $\mathbf{w}$, and $\mathbf{x}$ are elements of a vector space. Give an alternative for each of the following expressions other than that given in Theorem 1.7. In each case, prove that your expression is equal to the given expression.

    a. $\mathbf{v} - (\mathbf{w} + \mathbf{x})$

    b. $\mathbf{v} - (\mathbf{w} - \mathbf{x})$

    c. $\mathbf{v} - (-\mathbf{w} + \mathbf{x})$

    d. $\mathbf{v} - (-\mathbf{w} - \mathbf{x})$

1.5 Euclidean Spaces

In this section and the next two sections, we will examine some concrete examples of vector spaces. Each of these sections is devoted to a family of examples that will be extremely important throughout the course. Along the way we will also encounter a few curiosities and some spaces that are almost but not quite vector spaces. Each example will consist of a set of points and the two algebraic operations. Our primary task will be to check whether the eight axioms are true. The failure of any one of the axioms disqualifies the example from being a vector space. If all eight axioms hold, we will have an example of a vector space. In such a case, all the theorems we have proved for vector spaces will also hold.

The first example is based on the set of ordered pairs of real numbers,

$$\{(v_1, v_2) \mid v_1 \in \mathbb{R} \text{ and } v_2 \in \mathbb{R}\}.$$

This set is usually denoted $\mathbb{R}^2$. We are dealing with the familiar Cartesian coordinate system for the plane. You are undoubtedly familiar with its use in analytic geometry for plotting points and graphing functions. The addition of two elements of $\mathbb{R}^2$ is defined by adding the corresponding coordinates:

$$(v_1, v_2) + (w_1, w_2) = (v_1 + w_1, v_2 + w_2).$$

This equation defines the new use of the plus sign on the left side of the equation in terms of the ordinary addition of real numbers $v_1 + w_1$ and $v_2 + w_2$ on the right side of the equation. The multiplication of $(v_1, v_2) \in \mathbb{R}^2$ by a real number r is defined similarly:

$$r(v_1, v_2) = (rv_1, rv_2).$$

These operations have very simple geometric interpretations that we will investigate in Section 1.9. They also correspond to the operations physicists find so useful when dealing with vectors that represent forces or velocities in a plane.

A set has been given, and definitions for addition and scalar multiplication have been proposed. It is clear that the results of both of these operations yield elements in the set. Now we must verify the eight axioms of Definition 1.1. This is somewhat tedious, but in each case there is something to be checked. Once you catch on to what is happening and see the point of doing it, you will be able to breeze through such verifications.

Mathematical Strategy Session We will let $\mathbf{v} = (v_1, v_2)$, $\mathbf{w} = (w_1, w_2)$, and $\mathbf{x} = (x_1, x_2)$ denote the typical elements of $\mathbb{R}^2$ to be used in the verification of the eight axioms. We will use r and s to represent arbitrary real numbers. Each equality will be verified by the following six steps:

1. Start with the left-hand side.
2. Rewrite the vectors involved as ordered pairs of real numbers.
3. Use the vector addition and scalar multiplication as defined.
4. Use some rules of arithmetic of real numbers.
5. Write the result in terms of the vector space operations again.
6. Obtain the right-hand side of the equation.

Feel free to skim over some of these verifications. They are written out in full detail to serve as models for you to refer to in verifying the axioms in other situations.

1.
$$\begin{aligned}
\mathbf{v} + \mathbf{w} &= (v_1, v_2) + (w_1, w_2) \\
&= (v_1 + w_1, v_2 + w_2) \\
&= (w_1 + v_1, w_2 + v_2) \\
&= (w_1, w_2) + (v_1, v_2) \\
&= \mathbf{w} + \mathbf{v}.
\end{aligned}$$

2.
$$\begin{aligned}
(\mathbf{v} + \mathbf{w}) + \mathbf{x} &= ((v_1, v_2) + (w_1, w_2)) + (x_1, x_2) \\
&= (v_1 + w_1, v_2 + w_2) + (x_1, x_2) \\
&= ((v_1 + w_1) + x_1, (v_2 + w_2) + x_2) \\
&= (v_1 + (w_1 + x_1), v_2 + (w_2 + x_2)) \\
&= (v_1, v_2) + (w_1 + x_1, w_2 + x_2) \\
&= (v_1, v_2) + ((w_1, w_2) + (x_1, x_2)) \\
&= \mathbf{v} + (\mathbf{w} + \mathbf{x}).
\end{aligned}$$

3. For the zero vector of $\mathbb{R}^2$, let us try the ordered pair that has both coordinates equal to zero:
$$\begin{aligned}
\mathbf{v} + \mathbf{0} &= (v_1, v_2) + (0, 0) \\
&= (v_1 + 0, v_2 + 0) \\
&= (v_1, v_2) \\
&= \mathbf{v}.
\end{aligned}$$

4. The choice for $-\mathbf{v}$ should be clear. Let us take $-\mathbf{v} = (-v_1, -v_2)$:

$$
\begin{aligned}
\mathbf{v} + (-\mathbf{v}) &= (v_1, v_2) + (-v_1, -v_2) \\
&= (v_1 + (-v_1), v_2 + (-v_2)) \\
&= (0, 0) \\
&= \mathbf{0}.
\end{aligned}
$$

5.
$$
\begin{aligned}
r(\mathbf{v} + \mathbf{w}) &= r((v_1, v_2) + (w_1, w_2)) \\
&= r(v_1 + w_1, v_2 + w_2) \\
&= (r(v_1 + w_1), r(v_2 + w_2)) \\
&= (rv_1 + rw_1, rv_2 + rw_2) \\
&= (rv_1, rv_2) + (rw_1, rw_2) \\
&= r(v_1, v_2) + r(w_1, w_2) \\
&= r\mathbf{v} + r\mathbf{w}.
\end{aligned}
$$

6.
$$
\begin{aligned}
(r + s)\mathbf{v} &= (r + s)(v_1, v_2) \\
&= ((r + s)v_1, (r + s)v_2) \\
&= (rv_1 + sv_1, rv_2 + sv_2) \\
&= (rv_1, rv_2) + (sv_1, sv_2) \\
&= r(v_1, v_2) + s(v_1, v_2) \\
&= r\mathbf{v} + s\mathbf{v}.
\end{aligned}
$$

7.
$$
\begin{aligned}
r(s\mathbf{v}) &= r(s(v_1, v_2)) \\
&= r(sv_1, sv_2) \\
&= (r(sv_1), r(sv_2)) \\
&= ((rs)v_1, (rs)v_2) \\
&= (rs)(v_1, v_2) \\
&= (rs)\mathbf{v}.
\end{aligned}
$$

8.
$$
\begin{aligned}
1\mathbf{v} &= 1(v_1, v_2) \\
&= (1v_1, 1v_2) \\
&= (v_1, v_2) \\
&= \mathbf{v}.
\end{aligned}
$$

Because of the simplicity of the addition and scalar multiplication we have been working with on $\mathbb{R}^2$, it is easy to forget that there are lots of other ways to define these operations. For example, suppose we define addition and scalar multiplication on $\mathbb{R}^2$ by the rules

$$
(v_1, v_2) + (w_1, w_2) = (v_1 + w_2, v_2 + w_1),
$$
$$
r(v_1, v_2) = (rv_1, rv_2).
$$

Because of the peculiar pairings of the terms in the definition of the sum of two vectors, we might suspect trouble with an additive inverse. In order to check the additive inverse law, we must first locate the additive identity. In this case it is easy to verify that $\mathbf{0} = (0, 0)$ satisfies the condition given in Axiom 3 to be the additive identity:

$$\mathbf{v} + \mathbf{0} = (v_1, v_2) + (0, 0) = (v_1 + 0, v_2 + 0) = (v_1, v_2) = \mathbf{v}.$$

Now we are ready to track down an additive inverse for a given vector $\mathbf{v} = (v_1, v_2)$. If you write down the condition stated in the additive inverse law, you will quickly discover that $(-v_2, -v_1)$ is the only possibility for the inverse of $\mathbf{v} = (v_1, v_2)$. Of course, $(-v_2, -v_1)$ is not the standard inverse of (v_1, v_2); this is not surprising since we are not dealing with the standard notion of addition. Again it is easy to verify that this vector does the job:

$$(v_1, v_2) + (-v_2, -v_1) = (v_1 + (-v_1), v_2 + (-v_2))$$
$$= (0, 0)$$
$$= \mathbf{0}.$$

At this point we might go through the axioms from the top. The attempt to verify the commutative law of addition proceeds smoothly until the last step:

$$\mathbf{v} + \mathbf{w} = (v_1, v_2) + (w_1, w_2)$$
$$= (v_1 + w_2, v_2 + w_1)$$
$$= (w_2 + v_1, w_1 + v_2)$$
$$= (w_2, w_1) + (v_2, v_1).$$

The twisted definition of addition has interchanged the coordinates of the two vectors. A counterexample even as simple as $\mathbf{v} = (0, 0)$ and $\mathbf{w} = (1, 0)$ will yield $\mathbf{v} + \mathbf{w} = (0, 0) + (1, 0) = (0 + 0, 0 + 1) = (0, 1)$, clearly not equal to $\mathbf{w} + \mathbf{v} = (1, 0) + (0, 0) = (1 + 0, 0 + 0) = (1, 0)$. A single counterexample to a single one of the axioms is all that is needed to prevent a system from being a vector space.

Quick Example *With operations defined as in the previous example, determine whether Axiom 6 holds.*

Here is an attempt to verify Axiom 6:

$$(r + s)\mathbf{v} = (r + s)(v_1, v_2)$$
$$= ((r + s)v_1, (r + s)v_2)$$
$$= (rv_1 + sv_1, rv_2 + sv_2)$$
$$= (rv_1, rv_2) + (sv_2, sv_1)$$
$$= r(v_1, v_2) + s(v_2, v_1).$$

Now it is easy to choose values such as

$$r = 0, \quad s = 1, \quad v_1 = 0, \quad v_2 = 1$$

to give a specific counterexample. With these values the two sides of the equation in Axiom 6 are

$$(r + s)\mathbf{v} = (r + s)(v_1, v_2)$$
$$= (0 + 1)(0, 1)$$
$$= 1(0, 1)$$
$$= (0, 1)$$

and

$$rv + sv = r(v_1, v_2) + s(v_1, v_2)$$
$$= 0(0, 1) + 1(0, 1)$$
$$= (0, 0) + (0, 1)$$
$$= (0 + 1, 0 + 0) = (1, 0),$$

which are definitely not equal. ∎

Quick Example *Suppose we define addition on $\mathbb{R}^2$ by the rule*

$$(v_1, v_2) + (w_1, w_2) = (v_1 w_1, v_2 w_2).$$

Show that this operation satisfies the additive identity axiom, but not the additive inverse axiom.

In order for the sum of two vectors to equal the first vector, we must have $w_1 = 1$ and $w_2 = 1$. It is easy to check that $(1, 1)$ is indeed an additive identity for this operation:

$$(v_1, v_2) + (1, 1) = (v_1 \cdot 1, v_2 \cdot 1) = (v_1, v_2).$$

Thus, the additive identity axiom is satisfied.

The additive inverse axiom requires that for any vector $(v_1, v_2) \in \mathbb{R}^2$, we must produce a vector (w_1, w_2) so that

$$(v_1, v_2) + (w_1, w_2) = (v_1 w_1, v_2 w_2) = (1, 1).$$

We are forced to take $w_1 = \frac{1}{v_1}$ and $w_2 = \frac{1}{v_2}$. This is fine as long as $v_1 \neq 0$ and $v_2 \neq 0$. However, we are out of luck when either coordinate of (v_1, v_2) is zero. Hence, not every vector in $\mathbb{R}^2$ has an inverse when we use this operation. ∎

Other definitions of addition and scalar multiplication might seem even more bizarre, yet satisfy all the axioms and thus turn $\mathbb{R}^2$ into a vector space. Note, however, that this will not be the same vector space as $\mathbb{R}^2$ with the standard operations defined above. A set of points alone does not suffice to define a vector space; if we consider different notions of addition and scalar multiplication, we will get different vector spaces. Nevertheless, when we consider $\mathbb{R}^2$ as a vector space, unless mention is made to the contrary, we will assume that the standard operations are in effect.

Much more can be said about $\mathbb{R}^2$ as a vector space, but for now we want to note that $\mathbb{R}^2$ is merely one of an infinite family of vector spaces. These higher-dimensional analogs of $\mathbb{R}^2$ are known as the **Euclidean** vector spaces. Instead of considering pairs of real numbers, pick a positive integer n and consider ordered lists of n real numbers. The resulting set $\{(v_1, v_2, \ldots, v_n) \mid v_1 \in \mathbb{R}, v_2 \in \mathbb{R}, \ldots, v_n \in \mathbb{R}\}$ is denoted $\mathbb{R}^n$. It is the n-fold Cartesian product of the real numbers. If n is a low number, say 4, we can avoid subscripts by using letters of the alphabet:

$$\mathbb{R}^4 = \{(a, b, c, d) \mid a \in \mathbb{R}, b \in \mathbb{R}, c \in \mathbb{R}, d \in \mathbb{R}\}.$$

There are two points to be considered, however. There is no advantage to restricting our consideration to $n \leq 26$; in fact, there are advantages to leaving n unspecified. The subscripts and the three dots (meaning to fill in however many entries are necessary) are slightly awkward, but their use will pay off in the long run. We will follow the custom of using boldface letters for vectors and corresponding lowercase letters with subscripts to indicate the coordinates. Thus, $\mathbf{v} = (v_1, v_2, \ldots, v_n)$ represents a typical element of $\mathbb{R}^n$.

Notice that if n equals 1 or 2, our notation indicates more coordinates than actually should be there. In particular, the elements of $\mathbb{R}^1$ are simply the real numbers disguised in parentheses. Convince yourself that in this case, the vector space addition corresponds to ordinary addition of real numbers and scalar multiplication corresponds to ordinary multiplication.

Crossroads   A geometric or physical interpretation is often demanded for Euclidean spaces with $n > 3$. The three dimensions of length, width, and height are familiar interpretations for the first three coordinates. Time is often proposed as an interpretation of a fourth. Other parameters describing the attitude, mass, temperature, charge, color, spin, strangeness, and so forth can be chosen as physical interpretations of other coordinates. A Baskin-Robbins store might find that its daily sales totals are elements of $\mathbb{R}^{31}$; addition and scalar multiplication might even be of use in this context. Keep in mind that these are only interpretations for $\mathbb{R}^n$ as a mathematical model. To be on the safe side, even the three dimensions of physical space should be distinguished from the mathematical object $\mathbb{R}^3$, a set of ordered triples of real numbers. As soon as vector space operations of addition and scalar multiplication can be defined in any of these contexts, the theory of vector spaces is ready and waiting. As one of the fruits of our labors, we will eventually be able to do some geometry of lines and planes in n-dimensional space. For the present, however, we wish to treat the vector space $\mathbb{R}^n$ merely as an algebraic gadget.

The two operations are defined for $\mathbb{R}^n$ in strict analogy with the operations on $\mathbb{R}^2$. Thus, if $\mathbf{v} = (v_1, v_2, \ldots, v_n)$ and $\mathbf{w} = (w_1, w_2, \ldots, w_n)$ are elements of $\mathbb{R}^n$ and r is a real number, we define

$$\mathbf{v} + \mathbf{w} = (v_1 + w_1, v_2 + w_2, \ldots, v_n + w_n),$$
$$r\mathbf{v} = (rv_1, rv_2, \ldots, rv_n).$$

The verification that we have a vector space amounts to adapting the eight strings of equalities listed for $\mathbb{R}^2$ to cover the case of more than two coordinates. For example, to check Axiom 1, we write

$$\begin{aligned}
\mathbf{v} + \mathbf{w} &= (v_1, v_2, \ldots, v_n) + (w_1, w_2, \ldots, w_n) \\
&= (v_1 + w_1, v_2 + w_2, \ldots, v_n + w_n) \\
&= (w_1 + v_1, w_2 + v_2, \ldots, w_n + v_n) \\
&= (w_1, w_2, \ldots, w_n) + (v_1, v_2, \ldots, v_n) \\
&= \mathbf{w} + \mathbf{v}.
\end{aligned}$$

Crossroads The vector space $\mathbb{R}^n$ is the prototype for all vector spaces. The historical reason the eight axioms were selected was that they hold for $\mathbb{R}^n$ and they imply a large number of the important properties of $\mathbb{R}^n$. Also, we shall see in Section 6.8 that many vector spaces are disguised versions of $\mathbb{R}^n$. Nevertheless, we do not want to restrict our study of vector spaces to $\mathbb{R}^n$. As you encounter the vector spaces of matrices and functions in the next two sections, try to develop a sense of how these spaces are related to the Euclidean spaces. These examples (and many others) share the properties that derive from being instances of the abstract notion of vector space.

Exercises 1.5

1. Determine whether we obtain a vector space from $\mathbb{R}^2$ with operations defined by

$$(v_1, v_2) + (w_1, w_2) = (v_2 + w_2, v_1 + w_1),$$
$$r(v_1, v_2) = (rv_1, rv_2).$$

2. Determine whether we obtain a vector space from $\mathbb{R}^2$ with operations defined by

$$(v_1, v_2) + (w_1, w_2) = (v_1 + w_1, v_2 + w_2 + 1),$$
$$r(v_1, v_2) = (rv_1, rv_2).$$

Notice that the additive identity vector is $(0, -1)$ since

$$(v_1, v_2) + (0, -1) = (v_1 + 0, v_2 + (-1) + 1)$$
$$= (v_1, v_2).$$

as required by Axiom 3.

3. Determine whether we obtain a vector space from $\mathbb{R}^2$ with operations defined by

$$(v_1, v_2) + (w_1, w_2) = (v_1 + w_1 - 1, v_2 + w_2),$$
$$r(v_1, v_2) = (rv_1 - r + 1, rv_2).$$

4. Determine whether we obtain a vector space from $\mathbb{R}^2$ with operations defined by

$$(v_1, v_2) + (w_1, w_2) = (v_1 w_2 + v_2 w_1, v_2 w_2),$$
$$r(v_1, v_2) = (rv_1, v_2).$$

5. Determine whether we obtain a vector space from $\mathbb{R}^2$ with operations defined by

$$(v_1, v_2) + (w_1, w_2) = (v_1 + w_1, v_1 + w_1 + v_2 + w_2),$$
$$r(v_1, v_2) = (rv_1, rv_1 + rv_2).$$

6. Determine whether we obtain a vector space from $\mathbb{R}^2$ with operations defined by

$$(v_1, v_2) + (w_1, w_2) = (v_1 + w_1, v_2 + w_2),$$
$$r(v_1, v_2) = (0, 0).$$

It is especially interesting to have an example where exactly one of the axioms fails. This indicates that such an axiom is essential in the sense that there is no way to derive it from the other axioms.

7. Determine whether we obtain a vector space from $\mathbb{R}^2$ with operations defined by

$$(v_1, v_2) + (w_1, w_2) = (v_1 + w_1, v_2 + w_2),$$
$$r(v_1, v_2) = (rv_1, v_2).$$

8. Determine whether we obtain a vector space from the following subset of $\mathbb{R}^2$ with the standard operations:

$$S = \{(v_1, v_2) \in \mathbb{R}^2 \mid v_1 \text{ and } v_2 \text{ are integers}\}.$$

Here the identities stated in the eight axioms are not particularly in doubt. The real question is whether the definitions apply in as much generality as they should. The definition of a vector space requires that when we add together any pair of vectors or multiply any vector by any real number, the result is an element of the set. In the terminology of Section 1.1, the set must be closed under addition and scalar multiplication.

9. Determine whether we obtain a vector space from the following subset of $\mathbb{R}^2$ with the standard operations:

$$H = \{(v_1, v_2) \in \mathbb{R}^2 \mid v_1 \geq 0\}.$$

10. Determine whether we obtain a vector space from the following subset of $\mathbb{R}^3$ with the standard operations:

$$P = \{(v_1, v_2, v_3) \in \mathbb{R}^3 \mid v_1 = v_2 + v_3\}.$$

11. Show that the set $\{0\}$ consisting of only one element can be made into a vector space. Notice that Axiom 3 requires any vector space to have at least one element that will serve as the zero vector.

12. Verify Axioms 2 through 8 for $\mathbb{R}^n$ with the standard operations.

13. Find scalars a, b, and c so that

$$a(2, 3, -1) + b(1, 0, 4) + c(-3, 1, 2) = (7, 2, 5).$$

14. Find scalars a, b, c, and d so that

$$a(1, 0, 0, 0, 0) + b(1, 1, 0, 0, 0) + c(1, 1, 1, 0, 0) + d(1, 1, 1, 1, 0)$$
$$= (8, 5, -2, 3, 0).$$

15. Show that it is impossible to find scalars a, b, c, and d so that

$$a(1, 0, 0, 0, 0) + b(1, 1, 0, 0, 0) + c(1, 1, 1, 0, 0) + d(1, 1, 1, 1, 0)$$
$$= (8, 5, -2, 3, 1).$$

16. Give some useful interpretations for the vector space $\mathbb{R}^n$ with $n \geq 4$. What roles do the operations of addition and scalar multiplication play in your interpretations?

1.6 Matrices

In this section we will consider another versatile family of vector spaces. We will investigate the notion of matrices and the two operations needed to make a set of matrices into a vector space. We will consider additional matrix operations in later chapters. The new family of examples is defined as follows:

Definition 1.9

Let m and n be positive integers. We denote by $\mathbb{M}(m, n)$ the set of all rectangular arrays of real numbers with m horizontal **rows** and n vertical **columns**. Such a rectangular array is called a **matrix** of size m by n or, more simply, an $m \times n$ matrix. The number in row i and column j is the ij-**entry** of the matrix.

For example, the matrix

$$\begin{bmatrix} 3 & -5 & \frac{1}{2} \\ -4 & \pi & 0 \end{bmatrix}$$

has two rows and three columns; it is a 2×3 matrix. One way to denote the entries of an arbitrary $m \times n$ matrix is to use a variable with two subscripts. The first subscript tells which of the m rows contains the entry and the second tells which of the n columns. Thus, a_{ij} denotes the entry in row i and column j. In the preceding example $a_{13} = \frac{1}{2}$ and $a_{22} = \pi$. The general element of $\mathbb{M}(m, n)$ can be expressed as

$$A = \begin{bmatrix} a_{11} & a_{12} & \cdots & a_{1n} \\ a_{21} & a_{22} & \cdots & a_{2n} \\ \vdots & \vdots & & \vdots \\ a_{m1} & a_{m2} & \cdots & a_{mn} \end{bmatrix}.$$

The ith row looks a lot like the vector $(a_{i1}, a_{i2}, \ldots, a_{in})$ in $\mathbb{R}^n$. In fact, when $m = 1$, the matrices of $\mathbb{M}(1, n)$ are often identified with the points of $\mathbb{R}^n$. Notation is the only difference: a matrix is written with square brackets and without commas separating the components. Similarly, the jth column resembles a verticalized version of the vector $(a_{1j}, a_{2j}, \ldots, a_{mj}) \in \mathbb{R}^m$. Later it will be useful for us to identify these columns with vectors in $\mathbb{R}^m$. In fact, we will then begin to write elements of $\mathbb{R}^m$ as columns.

The next order of business is to define the two algebraic operations on $\mathbb{M}(m, n)$ that will complete the definition of $\mathbb{M}(m, n)$ as a vector space. Let

$$A = \begin{bmatrix} a_{11} & a_{12} & \cdots & a_{1n} \\ a_{21} & a_{22} & \cdots & a_{2n} \\ \vdots & \vdots & & \vdots \\ a_{m1} & a_{m2} & \cdots & a_{mn} \end{bmatrix} \quad \text{and} \quad B = \begin{bmatrix} b_{11} & b_{12} & \cdots & b_{1n} \\ b_{21} & b_{22} & \cdots & b_{2n} \\ \vdots & \vdots & & \vdots \\ b_{m1} & b_{m2} & \cdots & b_{mn} \end{bmatrix}.$$

We define addition by

$$
A + B =
\begin{bmatrix}
a_{11} + b_{11} & a_{12} + b_{12} & \cdots & a_{1n} + b_{1n} \\
a_{21} + b_{21} & a_{22} + b_{22} & \cdots & a_{2n} + b_{2n} \\
\vdots & \vdots & & \vdots \\
a_{m1} + b_{m1} & a_{m2} + b_{m2} & \cdots & a_{mn} + b_{mn}
\end{bmatrix}
$$

For $r \in \mathbb{R}$, we define scalar multiplication by

$$
rA =
\begin{bmatrix}
ra_{11} & ra_{12} & \cdots & ra_{1n} \\
ra_{21} & ra_{22} & \cdots & ra_{2n} \\
\vdots & \vdots & & \vdots \\
ra_{m1} & ra_{m2} & \cdots & ra_{mn}
\end{bmatrix}.
$$

That is, the entries of the sum $A + B$ are formed by adding the real numbers in the corresponding entries of A and B, and the entries of the product rA are obtained by multiplying the corresponding entries of A by the scalar r.

Note that the matrices A and B must be of the same size for us to add them. The result $A + B$ is the same size as the matrices A and B. Similarly, the scalar product rA is a matrix of the same size as A. In other words, the set $\mathbb{M}(m, n)$ is closed under addition and scalar multiplication.

Mathematical Strategy Session   When the size of the matrices under consideration is clearly understood, the entries of a matrix $A \in \mathbb{M}(m, n)$ can be symbolized by a typical element a_{ij}. We write $A = [a_{ij}]$ with the understanding that i ranges over the integers from 1 to m and j independently ranges over the integers from 1 to n. For example, with this notation we could write the defining rule for matrix addition as

$$[a_{ij}] + [b_{ij}] = [a_{ij} + b_{ij}].$$

Notice that this abbreviated notation follows the convention that the first subscript, often i, refers to the row, and the second, often j, refers to the column.

These operations are simply extensions of the operations defined for the vector spaces $\mathbb{R}^n$. The similarity will become apparent when you start checking the eight axioms. By now this should be a routine exercise that you can almost carry out mentally. With the abbreviated notation, it is especially easy to write down the verifications of these axioms. For example, to verify the distributive law of Axiom 5, let $r \in \mathbb{R}$ and let $A = [a_{ij}]$ and $B = [b_{ij}]$ be matrices in $\mathbb{M}(m, n)$. Then

$$
\begin{aligned}
r(A + B) &= r([a_{ij}] + [b_{ij}]) \\
&= r[a_{ij} + b_{ij}] \\
&= [r(a_{ij} + b_{ij})] \\
&= [ra_{ij} + rb_{ij}] \\
&= [ra_{ij}] + [rb_{ij}]
\end{aligned}
$$

$$= r[a_{ij}] + r[b_{ij}]$$
$$= rA + rB.$$

There is much more to be said about matrices, but the mere fact that $\mathbb{M}(m, n)$ forms a vector space means that we are instantly familiar with the basic properties of its algebraic structure. The eight axioms and the basic theorems for vector spaces enumerate many of the properties of matrix arithmetic.

Exercises 1.6

1. Verify that the operations of addition and scalar multiplication defined on $\mathbb{M}(m, n)$ satisfy the remaining seven vector space axioms (Axioms 1 through 4 and 6 through 8).

2. Write the 3×1 matrix $\begin{bmatrix} 5r + 3s \\ -r + 2s \\ s \end{bmatrix}$ as the scalar r times a 3×1 matrix plus the scalar s times another 3×1 matrix.

3. Suppose we know that the values of three variables x, y, z are of the form

$$\begin{aligned} x &= a + 4b + c \\ y &= 3a \quad\;\; - 2c \\ z &= \quad\;\; b + 5c \end{aligned}$$

Rewrite this result as an equation between $\begin{bmatrix} x \\ y \\ z \end{bmatrix}$ and a sum of the scalars $a, b,$ and c times three 3×1 matrices.

4. **a.** Find values of the scalars r and s so that
$$r\begin{bmatrix} 2 & -5 & 0 \\ -3 & 1 & 4 \end{bmatrix} + s\begin{bmatrix} 7 & 0 & 8 \\ 1 & -2 & 7 \end{bmatrix} = \begin{bmatrix} 0 & 0 & 0 \\ 0 & 0 & 0 \end{bmatrix}.$$

b. Find values of the scalars r and s so that
$$r\begin{bmatrix} 2 & -5 & 0 \\ -3 & 1 & 4 \end{bmatrix} + s\begin{bmatrix} 7 & 0 & 8 \\ 1 & -2 & 7 \end{bmatrix} = \begin{bmatrix} -1 & -15 & -8 \\ -10 & 5 & 5 \end{bmatrix}.$$

c. Are there solutions to the equations in parts a and b other than the ones you found?

5. **a.** Find values of the scalars a, b, c, d so that
$$a\begin{bmatrix} 1 & 2 \\ 3 & 4 \end{bmatrix} + b\begin{bmatrix} 0 & 1 \\ 2 & 3 \end{bmatrix} + c\begin{bmatrix} 0 & 0 \\ 1 & 2 \end{bmatrix} + d\begin{bmatrix} 0 & 0 \\ 0 & 1 \end{bmatrix} = \begin{bmatrix} 0 & 0 \\ 0 & 0 \end{bmatrix}.$$

b. Find values of the scalars a, b, c, d so that
$$a\begin{bmatrix} 1 & 2 \\ 3 & 4 \end{bmatrix} + b\begin{bmatrix} 0 & 1 \\ 2 & 3 \end{bmatrix} + c\begin{bmatrix} 0 & 0 \\ 1 & 2 \end{bmatrix} + d\begin{bmatrix} 0 & 0 \\ 0 & 1 \end{bmatrix} = \begin{bmatrix} 4 & 9 \\ 16 & 25 \end{bmatrix}.$$

c. Are there solutions to the equations in parts a and b other than the ones you found?

6. **a.** Find values of the scalars a and b so that

$$a\begin{bmatrix} 12 & 3 \\ 0 & -9 \end{bmatrix} + b\begin{bmatrix} 4 & 1 \\ 0 & -3 \end{bmatrix} = \begin{bmatrix} 0 & 0 \\ 0 & 0 \end{bmatrix}.$$

  b. Find values of the scalars a and b so that

$$a\begin{bmatrix} 12 & 3 \\ 0 & -9 \end{bmatrix} + b\begin{bmatrix} 4 & 1 \\ 0 & -3 \end{bmatrix} = \begin{bmatrix} -28 & -7 \\ 0 & 21 \end{bmatrix}.$$

  c. Are there solutions to the equations in parts a and b other than the ones you found?

7. Show that

$$S = \left\{ \begin{bmatrix} a & b \\ c & d \end{bmatrix} \in \mathbb{M}(2, 2) \;\middle|\; b = 0 \text{ and } c = 0 \right\}$$

is closed under the standard operations of addition and scalar multiplication of $\mathbb{M}(2, 2)$. Verify that with these operations S is a vector space.

8. Show that

$$S = \left\{ \begin{bmatrix} a & b \\ c & d \end{bmatrix} \in \mathbb{M}(2, 2) \;\middle|\; 2a + b = 0 \right\}$$

is closed under the standard operations of addition and scalar multiplication of $\mathbb{M}(2, 2)$. Verify that with these operations S is a vector space.

9. Notice that we can take a matrix $\begin{bmatrix} a_1 & a_2 & \cdots & a_n \end{bmatrix}$ in $\mathbb{M}(1, n)$, change the brackets to parentheses, and put commas between the entries to obtain an element $(a_1, a_2, \ldots, a_n)$ in $\mathbb{R}^n$.

  a. Show that this correspondence between $\mathbb{M}(1, n)$ and $\mathbb{R}^n$ is compatible with the operations of addition. That is, if $A, B \in \mathbb{M}(1, n)$ correspond to $\mathbf{a}, \mathbf{b}$ in $\mathbb{R}^n$, then the matrix sum $A + B$ corresponds to the Euclidean vector sum $\mathbf{a} + \mathbf{b}$.

  b. Show that this correspondence between $\mathbb{M}(1, n)$ and $\mathbb{R}^n$ is compatible with scalar multiplication.

10. Invent a rule for adding matrices that does not require the matrices to be the same size. Your rule should agree with the standard definition in the case that the matrices are the same size. When you use your definition of addition along with the standard definition of scalar multiplication, which of the eight vector space axioms does your system satisfy? Speculate as to whether or not your rules make the set of matrices of all sizes into a vector space.

1.7 Function Spaces

The final examples of vector spaces that we will consider are spaces of functions. This is ground common to linear algebra and differential equations. It will also provide us with a look at spaces that are infinite-dimensional. You may need a little time to become comfortable with these spaces, but they are not especially complicated if you recall the notion of function as used in your calculus course. We begin with an arbitrary nonempty set X. Often X will be an interval on the real number line, perhaps all of $\mathbb{R}$ (just as a set though—no algebraic operations are needed here). Consider a real-valued **function** f defined on the set X. We think of f as a rule that assigns to each element x of X a unique and unambiguous real number denoted $f(x)$. If X is a subset of $\mathbb{R}$, f is just an ordinary function like those you studied in beginning calculus. If X is $\mathbb{R}^n$, then f is a real-valued function of n real variables as studied in multivariate calculus.

> **Mathematical Strategy Session**   For the purpose of this family of vector spaces, we will consider only functions that map into $\mathbb{R}$. Speaking more generally, however, we can choose any sets X and Y and consider a function f that assigns to each element $x \in X$ an element $f(x) \in Y$. The only requirement is that the assignment be done in a well-defined manner. If x_1 and x_2 are distinct points of X, we allow the possibility that $f(x_1) = f(x_2)$. However, if x_1 and x_2 are two ways of writing the same point of X, then we require that $f(x_1) = f(x_2)$. For example, if X is the set of rational numbers, we require that $f(\frac{2}{3}) = f(\frac{4}{6})$.
>
> Here is some terminology you should be familiar with; we will refer to it later. A function f from X to Y will be denoted $f : X \to Y$. In this case, the set X is called the **domain** of f, and the set Y is called the **range** of f. We do not require every element of Y to be assigned to an element of X; the subset of all elements of Y that can be realized as $f(x)$ for some $x \in X$ is called the **image** of f. Some mathematicians refer to this set as the range, but that leaves no good word to refer to the set Y.

Now let $\mathbb{F}(X)$ denote the set of all possible real-valued functions defined on the nonempty set X. Since we are not even requiring the functions to be continuous, $\mathbb{F}(X)$ is certainly an enormously large set. We want to regard the functions in $\mathbb{F}(X)$ as mere points in a vector space. Granted these functions are more complicated than geometric points, but when we stand back far enough to see all of $\mathbb{F}(X)$, they will appear more like points. It is absolutely essential to distinguish between a function f and the real number $f(x)$ that is the image of some $x \in X$. The function f is an element in $\mathbb{F}(X)$; the value $f(x)$ is a point in $\mathbb{R}$. To know a single function $f \in \mathbb{F}(X)$ takes the knowledge of all the values of the real numbers $f(x)$ for the various points $x \in X$. To know that two functions f and g in $\mathbb{F}(X)$ are equal takes the knowledge that the real number $f(x)$ is equal to the corresponding real number $g(x)$ for all the various points $x \in X$.

> **Mathematical Strategy Session**   To minimize confusion about the objects under consideration, we will consistently follow the convention that letters such as x and y denote elements of sets such as X and Y, and f and g denote elements of a set such as $\mathbb{F}(X)$. We will also exercise due caution to distinguish between a function f and its value $f(x)$. In later chapters, however, when there is no danger of confusion, we will occasionally slip in a function defined by its

> formula. Thus, for example, the formality of introducing a function $f : \mathbb{R} \to \mathbb{R}$ defined by $f(x) = x^2$ will eventually give way to the simplicity of referring to the unnamed function x^2.

The next order of business is to define the two operations necessary to turn $\mathbb{F}(X)$ into a vector space. Suppose f and g are elements of $\mathbb{F}(X)$. The sum $f + g$ will be another real-valued function defined on X by the following rule: to each point $x \in X$, the function $f + g$ will assign the sum $f(x) + g(x)$ of the real numbers $f(x)$ and $g(x)$. In a similar fashion, if r is a real number and f is an element of $\underline{F}(X)$, we define the new function rf by saying that it assigns to any $x \in X$ the product $r(f(x))$ of the real numbers r and $f(x)$.

These two definitions can be written more briefly by the equations

$$(f + g)(x) = f(x) + g(x),$$
$$(rf)(x) = r(f(x)).$$

Notice that the expressions on the right side use ordinary addition and multiplication of real numbers to define the expressions on the left side. Also notice how parentheses are used for grouping as well as for part of the notation for function evaluation.

Quick Example *Let f and g be functions in $\mathbb{F}(\mathbb{R})$ defined by $f(x) = e^x$ and $g(x) = |4(x - 1)|$. Compute the real number $\left((-5)f + \frac{1}{2}g\right)(0)$.*

$$
\begin{aligned}
\left((-5)f + \tfrac{1}{2}g\right)(0) &= \left((-5)f\right)(0) + \left(\tfrac{1}{2}g\right)(0) && \text{\textit{definition of addition;}} \\
&= (-5)\left(f(0)\right) + \tfrac{1}{2}\left(g(0)\right) && \text{\textit{definition of scalar multiplication;}} \\
&= (-5)e^0 + \tfrac{1}{2}|4(0 - 1)| && \text{\textit{definitions of } f \text{ and } g;} \\
&= (-5) \cdot 1 + \tfrac{1}{2} \cdot 4 && \text{\textit{evaluation;}} \\
&= -3 && \text{\textit{simplification.}} \quad \blacksquare
\end{aligned}
$$

Although addition and scalar multiplication of functions are fairly easy to deal with algebraically, you should also try to develop a geometric intuition about these operations. The best way to do this is to think of a function in terms of its graph. Scalar multiplication will stretch the graph away from the x-axis (if the scalar is greater than 1 in absolute value) or shrink it toward the x-axis (if the scalar is less than 1 in absolute value), possibly with a reflection through the x-axis (if the scalar is negative). The graph of the sum of two functions is obtained by adding the y-values of the functions for each value of x. The following example illustrates these ideas.

Quick Example *Let f and g be functions in $\mathbb{F}(\mathbb{R})$ defined by $f(x) = \frac{1}{4}x - 1$ and $g(x) = \sin(2x)$. Sketch the graphs of $3f$, $-\frac{1}{2}g$, and $3f - \frac{1}{2}g$.*

The function $3f$ is defined by $(3f)(x) = (3)(f(x)) = 3(\frac{1}{4}x - 1) = \frac{3}{4}x - 3$. This scalar multiplication has the effect of stretching the graph of the line $y = \frac{1}{4}x - 1$, with slope $\frac{1}{4}$, away from the x-axis to yield a line with slope $\frac{3}{4}$. See Figure 1.1.

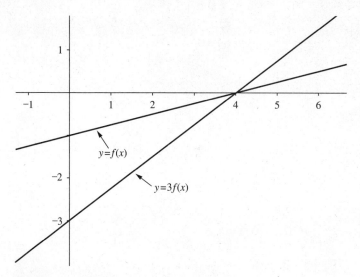

Figure 1.1 Multiplication by the scalar 3.

The function $-\frac{1}{2}g$ is defined by $(-\frac{1}{2}g)(x) = (-\frac{1}{2})(g(x)) = -\frac{1}{2}\sin(2x)$. Multiplication by the scalar $-\frac{1}{2}$ flips the graph across the x-axis and squeezes it toward the x-axis so that its oscillations have an amplitude of $\frac{1}{2}$. See Figure 1.2.

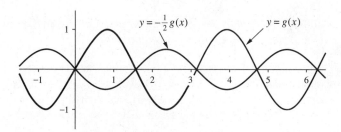

Figure 1.2 Multiplication by the scalar $-\frac{1}{2}$.

The function $3f - \frac{1}{2}g$ is defined by

$$\begin{aligned}
\left(3f - \tfrac{1}{2}g\right)(x) &= \left(3f\right)(x) - \left(\tfrac{1}{2}g\right)(x) \\
&= 3f(x) - \tfrac{1}{2}g(x) \\
&= 3(\tfrac{1}{4}x - 1) - \tfrac{1}{2}\sin(2x).
\end{aligned}$$

Adding the y-values of the two scalar multiples $3f$ and $-\frac{1}{2}g$ has the appearance of tilting the axis of the trigonometric function $y = -\frac{1}{2}\sin(2x)$ so that it oscillates about the line $y = \frac{3}{4}x - 3$. Although the oscillations are still vertical, they are no longer perpendicular to the axis. See Figure 1.3. ∎

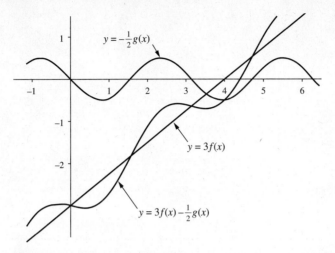

Figure 1.3 Addition of the functions $3f$ and $-\frac{1}{2}g$.

We must not neglect to verify that these function spaces are indeed examples of vector spaces. Once again, the details of verifying the eight axioms become somewhat routine after you see the pattern. Axiom 5 is verified in detail in the following example; you should try a few more and convince yourself that you can see how to do the others. Be sure to notice the function that is the **0** vector in $\mathbb{F}(X)$.

Quick Example   *Verify that the operations of addition and scalar multiplication defined on* $\mathbb{F}(X)$ *satisfy the distributive law of Axiom 5.*

Let f and g be arbitrary elements of $\mathbb{F}(X)$. Let r be an arbitrary real number. Axiom 5 requires that the functions $r(f + g)$ and $rf + rg$ be equal. As mentioned earlier, this means we must show that to each $x \in X$ the two functions assign the same real number. That is, for an arbitrary $x \in X$, we must establish the equality

$$\big(r\,(f + g)\big)(x) = \big(rf + rg\big)(x).$$

The parentheses around x are required by the notation for the value of a function; the other parentheses are for grouping. Starting with the left side we have the following string of equalities:

$$\big(r(f + g)\big)(x) = r\big((f + g)(x)\big) \qquad\qquad \text{definition of scalar multiplication,}$$
$$\text{r times } f + g;$$
$$= r\big(f(x) + g(x)\big) \qquad\qquad \text{definition of addition of } f \text{ and } g;$$
$$= r\big(f(x)\big) + r\big(g(x)\big) \qquad\qquad \text{distributive law of real numbers;}$$
$$= \big(rf\big)(x) + \big(rg\big)(x) \qquad\qquad \text{definition of scalar multiplication,}$$
$$\text{r times } f \text{ and r times } g;$$
$$= \big(rf + rg\big)(x) \qquad\qquad \text{definition of addition of } rf \text{ and } rg. \quad\blacksquare$$

Crossroads   As you may have anticipated, function spaces constitute a fairly versatile family of examples of vector spaces. They recur in many contexts to unify the concepts of differentiation, integration, differential equations, and calculus of several variables. With a little trickery, the examples $\mathbb{R}^n$ and $\mathbb{M}(m, n)$ can even be seen as special cases of $\mathbb{F}(X)$. For example, if we let X be any set with n elements, say $X = \{1, 2, \ldots, n\}$, then a function $f \in \mathbb{F}(X)$ is specified by a list of n real numbers $f(1), f(2), \ldots, f(n)$. Now just think of these as the n coordinates of a point in $\mathbb{R}^n$. Even the notions of addition and scalar multiplication correspond under this identification of $\mathbb{F}(X)$ and $\mathbb{R}^n$. You are asked to verify this in Exercise 8.

Exercises 1.7

1. For each of the following, either explain why the rule gives a well-defined function or find a number that has two values assigned to it.

    a. $f : \mathbb{R} \to \mathbb{R}$ where $f(x)$ is the third digit to the right of the decimal point in the decimal expansion of x.

    b. $g : \mathbb{R} \to \mathbb{R}$ defined by
    $$g(x) = \begin{cases} 0 & \text{if } x \text{ is irrational} \\ p + q & \text{if } x = \frac{p}{q}, \text{ where } p \text{ and } q \text{ are integers.} \end{cases}$$

    c. $h : \mathbb{R} \to \mathbb{R}$ defined by
    $$h(x) = \begin{cases} 0 & \text{if } x \text{ is irrational} \\ m - n & \text{if } x = \frac{2^m p}{2^n q}, \text{ where } p \text{ and } q \text{ are odd integers and} \\ & m \text{ and } n \text{ are nonnegative integers.} \end{cases}$$

    d. $s : [-1, 1] \to \mathbb{R}$ where $y = s(x)$ satisfies the equation $x^2 + y^2 = 1$.

2. Let f and g be functions in $\mathbb{F}(\mathbb{R})$ defined by $f(x) = x^2$ and $g(x) = \cos x$. Determine the following real numbers.

    a. $(f + g)(0)$ **b.** $(f + g)(5)$

    c. $(f + g)(\pi)$ **d.** $(3f)(-4)$

    e. $(-2g)(\pi)$ **f.** $((-1)f + 4g)(2)$

3. Let f and g be functions in $\mathbb{F}(\mathbb{R})$ defined by $f(x) = 1 - x$ and $g(x) = e^x$. Sketch the graphs of the following functions. Describe geometrically how the graphs relate to the graphs of f and g.

    a. $2f$ **b.** $-\frac{1}{2}f$

    c. $\frac{1}{10}g$ **d.** $-g$

    e. $g - f$ **f.** $2f + \frac{1}{10}g$

4. Rewrite the verification that Axiom 5 holds for $\mathbb{F}(\mathbb{R})$ using brackets for grouping and parentheses to denote function evaluation. Written with this distinction, the identity you need to verify looks like $[r[f + g]](x) = [rf + rg](x)$.

5. Let f, g, and h be functions in $\mathbb{F}(\mathbb{R})$ defined by $f(x) = (x+1)^3$, $g(x) = x^3 + 1$, and $h(x) = x^2 + x$. Let a and b denote scalars.

    a. Write the identity in the independent variable x that must hold if the function f is equal to the function $ag + bh$.

    b. Find scalars a and b so that $f = ag + bh$.

6. Let f be the function in $\mathbb{F}(\mathbb{R})$ defined by $f(x) = \sin(x + \frac{\pi}{3})$. Let a and b denote scalars.

    a. Write the identity in the independent variable x that must hold if the function f is equal to the function $a\sin + b\cos$.

    b. Find scalars a and b so that $f = a\sin + b\cos$.

7. Verify that the two operations defined on $\mathbb{F}(X)$ satisfy the remaining seven vector space axioms (Axioms 1 through 4 and 6 through 8).

8. Let $X = \{1, 2, \ldots, n\}$. Consider the correspondence between a function $f \in \mathbb{F}(X)$ and the point $(f(1), f(2), \ldots, f(n))$ in $\mathbb{R}^n$.

    a. Let $f, g \in \mathbb{F}(X)$. Add the two vectors in $\mathbb{R}^n$ corresponding to f and g. Show that the sum is the vector in $\mathbb{R}^n$ corresponding to the function $f + g \in \mathbb{F}(X)$.

    b. Let $f \in \mathbb{F}(X)$. Multiply the vector in $\mathbb{R}^n$ corresponding to f by any real number $r \in \mathbb{R}$. Show that the scalar product is the vector in $\mathbb{R}^n$ corresponding to the function $rf \in \mathbb{F}(X)$.

9. Let X be the set of ordered pairs of integers

$$X = \{(i, j) \mid i \text{ and } j \text{ are integers with } 1 \le i \le m \text{ and } 1 \le j \le n\}.$$

Think of the real number that a function $f \in \mathbb{F}(X)$ assigns to $(i, j) \in X$ as the entry a_{ij} of an $m \times n$ matrix.

    a. Add the matrices corresponding to two functions in $\mathbb{F}(X)$. Show that the sum corresponds to the sum of the functions.

    b. Multiply a real number by the matrix corresponding to a function in $\mathbb{F}(X)$. Show that the product corresponds to the product of the function by the same scalar.

10.   **a.** Show that the set of infinite sequences of real numbers with operations

$$(x_1, x_2, \ldots) + (y_1, y_2, \ldots) = (x_1 + y_1, x_2 + y_2, \ldots),$$
$$r(x_1, x_2, \ldots) = (rx_1, rx_2, \ldots)$$

    forms a vector space.

    b. Reformulate this as an instance of $\mathbb{F}(\mathbb{N})$ where $\mathbb{N} = \{1, 2, 3, \ldots\}$ denotes the set of natural numbers.

    c. Show that addition of infinite sequences is compatible with addition in $\mathbb{F}(\mathbb{N})$.

    d. Show that scalar multiplication of infinite sequences is compatible with scalar multiplication in $\mathbb{F}(\mathbb{N})$.

11.   a. Let $X = \{1, 2\}$ be a set with two elements and let $Y = \{a, b, c\}$ be a set
with three elements. How many functions are there with domain X and
range Y? (Suggestion: How many choices can be made for the value a
function assigns to 1? For each such choice, how many choices can be
made for the value a function assigns to 2?)

   b. Suppose X is a set with m elements and Y is a set with n elements. How
many functions are there with domain X and range Y?

   c. Does your formula for the number of functions from a set with m elements
to a set with n elements work (or even make sense) when $m = 0$ or $n = 0$?

12.   Let X be any nonempty set and let n be a positive integer. Let V denote
the set of all functions $f : X \to \mathbb{R}^n$. Any element of V corresponds to an
ordered list of functions $f_i : X \to \mathbb{R}$ for $i = 1, 2, \ldots, n$ such that $f(x) =
(f_1(x), f_2(x), \ldots, f_n(x))$ for all $x \in X$. Define addition and scalar multiplica-
tion on V in terms of the corresponding operations on the coordinate functions.
Show that with these operations, V is a vector space.

1.8 Subspaces

As you think about the relations among the examples of vector spaces presented in the
previous section, you will often find that one vector space can be seen as embedded
within a larger vector space. For example, it is very common to view $\mathbb{R}^2$ as the plane in $\mathbb{R}^3$
consisting of all elements whose third coordinate is zero. This identification of the ordered
pair $(v_1, v_2) \in \mathbb{R}^2$ with the ordered triple $(v_1, v_2, 0) \in \mathbb{R}^3$ is especially convenient since
the sum of two vectors in $\mathbb{R}^2$ corresponds to the sum of the vectors considered as lying
in $\mathbb{R}^3$. That is, $(v_1, v_2) + (w_1, w_2)$ corresponds to $(v_1, v_2, 0) + (w_1, w_2, 0)$. A similar relation
holds for the operation of scalar multiplication.

Although it would be stretching the truth to say that (v_1, v_2) is equal to $(v_1, v_2, 0)$,
we do want to investigate the situation of a vector space containing other vector spaces.
We will find, for example, that any plane in $\mathbb{R}^3$ passing through the origin is itself a
vector space. This notion of one vector space being contained in another is made precise
in the following definition. The definition takes careful notice of the compatibility of the
operations in the two vector spaces.

Definition 1.10

A **subspace** of a vector space V is a subset $S \subseteq V$ such that when the addition
and scalar multiplication of V are used to add and scalar-multiply the elements
of S, then S is a vector space.

Let us check that this definition applies to the plane $S = \{(v_1, v_2, 0) \mid v_1, v_2 \in \mathbb{R}\}$
considered above. First of all it is clear that $S \subseteq \mathbb{R}^3$, so $\mathbb{R}^3$ can play the role of the parent
vector space in the definition of subspace. The vector space operations are just the ordi-
nary operations of $\mathbb{R}^3$ restricted to the elements of S. Thus, $(v_1, v_2, 0) + (w_1, w_2, 0) =
(v_1 + w_1, v_2 + w_2, 0)$ and $r(v_1, v_2, 0) = (rv_1, rv_2, 0)$. Since the resulting vectors are in S,

we can now proceed to check the eight axioms. Unlike the examples in the previous sections, where we skipped most of the verifications out of sheer boredom, this time there is a mathematically legitimate reason for skipping all but the additive identity axiom and the additive inverse axiom.

Notice that each of the eight axioms requires a certain identity to hold in S. The fact that the identities hold for all elements of V automatically means that they will hold when the elements of S are plugged in. It is, of course, essential here for S to be a subset of V and for the operations in S to be inherited from those in V. The only trouble with the additive identity axiom is that $\mathbf{0}$ might not be in S. That is easily overcome in our example, because $\mathbf{0} = (0, 0, 0)$ clearly satisfies the defining condition to be an element of S. Since this is the zero vector in $\mathbb{R}^3$, it must be the zero vector in S; that is, it will satisfy the additive identity condition of Axiom 3 for all elements of S. The only trouble with the additive inverse axiom is that there might be some $\mathbf{v} \in S$ where $-\mathbf{v} \notin S$. Again we see there is nothing to worry about since if $\mathbf{v} = (v_1, v_2, 0)$ is in S, then $-\mathbf{v} = (-v_1, -v_2, 0)$ is also in S. Of course, $\mathbf{v} + (-\mathbf{v}) = \mathbf{0}$, as is required by Axiom 4. We conclude that S is a vector space and hence a subspace of $\mathbb{R}^3$.

Actually, the remarks made for this simple example hold in the setting of an arbitrary vector space. The following proof of the Subspace Theorem is just a matter of interpreting these remarks in the general setting. A slight bit of trickery even eliminates the need to check the additive identity and additive inverse axioms.

Recall from Section 1.1 the idea of a set being closed under an operation. To say S is **closed under addition** means that for any elements $\mathbf{v}$ and $\mathbf{w}$ in S, the sum $\mathbf{v} + \mathbf{w}$ is also in S. To say S is **closed under scalar multiplication** means that if $\mathbf{v}$ is in S and r is a real number, then $r\mathbf{v}$ is in S. For a subset S of a vector space V to be closed under the vector space operations of V just means that these operations define corresponding operations for S.

Theorem 1.11   Subspace Theorem

Suppose a subset S of a vector space V satisfies the following conditions:

1. S is nonempty.
2. S is closed under addition.
3. S is closed under scalar multiplication.

Then S is a subspace of V.

Proof   Since the identities in the eight axioms hold for all vectors in V, they hold automatically for the vectors in the subset S. The only difficulties are the additive identity axiom and the additive inverse axiom, which require a vector space to contain certain elements. Well, to show that S contains the $\mathbf{0}$ vector of V, just take any vector $\mathbf{v}$ in S (here is where it is vital that S be nonempty) and multiply it by the real number 0. We know that $0\mathbf{v} = \mathbf{0}$ and, since S is closed under scalar multiplication, this is an element of S. Hence, the additive identity axiom is satisfied. To show that $-\mathbf{v}$ is in S whenever $\mathbf{v}$ is in S, just multiply $\mathbf{v}$ by the real number -1. We know that $(-1)\mathbf{v} = -\mathbf{v}$. Again since S is closed under scalar multiplication, we know further that this is an element of S. So the additive inverse axiom is also satisfied. ■

The Subspace Theorem greatly simplifies the process of showing that a subset of a known vector space is a subspace. The following example illustrates the typical application of this theorem.

Consider the set $S = \{(x, y, z) \in \mathbb{R}^3 \mid 2x + 3y - z = 0\}$. Let us show that S with the standard notions of addition and scalar multiplication is a vector space. We will use the Subspace Theorem to show it is in fact a subspace of $\mathbb{R}^3$. Since S is defined to be a subset of $\mathbb{R}^3$, we need to check only that S is nonempty, that S is closed under addition, and that S is closed under scalar multiplication. To show that S is nonempty, let us check that the zero vector $\mathbf{0} = (0, 0, 0)$ in $\mathbb{R}^3$ is in S. (The zero vector must be in every vector space. Furthermore, it is usually the easiest vector to work with.) The condition for $(0, 0, 0)$ to be in S checks out: $2 \cdot 0 + 3 \cdot 0 - 0 = 0$, so $(0, 0, 0) \in S$. Now take two arbitrary elements $\mathbf{v} = (x, y, z)$ and $\mathbf{v}' = (x', y', z')$ of S. This means that $2x + 3y - z = 0$ and $2x' + 3y' - z' = 0$. We want to show that $\mathbf{v} + \mathbf{v}' = (x + x', y + y', z + z') \in S$. That is, we are required to show that $2(x + x') + 3(y + y') - (z + z') = 0$. This is easy to deduce from the two previous equations: simply add together the corresponding sides and rearrange some of the terms. More formally, we have the string of equalities

$$2(x + x') + 3(y + y') - (z + z') = (2x + 3y - z) + (2x' + 3y' - z')$$
$$= 0 + 0$$
$$= 0.$$

Similarly, take $r \in \mathbb{R}$ and $\mathbf{v} = (x, y, z) \in S$ (again this means that $2x + 3y - z = 0$). To show $r\mathbf{v} = (rx, ry, rz) \in S$, we need to show $2(rx) + 3(ry) - (rz) = 0$. Again this is easy:

$$2(rx) + 3(ry) - (rz) = r(2x + 3y - z)$$
$$= r0$$
$$= 0.$$

In the process of applying the Subspace Theorem, you may find that a nonempty subset of a vector space is not closed under one of the vector space operations. This means, of course, that the set fails to be a subspace. For example, consider the half plane

$$H = \{(v_1, v_2) \in \mathbb{R}^2 \mid v_1 \geq 0\}.$$

It is easy to check that H is closed under addition. Indeed, if (v_1, v_2) and (w_1, w_2) are elements of H, then $v_1 \geq 0$ and $w_1 \geq 0$, so $v_1 + w_1 \geq 0$. Hence, $(v_1, v_2) + (w_1, w_2) = (v_1 + w_1, v_2 + w_2)$ satisfies the condition to be an element of H. However, an element of H such as $(1, 0)$ multiplied by a negative scalar such as -1 gives the product $-1(1, 0) = (-1, 0)$, which is not in H. This specific counterexample to closure under scalar multiplication demonstrates that H is not a subspace of $\mathbb{R}^2$.

Mathematical Strategy Session   You will quickly become familiar with checking the three conditions of the Subspace Theorem for a given set S

to be a subspace of a vector space. The three hypotheses are: S is nonempty (it is usually trivial to show that $\mathbf{0} \in S$), S is closed under addition, and S is closed under scalar multiplication. Once these verifications are down to a routine, the commentary accompanying an application of the Subspace Theorem can be drastically reduced.

The following examples illustrate the Subspace Theorem in a variety of situations. In each example we are dealing with a subset of a known vector space, and, of course, we use the operations of addition and scalar multiplication inherited from that vector space.

Quick Example   *Show that* $S = \left\{ \begin{bmatrix} a & b \\ c & d \end{bmatrix} \in \mathbb{M}(2, 2) \ \middle| \ a + b + c + d = 0 \right\}$ *is a subspace of* $\mathbb{M}(2, 2)$.

The zero matrix $\mathbf{0} = \begin{bmatrix} 0 & 0 \\ 0 & 0 \end{bmatrix}$ has $a = b = c = d = 0$, so certainly $a + b + c + d = 0$. Thus, $\mathbf{0} \in S$, so $S \neq \varnothing$.

To show S is closed under addition, take two elements $\begin{bmatrix} a & b \\ c & d \end{bmatrix}$ and $\begin{bmatrix} a' & b' \\ c' & d' \end{bmatrix}$ in S. That is, the entries of these matrices satisfy $a + b + c + d = 0$ and $a' + b' + c' + d' = 0$. Now

$$\begin{bmatrix} a & b \\ c & d \end{bmatrix} + \begin{bmatrix} a' & b' \\ c' & d' \end{bmatrix} = \begin{bmatrix} a + a' & b + b' \\ c + c' & d + d' \end{bmatrix},$$

and it follows easily that this sum satisfies the condition to be in S:

$$(a + a') + (b + b') + (c + c') + (d + d')$$
$$= (a + b + c + d) + (a' + b' + c' + d')$$
$$= 0 + 0$$
$$= 0.$$

Hence, S is closed under addition.

To show S is closed under scalar multiplication, take a real number $r \in \mathbb{R}$ along with the element $\begin{bmatrix} a & b \\ c & d \end{bmatrix} \in S$. Again, the condition $a + b + c + d = 0$ makes it easy to show that the scalar product

$$r \begin{bmatrix} a & b \\ c & d \end{bmatrix} = \begin{bmatrix} ra & rb \\ rc & rd \end{bmatrix}$$

is in S. Indeed, $ra + rb + rc + rd = r(a + b + c + d) = r0 = 0$. Hence, S is closed under scalar multiplication.

By the Subspace Theorem, S is a subspace of $\mathbb{M}(2, 2)$.   ■

Quick Example   *Show that the set* $\mathbb{P}_2$ *of all polynomials of degree 2 or lower forms a subspace of* $\mathbb{F}(\mathbb{R})$.

The elements in $\mathbb{P}_2$ are functions p defined by formulas of the form $p(x) = a_2x^2 + a_1x + a_0$ for $a_2, a_1, a_0 \in \mathbb{R}$. We can choose $a_2 = a_1 = a_0 = 0$ to obtain the zero polynomial as a specific element in $\mathbb{P}_2$ and conclude that $\mathbb{P}_2 \neq \emptyset$.

Let p and q be two elements in $\mathbb{P}_2$. That is, $p(x) = a_2x^2 + a_1x + a_0$ and $q(x) = b_2x^2 + b_1x + b_0$ for some real numbers $a_2, a_1, a_0, b_2, b_1, b_0$. The function $p + q$ is defined by

$$
\begin{aligned}
(p + q)(x) &= p(x) + q(x) \\
&= (a_2x^2 + a_1x + a_0) + (b_2x^2 + b_1x + b_0) \\
&= (a_2 + b_2)x^2 + (a_1 + b_1)x + (a_0 + b_0).
\end{aligned}
$$

Since the coefficients $a_2 + b_2$, $a_1 + b_1$, and $a_0 + b_0$ are real numbers, this is certainly an element of $\mathbb{P}_2$. Hence, $\mathbb{P}_2$ is closed under addition.

Let r denote a real number. The product of r and the polynomial p is the function rp defined by

$$
\begin{aligned}
(rp)(x) &= r(a_2x^2 + a_1x + a_0) \\
&= (ra_2)x^2 + (ra_1)x + ra_0.
\end{aligned}
$$

Again, the coefficients ra_2, ra_1, and ra_0 are real numbers, so rp is an element of $\mathbb{P}_2$. Hence, $\mathbb{P}_2$ is closed under scalar multiplication.

By the Subspace Theorem, $\mathbb{P}_2$ is a subspace of $\mathbb{F}(\mathbb{R})$. ∎

Quick Example   *Show that the set* $\mathbb{C}(\mathbb{R}) = \{f \in \mathbb{F}(\mathbb{R}) \mid f \text{ is continuous}\}$ *of all continuous functions* $f : \mathbb{R} \to \mathbb{R}$ *is a vector space.*

The zero function is the constant function $\mathbf{0} : \mathbb{R} \to \mathbb{R}$ defined by $\mathbf{0}(x) = 0$ for all $x \in \mathbb{R}$. We know from calculus that such a constant function is continuous. Thus, $\mathbb{C}(\mathbb{R}) \neq \emptyset$. If f and g are continuous, we know from a theorem in calculus that $f + g$ is continuous. Thus, $\mathbb{C}(\mathbb{R})$ is closed under addition. If f is continuous and $r \in \mathbb{R}$, then we know again from a theorem in calculus that rf is continuous. Thus, $\mathbb{C}(\mathbb{R})$ is closed under scalar multiplication. Hence, by the Subspace Theorem, $\mathbb{C}(\mathbb{R})$ is a subspace of $\mathbb{F}(\mathbb{R})$. In particular, it is a vector space. ∎

Here is a more abstract example.

Quick Example   *Pick two vectors* $\mathbf{v}$ *and* $\mathbf{w}$ *from any vector space V. Let*

$$
S = \{a\mathbf{v} + b\mathbf{w} \mid a \in \mathbb{R} \text{ and } b \in \mathbb{R}\}.
$$

In words, S consists of all possible sums of scalar multiples of $\mathbf{v}$ *and* $\mathbf{w}$. *Prove that S is a subspace of V.*

To show S is nonempty, notice that $\mathbf{0} = 0\mathbf{v} + 0\mathbf{w} \in S$. To show S is closed under addition, take two arbitrary elements of S, say $a\mathbf{v} + b\mathbf{w}$ and $a'\mathbf{v} + b'\mathbf{w}$. Their sum is

$$
\begin{aligned}
(a\mathbf{v} + b\mathbf{w}) + (a'\mathbf{v} + b'\mathbf{w}) &= (a\mathbf{v} + a'\mathbf{v}) + (b\mathbf{w} + b'\mathbf{w}) \\
&= (a + a')\mathbf{v} + (b + b')\mathbf{w},
\end{aligned}
$$

which we see is in S since it is the sum of multiples of the vectors $\mathbf{v}$ and $\mathbf{w}$. To show S is closed under scalar multiplication, take $a\mathbf{v} + b\mathbf{w} \in S$ and $r \in \mathbb{R}$. The product is

$$
\begin{aligned}
r(a\mathbf{v} + b\mathbf{w}) &= r(a\mathbf{v}) + r(b\mathbf{w}) \\
&= (ra)\mathbf{v} + (rb)\mathbf{w},
\end{aligned}
$$

which we again see is in S. By the Subspace Theorem, S is a subspace of V. ∎

    The last example provides a particularly interesting subset in the case that the vector space V is $\mathbb{R}^3$. We shall see in Section 1.9 that such a set represents a plane through the origin provided $\mathbf{v}$ and $\mathbf{w}$ are suitably chosen to point in different directions. We can interpret the general result as saying that the vector space V contains the subspace S that is very much like a plane in $\mathbb{R}^3$.

    Here is a summary of some subspaces of $\mathbb{F}(X)$ that will be of importance in our work with function spaces. The domain X will typically be an interval of $\mathbb{R}$.

$$
\mathbb{C}(X) = \{f \in \mathbb{F}(X) \mid f \text{ is continuous on } X\}
$$
$$
\mathbb{D}(X) = \{f \in \mathbb{F}(X) \mid f \text{ is differentiable on } X\}
$$
$$
\mathbb{D}^{(n)}(X) = \{f \in \mathbb{F}(X) \mid \text{ the } n\text{th derivative of } f \text{ exists on } X\}
$$
$$
\mathbb{P}_n = \{f \in \mathbb{F}(\mathbb{R}) \mid f(x) = a_n x^n + \cdots + a_1 x + a_0 \text{ where } a_n, \ldots, a_0 \in \mathbb{R}\}
$$
$$
\mathbb{P} = \{f \in \mathbb{F}(\mathbb{R}) \mid f \text{ is a polynomial of any degree}\}
$$

Exercises 1.8

1.   Show that the line $S = \{(x, y) \in \mathbb{R}^2 \mid y = 3x\}$ is a subspace of $\mathbb{R}^2$.
2.   Show that the line $S = \{(x, y) \in \mathbb{R}^2 \mid y = 2x + 1\}$ is not a subspace of $\mathbb{R}^2$.
3.   Determine whether $S = \{(x, y) \in \mathbb{R}^2 \mid x \text{ and } y \text{ are integers}\}$ is a subspace of $\mathbb{R}^2$.
4.   Determine whether $S = \{(x, y) \in \mathbb{R}^2 \mid x^2 + y^2 < 0\}$ is a subspace of $\mathbb{R}^2$.
5.   Find a subset of $\mathbb{R}^2$ that is closed under scalar multiplication but is not closed under addition.
6.   Show that

$$
S = \left\{ \begin{bmatrix} a & b \\ c & d \end{bmatrix} \in \mathbb{M}(2, 2) \;\middle|\; a = b \text{ and } b + 2c = 0 \right\}
$$

    is a subspace of $\mathbb{M}(2, 2)$.

7.   Show that $\{f \in \mathbb{F}(\mathbb{R}) \mid f(0) = 0\}$ is a subspace of $\mathbb{F}(\mathbb{R})$.
8.   Show that $\{f \in \mathbb{C}([0, 1]) \mid \int_0^1 f(x)\, dx = 0\}$ is a subspace of $\mathbb{C}([0, 1])$.
9.   **a.** Show that $\mathbb{D}(\mathbb{R})$ is a subspace of $\mathbb{F}(\mathbb{R})$. Notice that $\mathbb{D}(\mathbb{R}) \subseteq \mathbb{C}(\mathbb{R}) \subseteq \mathbb{F}(\mathbb{R})$. This gives an interesting example of nesting of subspaces.

      b. Show that $\mathbb{D}^{(n)}(\mathbb{R})$ is a subspace of $\mathbb{F}(\mathbb{R})$.

    c. How are the subspaces $\mathbb{D}^{(1)}(\mathbb{R})$, $\mathbb{D}^{(2)}(\mathbb{R})$, ... nested among $\mathbb{D}(\mathbb{R})$, $\mathbb{C}(\mathbb{R})$, and $\mathbb{F}(\mathbb{R})$?

10. Find subspaces of $\mathbb{R}^2$ that are not nested; that is, neither is a subset of the other.

11. Determine whether $S = \{f \in \mathbb{D}(\mathbb{R}) \mid f'(x) + f(x) = 1 \text{ for all } x \in \mathbb{R}\}$ is a subspace of $\mathbb{D}(\mathbb{R})$.

12. Determine whether $S = \{f \in \mathbb{D}(\mathbb{R}) \mid 2f'(x) + x^2 f(x) = 0 \text{ for all } x \in \mathbb{R}\}$ is a subspace of $\mathbb{D}(\mathbb{R})$.

13. Generalize the example in this section to show for any nonnegative integer n, that $\mathbb{P}_n$ is a subspace of $\mathbb{F}(\mathbb{R})$.

14. Show that $\mathbb{P}$ is a subspace of $\mathbb{F}(\mathbb{R})$.

15. The following result has been attributed to Gertrude Stein: If S is a subspace of a vector space V, and T is a subspace of S, then T is a subspace of V. (Or, in the original formulation: A subspace of a subspace is a subspace.) Explain how this result is an immediate consequence of the definition of a subspace.

16. Suppose S is a subspace of a vector space V.

    a. Show that the additive identity of S is the additive identity of V.

    b. Show that the additive inverse of any vector in S is the same as its additive inverse in V.

17. If $X \subseteq Y$, what is the relation between $\mathbb{F}(X)$ and $\mathbb{F}(Y)$?

18. Let V be an arbitrary vector space. Pick n vectors $\mathbf{v}_1, \mathbf{v}_2, \ldots, \mathbf{v}_n$. Show that

$$S = \{a_1\mathbf{v}_1 + a_2\mathbf{v}_2 + \cdots + a_n\mathbf{v}_n \mid a_1, a_2, \ldots, a_n \in \mathbb{R}\}$$

is a subspace of V. This is a useful generalization of the example in the text.

19. Suppose S and T are subspaces of a vector space V. Show that the intersection $S \cap T$ is also a subspace of V.

20. Find subspaces S and T of $\mathbb{R}^2$ such that the union $S \cup T$ is not a subspace of $\mathbb{R}^2$.

21. In the proof of the Subspace Theorem, the existence of an additive identity and additive inverses are consequences of the closure properties of the vector space operations. Return to Definition 1.1, and try to adapt these arguments to derive the existence of an additive identity and additive inverses as consequences of the closure properties and the other six axioms. What difficulties do you encounter?

1.9 Lines and Planes

In working with a specific example of a vector space, we want an explicit description of the set of vectors that make up the space. Often the clearest description is in terms of a handful of selected vectors that generate the entire space. In this section we will examine how this works for lines and planes as subspaces of Euclidean vector spaces. In Chapter 3 we will find that this same process can be generalized to give us some key concepts for dealing with arbitrary vector spaces.

Mathematical Strategy Session   Have a pencil, an eraser, and plenty of scratch paper handy as you read this section. These items are helpful when you read any mathematics text, but they are especially important in this section. Here you will need to sketch diagrams of constructions that are described in the text and play around with variations of the examples as you read along.

Let us begin by reexamining the vector space $\mathbb{R}^2$ from a geometric point of view. We will represent a vector $\mathbf{v} = (v_1, v_2)$ in $\mathbb{R}^2$ by an arrow in the Cartesian coordinate plane from the origin to the point with coordinates (v_1, v_2). Envision a rectangle with one side along the x-axis from the origin to the point $(v_1, 0)$ and an adjacent side along the y-axis from the origin to the point $(0, v_2)$. (This rectangle will be degenerate if $v_1 = 0$ or $v_2 = 0$.) The arrow representing $\mathbf{v}$ is a diagonal of this rectangle.

You may be familiar with the **parallelogram rule** for adding two vectors $\mathbf{v} = (v_1, v_2)$ and $\mathbf{w} = (w_1, w_2)$ in $\mathbb{R}^2$. The two arrows that represent $\mathbf{v}$ and $\mathbf{w}$ form adjacent sides of a parallelogram. Move the arrow that represents $\mathbf{v}$ by parallel translation so that its tail is at the point with coordinates (w_1, w_2). This translated arrow forms a third side of the parallelogram. Its head coincides with the corner of the parallelogram opposite the origin. If we also translate the rectangle that encloses $\mathbf{v}$ and notice how it lines up with the rectangle that encloses $\mathbf{w}$, it is easy to see that this point has coordinates $(v_1 + w_1, v_2 + w_2) = \mathbf{v} + \mathbf{w}$. Indeed, since $\mathbf{v}$ is v_1 units to the right of the origin and v_2 units above the origin, the parallel translation will have its head v_1 units to the right of $\mathbf{w}$ (at $v_1 + w_1$ units to the right of the origin) and v_2 units above $\mathbf{w}$ (at $v_2 + w_2$ units above the origin). Thus, the arrow from the origin to the opposite corner of the parallelogram represents the sum of the vectors that define the parallelogram. Figure 1.4 shows a typical illustration of the parallelogram rule for adding vectors.

Scalar multiplication also has a simple geometric interpretation in $\mathbb{R}^2$. Choose a nonzero vector $\mathbf{v} = (v_1, v_2)$ in $\mathbb{R}$. Again consider the arrow that represents $\mathbf{v}$. For a positive scalar r, extend (or contract if $r < 1$) this arrow by a factor of r. Keep the tail of the new arrow at the origin. If we also magnify the rectangle that encloses the arrow for $\mathbf{v}$ by a factor of r, we see that the new arrow is the diagonal of the new rectangle. Hence, the head of the new arrow is at the point with coordinates $(rv_1, rv_2) = r\mathbf{v}$. For a negative value of r, extend the arrow in the opposite direction by a factor of $|r| = -r$. Again the head will be at the point with coordinates $(-|r|v_1, -|r|v_2) = (rv_1, rv_2) = r\mathbf{v}$. Figure 1.5 illustrates various multiples of a typical nonzero vector $\mathbf{v} \in \mathbb{R}^2$.

This interplay between geometry and algebra also works in $\mathbb{R}^3$. Vectors in $\mathbb{R}^3$ can be represented by arrows emanating from the origin. Two vectors determine a parallelogram, and the arrow from the vertex to the opposite corner of the parallelogram represents the sum of the two original vectors. Scalar multiples of a vector stretch, shrink, or reflect the vector according to the magnitude and sign of the scalar.

We can adapt these ideas to higher-dimensional Euclidean spaces to get a sense of geometry in $\mathbb{R}^n$ for $n > 3$. Since points corresponding to two vectors $\mathbf{v}$ and $\mathbf{w}$ will always lie in a plane through the origin, the geometric constructions of the parallelogram rule for addition can be carried out in this plane. Similarly, scalar multiples of a vector can be constructed in the line through the origin and the point corresponding to the vector.

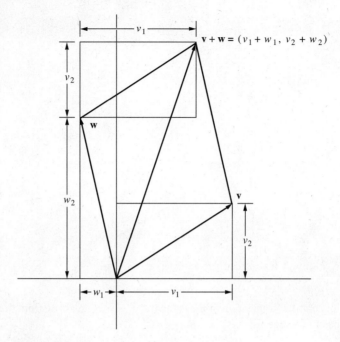

Figure 1.4 The parallelogram rule for addition.

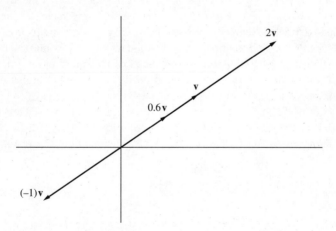

Figure 1.5 Scalar multiples of a vector **v**.

Quick Example *Suppose **v** and **w** are vectors in Euclidean space. Use the identity* $(\mathbf{v} - \mathbf{w}) + \mathbf{w} = \mathbf{v}$ *of Theorem 1.7, part g, to show that the arrow representing* $\mathbf{v} - \mathbf{w}$ *is a parallel translation of the arrow from the head of* **w** *to the head of* **v**.

The parallelogram rule for adding $\mathbf{v} - \mathbf{w}$ to **w** instructs us to move the arrow for $\mathbf{v} - \mathbf{w}$ by a parallel translation so that its tail is at the head of **w**. The head of this translated arrow determines the head of the arrow for the sum $(\mathbf{v} - \mathbf{w}) + \mathbf{w} = \mathbf{v}$.

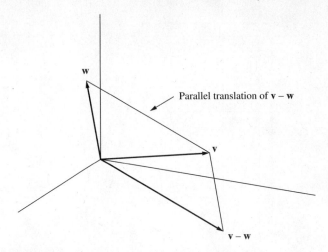

Figure 1.6 Geometric representation of $\mathbf{v} - \mathbf{w}$.

Thus, this parallel translation of the arrow for $\mathbf{v} - \mathbf{w}$ is an arrow from the head of $\mathbf{w}$ to the head of $\mathbf{v}$. See Figure 1.6. ∎

Here is how these ideas lead to a simple way of writing a line in Euclidean space as sets of vectors. Choose any nonzero vector $\mathbf{v}$ in $\mathbb{R}^2$ or $\mathbb{R}^3$. Consider all the vectors that result from multiplying $\mathbf{v}$ by different real numbers. As the value of the scalar r varies, the vectors $r\mathbf{v}$ trace out longer and shorter versions of $\mathbf{v}$. These vectors all lie on the line through $\mathbf{v}$ and the origin. Negative values of r also give vectors on this line. In fact, the set $\{r\mathbf{v} \mid r \in \mathbb{R}\}$ of all such multiples exactly coincides with this line. Notice that this set-theoretical description gives a unified approach to lines in $\mathbb{R}^2$ or $\mathbb{R}^3$. Furthermore, it suggests that we might use the same notation for defining lines in higher-dimensional Euclidean spaces and even for other vector spaces.

Let us examine another simple case. This time consider the two vectors $\mathbf{e}_1 = (1, 0)$ and $\mathbf{e}_2 = (0, 1)$ of the vector space $\mathbb{R}^2$. Take an arbitrary vector $(r, s) \in \mathbb{R}^2$. Multiply $\mathbf{e}_1$ by the real number r and $\mathbf{e}_2$ by the real number s. Then add the two resulting vectors. Clearly,

$$
\begin{aligned}
r\mathbf{e}_1 + s\mathbf{e}_2 &= r(1, 0) + s(0, 1) \\
&= (r, 0) + (0, s) \\
&= (r + 0, 0 + s) \\
&= (r, s).
\end{aligned}
$$

Thus, we see how to obtain any vector in $\mathbb{R}^2$ from $\mathbf{e}_1$ and $\mathbf{e}_2$ by use of the vector space operations. This is the sense in which we can say that $\mathbf{e}_1$ and $\mathbf{e}_2$ generate $\mathbb{R}^2$.

The two vectors $\mathbf{e}_1$ and $\mathbf{e}_2$ are particularly easy to use in generating all of $\mathbb{R}^2$, but other pairs of vectors would do just as well. For example, consider the two vectors $\mathbf{v}$ and $\mathbf{w}$ of $\mathbb{R}^2$ that are represented in Figure 1.7 as arrows in the Cartesian coordinate plane. For any point $\mathbf{x}$ in $\mathbb{R}^2$, draw lines through $\mathbf{x}$ that are parallel to the arrows representing $\mathbf{v}$ and $\mathbf{w}$. Now multiply $\mathbf{v}$ by the appropriate scalar r so that the product $r\mathbf{v}$ is represented by an arrow from the origin to the line parallel to the arrow for $\mathbf{w}$. Similarly, multiply

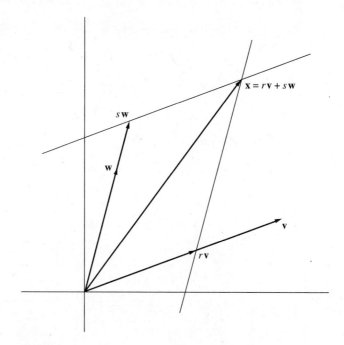

Figure 1.7   The two vectors **v** and **w** generate $\mathbb{R}^2$.

w by the appropriate scalar s so that the product $s\mathbf{w}$ is represented by an arrow from the origin to the line parallel to the arrow for **v**. Now $r\mathbf{v}$ and $s\mathbf{w}$ are represented as arrows along two sides of a parallelogram. The vertex opposite the origin is the point **x**. Thus, by the parallelogram rule of vector addition in $\mathbb{R}^2$, we see that $r\mathbf{v} + s\mathbf{w} = \mathbf{x}$. This geometric construction shows how to find the scalars r and s needed to obtain any vector **x** as a sum of scalar multiples of **v** and **w**. Other than the fact that **v** and **w** are vectors in different directions (that is, the arrows that represent them do not lie along the same line), there is really nothing special about these two vectors. Later we will reformulate this discussion in a more algebraic setting. For now, let us simply observe that the geometric argument works for any plane through the origin of $\mathbb{R}^3$. That is, we can generate any plane in $\mathbb{R}^3$ that passes through $(0, 0, 0)$ using two vectors. In fact, any two vectors **v** and **w** in the plane that have different directions will do the job: the set

$$\{r\mathbf{v} + s\mathbf{w} \mid r, s \in \mathbb{R}\}$$

of all vectors that can be written as a sum of scalar multiples of **v** and **w** will coincide with the plane. Notice also that **v**, **w**, and **0** are in this set (you are asked to verify this in Exercise 3 at the end of this section). Hence, this is the plane through these three points.

The fact that a single nonzero vector generates a line and that two vectors in different directions are needed to generate a plane corresponds to our geometric intuition that a line is somehow one-dimensional whereas a plane is two-dimensional.

What about lines and planes that do not contain the origin? Well, it is a simple matter to translate a line or plane that does contain the origin to a parallel copy passing through any prescribed point. For example, Figure 1.8 shows how adding the vector $\mathbf{x} \in \mathbb{R}^2$ to every vector in the line $\{r\mathbf{v} \mid r \in \mathbb{R}\}$ yields a parallel line that passes through **x**. This trick also works for lines and planes in $\mathbb{R}^3$. Thus, we can write the line through **x**

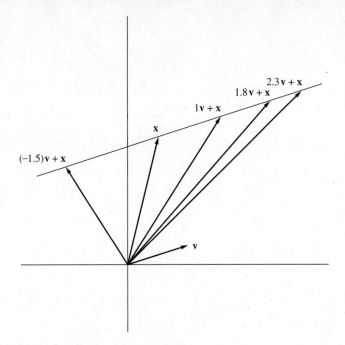

Figure 1.8 Generating a line through a vector **x**.

that is parallel to the line through **0** in the direction of **v** as the set $\{r\mathbf{v} + \mathbf{x} \mid r \in \mathbb{R}\}$. Notice that the difference between any two distinct vectors $r\mathbf{v} + \mathbf{x}$ and $s\mathbf{v} + \mathbf{x}$ on this line is $(r\mathbf{v}+\mathbf{x}) - (s\mathbf{v}+\mathbf{x}) = (r-s)\mathbf{v}$, a nonzero multiple of **v**. So if an arbitrary line is given in $\mathbb{R}^2$ or $\mathbb{R}^3$, any point on the line can be used as the translation vector **x**, and the difference between any two distinct vectors on the line can be used as the direction vector **v**. Once these vectors are determined, the line can be written as the set $\{r\mathbf{v} + \mathbf{x} \mid r \in \mathbb{R}\}$.

Similarly, the plane $\{r\mathbf{v} + s\mathbf{w} \mid r, s \in \mathbb{R}\}$ through the origin in $\mathbb{R}^3$ can be translated to pass through an arbitrary point **x** by adding **x** to each point of the plane. The set $\{r\mathbf{v} + s\mathbf{w} + \mathbf{x} \mid r, s \in \mathbb{R}\}$, then, is the resulting plane.

Quick Example   *Write the plane in $\mathbb{R}^3$ through the points $(1, 2, 3)$, $(0, 2, 1)$, and $(3, -1, 1)$ as a set in the form indicated above.*

Choose one of these points as the translation vector, say $\mathbf{x} = (1, 2, 3)$. The difference between any pair of points is a parallel translation of the arrow between the two points. Thus, the differences between any two pairs can serve as the direction vectors, for example, $\mathbf{v} = (1, 2, 3) - (0, 2, 1) = (1, 0, 2)$ and $\mathbf{w} = (1, 2, 3) - (3, -1, 1) = (-2, 3, 2)$. Notice that **v** is not a multiple of **w**, so the directions are distinct, and we can write the plane in the desired form:

$$\{r(1, 0, 2) + s(-2, 3, 2) + (1, 2, 3) \mid r, s \in \mathbb{R}\}. \quad \blacksquare$$

The only problem that could arise in the previous example is if **v** and **w** are in the same direction. However, this will happen only if the three original points lie on a single line in the plane. In this case the three points would not uniquely determine the plane of interest.

When n is greater than 3, we lose our geometric feel for the vector spaces $\mathbb{R}^n$. And, of course, we have other examples of vector spaces that further defy geometric intuition. Let us therefore use the geometric concepts developed in $\mathbb{R}^2$ and $\mathbb{R}^3$ to extend the notions of line and plane to arbitrary vector spaces.

Definition 1.12

Suppose V is a vector space. Let **v** be a nonzero vector in V. The **line** through a vector $\mathbf{x} \in V$ in the direction **v** is the set $\{r\mathbf{v} + \mathbf{x} \mid r \in \mathbb{R}\}$. If **v**, **w**, and **x** are three vectors in V with neither **v** nor **w** being a multiple of the other, then the **plane** through **x** in the directions **v** and **w** is the set $\{r\mathbf{v} + s\mathbf{w} + \mathbf{x} \mid r, s \in \mathbb{R}\}$.

Here are some examples to illustrate how geometric descriptions of lines and planes can be used to find the appropriate translation and direction vectors.

Suppose we want to describe the set of points (x, y) in $\mathbb{R}^2$ whose coordinates satisfy the equation $2x + 5y = 4$. If you already believe that this is a line, just find two points on it, say $(2, 0)$ and $(-3, 2)$. Compute the direction vector $(2, 0) - (-3, 2) = (5, -2)$. The line is $\{r(5, -2) + (-3, 2) \mid r \in \mathbb{R}\}$. If you want to be convinced that $\{(x, y) \in \mathbb{R}^2 \mid 2x + 5y = 4\}$ is a line, think of assigning to y an arbitrary real value r and solving for x. The two equations $y = r$ and $x = (4 - 5r)/2$ can be written as a single vector equation $(x, y) = (2 - \frac{5}{2}r, r)$ or, equivalently, $(x, y) = r(-\frac{5}{2}, 1) + (2, 0)$. Thus, the above set equals $\{r(-\frac{5}{2}, 1) + (2, 0) \mid r \in \mathbb{R}\}$. By choosing the arbitrary real number to be $2r$, we can eliminate the fractions and further simplify the expression for the line to

$$\{r(-5, 2) + (2, 0) \mid r \in \mathbb{R}\}.$$

Keep in mind that a line has many possible direction and translation vectors. As can be observed in the preceding example, the direction vectors $(5, -2)$, $(-\frac{5}{2}, 1)$, and $(-5, 2)$ are multiples of one another. The translation vector can be any point on the line.

Quick Example *Describe the set $\{(x, y, z) \in \mathbb{R}^3 \mid 6x + 2y - z = 6\}$ as a plane.*

Assign arbitrary real values to y and z, and solve for x. This gives the three equations

$$x = -\tfrac{1}{3}r + \tfrac{1}{6}s + 1$$
$$y = r$$
$$z = s$$

As a single vector equation, this is

$$(x, y, z) = \left(-\tfrac{1}{3}r + \tfrac{1}{6}s + 1,\ r,\ s\right)$$
$$= r\left(-\tfrac{1}{3}, 1, 0\right) + s\left(\tfrac{1}{6}, 0, 1\right) + (1, 0, 0).$$

Thus, we have the plane

$$\left\{ r\left(-\tfrac{1}{3}, 1, 0\right) + s\left(\tfrac{1}{6}, 0, 1\right) + (1, 0, 0) \mid r, s \in \mathbb{R} \right\}.$$

We can eliminate the fractions by choosing the scalar multiples to be $3r$ and $6s$. This gives a slightly simpler expression for this plane:

$$\{ r(-1, 3, 0) + s(1, 0, 6) + (1, 0, 0) \mid r, s \in \mathbb{R} \}. \quad \blacksquare$$

Exercises 1.9

1. Let (v_1, v_2) and (w_1, w_2) be points in $\mathbb{R}^2$ other than the origin.
   a. Verify that the line through $(0, 0)$ and (v_1, v_2) is parallel to the line through (w_1, w_2) and $(v_1 + w_1, v_2 + w_2)$.
   b. Verify that the distance between $(0, 0)$ and (v_1, v_2) is equal to the distance between (w_1, w_2) and $(v_1 + w_1, v_2 + w_2)$.
   c. Conclude that the points $(0, 0)$, (v_1, v_2), $(v_1 + w_1, v_2 + w_2)$, and (w_1, w_2) are the vertices of a parallelogram.

2. Let (v_1, v_2) be a point in $\mathbb{R}^2$ other than the origin. Let $r \in \mathbb{R}$.
   a. Verify that the line through $(0, 0)$ and (v_1, v_2) is the same as the line through $(0, 0)$ and (rv_1, rv_2).
   b. Verify that the distance between $(0, 0)$ and (rv_1, rv_2) is $|r|$ times the distance between $(0, 0)$ and (v_1, v_2).

3. Suppose **v** and **w** are elements of a vector space V. Show that **v**, **w**, and **0** are elements of the subset $\{r\mathbf{v} + s\mathbf{w} \mid r, s \in \mathbb{R}\}$.

4. Suppose **v** and **w** are elements of a vector space $V \neq \{\mathbf{0}\}$.
   a. Show that if one of these vectors is a multiple of the other, then there is a line through **0** containing both of them. Be careful of the cases where one or both of the vectors are equal to **0**.
   b. Show that if **v** and **w** are on a line through the origin, then one of these two vectors is a multiple of the other.

5. Write the line in $\mathbb{R}^2$ through the points $(1, -2)$ and $(2, 3)$ in the form $\{r\mathbf{v} + \mathbf{x} \mid r \in \mathbb{R}\}$.

6. Write the line in $\mathbb{R}^2$ that is tangent to the unit circle at the point $(0.6, 0.8)$ in the form $\{r\mathbf{v} + \mathbf{x} \mid r \in \mathbb{R}\}$.

7. Write the y-axis as a line in $\mathbb{R}^2$.

8. Write $\{(x, y) \in \mathbb{R}^2 \mid -3x + 4y = 1\}$ as a line in $\mathbb{R}^2$.

9. Show for any coefficients $a, b, c \in \mathbb{R}$, with not both a and b equal to zero, that $\{(x, y) \in \mathbb{R}^2 \mid ax + by = c\}$ is a line in $\mathbb{R}^2$.

10. Write the line in $\mathbb{R}^3$ through $(1, -2, 4)$ and $(-3, 0, 7)$ in the form $\{r\mathbf{v} + \mathbf{x} \mid r \in \mathbb{R}\}$.

11. Write the line through the origin and the opposite corner of the unit cube in the first octant of $\mathbb{R}^3$ in the form $\{r\mathbf{v} + \mathbf{x} \mid r \in \mathbb{R}\}$.

12. Write $\{(x, y, z) \in \mathbb{R}^3 \mid x + y + z = 3 \text{ and } z = 2\}$ as a line in $\mathbb{R}^3$.

13. Write the line in $\mathbb{R}^2$ that is tangent to the unit circle at the point $\left(\frac{1}{\sqrt{2}}, \frac{1}{\sqrt{2}}\right)$ in the form $\{r\mathbf{v} + \mathbf{x} \mid r \in \mathbb{R}\}$. (Suggestion: The tangent line will intersect the axes at points $(a, 0)$ and $(0, a)$ equidistant from the origin. You do not need to know the value of a to find a direction vector.)

14. Write the plane in $\mathbb{R}^3$ that is tangent to the unit sphere at the point $\left(\frac{1}{\sqrt{3}}, \frac{1}{\sqrt{3}}, \frac{1}{\sqrt{3}}\right)$ in the form $\{r\mathbf{v} + s\mathbf{w} + \mathbf{x} \mid r, s \in \mathbb{R}\}$. (Suggestion: To find direction vectors, use the fact that the plane intersects the three coordinate axes at points that are equidistant from the origin.)

15. Write $\{(x, y, z) \in \mathbb{R}^3 \mid x + 3y - 2z = 4\}$ as a plane in $\mathbb{R}^3$.

16. Write the line in $\mathbb{R}^4$ through $(3, 1, 0, -1)$ and $(1, -1, 3, 2)$ in the form $\{r\mathbf{v} + \mathbf{x} \mid r \in \mathbb{R}\}$.

17. Write the set $\left\{ \begin{bmatrix} a & 0 \\ 0 & b \end{bmatrix} \in \mathbb{M}(2, 2) \;\middle|\; a, b \in \mathbb{R} \right\}$ as a plane in $\mathbb{M}(2, 2)$.

18. Write the set $\left\{ \begin{bmatrix} 2a - b & b + 5 \\ a + 2 & a + b \end{bmatrix} \in \mathbb{M}(2, 2) \;\middle|\; a, b \in \mathbb{R} \right\}$ as a plane in $\mathbb{M}(2, 2)$.

19. Write the set of all solutions to the differential equation $y' = y$ as a line in $\mathbb{D}(\mathbb{R})$.

Project: Quotient Spaces

The topics in the concluding section of each chapter are presented as a challenge to students who want to explore some interesting mathematics beyond the basic material of the course. Here is your chance to undertake a project that is more comprehensive and open-ended than the regular exercise sets. You are encouraged to draw analogies and supply examples to develop your insight into these topics. Often your geometric intuition will suggest a route from a vague idea to a rigorous proof.

The first topic uses some basic set theory, the concept of a subspace, and the procedure for verifying the vector-space axioms.

1. Suppose S is a subspace of a vector space V. We want to define a relation between vectors of V. For any two vectors $\mathbf{v}, \mathbf{w} \in V$, we say that $\mathbf{v}$ is **equivalent** to $\mathbf{w}$, and write $\mathbf{v} \sim \mathbf{w}$, if and only if $\mathbf{v} - \mathbf{w} \in S$. Whether or not $\mathbf{v}$ is equivalent to $\mathbf{w}$ depends, of course, on the choice of the subspace S as well as the vectors $\mathbf{v}$ and $\mathbf{w}$. In any given context, however, we will assume that the subspace S is fixed for the duration of the discussion.

Try out the preceding definition. Choose the subspace S to be a line through the origin of $\mathbb{R}^2$ (you need to select a direction vector). Write down some vectors in $\mathbb{R}^2$ that are equivalent to $\mathbf{0}$. What is the set of all vectors in $\mathbb{R}^2$ that are equivalent to $\mathbf{0}$? Choose a vector $\mathbf{v}_0 \in \mathbb{R}^2$. What is the set of all vectors in $\mathbb{R}^2$ that are equivalent to $\mathbf{v}_0$? Consider cases where $\mathbf{v}_0 \in S$ and where $\mathbf{v}_0 \notin S$.

You may want to explore other examples, such as a plane in $\mathbb{R}^3$ or the subspace of $\mathbb{F}(\mathbb{R})$ consisting of all functions f that satisfy $f(1) = 0$.

2. Return now to the general situation of a given subspace S of a vector space V. Show that the equivalence relation satisfies the following conditions for all $\mathbf{v}, \mathbf{w}, \mathbf{x} \in V$.

    a. $\mathbf{v} \sim \mathbf{v}$ (reflexivity)
    b. if $\mathbf{v} \sim \mathbf{w}$, then $\mathbf{w} \sim \mathbf{v}$ (symmetry)
    c. if $\mathbf{v} \sim \mathbf{w}$ and $\mathbf{w} \sim \mathbf{x}$, then $\mathbf{v} \sim \mathbf{x}$ (transitivity)

3. To what extent do these three properties hold for other relations? Consider, for example, the *less-than-or-equal-to* relation among real numbers. How about the relation of divisibility among the positive integers? What happens with the relation on $\mathbb{R}^2$ defined by $(a, b) \sim (c, d)$ if and only if $ad = bc$? Also, be sure to try some genealogical relations on sets of people. Be the first on your block to collect all eight examples to show that the three properties may hold or fail in any combination.

4. For any vector $\mathbf{v}$ in the vector space V, define the **affine space** through $\mathbf{v}$ parallel to the subspace S, to be the set $\{\mathbf{v} + \mathbf{w} \mid \mathbf{w} \in S\}$. We will adopt the convenient and suggestive notation $\mathbf{v} + S$ for this set. Show for any $\mathbf{v} \in V$ that

$$\mathbf{v} + S = \{\mathbf{w} \in V \mid \mathbf{w} \sim \mathbf{v}\}.$$

Illustrate the definition of $\mathbf{v} + S$ with some simple examples. Find all affine subspaces of $\mathbb{R}^2$ parallel to a given line through the origin.

5. Adapt the preceding definition for adding a vector and a subspace to define the sum of two arbitrary subsets of a vector space. If $A = \{(x, 0) \in \mathbb{R}^2 \mid 0 \le x \le 1\}$ and $B = \{(0, y) \in \mathbb{R}^2 \mid 0 \le y \le 1\}$, what is $A + B$ according to your definition? Let C be the unit circle in $\mathbb{R}^2$ (centered at the origin and with radius equal to 1) and let D be the unit square in $\mathbb{R}^2$ (with vertices at $(0, 0)$, $(1, 0)$, $(0, 1)$, and $(1, 1)$). Sketch $C + D$. If E and F are finite sets, what can you say about the number of elements in $E + F$ in terms of the number of elements in E and the number of elements in F? Does there appear to be any relation between the dimensions of two sets and the dimension of their sum?

6. For any $\mathbf{v}, \mathbf{w} \in V$ show that if $\mathbf{v} \sim \mathbf{w}$, then $\mathbf{v} + S = \mathbf{w} + S$.
For any $\mathbf{v}, \mathbf{w} \in V$ show that if $\mathbf{v} + S = \mathbf{w} + S$, then $\mathbf{v} \sim \mathbf{w}$.
For any $\mathbf{v}, \mathbf{w} \in V$ show that if $\mathbf{v} + S \cap \mathbf{w} + S \ne \varnothing$, then $\mathbf{v} \sim \mathbf{w}$.
If S is a line through the origin of $\mathbb{R}^2$, find a way to choose one vector from each affine space parallel to S. Can you choose these vectors so that the set of all selected vectors forms a subspace of $\mathbb{R}^2$? What happens when S is a line through the origin of $\mathbb{R}^3$? What happens when S is a plane through the origin of $\mathbb{R}^3$?

7. Let V/S denote the set of all affine spaces parallel to the subspace S. This set of sets is known as the **quotient space** of V modulo the subspace S (or V mod S, for short). Consider the operation of addition defined on V/S by

$$(\mathbf{v} + S) + (\mathbf{w} + S) = (\mathbf{v} + \mathbf{w}) + S.$$

Show that this operation is well defined in the sense that the sum does not depend on the vectors $\mathbf{v}$ and $\mathbf{w}$ used to represent the affine spaces $\mathbf{v} + S$ and $\mathbf{w} + S$. That is, if $\mathbf{v}' + S = \mathbf{v} + S$ and $\mathbf{w}' + S = \mathbf{w} + S$, then

$$(\mathbf{v}' + S) + (\mathbf{w}' + S) = (\mathbf{v} + S) + (\mathbf{w} + S).$$

How does this definition of addition of affine subspaces compare with your general definition of the sum of two subsets of a vector space?

8. Give a definition for scalar multiplication on V/S that is similar in spirit to the preceding definition of addition of affine spaces. Show it is a well-defined operation. Show that with these operations, V/S is a vector space.

You will need the material in Chapter 3 and Chapter 6 to continue this project.

9. Continue to let S denote a subspace of a vector space V. Show that $P : V \to V/S$ defined by $P(\mathbf{v}) = \mathbf{v}+S$ is a linear function. Show that P is onto V/S. Suppose dim $S = k$ and dim $V = n$. What is dim V/S?

10. Suppose the linear map $T : V \to W$ is onto W. Let $K = \ker T$. Define $\theta : V/K \to W$ by $\theta(\mathbf{v} + K) = T(\mathbf{v})$.

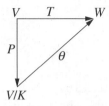

Show that θ is well defined. Show that θ is linear. Show that θ is an isomorphism.

Project: Vector Fields

In describing the motion of a freely falling object, we often assume for the sake of simplicity that the force of gravity is constant throughout the path of the object. Of course, this assumption breaks down if we want to model the motion of a satellite around a planet. Gravitational attraction is inversely proportional to the square of the distance between two objects; over large distances, this is definitely not a constant. A similar problem arises when physicists describe the forces created by a configuration of electrically charged objects. The force cannot be described by the magnitude and direction of a single vector. Instead, the force acting on a charged particle will depend on the position of the charged particle in relation to the configuration.

We need not just a single vector space, but an entire field of vector spaces, one attached to each point of the set in which the field acts. We can then select one vector out of each of the vector spaces to describe a force or other vector quantity that changes as we move around in the underlying space.

1. Start with a **vector field** defined on $\mathbb{R}^2$. That is, associate a vector space V with each point of the underlying set $\mathbb{R}^2$. A vector field assigns to each point of the underlying set $\mathbb{R}^2$ a vector in the vector space V associated with the point. This can be described by a function $\mathbf{F} : \mathbb{R}^2 \to V$. If V is $\mathbb{R}^2$, we can represent the vector $\mathbf{F}(\mathbf{v})$ associated with any point $\mathbf{v}$ by an arrow from $\mathbf{v}$ to $\mathbf{v} + \mathbf{F}(\mathbf{v})$. This arrow is a translation of the standard vector with its tail at the origin that usually depicts $\mathbf{F}(\mathbf{v})$.

Sketch some vectors to represent the vector field defined by $\mathbf{F}(x, y) = (-0.1x, -0.1y)$. Modify this function so that it might describe gravitational attraction in a planar universe. A. K. Dewdney describes such a world in great detail in his book *The Planiverse: Computer Contact with a Two-dimensional World*, published in 1984 by Poseidon Press. Sketch the vector field defined by $\mathbf{F}(x, y) = (-0.2y, 0.2x)$. Modify this to make the arrows circle clockwise. Can you modify the function so that the vector field has an elliptical look or a slight outward spiral? Can you cook up a vector field that spirals around two or more points?

2. Vector fields are often used to describe the motion a particle undergoes if it is free to drift through the field. In this context a vector field is called a **direction field** rather

than a force field. A direction field gives a global picture of motion. This is invaluable to meteorologists and oceanographers, who need to envision large bodies of air or water under the influence of a direction field. The constant direction vector field $F(x, y) = (2, 0)$ represents a constant flow in the positive x-direction. Experiment with modifications of this vector field to create a vector field that might more realistically describe the motion of a river, streamlines in a wind tunnel, ocean currents, or wind in a hurricane.

3. Once you have sketched a few vector fields by hand, you will probably appreciate the power of a computer in creating these diagrams. Look into the availability of a computer algebra system or other mathematical graphics package that can automate the sketching of vector fields from $\mathbb{R}^2$ to $\mathbb{R}^2$. You may even want to write your own program to explore the graphical world of vector fields.

4. A *linear* vector field is one given by a formula whose terms involve only constants multiplied by a variable. In particular, a linear vector field on $\mathbb{R}^2$ is of the form $F(x, y) = (ax + by, cx + dy)$. Plot the motion of an object affected by such a vector field. Try modest values of a, b, c, and d so that the motion in each step is relatively small compared with the sizes of the coordinates.

For the vector field $F(x, y) = (0.24x + 0.08y, 0.03x + 0.26y)$, notice that $F(4, 3) = (1.2, 0.9) = 0.3(4, 3)$. How does this affect the motion? Can you see another vector (x, y) such that $F(x, y)$ is a scalar multiple of (x, y)? How is the direction field for F fundamentally different from the direction field for $G(x, y) = (-y, x)$, where no point (x, y) has a direction vector in the same direction as (x, y) itself? Try other examples to see if you can classify the various kinds of qualitative behavior of linear vector fields.

5. Vector fields are ideal tools for describing force fields and the motion of a large mass of particles. They are also useful for describing change in other situations involving several variables. For example, the variables x and y might represent the population of two species of animals in a certain habitat. A vector field $F : \mathbb{R}^2 \to \mathbb{R}^2$ might describe the change $F(x, y)$ in (x, y) over the course of a year. Notice how the vector field easily gives changes in x and y that depend on the values of x and y. Vector addition describes the situation $(x_1, y_1) = (x_0, y_0) + F(x_0, y_0)$ after one year in terms of the starting values (x_0, y_0). Similarly, $(x_2, y_2) = (x_1, y_1) + F(x_1, y_1)$, and in general $(x_{n+1}, y_{n+1}) = (x_n, y_n) + F(x_n, y_n)$.

Mathematical models of ecological systems received a great deal of attention with the flourishing of mathematical biology in the 1970s. You will find that population models are typically based on differential equations (where the vector field describes instantaneous changes rather than discrete changes). However, these models can easily be converted to the discrete systems we have been considering. Some references are:

Colin Clark, *Mathematical Bioeconomics: The Optimal Management of Renewable Resources*, New York: John Wiley & Sons, 1976.

Raymond Greenwell, "Whales and Krill: A Mathematical Model," UMAP Module 610, *UMAP Journal*, vol. 3, no. 2, 1982, pp. 165–183.

J. Maynard Smith, *Models in Ecology*, New York: Cambridge University Press, 1974.

A classical predator-prey model is given by a vector field of the form $F(x, y) = (ax - bxy, cxy - dy)$, where a, b, c, and d are positive constants (typical values are $a = 0.06$, $b = 0.001$, $c = 0.0001$, and $d = 0.08$ for populations such as $x = 1000$ and $y = 50$). Plot a vector field for a predator-prey model. Compute the population of the two species over a number of years. How do the long-term trends depend on the populations you start with? What initial populations lead to no change? Explain the formula for **F** in terms of what is happening to the individuals in the two populations. Which variable corresponds to the predator species and which to the prey?

Modify the predator-prey vector field to obtain a vector field that models two competing populations. What does your model predict will happen to the populations? Try a model of populations in a symbiotic relation.

6. The base space for a vector field need not be a vector space. For example, the wind on the surface of the earth gives rise to a natural velocity vector field defined on a sphere. This leads to some surprising topological results. For example, at any instant there is at least one spot on the earth where the wind velocity is zero. Also, there always exists a pair of diametrically opposite points on the surface of the earth with the same temperature and barometric pressure. W. G. Chinn and N. E. Steenrod have presented ingenious proofs of these theorems in their book *First Concepts of Topology* (New York: Random House, 1966).

Chapter 1 Summary

Chapter 1 introduces the fundamental ideas of vector spaces. These ideas are generalizations of properties common to three families of classical examples of vector spaces. The axiomatic approach to vector spaces distinguishes the general concepts (such as algebraic laws and subspaces) that pertain to any vector space from specific properties (such as coordinates in $\mathbb{R}^n$, rows and columns of a matrix, domain and range of a function) that are important in individual examples.

Computations

Notation and standard vector space operations
Euclidean spaces, $\mathbb{R}^n$
Matrices, $\mathbb{M}(m, n)$
Function spaces, $\mathbb{F}(X)$
Domain, range, rule
Distinction between a function f and its value $f(x)$

Geometric interpretations
Parallelogram rule for addition in $\mathbb{R}^n$
Stretching, shrinking, reflecting for scalar multiplication
Graphs of sums and scalar products of functions

Theory

Sets
Notation
Operations and relations: element, union, intersection, subsets
Empty set, $\varnothing$
Natural numbers, $\mathbb{N}$
Real numbers, $\mathbb{R}$

Logic and proof techniques
Implication: hypothesis and conclusion
String of equalities
Substitution of equal quantities
Proofs of uniqueness theorems

Analogies
Set containment and ordering of real numbers
Operations on sets, real numbers, and vectors

Vector space axioms capturing the essence of a vector space
Closure of a set under an operation
Verification that a given operation satisfies an axiom
Counterexample when a system does not satisfy the axioms
Proving basic identities from the axioms

Subtraction
Definition in terms of addition and additive inverse
Cancellation Theorem
Tolerance of certain notational ambiguities

Subspaces
Definition as a vector space
Subspace Theorem to simplify verification of axioms
    Nonempty
    Closed under addition
    Closed under scalar multiplication
Common examples of subspaces
    Lines and planes through the origin
    Continuous functions, differentiable functions, polynomials

Applications

Euclidean vectors as lists of data

Geometry
Lines and planes through the origin
Lines and planes through arbitrary points

Qualitative meaning of magnitude and direction of a vector

Review Exercises

1. Explain why the empty set cannot be the underlying set for a vector space.

2. Explain why a subset of the real numbers that is closed under multiplication and contains more than one element must contain a positive number.

3. Suppose $\mathbf{v}$ and $\mathbf{w}$ are elements of a vector space.

    a. Prove that $(\mathbf{w} + \mathbf{v}) + \mathbf{w} = \mathbf{v} + 2\mathbf{w}$.

    b. Prove that $(\frac{1}{2}\mathbf{v} + \mathbf{w}) + \frac{1}{2}\mathbf{v} = \mathbf{v} + \mathbf{w}$.

4. Suppose **v** is an element of a vector space and a is a nonzero real number. Prove that $a\left(\frac{1}{a}\mathbf{v}\right) = \mathbf{v}$.

5. The 14 identities of Theorem 1.7 involve three different uses of the minus sign: the additive inverse of a vector, subtraction of vectors, and subtraction of real numbers. We can distinguish among these three uses if we let $\sim\mathbf{v}$ denote the additive inverse of a vector, $\mathbf{v}\ominus\mathbf{w}$ denote the vector obtained from subtracting **w** from **v**, and $a - b$ denote ordinary subtraction of real numbers. With this notation, Theorem 1.7, part e, becomes $\sim(\mathbf{v} + \mathbf{w}) = (\sim\mathbf{v})\ominus\mathbf{w}$. Rewrite the other identities of Theorem 1.7 with these symbols to distinguish the three different uses of minus signs.

6. Suppose **v**, **w**, and **x** are elements of a vector space. Prove that $\mathbf{v} + (\mathbf{w} - \mathbf{x}) = (\mathbf{v} + \mathbf{w}) - \mathbf{x}$.

7. Believe it or not, $\mathbb{R}^2$ with the operations of addition and scalar multiplication as defined below forms a vector space:

$$(v_1, v_2) + (w_1, w_2) = (v_1 + w_1 - 3, v_2 + w_2 + 5),$$
$$r(v_1, v_2) = (rv_1 - 3r + 3, rv_2 + 5r - 5).$$

a. Determine the additive identity and show that your candidate does indeed satisfy the required conditions to be an additive identity.

b. Verify that these operations on $\mathbb{R}^2$ satisfy the distributive law of Axiom 5:

$$r(\mathbf{v} + \mathbf{w}) = r\mathbf{v} + r\mathbf{w}.$$

8. Consider the nonstandard addition $\oplus$ and scalar multiplication $\odot$ defined on $\mathbb{P}_1$ by

$$(ax + b) \oplus (cx + d) = (b + d)x + (a + c),$$
$$r \odot (ax + b) = rbx + ra.$$

Determine whether $\mathbb{P}_1$ with these operations is a vector space. Either verify the axioms or give an explicit counterexample to one of them.

9. Determine values of real numbers a, b, c, and d so that

$$a\begin{bmatrix} 1 & 0 \\ 1 & 1 \\ 0 & 1 \end{bmatrix} + b\begin{bmatrix} 1 & 0 \\ 2 & 0 \\ 0 & 0 \end{bmatrix} + c\begin{bmatrix} 2 & 0 \\ 1 & -1 \\ 0 & 2 \end{bmatrix} + d\begin{bmatrix} 1 & 0 \\ 0 & 0 \\ 0 & 0 \end{bmatrix} = \begin{bmatrix} -1 & 0 \\ 4 & 0 \\ 0 & 3 \end{bmatrix}$$

10. A *quadratic* polynomial is an element of $\mathbb{P}_2$ of the form $p(x) = ax^2 + bx + c$, where $a, b, c \in \mathbb{R}$ and $a \neq 0$.

a. Is the set of quadratic polynomials closed under addition?

b. Is the set of quadratic polynomials closed under scalar multiplication?

c. Does the set of quadratic polynomials form a subspace of $\mathbb{P}_2$?

11. Let a and b denote scalars and let f, g, h, and **1** denote the functions in $\mathbb{F}(\mathbb{R})$ defined by

$$f(x) = \cos^2 x, \quad g(x) = \sin^2 x, \quad h(x) = \cos 2x, \quad \mathbf{1}(x) = 1.$$

    a. Write the identity in the independent variable x that must hold if the function h is equal to the function $af + bg$.

    b. Find scalars a and b so that $h = af + bg$.

    c. Write the identity in the independent variable x that must hold if the function **1** is equal to the function $af + bg$.

    d. Find scalars a and b so that $1 = af + bg$.

12. Suppose S and T are subspaces of a vector space V with $S \subseteq T$. Explain why the conclusion that S is a subspace of T is an immediate consequence of the definition of a subspace.

13. Determine whether the set $\{(x, y, z) \in \mathbb{R}^3 \mid |x| = |y| = |z|\}$ is a subspace of $\mathbb{R}^3$.

14. **a.** Show that the set $E = \{f \in \mathbb{F}(\mathbb{R}) \mid f(-x) = f(x) \text{ for all } x \in \mathbb{R}\}$ of all even functions in $\mathbb{F}(\mathbb{R})$ is a subspace of $\mathbb{F}(\mathbb{R})$.

    b. Show that the set $O = \{f \in \mathbb{F}(\mathbb{R}) \mid f(-x) = -f(x) \text{ for all } x \in \mathbb{R}\}$ of all odd functions in $\mathbb{F}(\mathbb{R})$ is a subspace of $\mathbb{F}(\mathbb{R})$.

15. Consider the unit cube in the first octant of $\mathbb{R}^3$. One vertex is at the origin, three other vertices lie on the axes of the coordinate system, three other vertices lie in the coordinate planes, and the remaining vertex is opposite the origin.

    a. Consider the plane passing through the three vertices that lie on the axes. Write this set in the standard form of a plane in $\mathbb{R}^3$.

    b. Consider the plane passing through the three vertices that lie in the coordinate planes but not on the axes. Write this set in the standard form of a plane in $\mathbb{R}^3$.

    c. Determine the scalar multiples of the vector $(1, 1, 1)$ that lie in these two planes.

    d. What is the distance between these two planes?

16. Let V be a vector space.

    a. For any $\mathbf{v} \in V$, show that the set $S = \{r\mathbf{v} \mid r \in \mathbb{R}\}$ is a subspace of V.

    b. Let $\mathbf{v}$ be a nonzero element of V and let $\mathbf{x}$ be any vector in V. Show that if the line $L = \{r\mathbf{v} + \mathbf{x} \mid r \in \mathbb{R}\}$ passes through $\mathbf{0}$, then L is equal to $\{r\mathbf{v} \mid r \in \mathbb{R}\}$.

    c. Prove that a line in V is a subspace of V if and only if the line passes through $\mathbf{0}$.

17. Let V be a vector space.

    a. Suppose $\mathbf{v}$, $\mathbf{w}$, and $\mathbf{x}$ are vectors in V with neither $\mathbf{v}$ nor $\mathbf{w}$ being a multiple of the other. Show that if the plane $P = \{r\mathbf{v} + s\mathbf{w} + \mathbf{x} \mid r, s \in \mathbb{R}\}$ passes through $\mathbf{0}$, then P is equal to $\{r\mathbf{v} + s\mathbf{w} \mid r, s \in \mathbb{R}\}$.

    b. Prove that a plane in V is a subspace of V if and only if the plane passes through $\mathbf{0}$.

Systems of Linear Equations

Before continuing the study of vector spaces, we want to develop a systematic computational technique for solving certain systems of simultaneous equations. This chapter presents a simple but powerful algorithm for determining all solutions of systems of linear equations. Matrices will provide a convenient notation for conceptualizing the massive arrays of coefficients. Watch for the vector space operations of addition and scalar multiplication lurking in the background at all times.

2.1 Notation and Terminology

You are undoubtedly familiar with methods for solving systems of two simultaneous equations in two unknowns, such as

$$2x + 3y = 7$$
$$6x - y = 2$$

You may also have considered systems of three or more equations in three or more unknowns, such as

$$x + y - z = 1$$
$$2x + y \quad\quad = 2$$
$$x - 2y + 4z = 0$$

The technique we will develop is based on the familiar method of using one equation to eliminate the occurrence of a variable in another of the equations and gradually accumulating values of all the variables.

♜ **Mathematical Strategy Session**   There are many advantages to proceeding systematically:

- We will be able to deal with any number of equations in any number of unknowns.
- If there is a solution, the method will find it.
- If there is more than one solution, the method will find all of them.
- If there are no solutions, the method will let us know.
- We will be able to adapt the method to solve related problems.
- We will understand how a computer program can assist with the routine arithmetic.

Let us begin with an example of three linear equations in three unknowns:

$$\begin{aligned} x + y - z &= -1 \\ 2x + 4y \phantom{{}+2z} &= -4 \\ -x + 3y + 2z &= -6 \end{aligned}$$

The term **linear** refers to the fact that the only operations applied to the variables are multiplication by various constants and addition of the resulting terms. The analogy with the two vector space operations gives a hint as to why linear equations are so useful in linear algebra.

We are interested in triples of real numbers corresponding to values for x, y, and z. Let us define a set consisting of those triples of real numbers that make all three equations true when substituted for the three corresponding variables. Taking advantage of our notation for $\mathbb{R}^3$ as the set of all ordered triples of real numbers, we can write

$$S = \{(x, y, z) \in \mathbb{R}^3 \mid x + y - z = -1, \ 2x + 4y = -4, \ \text{and} \ -x + 3y + 2z = -6\}.$$

This set S is called the **solution set** of the system of equations. An element of S is called a **solution** of the system. To check whether a point of $\mathbb{R}^3$ such as $(-4, 1, -2)$ is in S, we plug in -4 for x, 1 for y, and -2 for z, and see whether all three equations hold. For the point $(-4, 1, -2)$, the first two equations hold, but $-(-4) + 3 \cdot 1 + 2 \cdot (-2) = -6$ is false. We conclude that $(-4, 1, -2) \notin S$. Apparently, it is easy to stumble across elements of $\mathbb{R}^3$ that are not in S. We want to know if S contains any points of $\mathbb{R}^3$ at all.

Suppose we simplify the system by replacing the third equation with the sum of the first and the third. That is, consider the new system

$$\begin{aligned} x + y - z &= -1 \\ 2x + 4y \phantom{{}+z} &= -4 \\ 4y + z &= -7 \end{aligned}$$

Notice that any solution of the original system will be a solution of the new system. Indeed, the first two equations are exactly the same, and the third holds since it is the result of adding the equal quantities in the first equation to the equal quantities in the third equation of the original system. Also note that any solution of the new system is a solution of the original system. This is because we can recover the third equation of the original system by subtracting the first equation from the new third equation. It follows, then, that the new system has exactly the same solution set as the old system. The new system has the advantage of being slightly simpler.

To make further progress toward simplification, we can eliminate the variable x in the second equation by adding -2 times the first equation to the second equation. This produces the system

$$
\begin{array}{rcl}
x + y - z &=& -1 \\
2y + 2z &=& -2 \\
4y + z &=& -7
\end{array}
$$

Notice that we can recover the previous system by adding 2 times the first equation to the second. Thus, we can argue as above that no solutions have been lost and no new solutions have appeared. Hence, we still have exactly the same solution set.

Next we would like to use the y in the second equation to eliminate the occurrences of y in the other equations. First we will multiply the second equation by $\frac{1}{2}$. The system is changed again:

$$
\begin{array}{rcl}
x + y - z &=& -1 \\
y + z &=& -1 \\
4y + z &=& -7
\end{array}
$$

Take a minute to convince yourself that this did not affect the solution set. Now add -4 times the second equation to the third to give the following system (again with the same solution set):

$$
\begin{array}{rcl}
x + y - z &=& -1 \\
y + z &=& -1 \\
-3z &=& -3
\end{array}
$$

Multiply the third equation by $-\frac{1}{3}$ to obtain $z = 1$. This gives the following system, where it is clear what value z must have in order for the equations to hold:

$$
\begin{array}{rcl}
x + y - z &=& -1 \\
y + z &=& -1 \\
z &=& 1
\end{array}
$$

Subtracting the third equation from the second and then adding the third equation to the first will further simplify things to

$$
\begin{array}{rcl}
x + y & = & 0 \\
y & = & -2 \\
z & = & 1
\end{array}
$$

Finally, subtract the second equation from the first to product the simplest possible system:

$$
\begin{array}{rcl}
x & = & 2 \\
y & = & -2 \\
z & = & 1
\end{array}
$$

There is obviously only the one point $(2, -2, 1)$ in the solution set of this system. But we have been careful not to change this solution set as we simplified the system of equations. Hence, $S = \{(2, -2, 1)\}$ is the entire solution set of the original system as well.

Before we consider some other examples to explain more thoroughly this technique for solving systems of equations, there is a simple observation that will save a lot of writing. We can abbreviate the system of equations we started with by recording only the coefficients and the constants on the right side of the equations. Thus, the 3×4 matrix

$$\begin{bmatrix} 1 & 1 & -1 & -1 \\ 2 & 4 & 0 & -4 \\ -1 & 3 & 2 & -6 \end{bmatrix}$$

will represent the original system of three equations in three unknowns. The vertical line between the third and fourth columns is not part of the matrix; it appears here to indicate that the columns on the left are coefficients for x, y, and z, and the final column contains the constants on the right side of the equations. This matrix is called the **augmented matrix** of the system. The 3×3 matrix of coefficients

$$\begin{bmatrix} 1 & 1 & -1 \\ 2 & 4 & 0 \\ -1 & 3 & 2 \end{bmatrix}$$

has been augmented by the column $\begin{bmatrix} -1 \\ -4 \\ -6 \end{bmatrix}$ of constants. Now, instead of performing operations on the equations to simplify the system, we will just work with the rows of the augmented matrix. For example, the first step in the process described above can be performed by replacing the third row of the original matrix by the sum of the first and third rows. We let R_i denote the ith row of a matrix. This operation can be symbolized by $R_1 + R_3 \rightarrow R_3$. We suppress the vertical line in the augmented matrix and write the operation as follows:

$$\begin{bmatrix} 1 & 1 & -1 & -1 \\ 2 & 4 & 0 & -4 \\ -1 & 3 & 2 & -6 \end{bmatrix} \quad R_1 + R_3 \rightarrow R_3 \quad \begin{bmatrix} 1 & 1 & -1 & -1 \\ 2 & 4 & 0 & -4 \\ 0 & 4 & 1 & -7 \end{bmatrix}.$$

The entire process can be written

$$\begin{bmatrix} 1 & 1 & -1 & -1 \\ 2 & 4 & 0 & -4 \\ -1 & 3 & 2 & -6 \end{bmatrix} \quad \begin{matrix} -2R_1 + R_2 \rightarrow R_2 \\ 1R_1 + R_3 \rightarrow R_3 \end{matrix} \quad \begin{bmatrix} 1 & 1 & -1 & -1 \\ 0 & 2 & 2 & -2 \\ 0 & 4 & 1 & -7 \end{bmatrix}$$

$$\tfrac{1}{2}R_2 \rightarrow R_2 \quad \begin{bmatrix} 1 & 1 & -1 & -1 \\ 0 & 1 & 1 & -1 \\ 0 & 4 & 1 & -7 \end{bmatrix}$$

$$-4R_2 + R_3 \rightarrow R_3 \quad \begin{bmatrix} 1 & 1 & -1 & -1 \\ 0 & 1 & 1 & -1 \\ 0 & 0 & -3 & -3 \end{bmatrix}$$

$$-\tfrac{1}{3}R_3 \rightarrow R_3 \quad \begin{bmatrix} 1 & 1 & -1 & -1 \\ 0 & 1 & 1 & -1 \\ 0 & 0 & 1 & 1 \end{bmatrix}$$

$$
\begin{array}{c}
-1R_3 + R_2 \rightarrow R_2 \\
1R_3 + R_1 \rightarrow R_1
\end{array}
\qquad
\begin{bmatrix}
1 & 1 & 0 & 0 \\
0 & 1 & 0 & -2 \\
0 & 0 & 1 & 1
\end{bmatrix}
$$

$$
-1R_2 + R_1 \rightarrow R_1
\qquad
\begin{bmatrix}
1 & 0 & 0 & 2 \\
0 & 1 & 0 & -2 \\
0 & 0 & 1 & 1
\end{bmatrix}.
$$

Reinterpreting the last matrix in terms of equations gives the system

$$
\begin{array}{rcl}
x & = & 2 \\
y & = & -2 \\
z & = & 1
\end{array}
$$

Here is another example of a system of three equations in three unknowns:

$$
\begin{array}{rcr}
-3y + z & = & 1 \\
x + y - 2z & = & 2 \\
x - 2y - z & = & -1
\end{array}
$$

The augmented matrix that encodes this system is

$$
\begin{bmatrix}
0 & -3 & 1 & 1 \\
1 & 1 & -2 & 2 \\
1 & -2 & -1 & -1
\end{bmatrix}.
$$

Again we would like to simplify this matrix using the operations as in the previous example. The first step is to eliminate the appearance of x in all but one equation. In the previous example we added appropriate multiples of the first equation to the other equation until the first column of the matrix looked like

$$
\begin{bmatrix}
1 & ? & ? & ? \\
0 & ? & ? & ? \\
0 & ? & ? & ?
\end{bmatrix}.
$$

The 0 in the upper left corner of our matrix prevents us from beginning in exactly the same way, but it is obvious how to avoid this difficulty. Let us interchange the first and second rows of the matrix. This corresponds to rewriting the equations in a different order and, of course, does not change the solution set. Thus, the first step can be written

$$
\begin{bmatrix}
0 & -3 & 1 & 1 \\
1 & 1 & -2 & 2 \\
1 & -2 & -1 & -1
\end{bmatrix}
\qquad R_1 \leftrightarrow R_2 \qquad
\begin{bmatrix}
1 & 1 & -2 & 2 \\
0 & -3 & 1 & 1 \\
1 & -2 & -1 & -1
\end{bmatrix}.
$$

Now we can proceed as in the previous example. We use the first equation to eliminate the x variable from the other equations and then the second equation to eliminate the y variable from the equation below it:

$$
\begin{bmatrix}
1 & 1 & -2 & 2 \\
0 & -3 & 1 & 1 \\
1 & -2 & -1 & -1
\end{bmatrix}
\qquad -1R_1 + R_3 \rightarrow R_3 \qquad
\begin{bmatrix}
1 & 1 & -2 & 2 \\
0 & -3 & 1 & 1 \\
0 & -3 & 1 & -3
\end{bmatrix}
$$

$$\frac{1}{3}R_2 \rightarrow R_2 \quad \begin{bmatrix} 1 & 1 & -2 & 2 \\ 0 & 1 & -\frac{1}{3} & -\frac{1}{3} \\ 0 & -3 & 1 & -3 \end{bmatrix}$$

$$3R_2 + R_3 \rightarrow R_3 \quad \begin{bmatrix} 1 & 1 & -2 & 2 \\ 0 & 1 & -\frac{1}{3} & -\frac{1}{3} \\ 0 & 0 & 0 & -4 \end{bmatrix}.$$

Although we could further simplify this matrix, let us stop to look at the third row. It represents the equation $0x + 0y + 0z = -4$. There are no earthly values of x, y, and z that will make 0 equal to -4. We conclude that the solution set S of this last system of equations is empty. Of course, $S = \varnothing$ is also the solution set for the original system of equations. This example illustrates the typical way the simplification procedure leads to the conclusion that there are no solutions to a system of linear equations.

Mathematical Strategy Session   Notice that after the first step in the preceding reduction, we could have subtracted the second row from the third and come to the same conclusion. The savings in time and effort offered by such shortcuts must be weighed against the benefits of having a uniform, systematic procedure that is simple to describe, easy to implement on a computer, and convenient to work with.

Before describing the procedure in detail, let us consider one more example. This example illustrates a system of equations that has an infinite number of solutions:

$$\begin{aligned} -3y + z &= 1 \\ x + y - 2z &= 2 \\ x - 2y - z &= 3 \end{aligned}$$

This is just a minor modification of the system considered in the previous example; therefore, the reduction process begins as before:

$$\begin{bmatrix} 0 & -3 & 1 & 1 \\ 1 & 1 & -2 & 2 \\ 1 & -2 & -1 & 3 \end{bmatrix} \quad R_1 \leftrightarrow R_2 \quad \begin{bmatrix} 1 & 1 & -2 & 2 \\ 0 & -3 & 1 & 1 \\ 1 & -2 & -1 & 3 \end{bmatrix}$$

$$-1R_1 + R_3 \rightarrow R_3 \quad \begin{bmatrix} 1 & 1 & -2 & 2 \\ 0 & -3 & 1 & 1 \\ 0 & -3 & 1 & 1 \end{bmatrix}$$

$$-\frac{1}{3}R_2 \rightarrow R_2 \quad \begin{bmatrix} 1 & 1 & -2 & 2 \\ 0 & 1 & -\frac{1}{3} & -\frac{1}{3} \\ 0 & -3 & 1 & 1 \end{bmatrix}$$

$$3R_2 + R_3 \rightarrow R_3 \qquad \begin{bmatrix} 1 & 1 & -2 & 2 \\ 0 & 1 & -\frac{1}{3} & -\frac{1}{3} \\ 0 & 0 & 0 & 0 \end{bmatrix}$$

$$-1R_2 + R_1 \rightarrow R_1 \qquad \begin{bmatrix} 1 & 0 & -\frac{5}{3} & \frac{7}{3} \\ 0 & 1 & -\frac{1}{3} & -\frac{1}{3} \\ 0 & 0 & 0 & 0 \end{bmatrix}$$

The third equation has been reduced to $0x + 0y + 0z = 0$. This puts no restriction at all in the solution set. Thus, we need to consider only the equations corresponding to the first two rows of the simplified matrix. They are:

$$x \qquad - \tfrac{5}{3}z = \tfrac{7}{3}$$
$$y - \tfrac{1}{3}z = -\tfrac{1}{3}$$

At this point it should be clear why there is an infinite number of solutions. If any real number is chosen for any one of the unknowns, there are values of the other two unknowns that will give solutions to the two equations. For example, if we choose $x = 1$, then from the first equation we must have

$$z = -\tfrac{3}{5}\left(\tfrac{7}{3} - x\right) = -\tfrac{3}{5}\left(\tfrac{7}{3} - 1\right) = \left(-\tfrac{3}{5}\right)\left(\tfrac{4}{3}\right) = -\tfrac{4}{5},$$

and from the second equation

$$y = -\tfrac{1}{3} + \tfrac{1}{3}z = -\tfrac{1}{3} + \tfrac{1}{3}\left(-\tfrac{4}{5}\right) = -\tfrac{9}{15} = -\tfrac{3}{5}.$$

The way the two equations are written, however, it is much easier to assign an arbitrary value to z and let that value determine x and y. In fact, let us introduce a new variable r to stand for the specific (but arbitrary) value of z. Then

$$x = \tfrac{5}{3}r + \tfrac{7}{3}$$

and

$$y = \tfrac{1}{3}r - \tfrac{1}{3}.$$

We can write the solution set as follows:

$$S = \{(x, y, z) \in \mathbb{R}^3 \mid x = \tfrac{5}{3}r + \tfrac{7}{3}, y = \tfrac{1}{3}r - \tfrac{1}{3}, z = r \quad \text{for some } r \in \mathbb{R}\}.$$

A slightly neater way to arrive at this solution set is to take advantage of the notation for vector space algebra. The three equations

$$x = \tfrac{5}{3}r + \tfrac{7}{3}$$
$$y = \tfrac{1}{3}r - \tfrac{1}{3}$$
$$z = r$$

can be written as one equation in terms of 3×1 matrices:

$$\begin{bmatrix} x \\ y \\ z \end{bmatrix} = r\begin{bmatrix} \frac{5}{3} \\ \frac{1}{3} \\ 1 \end{bmatrix} + \begin{bmatrix} \frac{7}{3} \\ -\frac{1}{3} \\ 0 \end{bmatrix}.$$

Since r is allowed to take on all real values, we can even eliminate some fractions by replacing r by $3r$. The vector equation becomes

$$\begin{bmatrix} x \\ y \\ z \end{bmatrix} = r \begin{bmatrix} 5 \\ 1 \\ 3 \end{bmatrix} + \begin{bmatrix} \frac{7}{3} \\ -\frac{1}{3} \\ 0 \end{bmatrix}.$$

Thus, we can write the solution set as

$$S = \left\{ r(5, 1, 3) + \left(\tfrac{7}{3}, -\tfrac{1}{3}, 0 \right) \mid r \in \mathbb{R} \right\}.$$

We instantly recognize this set as a line in $\mathbb{R}^3$.

You will find that you can use this example as a pattern whenever a system has an infinite number of solutions.

♖ Mathematical Strategy Session   In the previous example we can read the 3×1 matrices as vectors in $\mathbb{R}^3$ with their coordinates written vertically. This column notation for vectors in $\mathbb{R}^n$ is often convenient, and we will use it freely. At the end of the example we switched back to row notation simply to conserve space in writing the solution set. Although it would not be appropriate to mix notation in a single expression such as

$$\begin{bmatrix} 1 \\ 1 \end{bmatrix} + (0, 2),$$

you should feel free to choose the notation that suits your immediate needs.

Exercises 2.1

1. In each of the following systems, add a multiple of one equation to another equation to obtain an equivalent system with no more than one unknown in each equation. Write the solution set.

    a. $x + 2y = 7$
    $y = 2$

    b. $x = 2$
    $-2x + 6y = 5$

    c. $x - 3y = 2$
    $y = -1$
    $z = 5$

    d. $x + 7z = 0$
    $y = -3$
    $z = 2$

2. The following systems of equations in three unknowns x, y, and z have an infinite number of solutions. Assign an arbitrary variable to one of the unknowns and solve for all the unknowns in terms of the arbitrary variable. Write the solution set in the form of a line in $\mathbb{R}^3$.

    a. $x + 2y = 2$
    $z = 3$

    b. $x - 5z = 1$
    $y + 2z = -1$

    c. $\begin{aligned} y &= 4 \\ z &= 2 \end{aligned}$ **d.** $\begin{aligned} x - 2y + 4z &= 1 \\ y - 2z &= -1 \end{aligned}$

3. The following systems have an infinite number of solutions. Assign arbitrary variables to two of the unknowns and solve for all the unknowns in terms of these two arbitrary variables. Write the solution set in the form of a plane in Euclidean space.

    a. $\begin{aligned} w + 2x + 3z &= 1 \\ y - z &= 2 \end{aligned}$ **b.** $\begin{aligned} x_1 + 3x_2 - x_3 &= -6 \\ x_4 &= 5 \end{aligned}$

    c. $\begin{aligned} x_1 - x_2 + x_3 - x_4 &= 1 \\ x_2 - 2x_3 - x_4 &= 3 \\ 2x_2 - 4x_3 - 2x_4 &= 6 \end{aligned}$ **d.** $\begin{aligned} x_1 + 3x_2 - x_5 &= 0 \\ x_1 + 3x_2 - x_5 &= 0 \\ x_3 + 2x_5 &= -1 \\ x_4 + x_5 &= 2 \end{aligned}$

4.   **a.** Show that the solution set of the equation $ax + by + cz = d$ is a plane in $\mathbb{R}^3$ as long as one or more of the coefficients a, b, c is nonzero.

    b. Under what conditions on the coefficients does this plane pass through the origin?

    c. What are the possible solution sets if $a = b = c = 0$?

5. Geometrically determine the possible solution sets to a system of two equations in three unknowns:

$$a_1 x + b_1 y + c_1 z = d_1$$
$$a_2 x + b_2 y + c_2 z = d_2$$

Sketch pictures to illustrate the possibilities.

6. Geometrically determine the possible solution sets to a system of three equations in three unknowns:

$$a_1 x + b_1 y + c_1 z = d_1$$
$$a_2 x + b_2 y + c_2 z = d_2$$
$$a_3 x + b_3 y + c_3 z = d_3$$

Sketch pictures to illustrate the possible ways each of these solution sets can arise.

2.2 Gaussian Elimination

The process we have been using to simplify matrices is known as Gauss-Jordan reduction or, more simply, Gaussian elimination. The method will be described in two parts. First is a description of the three legal row operations. This is followed by the strategy employed in simplifying the matrix.

Crossroads The article "Gauss-Jordan Reduction: A Brief History" by Steven Althoen and Renate McLaughlin in the February 1987 issue of the *American Mathematical Monthly* gives some interesting background into the origins of this method.

We will find that Gaussian elimination is useful in solving many types of problems that can be formulated in terms of systems of linear equations. In some of these applications we will not be dealing with an augmented matrix that directly represents the system of equations. Thus, it is convenient to describe the process as applied to an arbitrary matrix rather than thinking of the matrix necessarily as the augmented matrix for some system of equations.

Definition 2.1

The three **elementary row operations** are:

1. Interchange two rows: $R_i \leftrightarrow R_j$. Interchange the entries in row i with the corresponding entries in another row j. Leave all other rows unchanged.
2. Multiply a row by a nonzero constant: $cR_i \to R_i$. Replace each entry in row i by the product of that entry with a nonzero real number c. Leave all other rows unchanged.
3. Add a multiple of one row to another row: $cR_i + R_j \to R_j$. Replace each entry in row j by the sum of that entry and a constant c times the corresponding entry in row i. Leave other rows, including row i, unchanged.

The following definition will be used in describing the strategy for reducing a matrix.

Definition 2.2

A **leading entry** of a matrix is the first nonzero entry of a row of the matrix as you read across the row from left to right. A **leading 1** of a matrix is a leading entry that is equal to 1.

Mathematical Strategy Session Here is the Gauss-Jordan reduction algorithm for using the three row operations to simplify a matrix. The idea consists of building up a pattern of leading 1s in columns that otherwise contain only 0s. It is important to notice that row operations performed in this process do not disturb the pattern of 1s and 0s obtained in previous steps.

1. Interchange rows so that among all the rows the first row begins with the fewest zeros. Often you will begin with a matrix that has a nonzero entry in the upper left corner; in such a case no row interchanges need to be done for the first step.

2. Now we want the first row to have a leading 1. To accomplish this, multiply the first row by the reciprocal of its leading entry.

3. In the column below the leading 1 created in the previous step, there may be some nonzero entries. Add the appropriate multiples of the first row to each of the other rows to replace these entries in this column by zeros. At this point all entries below and in any columns to the left of the leading 1 in the first row will be zeros. (Possibly there will be no entries at all to the left of this leading 1.)

4. Next apply steps 1, 2, and 3 to the submatrix formed by ignoring the first row of the matrix. This will produce a leading 1 in the next row with only zero entries below and to the left of it.

5. Continue to repeat this process until you either reach the bottom of the matrix or find that all the remaining rows have only zero entries.

Quick Example   *Apply the first five steps of the Gauss-Jordan reduction algorithm*

$$\text{to the matrix } \begin{bmatrix} 0 & 2 & -6 & 4 \\ -1 & 1 & 1 & 0 \\ 2 & 1 & -13 & 2 \\ 1 & -1 & 0 & 2 \end{bmatrix}.$$

$$\begin{bmatrix} 0 & 2 & -6 & 4 \\ -1 & 1 & 1 & 0 \\ 2 & 1 & -13 & 2 \\ 1 & -1 & 0 & 2 \end{bmatrix} \quad \begin{matrix} \textbf{Step 1} \\ R_1 \leftrightarrow R_2 \end{matrix} \quad \begin{bmatrix} -1 & 1 & 1 & 0 \\ 0 & 2 & -6 & 4 \\ 2 & 1 & -13 & 2 \\ 1 & -1 & 0 & 2 \end{bmatrix}$$

$$\begin{matrix} \textbf{Step 2} \\ -1R_1 \to R_1 \end{matrix} \quad \begin{bmatrix} 1 & -1 & -1 & 0 \\ 0 & 2 & -6 & 4 \\ 2 & 1 & -13 & 2 \\ 1 & -1 & 0 & 2 \end{bmatrix}$$

$$\begin{matrix} \textbf{Step 3} \\ -2R_1 + R_3 \to R_3 \\ -1R_1 + R_4 \to R_4 \end{matrix} \quad \begin{bmatrix} 1 & -1 & -1 & 0 \\ 0 & 2 & -6 & 4 \\ 0 & 3 & -11 & 2 \\ 0 & 0 & 1 & 2 \end{bmatrix}$$

$$\begin{matrix} \textbf{Step 4} \\ \textit{(applying steps 1 and 2} \\ \textit{to the bottom three rows)} \\ \tfrac{1}{2}R_2 \to R_2 \end{matrix} \quad \begin{bmatrix} 1 & -1 & -1 & 0 \\ 0 & 1 & -3 & 2 \\ 0 & 3 & -11 & 2 \\ 0 & 0 & 1 & 2 \end{bmatrix}$$

$$\begin{matrix} \textbf{Step 4, continued} \\ \textit{(applying step 3 to the} \\ \textit{bottom three rows)} \\ -3R_2 + R_3 \to R_3 \end{matrix} \quad \begin{bmatrix} 1 & -1 & -1 & 0 \\ 0 & 1 & -3 & 2 \\ 0 & 0 & -2 & -4 \\ 0 & 0 & 1 & 2 \end{bmatrix}$$

Step 5
(applying steps 1, 2, and 3 to
the bottom two rows)
$-\frac{1}{2}R_3 \to R_3$
$-1R_3 + R_4 \to R_4$

$$\begin{bmatrix} 1 & -1 & -1 & 0 \\ 0 & 1 & -3 & 2 \\ 0 & 0 & 1 & 2 \\ 0 & 0 & 0 & 0 \end{bmatrix}$$ ∎

After these first five steps, the matrix will be in row-echelon form; that is, it will satisfy the conditions of the following definition.

Definition   2.3

A matrix is in **row-echelon form** when:

1. Each row that is not entirely zero has a leading 1.
2. The leading 1s appear in columns farther to the right as you consider the rows of the matrix from top to bottom.
3. Any rows of zeros are together at the bottom of the matrix.

Quick Example   *Verify that the matrix*

$$\begin{bmatrix} 0 & 1 & 5 & 2 & 1 & 0 \\ 0 & 0 & 0 & 1 & -2 & 1 \\ 0 & 0 & 0 & 0 & 1 & 3 \\ 0 & 0 & 0 & 0 & 0 & 0 \end{bmatrix}$$

is in row-echelon form.

The first three rows are not entirely zero. Each has a leading 1. The leading 1 in the first row is in column 2, the leading 1 in the second row is in column 4, and the leading 1 in the third row is in column 5. Thus, the leading 1s appear farther to the right as we move down the matrix. Finally, there is one row of zeros at the bottom of the matrix.   ∎

Once a matrix has been transformed to have leading 1s in the staircase pattern characteristic of row-echelon form, one further step will produce a matrix that satisfies an additional condition. Our goal is to obtain a matrix that satisfies the four conditions in the following definition.

Definition   2.4

A matrix is in **reduced row-echelon form** when:

1. Each row that is not entirely zero has a leading 1.
2. The leading 1s appear in columns farther to the right as you consider the rows of the matrix from top to bottom.
3. Any rows of zeros are together at the bottom of the matrix.
4. Only zeros appear in the portions of the columns above the leading 1s.

Mathematical Strategy Session Here is the final step that will transform a matrix in row-echelon form into a matrix in reduced row-echelon form. It involves using each leading 1 in turn to clear out the nonzero entries in the portion of the column above it. By starting at the bottom, you will be able to create zeros that will reduce the amount of arithmetic later.

6. Beginning with the bottommost row that contains a leading 1 and working upward, add appropriate multiples of this row to the rows above it to introduce zeros in the column above the leading 1.

Quick Example *Apply the sixth step in the Gauss-Jordan reduction algorithm to put the matrix*

$$\begin{bmatrix} 1 & -1 & -1 & 0 \\ 0 & 1 & -3 & 2 \\ 0 & 0 & 1 & 2 \\ 0 & 0 & 0 & 0 \end{bmatrix}$$

in reduced row-echelon form.

$$\begin{bmatrix} 1 & -1 & -1 & 0 \\ 0 & 1 & -3 & 2 \\ 0 & 0 & 1 & 2 \\ 0 & 0 & 0 & 0 \end{bmatrix}$$

Step 6
(using the leading 1 in the third row)
$3R_3 + R_2 \rightarrow R_2$
$1R_3 + R_1 \rightarrow R_1$

$$\begin{bmatrix} 1 & -1 & 0 & 2 \\ 0 & 1 & 0 & 8 \\ 0 & 0 & 1 & 2 \\ 0 & 0 & 0 & 0 \end{bmatrix}$$

Step 6
(using the leading 1 in the second row)
$1R_2 + R_1 \rightarrow R_1$

$$\begin{bmatrix} 1 & 0 & 0 & 10 \\ 0 & 1 & 0 & 8 \\ 0 & 0 & 1 & 2 \\ 0 & 0 & 0 & 0 \end{bmatrix} \ \blacksquare$$

As you practice putting a few matrices in reduced row-echelon form, keep in mind the relation between the row operations you are performing and the idea of eliminating unknowns in a system of linear equations. This will help you keep track of the steps to be performed.

It should become clear that the Gauss-Jordan reduction algorithm is an efficient, mechanical, straightforward way of transforming any matrix to reduced row-echelon form. The most annoying feature of the process is the necessity of dealing with fractions. Most of the examples and exercises in this text have been rigged to minimize the amount of complicated arithmetic. You should feel free to take advantage of any available computer facilities that enable you to concentrate on the reduction algorithm rather than the arithmetic.

Crossroads As you apply the Gauss-Jordan reduction algorithm, you may occasionally notice a shortcut that will save a few steps or make the arithmetic easier. For example, you may be able to reduce the number of computations involving fractions if you delay changing a leading entry to a leading 1 until you have cleared out the nonzero entries below the leading entry. Of course, these different sequences of row operations will not change the solution set if the matrix comes from a system of equations (no row operation ever changes the solution set). Yet what if the two sequences of operations result in two different row-echelon forms? It is reassuring to find that the resulting reduced row-echelon forms will be the same. The proof that there is only one possible result when a given matrix is put in reduced row-echelon form requires some ideas that will be developed soon. The following theorem is for background information; it will not be used in this text. A one-paragraph proof of this result can be found in the article "The Reduced Row Echelon Form of a Matrix Is Unique: A Simple Proof" by Thomas Yuster in the March 1984 issue of *Mathematics Magazine*. (You may also enjoy the cartoon and riddle that immediately follow the article.)

Theorem   2.5

Suppose a sequence of elementary row operations transforms a matrix M into a matrix M' in reduced row-echelon form. Suppose another sequence transforms M into a matrix M'', also in reduced row-echelon form. Then $M' = M''$.

Exercises 2.2

1.   How many columns of the matrix

$$\begin{bmatrix} 1 & 0 & 1 & 0 \\ 0 & 1 & 2 & 0 \\ 0 & 0 & 0 & 1 \end{bmatrix}$$

contain leading 1s? How many rows of this matrix contain leading 1s? If your answers to these two questions are different, be sure to check very carefully the definition of leading 1s.

2.   For each type of row operation, show that there is a row operation that will undo it. That is, if M is transformed into M' by a certain row operation, determine a row operation that can be applied to M' to yield M.

3.   If two row operations are applied in succession to transform the matrix M into the matrix M', describe the row operations that will transform M' back to M. What if n row operations are used?

4.   What is the next step in applying the Gauss-Jordan reduction algorithm to each of the following matrices?

a. $\begin{bmatrix} 1 & 5 & -1 & 2 \\ 0 & 0 & 1 & 3 \\ 0 & -5 & 1 & 0 \end{bmatrix}$
  b. $\begin{bmatrix} 1 & 2 \\ 3 & 1 \\ 2 & 3 \\ 0 & 1 \\ 4 & 3 \end{bmatrix}$

c. $\begin{bmatrix} 1 & -1 & -\frac{4}{7} & 0 & 6 \\ 0 & 1 & \frac{4}{7} & 0 & -6 \\ 0 & 0 & 1 & \frac{2}{3} & \frac{9}{8} \end{bmatrix}$
  d. $\begin{bmatrix} 0 & 0 & 0 & 0 \\ 0 & 0 & 0 & 0 \\ 0 & 0 & 0 & 0 \end{bmatrix}$

5. Put the following matrices in reduced row-echelon form.

a. $\begin{bmatrix} 1 & -2 & 3 \\ 2 & -3 & 6 \\ -1 & 2 & -2 \end{bmatrix}$
  b. $\begin{bmatrix} 2 & 1 & -2 & -5 \\ 1 & 1 & -1 & -3 \\ 3 & 2 & -2 & -4 \end{bmatrix}$

c. $\begin{bmatrix} 1 & 0 & -2 & 1 \\ 3 & -1 & -7 & 0 \\ 2 & -3 & -7 & -7 \end{bmatrix}$
  d. $\begin{bmatrix} 1 & -2 & 2 & 11 \\ -1 & 2 & 3 & -1 \\ -2 & 4 & 0 & -14 \end{bmatrix}$

6. Put the following matrices in reduced row-echelon form.

a. $\begin{bmatrix} 2 & -4 \\ -3 & 6 \\ 1 & 2 \\ -2 & 4 \end{bmatrix}$
  b. $\begin{bmatrix} 0 & 0 & 0 & 1 \\ 0 & 1 & 1 & 0 \\ 1 & 0 & 1 & 0 \end{bmatrix}$

c. $\begin{bmatrix} 0 & 1 & 2 & 0 \\ -2 & 0 & -6 & -1 \\ 4 & -1 & 10 & 2 \\ 1 & 0 & 3 & 0 \end{bmatrix}$
  d. $\begin{bmatrix} 0 & 1 & 4 & 1 & 1 & 1 \\ 0 & -1 & -4 & 1 & -1 & 3 \\ 0 & 1 & 4 & 0 & 1 & -1 \\ 0 & 2 & 8 & 3 & 2 & 4 \end{bmatrix}$

7. Regard two row-reduced matrices as having the same echelon pattern if the leading 1s occur in the same positions. If we let an asterisk act as a wild card to denote any number, the four echelon patterns for a 2×2 matrix are

$$\begin{bmatrix} 1 & 0 \\ 0 & 1 \end{bmatrix}, \quad \begin{bmatrix} 1 & * \\ 0 & 0 \end{bmatrix}, \quad \begin{bmatrix} 0 & 1 \\ 0 & 0 \end{bmatrix}, \quad \begin{bmatrix} 0 & 0 \\ 0 & 0 \end{bmatrix}.$$

a. Write down the possible echelon patterns for a 2×3 matrix.

b. How many patterns are possible for a 3×5 matrix?

8. Give an example to show that the row-echelon form of a matrix is not unique. (Suggestion: When you perform step 6 of the Gauss-Jordan reduction process, all the matrices you create are in row-echelon form.

9. **a.** Show that the system

$$ax + by = r$$
$$cx + dy = s$$

has a unique solution if $ad - bc \neq 0$. (Suggestion: Consider separately the cases where $a = 0$ and where $a \neq 0$.)

b. Show that the system does not have a unique solution if $ad - bc = 0$. What are the possibilities in this case?

2.3 Solving Linear Systems

Since the applications of Gaussian elimination tend to be related to solving systems of linear equations, let us return to that problem and formulate a standard way to solve such systems and to write the solution set S.

The first step is to form the augmented matrix of the system. Each row of this matrix corresponds to one of the equations. The rightmost column contains the constants that appear on the right side of the equations. Each of the other columns corresponds to one of the unknowns; the entries are the coefficients of that unknown. The next step is to apply the Gauss-Jordan reduction process to put the matrix in reduced row-echelon form.

We know from examples in Section 2.1 that there are three possibilities. If the rightmost column contains a leading 1, the row it is in corresponds to the equation $0 = 1$. Thus, the system has no solution, and we can describe the solution set S by writing

$$S = \emptyset.$$

If every column but the rightmost contains a leading 1, then the system has a unique solution. The coordinates of this single vector can be read in the rightmost column of the reduced matrix. We can write the solution set as

$$S = \{(b_1, b_2, \ldots, b_n)\}.$$

Finally, suppose there are columns other than rightmost that do not contain leading 1s of any row. The corresponding unknowns are called **free variables**, and we can assign arbitrary values to them. We will typically use parameters such as r, s, and t to denote the real numbers assigned to these variables. The **leading variables**, those unknowns corresponding to columns with leading 1s, will be determined by the values assigned to the free variables. For example, suppose the reduced matrix of a system turns out to be

$$\begin{bmatrix} 1 & 1 & 0 & 0 & -2 & 4 \\ 0 & 0 & 1 & 0 & 5 & -1 \\ 0 & 0 & 0 & 1 & 3 & 0 \\ 0 & 0 & 0 & 0 & 0 & 0 \end{bmatrix}.$$

Notice that the second and fifth columns do not contain a leading 1 of any row. We assign the values r and s to the corresponding free variables x_2 and x_5, and allow r and s independently to take on all real values. Then the first row corresponds to the equation

$$x_1 = -r + 2s + 4.$$

The second and third rows give equations for the other two leading variables x_3 and x_5 in terms of the parameters r and s. Altogether, we have one equation for each variable:

$$
\begin{aligned}
x_1 &= -r + 2s + 4 \\
x_2 &= r \\
x_3 &= -5s - 1 \\
x_4 &= -3s \\
x_5 &= s
\end{aligned}
$$

Using the notation of column vectors, we have

$$\begin{bmatrix} x_1 \\ x_2 \\ x_3 \\ x_4 \\ x_5 \end{bmatrix} = r \begin{bmatrix} -1 \\ 1 \\ 0 \\ 0 \\ 0 \end{bmatrix} + s \begin{bmatrix} 2 \\ 0 \\ -5 \\ -3 \\ 1 \end{bmatrix} + \begin{bmatrix} 4 \\ 0 \\ -1 \\ 0 \\ 0 \end{bmatrix}.$$

Thus, the set of all solutions is a plane in $\mathbb{R}^5$. We can switch back to row notation and write the solution set as

$$S = \{r(-1, 1, 0, 0, 0) + s(2, 0, -5, -3, 1) + (4, 0, -1, 0, 0) \mid r, s \in \mathbb{R}\}.$$

Now that we have seen how to apply the Gauss-Jordan reduction technique to solve any system of linear equations, let us concentrate on the special situation where the constants on the right side of the equations are all equal to zero. Such a system is called a **homogeneous** system. That is, a homogeneous system of m equations in n unknowns will be of the form

$$a_{11}x_1 + a_{12}x_2 + \cdots + a_{1n}x_n = 0$$
$$a_{21}x_1 + a_{22}x_2 + \cdots + a_{2n}x_n = 0$$
$$\vdots \qquad \vdots \qquad \qquad \vdots \qquad \vdots$$
$$a_{m1}x_1 + a_{m2}x_2 + \cdots + a_{mn}x_n = 0$$

The first fact you should notice about a homogeneous system is that the solution set will never be empty. Indeed, $\mathbf{0} = (0, 0, \ldots, 0)$ will always be a solution. This is known as the **trivial solution**. For a homogeneous system, the question is whether the solution set S will contain any nontrivial solutions (solutions other than $\mathbf{0}$).

Next, notice that the augmented matrix of a homogeneous system will have a column of zeros as its rightmost column. Furthermore, the three row operations used in the Gauss-Jordan reduction process will always preserve this column of zeros. Hence, when we apply the reduction process to a homogeneous system, we will frequently omit the augmentation column. We will apply the standard reduction process to the coefficient matrix with the understanding that the constants on the right side of the equations are all equal to zero.

Quick Example   *Solve the homogeneous system*

$$x - 2y - z = 0$$
$$2x - y + 2z = 0$$
$$x + y + 3z = 0$$

We begin by reducing the coefficient matrix:

$$\begin{bmatrix} 1 & -2 & -1 \\ 2 & -1 & 2 \\ 1 & 1 & 3 \end{bmatrix} \longrightarrow \begin{bmatrix} 1 & -2 & -1 \\ 0 & 3 & 4 \\ 0 & 3 & 4 \end{bmatrix}$$

$$\longrightarrow \begin{bmatrix} 1 & -2 & -1 \\ 0 & 1 & \frac{4}{3} \\ 0 & 0 & 0 \end{bmatrix}$$

$$\longrightarrow \begin{bmatrix} 1 & 0 & \frac{5}{3} \\ 0 & 1 & \frac{4}{3} \\ 0 & 0 & 0 \end{bmatrix}$$

This final coefficient matrix corresponds to the system

$$
\begin{aligned}
x \quad\; + \tfrac{5}{3}z &= 0 \\
y + \tfrac{4}{3}z &= 0 \\
0 &= 0
\end{aligned}
$$

If we write our choice of an arbitrary value for the free variable z as $3r$, we have

$$\begin{bmatrix} x \\ y \\ z \end{bmatrix} = r \begin{bmatrix} -5 \\ -4 \\ 3 \end{bmatrix}.$$

as the general form of a solution. Thus, the solution set is a line in $\mathbb{R}^3$. In the standard form for a line, the solution set looks like

$$S = \{r(-5, -4, 3) \mid r \in \mathbb{R}\}. \quad \blacksquare$$

Crossroads   Notice that the solution set to the linear system in the previous example contains not just the trivial solution, but an infinite number of other points. The solution set is a line in $\mathbb{R}^3$ through the origin. In particular (see Exercise 18 of Section 1.8), this set is a subspace of $\mathbb{R}^3$. In Chapter 5 we will develop the tools to make it easy to show that the solution set of any homogeneous system of linear equations in n unknowns is a subspace of $\mathbb{R}^n$. In fact, by the time you reach Exercise 16 of Section 5.1, this result will be a straightforward application of the Subspace Theorem.

You may be familiar with the general result that a homogeneous system with more unknowns than equations will always have a nontrivial solution. This result will play a key role in our development of the concept of the dimension of a vector space.

Theorem   2.6 Fundamental Theorem of Homogeneous Systems

A homogeneous system of m linear equations in n unknowns with $n > m$ has at least one nontrivial solution.

Proof   The coefficient matrix of such a system will be an $m \times n$ matrix. In particular, it will have more columns than rows. Thus, when we apply the Gauss-Jordan reduction process to this matrix, the number of leading 1s—which cannot exceed m, the number of rows—will be less than n, the number of columns. It follows that at least one of the columns will not have a leading 1. The unknown corresponding to such a column can be assigned an arbitrary real number. In particular, any non-zero value for this free variable will lead to a nontrivial solution of the system of equations.   ■

Exercises 2.3

Solve the systems of linear equations. For each system, write the solution as a set with zero or one point, or as a line, plane, or higher-dimensional analog in $\mathbb{R}^n$. Once you have mastered the steps in the Gauss-Jordan reduction algorithm, you are encouraged to take advantage of any available computer facilities for reducing the matrices associated with these systems of equations.

1. a.
$$\begin{aligned} x + y - z &= 1 \\ 2x + 4y &= 4 \\ -x + 3y + 2z &= 6 \end{aligned}$$

b.
$$\begin{aligned} x + y - z &= 3 \\ 2x + 4y &= 12 \\ -x + 3y + 2z &= 18 \end{aligned}$$

c.
$$\begin{aligned} x + y - z &= 0 \\ 2x + 4y &= 0 \\ -x + 3y + 2z &= 0 \end{aligned}$$

d. Compare your solutions of these systems with the solution of the example near the beginning of Section 2.1. Make a conjecture as to the solution of

$$\begin{aligned} x + y - z &= \tfrac{1}{2} \\ 2x + 4y &= 2 \\ -x + 3y + 2z &= 3 \end{aligned}$$

e. Formulate a general principle about what seems to be going on here.

2. a.
$$\begin{aligned} x + y - 2z &= 1 \\ 3x + 2y + z &= -5 \\ -x + y + z &= -11 \end{aligned}$$

b.
$$\begin{aligned} 3x - 5y + 2z &= -8 \\ x + y &= -3 \\ 2y - 4z &= -2 \end{aligned}$$

c.
$$\begin{aligned} x + 2y + z &= 1 \\ x - y + 2z &= 2 \\ x + 8y - z &= -1 \end{aligned}$$

d.
$$\begin{aligned} x + 2y + z &= 1 \\ x - y + 2z &= 2 \\ x + 8y - z &= -2 \end{aligned}$$

3. a.
$$\begin{aligned} 2x - 4y + z &= 1 \\ 4x + 2y - z &= 1 \end{aligned}$$

b.
$$\begin{aligned} 3x + 6y + z &= 0 \\ x + 2y + z &= 2 \end{aligned}$$

4. a.
$$\begin{aligned} 4x - 6y + \tfrac{2}{3}z &= 2 \\ 6x - 9y + z &= 3 \end{aligned}$$

b.
$$\begin{aligned} 4x - 6y + \tfrac{2}{3}z &= 2 \\ 6x - 9y + z &= 2 \end{aligned}$$

5. a.
$$\begin{aligned} w + 3x + y - z &= 0 \\ w + 3x + 2y - z &= -2 \\ 2w + x + y + z &= 0 \\ -w + 2x - y + z &= 1 \end{aligned}$$

b.
$$\begin{aligned} x_1 + x_2 + x_3 &= 3 \\ -x_1 + x_2 + x_4 &= 0 \\ x_1 - x_2 + x_3 + x_4 &= -4 \end{aligned}$$

c.
$$\begin{aligned} x_3 + x_4 &= 1 \\ 2x_1 + x_2 + 2x_3 &= 0 \\ x_2 - 4x_4 &= -4 \end{aligned}$$

    d. $x_1 + x_2 + x_3 + x_4 + x_5 + x_6 = 6$
$$x_1 - x_2 + x_3 - x_4 + x_5 - x_6 = -6$$
$$x_1 + x_2 + x_3 - x_4 - x_5 - x_6 = -2$$
$$x_2 - x_3 + x_4 - x_5 + x_6 = 8$$

6.   **a.** $x + 5y + 3z = -7$
$$-x - 5y - 2z = 4$$
$$2x + 10y + 5z = -11$$

    b. $w \quad + y - 2z = 1$
$$2w + x \qquad\qquad = 4$$
$$w + x - y + z = 3$$

    c. $2x_1 + x_2 + 2x_3 - 3x_4 - 2x_5 = 5$
$$2x_3 + x_4 + x_5 = 1$$
$$-4x_1 - 2x_2 \qquad + 7x_4 + 8x_5 = -9$$

7.   **a.** $x_1 + 2x_2 - x_3 + x_4 - 2x_5 = 7$
$$2x_1 - x_2 + x_3 + x_4 \qquad = 3$$
$$x_1 - 3x_2 + 2x_3 \qquad + 2x_5 = -4$$

    b. $x_1 + 2x_2 - x_3 + x_4 - 2x_5 = 6$
$$2x_1 - x_2 + x_3 + x_4 \qquad = 3$$
$$x_1 - 3x_2 + 2x_3 \qquad + 2x_5 = -4$$

    c. $x_1 + 2x_2 - x_3 + x_4 - 2x_5 = 7$
$$2x_1 - x_2 + x_3 + x_4 \qquad = 3$$
$$x_1 - 3x_2 + 2x_3 + x_4 + 2x_5 = -4$$

    d. $x_1 + 2x_2 - x_3 + x_4 - 2x_5 = 0$
$$2x_1 - x_2 + x_3 + x_4 \qquad = 0$$
$$x_1 - 3x_2 + 2x_3 \qquad + 2x_5 = 0$$

8.   **a.** $2x_1 + x_2 + 3x_3 = 0$ **b.** $5x_1 - 2x_2 + x_3 = 0$
$$x_1 - x_2 + x_3 = 0 \qquad\qquad\qquad\quad 2x_1 + 4x_2 + x_3 = 0$$
$$x_1 + 2x_2 - x_3 = 0$$

    c. $12x - 3y - 2z = 0$ **d.** $x + y + z = 0$
$$x - y - z = 0 \qquad\qquad\qquad\qquad x - y + 2z = 0$$
$$6x + 3y + 4z = 0 \qquad\qquad\qquad\quad 2x \qquad + 3z = 0$$
$$x - 3y + 3z = 0$$
$$x + 3y \qquad = 0$$

9.   Prove that if a homogeneous system of linear equations has a nontrivial solution, then it has an infinite number of solutions.

2.4 Applications

This section is devoted to sampling the variety of applications of systems of linear equations that arise in problems in mathematics and other quantitative disciplines.

    In Section 1.9 we saw how systems of linear equations are intimately related to the study of lines and planes in Euclidean spaces. We will frequently encounter important results in linear algebra that we will be able to reformulate in terms of systems of linear equations.

You have probably also used systems of linear equations in other mathematics courses, perhaps in the integration technique involving partial fractions decompositions or in the method of undetermined coefficients for solving differential equations. You may be surprised to see how naturally linear systems arise in other areas of the physical, biological, and social sciences.

Let us begin with an example of curve fitting. The general problem is to find a mathematical expression for a curve that satisfies certain specified conditions. Mathematicians have studied the problem of curve fitting as a basic technique for approximating a complicated function by a simpler function. You may recall, for example, that Simpson's rule is based on integrating quadratic polynomials that approximate a more complicated function. The development of computer graphics has further increased the interest in finding formulas for curves that pass through predetermined points in predetermined directions.

Quick Example   *Determine a polynomial p whose graph passes through (0, 2) with a slope of −1 and through (1, 1) with a slope of 2.*

It seems likely that the four coefficients of a cubic polynomial defined by $p(x) = ax^3 + bx^2 + cx + d$ will provide sufficient flexibility to meet the four requirements. For the graph of p to pass through (0, 2) and (1, 1), we must have $p(0) = 2$ and $p(1) = 1$. To meet the conditions on the slope, p must also satisfy $p'(0) = -1$ and $p'(1) = 2$. An instant after we compute the derivative $p'(x) = 3ax^2 + 2bx + c$, we can write down four linear equations in four unknowns:

$$\begin{aligned} d &= 2 \\ a + b + c + d &= 1 \\ c &= -1 \\ 3a + 2b + c &= 2 \end{aligned}$$

The single solution $(a, b, c, d) = (3, -3, -1, 2)$ gives the cubic polynomial defined by $p(x) = 3x^3 - 3x^2 - x + 2$ that satisfies the four conditions.   ■

The field of twentieth-century mathematics known as linear programming deals with linear systems involving both equalities and inequalities. Problems of allocating limited resources frequently lead to linear programming models. The following example illustrates such a situation in which most of the linear constraints are equalities.

Quick Example   *A manufacturer makes three kinds of plastics: A, B, and C. Each kilogram of A produces 10 grams of particulate matter discharged into the air and 30 liters of liquid waste discharged into the river. Each kilogram of B produces 20 grams of particulate matter and 50 liters of liquid waste. Each kilogram of C produces 20 grams of particulate matter and 110 liters of liquid waste. The Environmental Protection Agency limits the company to 2550 grams of particulate matter per day and 7950 liters of liquid waste discharge per day. Determine the production levels of the three plastics that will result in emission levels that reach the maximum values for both particulate matter and liquid waste.*

We begin by introducing variables x, y, and z to represent the number of kilograms of the three kinds of plastics that can be made. The constraints on the particulate and liquid waste levels translate into the two equations

$$10x + 20y + 20z = 2550$$
$$30x + 50y + 110z = 7950$$

Our standard techniques yield the solution set

$$\{r(-12, 5, 1) + (315, -30, 0) \mid r \in \mathbb{R}\}.$$

Keeping in touch with reality, we need to avoid specifying negative production levels for any of the three plastics. The condition $x = -12r + 315 \geq 0$ results in the restriction $r \leq \frac{315}{12} = 26.25$. The condition $y = 5r - 30 \geq 0$ results in the restriction $r \geq 6$. This also ensures that $z = r$ will be safely positive. Thus, the set of possible values for x, y, and z is the line segment

$$\{r(-12, 5, 1) + (315, -30, 0) \mid 6 \leq r \leq 26.25\}. \quad \blacksquare$$

The final example in this section deals with the problem of determining the distribution of material in a system at equilibrium. Although it is stated as a physiological problem, simple reformulations result in problems as diverse as transfer of energy in thermodynamics and migration patterns of insect colonies in entomology.

Quick Example   *A patient is injected with 1 gram of a long-lasting drug. At any time thereafter, a portion of the drug is in use by the body, a portion is stored in the liver, and the remainder is in circulation throughout the body. Each day, suppose that of the amount in the circulatory system on the previous day, 20% is stored in the liver, 30% goes into use, and the rest remains in circulation. Suppose that 20% of the amount in the liver and 10% of the amount in use on the previous day are released into the bloodstream, with no direct transfer between the liver and the sites of utilization. Find the amount of the drug in the various locations when the distribution reaches equilibrium.*

Let x denote the number of grams of the drug in the circulatory system, let y denote the amount stored in the liver, and let z denote the amount in use by the body. The condition can be concisely formulated as the system

$$
\begin{aligned}
x + \quad y + \quad z &= 1 \\
0.5x + 0.2y + 0.1z &= x \\
0.2x + 0.8y \quad\quad\;\; &= y \\
0.3x \quad\quad + 0.9z &= z
\end{aligned}
$$

After rewriting the system with all the variables on the left of the equal signs and performing Gaussian elimination, we determine a unique solution $(x, y, z) = (0.2, 0.2, 0.6)$ for the system. Under the assumption that the drug does reach an equilibrium distribution, we expect to find 20% in the circulatory system, 20% stored in the liver, and 60% in use.

Exercises 2.4

1.   Find the coefficients of a cubic polynomial $p(x) = ax^3 + bx^2 + cx + d$ whose graph passes through the four points $(-1, -3)$, $(0, -4)$, $(1, -7)$, $(2, 6)$.

2. Find an equation of a parabola that coincides with the graph of the sine function at $x = 0$, $x = \pi/2$, and $x = \pi$.

3. **a.** Find the polynomial of minimal degree whose graph coincides with the graph of the exponential function at $x = 0$, $x = 1$, and $x = 2$.

    b. What if we additionally require the slopes of the two graphs to agree at $x = 0$ and $x = 2$?

4. Find values of a, b, and c so that the circle with equation

$$x^2 + y^2 + ax + by + c = 0$$

will pass through the three points $(1, 1)$, $(-2, 0)$, and $(0, 1)$.

5. Find values of a, b, and c so that the function defined by $f(x) = a + b\sin x + c\cos x$ satisfies $f(0) = 1$, $f(\frac{\pi}{2}) = 1$, and $\int_0^{\pi/2} f(x)\,dx = 1$.

6. A city has 700 teachers, 80 firefighters, 160 police officers, and 290 clerical workers. Within each of these four groups, all employees are to receive a salary increase of the same number of dollars. A total of \$500,000 is available for raises for these four groups of employees. The raise for a firefighter is to be \$100 larger than the raise for a police officer. A teacher must get a raise equal to the average raises of a firefighter and a police officer. What are the possible raises?

7. Suppose a certain virus exists in forms A, B, and C. Half of the offspring of type A will also be of type A, with the rest equally divided between types B and C. Of the offspring of type B, 5% will mutate to type A, and the rest will remain of type B. Of the offspring of type C, 10% will mutate to type A, 20% will mutate to type B, and the other 70% will be of type C. What will be the distribution of the three types when the population of the virus reaches equilibrium?

8. (Partial fractions decomposition) Find values of the constants a, b, c, and d so that

$$\frac{11x^3 - x + 2}{(x^2 - 1)(x^2 + 2x)} = \frac{a}{x - 1} + \frac{b}{x + 1} + \frac{c}{x + 2} + \frac{d}{x}.$$

9. (Method of undetermined coefficients) Find values of the coefficients a, b, and c so that

$$y = ae^x + bxe^x + cx^2e^x$$

is a solution to the differential equation

$$y' + y = e^x + xe^x + x^2e^x.$$

Project: Numerical Methods

The Gauss-Jordan reduction algorithm is a tool of amazing versatility. We will see it play a crucial role in the development of certain theoretical results, and it will repeatedly appear as a standard technique in working with concrete examples. Outside of textbook exercises, the reduction of matrices commonly relies on the speed and accuracy of a digital computer. However, since computers typically represent numbers with a limited number of significant figures, a certain amount of error is inherent in such computations. The project in this section concerns some numerical difficulties that arise as small initial errors are propagated through a sequence of computations.

1. Determine the solution to the system

$$(1/a)x + y = 1$$
$$-x + y = 0$$

in terms of the value of the parameter a. What happens to the system and the solution as a increases to infinity?

2. Investigate the effect of error introduced when calculations are limited in precision. Take $a = 1000$ in the preceding system and apply the Gauss-Jordan reduction algorithm, rounding after each step to three significant figures. How does the approximate solution compare with the true solution rounded to three figures? Trace through the steps of the reduction process to locate the causes of this trouble.

Interchange the rows of the augmented matrix and repeat this investigation. The leading entry that is used to create zeros elsewhere in its column is known as the **pivot entry**. The process of interchanging rows prior to creating these zeros is known as **pivoting.**

What do you suggest should be the pivot entry for $\begin{bmatrix} 100 & 100a & 100a \\ -1 & 1 & 0 \end{bmatrix}$ if a is large?

This example shows that the pivot entry should be chosen by looking at its size relative to the other entries in its row. Suggest some formulas for measuring the size of a row of a matrix. Which of these is simplest from the standpoint of implementation on a computer? Devise a strategy for choosing pivot entries to reduce the effect of round-off errors.

3. Test your pivoting strategy. Create some systems of equations for which you know the exact solutions. Be sure to use some systems where the coefficients are not all of the same order of magnitude. Take advantage of any computer facilities available to you in carrying out these tests. Is the accuracy of your results comparable to the precision used in the intermediate steps? How do your results compare with those obtained when pivoting is used only to avoid division by zero?

4. You may be interested in reading the article "Why Should We Pivot in Gaussian Elimination?" by Edward Rozema in the January 1988 issue of the *College Mathematics Journal.* This article discusses pivoting from the point of view outlined here. It also discusses some ill-conditioned systems where pivoting may lead to increased errors.

5. Techniques for dealing with linear systems are important topics in numerical analysis. You may wish to investigate other aspects of the Gauss-Jordan reduction process (total pivoting or LU-decompositions, for example), iterative methods for repeatedly reducing the error in an approximate solution, or the condition number associated with a matrix that leads to an estimate of the size of the error in an approximate solution. Most textbooks on numerical analysis contain a wealth of information on these topics.

Project: Further Applications of Linear Systems

There is an inexhaustible supply of applications of linear systems. Whenever a relation exists between two sets of variables, the simplest model to try is one in which the dependent variables are sums of constants times the independent variables. When we want to determine the values of the constants or to reverse the roles of dependent and independent variables, we must solve a system of linear equations. Here are brief descriptions of three broad topics where linear systems are crucial. You may wish to explore the mathematical background involved in the topic, or you may prefer to investigate a specific application to a model of particular interest to you.

1. Linear programming is a technique for allocating limited resources under linear constraints. The constraints typically involve a system of inequalities. The set of points that satisfy the inequalities is called the **feasible region**. The **objective function** measures the success of the allocation program. Like the constraints, the objective function involves a sum of constants times the variables. One of the basic results of linear programming is that the extreme values of an objective function occur at the vertices of the feasible region.

You may want to take a typical linear programming problem and try a brute-force approach to solving it. Find all vertices determined by subsets of the constraints, determine which are in the feasible region, and choose one that optimizes the objective function. The simplex method is a very successful technique for reducing the amount of trial and error involved in solving a linear programming problem. Some references to linear programming, the simplex method, and related results are:

Frederick Hillier and Gerald Lieberman, *Introduction to Operations Research*, 5th ed., New York: McGraw-Hill, 1990.

Margaret Lial, Charles Miller, and Raymond Greenwell, *Finite Mathematics*, 5th ed., New York: HarperCollins College Publishers, 1993.

Paul Long and Jay Graening, *Finite Mathematics: An Applied Approach*, New York: Harper-Collins College Publishers, 1993.

Daniel Maki and Maynard Thompson, *Mathematical Models and Applications*, Englewood Cliffs, NJ: Prentice-Hall, 1973.

Wayne Winston, *Operations Research: Applications and Algorithms*, 2nd ed., Boston: PWS-Kent, 1991.

2. Data gathered in the physical world are seldom perfectly consistent. The discrepancies are usually attributed to inaccurate measurements and other sources of error. Regression analysis attempts to estimate the values of parameters in a general model so that it best fits the observed data. The most common measure of how well a model fits a set of data is the sum of squares of the deviation at each data point. If we differentiate this quantity with respect to each of the parameters in the model, the minimum value will occur at a point where these derivatives are all zero. This gives a system of equations. For a simple model, the system is often linear.

Consult any statistics textbook for examples of regression analysis. A good starting point is

Robert Hogg and Elliot Tanis, *Probability and Statistical Inference*, 4th ed., New York: Macmillan, 1993.

You are likely to find a discussion of the distributions of the parameters as well as formulas for estimating their most likely values. You may want to try the technique in more complicated models or for other measures of goodness of fit. If you discover a new probability distribution, you can name it after yourself.

3. If a soap film or cell membrane is supported by a rigid frame, the surface quickly relaxes to an equilibrium shape. The position of each point will be determined by the positions of points immediately surrounding it. Imagine a grid of points superimposed on the surface. The position of each of the grid points can be approximated by some kind of average of the positions of neighboring points on the grid. This relaxation method can be adapted to many other situations where the value of a quantity is determined as an average of values of the quantity at neighboring points. Temperature distribution on a conductive plate with fixed boundary conditions is a classic case; the distribution of pollution in a trapped air mass is a three-dimensional version.

Create a relaxation model for some phenomenon of interest to you. Define a grid over a region, assign values to certain grid points (perhaps along the boundary of the region), and decide how you want to approximate the value at each grid point in terms of the values of neighboring points. You should obtain a system of linear equations. There will be one equation and one unknown for each grid point whose value is not predetermined.

The size of the system of equations increases quite rapidly as a finer grid is used to attain more accuracy. You might want to solve a large system with an iterative technique: initialize all variables with some reasonable values, and apply your averaging formula to compute new values of the variables one at a time while making repeated passes through the grid. Try to speed up the convergence of this iterative process. Think of a good strategy for making the initial assignment of values. Try different averaging formulas. Extrapolate the change in a value during one iteration to anticipate what its change might be in the next iteration.

Chapter 2 Summary

Chapter 2 presents the Gauss-Jordan reduction algorithm as a tool for solving systems of linear equations. Three types of elementary row operations allow us to simplify the coefficient matrix of a linear system of equations. Then we can easily determine the solution set of the system.

Computations

The three elementary row operations

Recognize leading entries of a matrix

Gauss-Jordan reduction algorithm
Systematic use of the three row operations
Transform a matrix to row-echelon form
Transform a matrix to reduced row-echelon form

Systems of linear equations
Code a system as an augmented matrix
Perform the Gauss-Jordan reduction algorithm
Read off the solution set from the reduced matrix
Three possibilities: $\varnothing$, singleton, infinite set of solutions
Geometric interpretation of infinite solution sets
Lines, planes, higher-dimensional analogs in $\mathbb{R}^n$

Theory

Advantages of a systematic and effective solution technique

Homogeneous systems of linear equations
    The trivial solution always exists
    Ignore augmentation column
    Possible nontrivial solutions
    Fundamental Theorem of Homogeneous Systems

Applications

Curve fitting and the method of undetermined coefficients

Linear constraints

Equilibrium distributions

Review Exercises

1. **a.** The second type of elementary row operation allows us to multiply a row of a matrix by a nonzero constant. Why is it important that the constant be nonzero?

   b. The third type of elementary row operation allows us to add a multiple of one row to another row of a matrix. What happens if the multiplier is zero? Why is it not important to prevent this from happening?

2. Consider the matrix

$$A = \begin{bmatrix} 0 & 1 & 1 & 0 & 1 \\ 0 & 0 & 1 & 1 & 1 \\ 0 & 0 & 0 & 0 & 1 \\ 0 & 0 & 0 & 0 & 0 \end{bmatrix}.$$

   a. How many rows of A have leading 1s?

   b. How many columns of A have leading 1s?

   c. How many leading 1s does A have?

3. Illustrate the steps in the Gauss-Jordan reduction algorithm in putting the matrix

$$\begin{bmatrix} 0 & 2 & 4 & 0 & 6 \\ 1 & -1 & 2 & 1 & 1 \\ 2 & -2 & 4 & 2 & 2 \\ -2 & 3 & -2 & 1 & 4 \end{bmatrix}$$

   in reduced row-echelon form.

4. Suppose a system of nine linear equations in seven variables has an augmented matrix that, when put in reduced row-echelon form, has four leading 1s.

   a. Describe the solution set if one of the leading 1s is in the rightmost column.

   b. Describe the solution set if none of the leading 1s is in the rightmost column.

5. Suppose a system of m equations in n variables has an augmented matrix that, when put in reduced row-echelon form, has r leading 1s.

a. Describe the solution set if one of the leading 1s is in the rightmost column.

b. Describe the solution set if none of the leading 1s is in the rightmost column.

6. For each of the following systems of linear equations, give the augmented matrix corresponding to the system. Put the matrix in reduced row-echelon form. Write the solution set of the system as a set with zero or one point, or as a line, plane, or higher-dimensional analog.

a.
$$
\begin{aligned}
-x_1 + x_2 - 2x_3 \quad\;\; &= -2 \\
x_1 - x_2 + x_3 + x_4 &= 1 \\
2x_1 - 2x_2 + x_3 + 3x_4 &= 1
\end{aligned}
$$

b.
$$
\begin{aligned}
x_2 + 2x_3 + \quad x_4 &= -3 \\
2x_1 \quad\;\; + 4x_3 - 6x_4 &= 6 \\
5x_1 + 3x_2 + 16x_3 - 12x_4 &= 6
\end{aligned}
$$

7. In applying the Gauss-Jordan reduction algorithm to transform each of the following matrices to reduced row-echelon form, what elementary row operation would be applied first? Write down the matrix resulting from this operation.

a. $\begin{bmatrix} 4 & 2 & 0 & -8 \\ 1 & 3 & 2 & 2 \\ 3 & 0 & 7 & -1 \end{bmatrix}$
　　b. $\begin{bmatrix} 0 & 1 & 3 & 2 & 2 & 0 & 0 \\ 0 & 0 & 0 & 1 & 1 & 1 & 0 \\ 0 & 0 & 0 & 0 & 1 & 5 & 1 \end{bmatrix}$

c. $\begin{bmatrix} 0 & 1 & 5 & 2 \\ -3 & 7 & 2 & 1 \\ -2 & 1 & 3 & -2 \end{bmatrix}$
　　d. $\begin{bmatrix} 1 & 0 & 4 & 0 & -2 \\ 0 & 1 & -2 & 0 & 1 \\ 0 & 0 & 0 & 1 & 0 \\ 0 & 0 & 0 & 0 & 0 \end{bmatrix}$

8. Which of the following matrices are in row-echelon form? Which are in reduced row-echelon form?

a. $\begin{bmatrix} 1 & 0 \\ 0 & 0 \\ 0 & 1 \end{bmatrix}$
　　b. $\begin{bmatrix} 1 & 0 & -2 & 3 & 0 \\ 0 & 1 & 1 & 0 & 0 \\ 0 & 0 & 0 & 1 & 0 \\ 0 & 0 & 0 & 0 & 0 \end{bmatrix}$

c. $\begin{bmatrix} 0 & 1 & 9 & 0 \\ 0 & 0 & 0 & 1 \end{bmatrix}$
　　d. $\begin{bmatrix} 1 & 3 & 0 & 4 & 0 \\ 0 & 0 & 2 & 7 & 0 \\ 0 & 0 & 0 & 0 & 1 \end{bmatrix}$

9. Solve the following systems of linear equations. Write the solution set as a set with zero or one element, or in the standard form of a line or plane.

a.
$$
\begin{aligned}
-w + 2x + y + z &= -4 \\
w + x - y + z &= 3 \\
2w - 3x \quad\;\; + 2z &= 3 \\
x + 4y - z &= 7
\end{aligned}
$$

b.
$$
\begin{aligned}
-w + 2x + y + z &= -4 \\
w + x - y + z &= 3 \\
2w - 3x \quad\;\; + 2z &= 3 \\
3x \quad\;\; + 2z &= -1
\end{aligned}
$$

$$\text{c.} \quad \begin{aligned} -w + 2x + y + z &= -4 \\ w + x - y + z &= 3 \\ 2w - 3x \quad + 2z &= 3 \\ 3x \quad + 2z &= 1 \end{aligned}$$

10. Solve the following homogeneous system of linear equations:

$$\begin{aligned} -x_1 - x_2 + 2x_3 - 3x_4 &= 0 \\ 2x_1 + 2x_2 - x_3 + 9x_4 &= 0 \\ x_1 + x_2 - 2x_3 + 3x_4 &= 0 \\ x_3 + x_4 &= 0 \end{aligned}$$

11. **a.** Suppose the solution set of a system of linear equations is known to be the empty set. Explain why the system cannot be homogeneous.

    b. Suppose **0** is known to be a solution of a system of linear equations. Explain why the system must be homogeneous.

12. Consider a system of five linear equations in six unknowns.

    a. Is it possible for the solution set to be $\emptyset$?

    b. Must the solution set be $\emptyset$?

    c. Is it possible for the solution set to consist of a single point?

    d. Must the solution set consist of a single point?

    e. Is it possible for the solution set to be a line?

    f. Must the solution set be a line?

    g. Is it possible for the solution set to be a plane?

    h. Must the solution set be a plane?

13. Consider a system of six linear equations in five unknowns.

    a. Is it possible for the solution set to be $\emptyset$?

    b. Must the solution set be $\emptyset$?

    c. Is it possible for the solution set to consist of a single point?

    d. Must the solution set consist of a single point?

    e. Is it possible for the solution set to be a line?

    f. Must the solution set be a line?

    g. Is it possible for the solution set to be a plane?

    h. Must the solution set be a plane?

14. Determine the coefficients of the polynomial $p \in P_n$ that satisfies $p(a) = c_0, p'(a) = c_1, \ldots, p^{(n)}(a) = c_n$ for some point $a \in \mathbb{R}$ and constants $c_0, c_1, \ldots, c_n$. Show that there is only one polynomial in P_n that satisfies these conditions. Where have you run across such polynomials in your calculus course?

15. Find a cubic polynomial $p(x)$ that satisfies $p(0) = -1, p'(0) = 5, p(1) = 3, p'(1) = 1$.

16. Suppose a country has three political parties: Left, Center, and Right. Every election one-fifth of the Leftists and one-fourth of the Rightists switch to the Center party, and one-third of the Centrists move to the extreme groups, equally divided between the Left and Right. If the number of members in each of the three parties remains constant from one year to the next, determine the fractions of the politically active citizens that are members of each of the three parties.

17. Find values of the coefficients a and b so that $y = a \sin 2x + b \cos 2x$ is a solution to the differential equation $y'' + 3y' - y = \sin 2x$.

Dimension Theory

Every man and woman in the street has an intuitive idea of dimension and would agree, for example, that a line is one-dimensional whereas a plane is two-dimensional. But what about a circle, such as the following set?

$$\{(x, y) \in \mathbb{R}^2 \mid x^2 + y^2 = 1\}$$

On one hand, it seems to be one-dimensional since it is more or less a bent version of a line. On the other hand, its very definition makes reference to a two-dimensional space to contain it. On yet another hand, if we put a knot in a circle so that it doesn't lie in a plane, might that make the resulting curve three-dimensional? It is interesting to speculate as to whether every set has a dimension and whether the number of dimensions is always a positive integer.

One of the triumphs of linear algebra is a geometrically appealing definition of dimension. Of course, this definition applies only in the context of vector spaces. But whenever a vector space turns out to have a finite dimension, that single number (the dimension) completely describes all the essential features of the vector space.

The definition of the dimension of a vector space is built up in several stages. Along the way, we will look at examples and theorems that demonstrate the reasonableness of the definitions, simplify the verifications of the defining conditions, and indicate some useful consequences of the definitions.

3.1 Linear Combinations

In this section we will take the first step toward developing the idea of the dimension of a vector space. We simply apply the two vector space operations to obtain new vectors from old ones.

Take a finite set of vectors $\{v_1, v_2, \ldots, v_n\}$ in some vector space V and apply the two vector space operations of addition and scalar multiplication. It should be clear that any vector that results can be written $r_1v_1 + r_2v_2 + \cdots + r_nv_n$. Here $r_1, r_2, \ldots, r_n$ are n real numbers used to multiply the n vectors; the scalar products are then added together. This leads to the first of our definitions.

Definition 3.1

Let $\{v_1, v_2, \ldots, v_n\}$ be a finite set of vectors in a vector space V. A **linear combination** of $v_1, v_2, \ldots, v_n$ is any vector of the form

$$r_1v_1 + r_2v_2 + \cdots + r_nv_n,$$

where $r_1, r_2, \ldots, r_n \in \mathbb{R}$. The real numbers $r_1, r_2, \ldots, r_n$ are called the **coefficients** of the linear combination.

This definition requires the use of all the vectors in the set. If, for some reason, you prefer to omit any of the vectors from the summation, simply choose the corresponding coefficient to be 0. We have already verified that $0v = 0$ for any vector v, and we know that the addition of the zero vector can be deleted without changing the sum. It will also be convenient for later purposes if we agree that the vector 0 is a linear combination of the vectors in any set, even if the set is empty.

Crossroads The idea of forming linear combinations is actually very common. A recipe that says to mix 5 gallons of water, 4 pounds of sugar, 3 ounces of extract, and $\frac{1}{2}$ teaspoon of yeast is talking about a linear combination in an informal sense (it isn't clear what vector space we are dealing with). The left side of the linear equation

$$a_1x_1 + a_2x_2 + \cdots + a_nx_n = b$$

is just a linear combination of the unknowns $x_1, x_2, \ldots, x_n$. A related example that fits the definition precisely stems from a system of linear equations:

$$
\begin{aligned}
a_{11}x_1 + a_{12}x_2 + \cdots + a_{1n}x_n &= b_1 \\
a_{21}x_1 + a_{22}x_2 + \cdots + a_{2n}x_n &= b_2 \\
\vdots \qquad\quad \vdots \qquad\qquad \vdots \qquad\quad \vdots \\
a_{m1}x_1 + a_{m2}x_2 + \cdots + a_{mn}x_n &= b_m
\end{aligned}
$$

Here the left side can be written as the linear combination

$$
x_1 \begin{bmatrix} a_{11} \\ a_{21} \\ \vdots \\ a_{m1} \end{bmatrix} + x_2 \begin{bmatrix} a_{12} \\ a_{22} \\ \vdots \\ a_{m2} \end{bmatrix} + \cdots + x_n \begin{bmatrix} a_{1n} \\ a_{2n} \\ \vdots \\ a_{mn} \end{bmatrix},
$$

where the unknowns of the linear system are the coefficients of the linear combination, and the coefficients of the linear system form vectors in $\mathbb{R}^m$ written as columns. Our techniques for finding the vectors in $\mathbb{R}^n$ that are solutions to such a system will be extremely useful in our study of the dimension of a vector space.

We encountered the geometric idea behind linear combinations in Section 1.9 in the discussion of lines and planes through the origin. Figure 1.6 illustrates a typical vector **x** as a linear combination of two vectors **v** and **w**.

If we take specific vectors, say the three vectors $(1, 1, 1)$, $(-2, 0, 1)$, and $(-1, 3, 5)$ in $\mathbb{R}^3$, it is a simple matter to choose various triples of scalars to use in forming linear combinations of the three vectors. You might want to grab your pencil and write down a few just to solidify your comprehension of the concept of linear combinations. About midway through your second example, you are likely to become bored with this rather trivial process. At this point, we want to turn our attention to a more interesting question: Given some specific vector, can we write it as a linear combination of the three vectors?

Quick Example   *Determine whether $(8, -2, -7)$ is a linear combination of the vectors $(1, 1, 1)$, $(-2, 0, 1)$, and $(-1, 3, 5)$.*

We need to decide whether

$$
a(1, 1, 1) + b(-2, 0, 1) + c(-1, 3, 5) = (8, -2, -7)
$$

for some values of the scalars a, b, and c. Reformulating this equation as

$$
(a - 2b - c, \ a + 3c, \ a + b + 5c) = (8, -2, -7),
$$

we see that we are dealing with a system of three linear equations

$$
\begin{aligned}
a - 2b - c &= 8 \\
a \phantom{{}- 2b} + 3c &= -2 \\
a + b + 5c &= -7
\end{aligned}
$$

whose unknowns are the three coefficients we are looking for. Our standard technique of Gaussian elimination goes as follows:

$$
\begin{bmatrix} 1 & -2 & -1 & 8 \\ 1 & 0 & 3 & -2 \\ 1 & 1 & 5 & -7 \end{bmatrix}
\quad
\begin{matrix} -1R_1 + R_2 \to R_2 \\ -1R_1 + R_3 \to R_3 \end{matrix}
\quad
\begin{bmatrix} 1 & -2 & -1 & 8 \\ 0 & 2 & 4 & -10 \\ 0 & 3 & 6 & -15 \end{bmatrix}
$$

$$
\tfrac{1}{2}R_2 \to R_2
\quad
\begin{bmatrix} 1 & -2 & -1 & 8 \\ 0 & 1 & 2 & -5 \\ 0 & 3 & 6 & -15 \end{bmatrix}
$$

$$-3R_2 + R_3 \rightarrow R_3 \qquad \begin{bmatrix} 1 & -2 & -1 & 8 \\ 0 & 1 & 2 & -5 \\ 0 & 0 & 0 & 0 \end{bmatrix}$$

$$2R_2 + R_1 \rightarrow R_1 \qquad \begin{bmatrix} 1 & 0 & 3 & -2 \\ 0 & 1 & 2 & -5 \\ 0 & 0 & 0 & 0 \end{bmatrix}$$

We see that we are actually free to choose one of the coefficients to be any real number and still come up with values for the other two coefficients that will give $(8, -2, -7)$ as the value of the linear combination. For simplicity, let us choose $c = 0$. This forces $a = -2$ and $b = -5$. It is an easy matter now to confirm that $-2(1, 1, 1) + (-5)(-2, 0, 1) + 0(-1, 3, 5)$ does indeed equal $(8, -2, -7)$. ∎

From our work with systems of linear equations, we know that the existence of a solution to the system in the preceding example is not guaranteed a priori. For example, if we try to show that $(8, -2, -6)$ is a linear combination of the same three vectors, we are faced with a system of equations

$$\begin{aligned} a - 2b - c &= 8 \\ a + 3c &= -2 \\ a + b + 5c &= -6 \end{aligned}$$

that has no solution (as can readily be verified by our row reduction technique). In this case we conclude that $(8, -2, -6)$ cannot be expressed as a linear combination of $(1, 1, 1), (-2, 0, 1)$, and $(-1, 3, 5)$.

These same simple ideas work in many of our other examples of vector spaces. You can use the coefficients of a polynomial and the entries of a matrix very much like the coordinates of a vector in Euclidean space.

Quick Example   *Determine whether the polynomial defined by $q(x) = x^2 + x + 2$ is a linear combination of the polynomials defined by $p_1(x) = x^2 + 5$ and $p_2(x) = x^2 + 2x - 1$.*

We are trying to find real values of a and b so that

$$a(x^2 + 5) + b(x^2 + 2x - 1) = x^2 + x + 2.$$

Equating coefficients of corresponding powers of x, we see that we need to solve the system

$$\begin{aligned} a + b &= 1 \\ 2b &= 1 \\ 5a - b &= 2 \end{aligned}$$

In this case, we find that there is exactly one linear combination (with $a = \frac{1}{2}$ and $b = \frac{1}{2}$) of p_1 and p_2 that will give the polynomial q. ∎

You cannot always rely on coordinates or coefficients. When you work with linear combinations of functions other than polynomials or of vectors in an abstract setting, look for special features of the vectors to exploit in connection with the definition of linear combination.

Quick Example   *Determine whether the exponential function* $\exp(x) = e^x$ *is a linear combination of the functions defined by* $f(x) = e^{2x}$ *and* $g(x) = e^{3x}$.

We are looking for scalars a and b so that $\exp = af + bg$. To an arbitrary real number x, the function on the left side of the equation assigns the value $\exp(x)$, and the function on the right side assigns the value $(af + bg)(x) = af(x) + bg(x)$. For the two functions to be equal, the two values must be equal for any given value of x. Experience with the exponential function suggests that such an identity is unlikely. Let us plug some simple values of x into both functions to try to arrive at a contradiction.

$$x = -1 \quad \text{gives} \quad e^{-1} = ae^{-2} + be^{-3};$$
$$x = 0 \quad \text{gives} \quad 1 = a \phantom{e^{-2}} + b;$$
$$x = 1 \quad \text{gives} \quad e = ae^{2} + be^{3}.$$

The first two equations force $a = e + 1$ and $b = -e$. These values do not satisfy the third equation. Hence, exp is not a linear combination of f and g.   ∎

Exercises 3.1

1. Write the following vectors as linear combinations of $(2, 1, -1)$, $(3, 1, 1)$, $(-4, 0, 6)$.

   a. $(1, 2, 6)$ **b.** $(9, 3, -5)$

   c. $(-12, -1, 9)$ **d.** $(1, \frac{2}{3}, 1)$

2. Write the following vectors as linear combinations of $(1, 0, 0)$, $(1, 1, 0)$, $(1, 1, 1)$.

   a. $(3, 0, 0)$ **b.** $(3, 5, 0)$

   c. $(3, 5, -2)$ **d.** (r, s, t)

3. Determine which of the following vectors are linear combinations of $(1, 1, 1, 0)$, $(1, 0, 0, 1)$, $(0, 1, 0, 1)$.

   a. $(1, 5, 3, 0)$ **b.** $(0, 0, 1, -2)$

   c. $(1, 1, 1, -3)$ **d.** $(-3, -5, -5, 2)$

   e. $(1, 0, 2, -1)$ **f.** $(1, 1, \frac{1}{2}, 1)$

4. Write the polynomials defined by the following formulas as linear combinations of p_1, p_2, p_3, where

   $$p_1(x) = x + 1, \qquad p_2(x) = x^2 + x, \qquad p_3(x) = x^3 + x^2.$$

   a. $x^3 + x^2 + x + 1$ **b.** $x^2 + 2x + 1$

   c. $4x^3 - 7x^2 - 3x + 8$ **d.** $(x + 1)(x^2 - x + 1)$

   e. $(x + 1)(x^2 + x)$ **f.** $(x + 1)^3$

5. Determine which of the following matrices are linear combinations of

$$\begin{bmatrix} 2 & 1 & 0 \\ 1 & 0 & 1 \end{bmatrix}, \quad \begin{bmatrix} 1 & -1 & 1 \\ 0 & 1 & 1 \end{bmatrix}, \quad \begin{bmatrix} 1 & 0 & 2 \\ 1 & 0 & 2 \end{bmatrix}, \quad \begin{bmatrix} 1 & 0 & 1 \\ 0 & 1 & 0 \end{bmatrix}.$$

$$\textbf{a.} \begin{bmatrix} 4 & 3 & 3 \\ 2 & 1 & 1 \end{bmatrix} \qquad\qquad \textbf{b.} \begin{bmatrix} 0 & 1 & 0 \\ 0 & 1 & 0 \end{bmatrix}$$

$$\textbf{c.} \begin{bmatrix} 0 & 2 & 3 \\ 1 & 0 & -1 \end{bmatrix} \qquad\qquad \textbf{d.} \begin{bmatrix} 3 & 1 & 2 \\ 2 & 0 & 3 \end{bmatrix}$$

6. Show that the exponential function $\exp(x) = e^x$ is not a linear combination of the functions sin and cos. (Suggestion: Write down the identity in the variable x that would hold if such a linear combination were possible. Derive some consequences of this identity by choosing specific values of x. Show that these consequences are incompatible.)

7. Determine whether the function defined by $x^2 e^x$ is a linear combination of the functions defined by e^x and xe^x.

8. The last example in this section proposed an identity between two functions. If this identity were to hold, the derivatives of the two functions would also be equal for all values of the variable x. Take the first and second derivatives of both sides of this equation. Plug $x = 0$ into both sides of the original equation and the two new equations. Obtain a contradiction, again showing that no such linear combination is possible.

9. **a.** Show that the function $f \in F(\mathbb{R})$ defined by $f(x) = \cos 2x$ is a linear combination of the functions defined by $\sin^2 x$ and 1.

   b. Show that cos is not a linear combination of these two functions.

10. Consider a finite set $\{\mathbf{v}_1, \mathbf{v}_2, \ldots, \mathbf{v}_n\}$ of vectors in a vector space. Show that any vector $\mathbf{v}_i$ for $1 \le i \le n$ is a linear combination of $\mathbf{v}_1, \mathbf{v}_2, \ldots, \mathbf{v}_n$.

11. Suppose that $\mathbf{v} = 5\mathbf{w}_1 - 2\mathbf{w}_2$ and that $\mathbf{w}_1 = 8\mathbf{x}_1 + \mathbf{x}_2$ and $\mathbf{w}_2 = -3\mathbf{x}_1 + 2\mathbf{x}_2$. Show that $\mathbf{v}$ is a linear combination of $\mathbf{x}_1$ and $\mathbf{x}_2$.

12. Suppose that $\mathbf{v}$ is a linear combination of $\mathbf{w}_1$ and $\mathbf{w}_2$ and that $\mathbf{w}_1$ and $\mathbf{w}_2$ are linear combinations of $\mathbf{x}_1$ and $\mathbf{x}_2$. Show that $\mathbf{v}$ is a linear combination of $\mathbf{x}_1$ and $\mathbf{x}_2$.

13. If $\mathbf{v}$ is a linear combination of $\mathbf{w}$ and $\mathbf{x}$, under what conditions is $\mathbf{w}$ a linear combination of $\mathbf{v}$ and $\mathbf{x}$?

14. **a.** Draw an equilateral triangle with sides 1 unit in length. On each side of this triangle, use the middle third as the base of an outward-pointing equilateral triangle. Replace the segment where the small triangle intersects the large triangle by the two other sides of the new triangle. What is the length of the resulting 12-sided star? What is the area enclosed by the curve?

   b. Repeat the replacement process on each of the 12 sides of this star. What is the length of the resulting 48-sided snowflake? What area does it enclose?

   c. After this replacement process has been repeated n times, how many sides does the curve have? What is its length? What area does it enclose?

   d. What is the length of the limiting **snowflake curve**? What area does it enclose? Speculate on the dimension of the snowflake curve.

   e. For further information and fascinating illustrations concerning sets that in some sense have fractional dimension, see the books:

Benoit Mandelbrot, *The Fractal Geometry of Nature,* New York: W. H. Freeman, 1982.

Heinz-Otto Peitgen and Peter H. Richter, *The Beauty of Fractals*, New York: Springer-Verlag, 1986.

Michael Barnsley, *Fractals Everywhere,* San Diego: Academic Press, 1988.

3.2 Span

The next step in the development of the concept of the dimension of a vector space is built on the idea of linear combinations introduced in the previous section. We now want to consider the set of all linear combinations of a given set of vectors.

Definition 3.2

Let $\{\mathbf{v}_1, \mathbf{v}_2, \ldots, \mathbf{v}_n\}$ be a finite set of vectors in a vector space V. The subset of V **spanned** by $\{\mathbf{v}_1, \mathbf{v}_2, \ldots, \mathbf{v}_n\}$ is the set of all linear combinations of $\mathbf{v}_1, \mathbf{v}_2, \ldots, \mathbf{v}_n$. This set is called the **span** of $\{\mathbf{v}_1, \mathbf{v}_2, \ldots, \mathbf{v}_n\}$ and is denoted

$$\text{span}\{\mathbf{v}_1, \mathbf{v}_2, \ldots, \mathbf{v}_n\}.$$

The concept of the span of a set of vectors is an extension of the idea of using a nonzero vector to generate a line or using two vectors in different directions to generate a plane. The following figures illustrate some typical examples of sets of vectors spanning lines or planes in Euclidean spaces.

In Figure 3.1 the vectors $\mathbf{v}$ and $\mathbf{w}$ are in different directions, so any vector in $\mathbb{R}^2$ is a linear combination of $\mathbf{v}$ and $\mathbf{w}$. Hence, the set $\{\mathbf{v}, \mathbf{w}\}$ spans the vector space $\mathbb{R}^2$. Similarly, $\{\mathbf{w}, \mathbf{x}\}$ spans $\mathbb{R}^2$. Of course, the larger set $\{\mathbf{v}, \mathbf{w}, \mathbf{x}\}$ also spans $\mathbb{R}^2$. However, $\mathbf{x}$ is a multiple of $\mathbf{v}$, so the set $\{\mathbf{v}, \mathbf{x}\}$ spans a line through the origin.

In Figure 3.2 the three vectors $\mathbf{e}_1$, $\mathbf{e}_2$, and $\mathbf{e}_3$ along the axes form the prototypical spanning set for $\mathbb{R}^3$. Other sets such as $\{\mathbf{e}_1, \mathbf{e}_2, \mathbf{w}\}$ and $\{\mathbf{e}_2, \mathbf{e}_3, \mathbf{v}\}$ also span $\mathbb{R}^3$. However, $\mathbf{e}_1$, $\mathbf{v}$, and $\mathbf{w}$ all lie in the same plane through the origin. Hence, $\{\mathbf{e}_1, \mathbf{v}, \mathbf{w}\}$ spans this plane and does not span all of $\mathbb{R}^3$.

We are frequently interested in the situation where a subset $\{\mathbf{v}_1, \mathbf{v}_2, \ldots, \mathbf{v}_n\}$ spans the entire vector space V. This is just an easy way of stating that every vector in V is a linear combination of $\mathbf{v}_1, \mathbf{v}_2, \ldots, \mathbf{v}_n$. We know, of course, that the span of a finite set of vectors is always a subset of the vector space that contains them. The following theorem should be no surprise when you consider similar results about lines and planes through the origin in $\mathbb{R}^n$.

Theorem 3.3

The span of any finite set $\{\mathbf{v}_1, \mathbf{v}_2, \ldots, \mathbf{v}_n\}$ of vectors in a vector space V is a subspace of V.

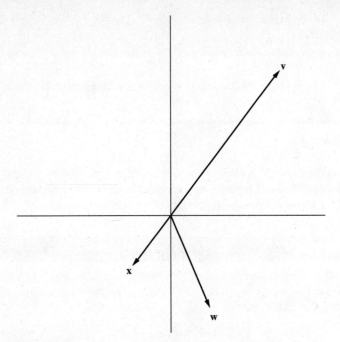

Figure 3.1 $\{\mathbf{v}, \mathbf{w}\}$ spans $\mathbb{R}^2$; $\{\mathbf{w}, \mathbf{x}\}$ spans $\mathbb{R}^2$, $\{\mathbf{v}, \mathbf{x}\}$ spans a line through the origin.

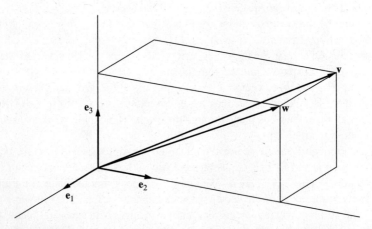

Figure 3.2 $\{\mathbf{e}_1, \mathbf{e}_2, \mathbf{w}\}$ spans $\mathbb{R}^3$; $\{\mathbf{e}_2, \mathbf{e}_3, \mathbf{v}\}$ spans $\mathbb{R}^3$; $\{\mathbf{e}_1, \mathbf{v}, \mathbf{w}\}$ spans a plane through the origin.

Proof The proof consists of a straightforward application of the Subspace Theorem. Let $S = \text{span}\{\mathbf{v}_1, \mathbf{v}_2, \ldots, \mathbf{v}_n\}$; that is,

$$S = \{a_1\mathbf{v}_1 + a_2\mathbf{v}_2 + \cdots + a_n\mathbf{v}_n \mid a_1, a_2, \ldots, a_n \in \mathbb{R}\}.$$

First, $S \neq \varnothing$ since $\mathbf{0} \in S$. Indeed, letting $a_1 = a_2 = \cdots = a_n = 0$, we have

$$0\mathbf{v}_1 + 0\mathbf{v}_2 + \cdots + 0\mathbf{v}_n = \mathbf{0} + \mathbf{0} + \cdots + \mathbf{0}$$
$$= \mathbf{0}.$$

(Also, recall our agreement that $\mathbf{0}$ is a linear combination of any set of vectors, even if $n = 0$.)

To show S is closed under addition, we simply observe that the sum of two arbitrary linear combinations $a_1\mathbf{v}_1 + a_2\mathbf{v}_2 + \cdots + a_n\mathbf{v}_n$ and $b_1\mathbf{v}_1 + b_2\mathbf{v}_2 + \cdots + b_n\mathbf{v}_n$ can again be written as a linear combination

$$(a_1 + b_1)\mathbf{v}_1 + (a_2 + b_2)\mathbf{v}_2 + \cdots + (a_n + b_n)\mathbf{v}_n.$$

Similarly, to show S is closed under scalar multiplication, notice that the product of any real number r with $a_1\mathbf{v}_1 + a_2\mathbf{v}_2 + \cdots + a_n\mathbf{v}_n$ again gives a linear combination $(ra_1)\mathbf{v}_1 + (ra_2)\mathbf{v}_2 + \cdots + (ra_n)\mathbf{v}_n$, as desired.

By the Subspace Theorem, then, S is a subspace of the vector space V. ∎

Thus, we see that a finite set of vectors in a vector space V spans some subspace of V. An interesting problem is to describe this subspace in some way, and in particular to determine if it is all of V. That is, given a finite set $\{\mathbf{v}_1, \mathbf{v}_2, \ldots, \mathbf{v}_n\}$ of vectors in a vector space V, we want to know whether every vector of V is a linear combination of $\{\mathbf{v}_1, \mathbf{v}_2, \ldots, \mathbf{v}_n\}$. Here are some examples showing how to solve this problem.

Quick Example *Show that* $\left\{\begin{bmatrix} 1 & 0 \\ 0 & 0 \end{bmatrix}, \begin{bmatrix} 0 & 1 \\ 0 & 0 \end{bmatrix}, \begin{bmatrix} 0 & 0 \\ 1 & 0 \end{bmatrix}, \begin{bmatrix} 0 & 0 \\ 0 & 1 \end{bmatrix}\right\}$ *spans the vector space* $\mathbb{M}(2, 2)$.

Observe that any vector $\begin{bmatrix} a & b \\ c & d \end{bmatrix}$ in $\mathbb{M}(2, 2)$ can be written

$$a\begin{bmatrix} 1 & 0 \\ 0 & 0 \end{bmatrix} + b\begin{bmatrix} 0 & 1 \\ 0 & 0 \end{bmatrix} + c\begin{bmatrix} 0 & 0 \\ 1 & 0 \end{bmatrix} + d\begin{bmatrix} 0 & 0 \\ 0 & 1 \end{bmatrix}$$

using the four entries of the matrix as scalars. ∎

Next is a slightly more complicated example.

Quick Example *Show that the set* $\{(1, -2, 3), (1, 0, 1), (0, 1, -2)\}$ *of three vectors spans* $\mathbb{R}^3$.

For any vector $(b_1, b_2, b_3) \in \mathbb{R}^3$ we want to find coefficients r_1, r_2, and r_3 so that

$$r_1(1, -2, 3) + r_2(1, 0, 1) + r_3(0, 1, -2) = (b_1, b_2, b_3).$$

This amounts to solving a system of three linear equations in three unknowns:

$$r_1 + r_2 \qquad = b_1$$
$$-\,2r_1 \qquad + \ r_3 = b_2$$
$$3r_1 + r_2 - 2r_3 = b_3$$

Keep in mind that the variables b_1, b_2, and b_3 represent given but arbitrary real numbers, and r_1, r_2, and r_3 are the unknowns. We proceed with Gaussian elimination of the augmented matrix. Of course, the entries in the rightmost column will be expressed in terms of b_1, b_2, and b_3, rather than specific real values.

$$\begin{bmatrix} 1 & 1 & 0 & b_1 \\ -2 & 0 & 1 & b_2 \\ 3 & 1 & -2 & b_3 \end{bmatrix} \longrightarrow \begin{bmatrix} 1 & 1 & 0 & b_1 \\ 0 & 2 & 1 & 2b_1 + b_2 \\ 0 & -2 & -2 & -3b_1 + b_3 \end{bmatrix}$$

$$\longrightarrow \begin{bmatrix} 1 & 1 & 0 & b_1 \\ 0 & 1 & \frac{1}{2} & b_1 + \frac{1}{2}b_2 \\ 0 & -2 & -2 & -3b_1 + \ b_3 \end{bmatrix}$$

$$\longrightarrow \begin{bmatrix} 1 & 1 & 0 & b_1 \\ 0 & 1 & \frac{1}{2} & b_1 + \frac{1}{2}b_2 \\ 0 & 0 & -1 & -b_1 + \ b_2 + b_3 \end{bmatrix}$$

$$\longrightarrow \begin{bmatrix} 1 & 1 & 0 & b_1 \\ 0 & 1 & 0 & \frac{1}{2}b_1 + b_2 + \frac{1}{2}b_3 \\ 0 & 0 & 1 & b_1 - b_2 - \ b_3 \end{bmatrix}$$

$$\longrightarrow \begin{bmatrix} 1 & 0 & 0 & \frac{1}{2}b_1 - b_2 - \frac{1}{2}b_3 \\ 0 & 1 & 0 & \frac{1}{2}b_1 + b_2 + \frac{1}{2}b_3 \\ 0 & 0 & 1 & b_1 - b_2 - \ b_3 \end{bmatrix}$$

This reduced matrix gives more information than we bargained for. It actually tells us the values of the coefficients in terms of the coordinates of the vector (b_1, b_2, b_3) we are trying to obtain:

$$r_1 = \tfrac{1}{2}b_1 - b_2 - \tfrac{1}{2}b_3$$
$$r_2 = \tfrac{1}{2}b_1 + b_2 + \tfrac{1}{2}b_3$$
$$r_3 = \ b_1 - b_2 - \ b_3$$

Actually, once the matrix is in row-echelon form (before we perform the last steps to put it in reduced row-echelon form), it is clear that the system has a unique solution. At that point, then, we can stop and claim that the three vectors do indeed span all of $\mathbb{R}^3$. ∎

In the first section of this chapter we determined that $(8, -2, -6)$ is not a linear combination of the three vectors $(1, 1, 1)$, $(-2, 0, 1)$, and $(-1, 3, 5)$. Thus, the three vectors span a subspace of $\mathbb{R}^3$ that may be a line or a plane through the origin, but it is definitely not all of $\mathbb{R}^3$. Here is how to see what is going on.

Quick Example   *Write the subspace spanned by $\{(1, 1, 1), (-2, 0, 1), (-1, 3, 5)\}$ in the standard form of a line or plane through the origin.*

Let (b_1, b_2, b_3) denote a vector in the subspace spanned by the three vectors. That is, for some real numbers $r_1, r_2,$ and r_3 we have the equation $r_1(1, 1, 1) + r_2(-2, 0, 1) + r_3(-1, 3, 5) = (b_1, b_2, b_3)$. We can rewrite this as a system of equations:

$$
\begin{aligned}
r_1 - 2r_2 - r_3 &= b_1 \\
r_1 \qquad\quad + 3r_3 &= b_2 \\
r_1 + r_2 + 5r_3 &= b_3
\end{aligned}
$$

We next put the augmented matrix of this system in row-echelon form:

$$
\begin{bmatrix}
1 & -2 & -1 & b_1 \\
1 & 0 & 3 & b_2 \\
1 & 1 & 5 & b_3
\end{bmatrix}
\longrightarrow
\begin{bmatrix}
1 & -2 & -1 & b_1 \\
0 & 2 & 4 & -b_1 + b_2 \\
0 & 3 & 6 & -b_1 + b_3
\end{bmatrix}
$$

$$
\longrightarrow
\begin{bmatrix}
1 & -2 & -1 & b_1 \\
0 & 1 & 2 & -\frac{1}{2}b_1 + \frac{1}{2}b_2 \\
0 & 3 & 6 & -b_1 + b_3
\end{bmatrix}
$$

$$
\longrightarrow
\begin{bmatrix}
1 & -2 & -1 & b_1 \\
0 & 1 & 2 & -\frac{1}{2}b_1 + \frac{1}{2}b_2 \\
0 & 0 & 0 & \frac{1}{2}b_1 - \frac{3}{2}b_2 + b_3
\end{bmatrix}
$$

Now it is clearly possible to pick a vector (b_1, b_2, b_3) so that $\frac{1}{2}b_1 - \frac{3}{2}b_2 + b_3$ is not zero. For such values of b_1, b_2, and b_3, the system will have no solutions. The resulting vector (b_1, b_2, b_3) will not be in the subspace of $\mathbb{R}^3$ spanned by $(1, 1, 1)$, $(-2, 0, 1)$, and $(-1, 3, 5)$. More precisely, we can say that (b_1, b_2, b_3) will be in the span of the three given vectors if and only if $\frac{1}{2}b_1 - \frac{3}{2}b_2 + b_3 = 0$. The last step is to write the set

$$
\{(b_1, b_2, b_3) \mid \tfrac{1}{2}b_1 - \tfrac{3}{2}b_2 + b_3 = 0\}
$$

of all such vectors in the standard form for a line or plane through the origin. If we choose arbitrary values r and s for b_2 and b_3, respectively, we can write $b_1 = 3b_2 - 2b_3 = 3r - 2s$. Then

$$
\begin{bmatrix}
b_1 \\
b_2 \\
b_3
\end{bmatrix}
=
\begin{bmatrix}
3r - 2s \\
r \\
s
\end{bmatrix}
= r
\begin{bmatrix}
3 \\
1 \\
0
\end{bmatrix}
+ s
\begin{bmatrix}
-2 \\
0 \\
1
\end{bmatrix}.
$$

Thus, we have

$$
\{r(3, 1, 0) + s(-2, 0, 1) \mid r, s \in \mathbb{R}\}
$$

as an explicit representation of the subspace spanned by the three vectors. We see that it is in fact a plane through the origin. ∎

Here is one final example to illustrate this use of the definitions of linear combination and span in the setting of an abstract vector space.

Quick Example   *In a vector space V, suppose that* **v** *and* **w** *are both linear combinations of* **x** *and* **y**. *Prove that if* {**v**, **w**} *spans V, then* {**x**, **y**} *spans V.*

The statement that **v** and **w** are linear combinations of **x** and **y** allows us to set up equations

$$\mathbf{v} = a\mathbf{x} + b\mathbf{y}$$
$$\mathbf{w} = c\mathbf{x} + d\mathbf{y}$$

where we introduce a, b, c, and d as the scalar multiples in the linear combinations. Now we are trying to prove an implication. The hypothesis is the statement that $\{\mathbf{v}, \mathbf{w}\}$ spans V. We interpret this assumption according to the definition of span: any vector in V can be written as a linear combination of **v** and **w**. We need to prove that any vector in V can be written as a linear combination of **x** and **y**.

Everything is nearly ready. Let **u** denote an arbitrary element of V. The hypothesis says we can write $\mathbf{u} = r\mathbf{v} + s\mathbf{w}$ for some scalars r and s. We can use the equations for **v** and **w** to rewrite **u** in terms **x** and **y**:

$$
\begin{aligned}
\mathbf{u} &= r\mathbf{v} + s\mathbf{w} \\
&= r(a\mathbf{x} + b\mathbf{y}) + s(c\mathbf{x} + d\mathbf{y}) \\
&= (ra + sc)\mathbf{x} + (rb + sd)\mathbf{y}.
\end{aligned}
$$

Thus, we see that we can write an arbitrary element $\mathbf{u} \in V$ as a linear combination of **x** and **y**. ■

Exercises 3.2

1. Consider the five vectors depicted in Figure 3.2. Determine whether each of the following subspaces is a line, a plane, or all of $\mathbb{R}^3$.

    a. span$\{\mathbf{e}_1, \mathbf{e}_2, \mathbf{v}\}$ **b.** span$\{\mathbf{e}_2, \mathbf{e}_3, \mathbf{w}\}$

    c. span$\{\mathbf{e}_2, \mathbf{e}_3, \mathbf{v}, \mathbf{w}\}$ **d.** span$\{\mathbf{v}, -\mathbf{v}, 5\mathbf{v}\}$

    e. span$\{\mathbf{v}, \mathbf{w}\}$ **f.** span$\{\mathbf{e}_1, \mathbf{e}_2 + \mathbf{e}_3\}$

    g. span$\{\mathbf{e}_1, \mathbf{e}_1 + \mathbf{w}, \mathbf{v}, \mathbf{w}\}$ **h.** span$\{\mathbf{e}_1 + \mathbf{e}_2, \mathbf{e}_1 + \mathbf{e}_3, \mathbf{e}_2 + \mathbf{e}_3\}$

2. **a.** Show that $\{(2, 4, -2), (3, 2, 0), (1, -2, -2)\}$ spans $\mathbb{R}^3$.

    b. Show that $\{(2, 4, -2), (3, 2, 0), (1, -2, 2)\}$ does not span $\mathbb{R}^3$.

3. Show that the set $\{(3, 1, 1), (1, 5, 3), (-2, 2, 1), (5, 4, 3), (1, 1, 1)\}$ spans $\mathbb{R}^3$.

4. Show that $\{(1, -1, 5), (3, 1, 3), (1, 2, -4)\}$ does not span $\mathbb{R}^3$. Write the subspace

    $$\text{span}\{(1, -1, 5), (3, 1, 3), (1, 2, -4)\}$$

    in the standard form of a plane through the origin.

5. Show that the set

    $$\left\{ \begin{bmatrix} 2 & 1 & 0 \\ -1 & 1 & 0 \end{bmatrix}, \begin{bmatrix} 1 & 0 & 1 \\ 1 & 1 & 1 \end{bmatrix}, \begin{bmatrix} -1 & 2 & 1 \\ 0 & 0 & 1 \end{bmatrix}, \begin{bmatrix} 0 & -2 & 1 \\ 1 & -1 & 2 \end{bmatrix} \right\}$$

    does not span $\mathbb{M}(2, 3)$.

6. Determine whether $\left\{ \begin{bmatrix} 1 & 1 \\ 0 & 0 \end{bmatrix}, \begin{bmatrix} 0 & 1 \\ 0 & 1 \end{bmatrix}, \begin{bmatrix} 0 & 0 \\ 1 & 1 \end{bmatrix}, \begin{bmatrix} 1 & 0 \\ 1 & 0 \end{bmatrix} \right\}$ spans $\mathbb{M}(2, 2)$.

7.   Determine whether the set of four polynomials defined by

$$p_1(x) = x^2 + 1, \qquad p_2(x) = x^2 + x + 1,$$
$$p_3(x) = x^3 + x, \qquad p_4(x) = x^3 + x^2 + x + 1$$

spans $\mathbb{P}_3$.

8.   **a.** If possible, give a set with five elements that spans $\mathbb{P}_3$.

   b. If possible, give a set with four elements that spans $\mathbb{P}_3$.

   c. If possible, give a set with three elements that spans $\mathbb{P}_3$.

9.   Prove that no finite set of polynomials spans $\mathbb{P}$.

10.   Give a geometric argument to show that it is impossible for a set with two elements to span $\mathbb{R}^3$.

11.   **a.** Suppose $\{v_1, v_2, v_3, v_4\}$ is a set of four vectors in a vector space V. Show that if $\{v_1, v_2, v_3\}$ spans V, then $\{v_1, v_2, v_3, v_4\}$ also spans V.

   b. Suppose $\{v_1, \ldots, v_m, v_{m+1}, \ldots, v_{m+k}\}$ is a set of $m + k$ vectors in a vector space V. Show that if $\{v_1, \ldots, v_m\}$ spans V, then

$$\{v_1, \ldots, v_m, v_{m+1}, \ldots, v_{m+k}\}$$

   also spans V.

12.   Suppose S is a subspace of a vector space V. Prove that if $v_1, \ldots, v_n \in S$, then $\text{span}\{v_1, \ldots, v_n\} \subseteq S$.

13.   Suppose $v_1, \ldots, v_m$ and $w_1, \ldots, w_n$ are vectors in a vector space. Prove that if

$$w_1, \ldots, w_n \in \text{span}\{v_1, \ldots, v_m\} \quad \text{and} \quad v_1, \ldots, v_m \in \text{span}\{w_1, \ldots, w_n\},$$

   then $\text{span}\{v_1, \ldots, v_m\} = \text{span}\{w_1, \ldots, w_n\}$.

14.   Prove that if $\{v, w\}$ spans V, then $\{v + w, w\}$ also spans V.

15.   In a vector space V suppose that v and w are both linear combinations of x and y. Prove or give a counterexample to the claim that if $\{x, y\}$ spans V, then $\{v, w\}$ spans V.

16.   By Exercise 11 of Section 1.5 the set $\{0\}$ containing only the zero vector forms a vector space. What sets span this vector space?

3.3 Linear Independence

A single nonzero vector spans a line, but two nonzero vectors will not span a plane unless they are in different directions. Similarly, three vectors in a plane through the origin cannot span a space larger than the plane that contains them. To span a larger subspace, one of the vectors must in some sense determine a direction that is independent of the directions of the vectors in the plane through the origin. The following definition makes this notion precise. Theorem 3.5 in this section shows how this definition is an algebraic adaptation of some of these geometric ideas.

Definition 3.4

A finite set $\{v_1, v_2, \ldots, v_n\}$ of vectors in a vector space is **linearly independent** if and only if the equation

$$r_1 v_1 + r_2 v_2 + \cdots + r_n v_n = 0$$

implies that $r_1 = r_2 = \cdots = r_n = 0$. If it is possible for the equation to hold when one or more of the coefficients is nonzero, the set is **linearly dependent**.

Mathematical Strategy Session   To show that a set is linearly independent, we must establish a logical implication. The hypothesis is that we have a linear combination of the vectors that equals 0, the additive identity vector in the vector space. Of course, we know this will happen if all the coefficients $r_1, r_2, \ldots, r_n$ are chosen to be zero. The force of the definition is that from this one equation we can conclude that this is the only linear combination of $\{v_1, v_2, \ldots, v_n\}$ that gives the zero vector. In other words, we start by assuming that the vector equation $r_1 v_1 + r_2 v_2 + \cdots + r_n v_n = 0$ holds. From this equation we must deduce the n scalar equations $r_1 = 0$, $r_2 = 0$, $\ldots$, $r_n = 0$. If any other values for any of $r_1, r_2, \ldots, r_n$ are possible, then the set of vectors is linearly dependent.

Whenever you deal with an implication, be careful to keep straight which condition is the hypothesis and which is the conclusion. Interchanging these conditions in an implication $P \Longrightarrow Q$ produces a new implication $Q \Longrightarrow P$. This is known as the **converse** of the original implication. The truth of the converse is not necessarily related to the truth of the original implication. For example, you will undoubtedly agree with the implication

$$\text{if } x > 0, \text{ then } x^2 > 0;$$

whereas you can easily find a negative value of x that serves as a counterexample to the converse,

$$\text{if } x^2 > 0, \text{ then } x > 0.$$

Another way to form the converse of an implication $P \Longrightarrow Q$ is to rephrase it

$$Q \text{ if } P$$

and change the word *if* to *only if*. The resulting statement of the converse,

$$Q \text{ only if } P,$$

is usually seen in conjunction with the original implication. These two directions of the implication are combined by stating

$$P \text{ if and only if } Q.$$

> This is known as a **biconditional** statement or a **logical equivalence**. It is de-
> noted $P \Longleftrightarrow Q$ and means that the conditions P and Q are both true or else they
> are both false.

As was the case with the definitions in the previous sections, we need to agree on a
special convention for $n = 0$ (that is, for the empty set). To make things work out nicely
later on, let us agree that the empty set of vectors is linearly independent. This is in line
with the definition of linear independence, provided you accept an implication with a
vacuous conclusion as being automatically true.

As you will see from the following examples, the verification of the definition of
linear independence can usually be done by formulating and solving a problem involving
a system of linear equations.

Quick Example   *Show that $\{(1, 1, 0), (1, 0, 1), (0, 1, 1)\}$ is a linearly independent
subset of $\mathbb{R}^3$.*

We start by assuming we have three scalars a, b, and c for which

$$a(1, 1, 0) + b(1, 0, 1) + c(0, 1, 1) = (0, 0, 0).$$

We want to find all values possible for the three scalars. This vector equation is
equivalent to the homogeneous system of three equations in three unknowns:

$$
\begin{aligned}
a + b \quad\;\; &= 0 \\
a \quad\;\; + c &= 0 \\
b + c &= 0
\end{aligned}
$$

The coefficient matrix of this system can quickly be reduced:

$$
\begin{bmatrix} 1 & 1 & 0 \\ 1 & 0 & 1 \\ 0 & 1 & 1 \end{bmatrix}
\longrightarrow
\begin{bmatrix} 1 & 1 & 0 \\ 0 & -1 & 1 \\ 0 & 1 & 1 \end{bmatrix}
\longrightarrow
\begin{bmatrix} 1 & 1 & 0 \\ 0 & 1 & -1 \\ 0 & 1 & 1 \end{bmatrix}
$$

$$
\longrightarrow
\begin{bmatrix} 1 & 1 & 0 \\ 0 & 1 & -1 \\ 0 & 0 & 2 \end{bmatrix}
\longrightarrow
\begin{bmatrix} 1 & 1 & 0 \\ 0 & 1 & -1 \\ 0 & 0 & 1 \end{bmatrix}
$$

$$
\longrightarrow
\begin{bmatrix} 1 & 1 & 0 \\ 0 & 1 & 0 \\ 0 & 0 & 1 \end{bmatrix}
\longrightarrow
\begin{bmatrix} 1 & 0 & 0 \\ 0 & 1 & 0 \\ 0 & 0 & 1 \end{bmatrix}
$$

We see that the only possible solution is $a = 0$, $b = 0$, and $c = 0$. Thus, the set is
linearly independent.   ∎

Quick Example   *Determine if the set of matrices*

$$
\left\{ \begin{bmatrix} 1 & 1 \\ 1 & -1 \end{bmatrix}, \begin{bmatrix} 1 & -1 \\ 0 & 0 \end{bmatrix}, \begin{bmatrix} 0 & 2 \\ 1 & -1 \end{bmatrix} \right\}
$$

is linearly independent.

We start by writing down a linear combination of the matrices and assuming it is equal to the zero matrix:

$$a\begin{bmatrix} 1 & 1 \\ 1 & -1 \end{bmatrix} + b\begin{bmatrix} 1 & -1 \\ 0 & 0 \end{bmatrix} + c\begin{bmatrix} 0 & 2 \\ 1 & -1 \end{bmatrix} = \begin{bmatrix} 0 & 0 \\ 0 & 0 \end{bmatrix}.$$

This gives four equations in three unknowns:

$$\begin{aligned} a + b &= 0 \\ a - b + 2c &= 0 \\ a \quad + c &= 0 \\ -a \quad - c &= 0 \end{aligned}$$

The coefficient matrix of this system reduces as follows:

$$\begin{bmatrix} 1 & 1 & 0 \\ 1 & -1 & 2 \\ 1 & 0 & 1 \\ -1 & 0 & -1 \end{bmatrix} \longrightarrow \begin{bmatrix} 1 & 1 & 0 \\ 0 & -2 & 2 \\ 0 & -1 & 1 \\ 0 & 1 & -1 \end{bmatrix}$$

$$\longrightarrow \begin{bmatrix} 1 & 1 & 0 \\ 0 & 1 & -1 \\ 0 & 0 & 0 \\ 0 & 0 & 0 \end{bmatrix} \longrightarrow \begin{bmatrix} 1 & 0 & 1 \\ 0 & 1 & -1 \\ 0 & 0 & 0 \\ 0 & 0 & 0 \end{bmatrix}$$

Since the third column does not contain a leading 1, we see that there are solutions with arbitrary real values assigned to the free variable c. In particular, we are not forced to use $a = b = c = 0$ in the previous equation in $\mathbb{M}(2, 2)$. Thus, the set of matrices is linearly dependent. ∎

In the preceding example, notice that the matrix $\begin{bmatrix} 0 & 2 \\ 1 & -1 \end{bmatrix}$ introduces the third column, which ultimately led to the nonuniqueness of the solution. It appears that if we delete this matrix, the resulting set

$$\left\{ \begin{bmatrix} 1 & 1 \\ 1 & -1 \end{bmatrix}, \begin{bmatrix} 1 & -1 \\ 0 & 0 \end{bmatrix} \right\}$$

is linearly independent. You can readily verify this by following the reduction operations with the third column covered up. Actually, in this particular example any two of the three matrices yield a linearly independent set.

To determine whether a set of polynomials is linearly independent, we can use the coefficients of the various powers of x in much the same way as we used coordinates of Euclidean vectors and entries in matrices to obtain a homogeneous system of linear equations. In dealing with a set of functions other than polynomials, it is often possible to plug values in for the variable to obtain a homogeneous system that has only the trivial solution. If this doesn't work, it may be that the set is linearly dependent. An identity may be lurking around to show that a nontrivial linear combination of the functions is equal to zero.

Quick Example   *Determine whether the functions defined by* $\sin^2 x$, $\cos 2x$, *and* 1 *form a linearly independent set.*

Suppose $a \sin^2 x + b \cos 2x + c = 0$ for all real numbers x. Let us choose some convenient values to plug in for x:

$$
\begin{array}{lll}
x = 0 & \text{gives} & a \cdot 0^2 + \quad b \cdot 1 + c = 0; \\[4pt]
x = \frac{\pi}{4} & \text{gives} & a\left(\frac{\sqrt{2}}{2}\right)^2 + \quad b \cdot 0 + c = 0; \\[4pt]
x = \frac{\pi}{3} & \text{gives} & a\left(\frac{\sqrt{3}}{2}\right)^2 + \quad b(-\frac{1}{2}) + c = 0; \\[4pt]
x = \frac{\pi}{2} & \text{gives} & a \cdot 1^2 + b \cdot (-1) + c = 0; \\[4pt]
x = -\frac{\pi}{6} & \text{gives} & a(-\frac{1}{2})^2 + \quad b \cdot \frac{1}{2} + c = 0.
\end{array}
$$

Clearly, $b = -c$ and $a = -2c$. But nothing indicates that the three variables must all be 0. In fact, $c = -1$ brings to memory the trigonometric identity $\cos 2x = 1 - 2 \sin^2 x$. Thus, the functions form a linearly dependent set. ∎

The following theorem provides a useful criterion for determining whether or not a set of vectors is linearly independent.

Theorem 3.5

The set $\{\mathbf{v}_1, \mathbf{v}_2, \ldots, \mathbf{v}_n\}$ of vectors in a vector space is linearly dependent if and only if one of the vectors can be written as a linear combination of the other vectors in the set.

♛ **Mathematical Strategy Session** This theorem is claiming that two conditions are logically equivalent. That is, if the first condition holds, then so does the second; and if the second condition holds, then so does the first. The proof will break into two separate parts, each assuming one of the conditions as its hypothesis and deriving the other condition from it.

Proof First suppose that $\{\mathbf{v}_1, \mathbf{v}_2, \ldots, \mathbf{v}_n\}$ is linearly dependent. That means that there are n real numbers $r_1, r_2, \ldots, r_n$, not all of which are zero (although some of them may be zero), such that

$$r_1 \mathbf{v}_1 + r_2 \mathbf{v}_2 + \cdots + r_n \mathbf{v}_n = \mathbf{0}.$$

Here is how to use this equation and the fact that at least one of the scalars, say r_i, is not equal to zero to write $\mathbf{v}_i$ as a linear combination of the other vectors $\mathbf{v}_1, \mathbf{v}_2, \ldots, \mathbf{v}_{i-1}, \mathbf{v}_{i+1}, \ldots, \mathbf{v}_n$. Simply add $-r_i \mathbf{v}_i$ to both sides of the equation. This gives

$$-r_i \mathbf{v}_i + \mathbf{0} = -r_i \mathbf{v}_i + (r_1 \mathbf{v}_1 + \cdots + r_{i-1} \mathbf{v}_{i-1} + r_i \mathbf{v}_i + r_{i+1} \mathbf{v}_{i+1} + \cdots + r_n \mathbf{v}_n).$$

After some standard use of the vector space axioms, we have

$$-r_i \mathbf{v}_i = r_1 \mathbf{v}_1 + \cdots + r_{i-1} \mathbf{v}_{i-1} + r_{i+1} \mathbf{v}_{i+1} + \cdots + r_n \mathbf{v}_n.$$

Now since $r_i \neq 0$, we can divide both sides of this equation by $-r_i$. More precisely, we can multiply both sides by the scalar $-\frac{1}{r_i}$. This gives

$$\mathbf{v}_i = \left(-\frac{r_1}{r_i}\right)\mathbf{v}_1 + \cdots + \left(-\frac{r_{i-1}}{r_i}\right)\mathbf{v}_{i-1} + \left(-\frac{r_{i+1}}{r_i}\right)\mathbf{v}_{i+1} + \cdots + \left(-\frac{r_n}{r_i}\right)\mathbf{v}_n,$$

and we have the desired linear combination.

Next let us work on the proof of the converse implication. Suppose that one of the vectors can be written as a linear combination of the other vectors. Let us say

$$\mathbf{v}_i = r_1\mathbf{v}_1 + \cdots + r_{i-1}\mathbf{v}_{i-1} + r_{i+1}\mathbf{v}_{i+1} + \cdots + r_n\mathbf{v}_n.$$

Add $-\mathbf{v}_i$ (alias $(-1)\mathbf{v}_i$) to both sides of this equation to obtain

$$r_1\mathbf{v}_1 + \cdots + r_{i-1}\mathbf{v}_{i-1} + (-1)\mathbf{v}_i + r_{i+1}\mathbf{v}_{i+1} + \cdots + r_n\mathbf{v}_n = \mathbf{0}.$$

The scalars $r_1, r_2, \ldots, r_{i-1}, r_{i+1}, \ldots, r_n$ in this part of the proof are unrelated to the scalars in the first part (of course, the vectors are the same). So there is no reason to conclude that any of them are nonzero. Nevertheless, the scalar multiple of $\mathbf{v}_i$ is the real number -1, which beyond all doubt is not equal to zero. Hence, the set $\{\mathbf{v}_1, \mathbf{v}_2, \ldots, \mathbf{v}_n\}$ is linearly dependent. ∎

As a consequence of this theorem, it becomes very easy to check whether a set of two vectors is linearly independent. Simply observe whether or not one vector is a scalar multiple of the other. After all, this is the only way to form a linear combination with the remaining vector. Thus, $\{(-3, 8, 7), (0, 0, 0)\}$ is linearly dependent in $\mathbb{R}^3$ since the second vector is 0 times the first. Also, $\left\{\begin{bmatrix} 8 & -6 & 0 \\ 2 & -2 & 4 \end{bmatrix}, \begin{bmatrix} 12 & -9 & 0 \\ 3 & -3 & 6 \end{bmatrix}\right\}$ is linearly dependent in $\mathbb{M}(2, 3)$ since the first matrix is $\frac{2}{3}$ times the second. Similarly, since $\ln x^2 = 2\ln x$, we see that the functions defined by $\ln x$ and $\ln x^2$ form a linearly dependent subset of $\mathbb{F}((0, \infty))$, the vector space of real-valued functions defined on the interval $(0, \infty)$.

On the other hand, the set $\{(7, 8, 9, 10), (14, 16, 18, 19)\}$ is linearly independent in $\mathbb{R}^4$, $\left\{\begin{bmatrix} 5 & 10 \\ 15 & 20 \end{bmatrix}, \begin{bmatrix} 5 & 15 \\ 10 & 20 \end{bmatrix}\right\}$ is linearly independent in $\mathbb{M}(2, 2)$, and $\{\sin, \cos\}$ is linearly independent in $\mathbb{F}(\mathbb{R})$.

To use Theorem 3.5 to show that a set of three vectors is linearly independent requires a little more checking. The condition is that none of the vectors is a linear combination of the other two. It is not enough to show that no vector is a multiple of any other. This is the case with the set $\{(1, 0), (0, 1), (1, 1)\}$ of three vectors in $\mathbb{R}^2$, yet the set is linearly dependent since the third vector is the sum of the first two.

We can derive a nice geometric interpretation of this theorem as it applies to a set of three vectors. The set is linearly dependent if and only if one of the vectors is a linear combination of the other two, that is, if and only if one vector lies in the subspace (typically a plane through the origin) spanned by the other two. To put it another way, a set of three vectors is linearly independent if and only if the three vectors do not all lie in the same plane through the origin.

The diagrams from the previous section are repeated here to illustrate these geometric interpretations of linear dependence and linear independence. In Figure 3.3, $\mathbf{x}$ is a scalar multiple of $\mathbf{v}$, so the set $\{\mathbf{x}, \mathbf{v}\}$ is linearly dependent. On the other hand, $\mathbf{w}$ is not a multiple of $\mathbf{v}$ or of $\mathbf{x}$. Hence, the set $\{\mathbf{v}, \mathbf{w}\}$ is linearly independent, as is the set $\{\mathbf{w}, \mathbf{x}\}$.

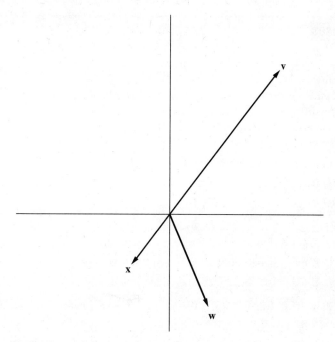

Figure 3.3 The set $\{\mathbf{x}, \mathbf{v}\}$ is linearly dependent; $\{\mathbf{v}, \mathbf{w}\}$ and $\{\mathbf{w}, \mathbf{x}\}$ are linearly independent.

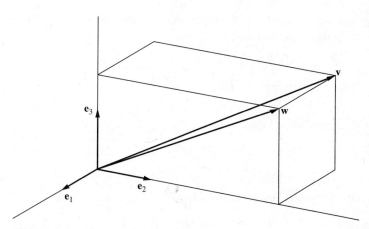

Figure 3.4 The set $\{\mathbf{e}_1, \mathbf{v}, \mathbf{w}\}$ is linearly dependent; $\{\mathbf{e}_2, \mathbf{v}, \mathbf{w}\}$ is linearly independent.

In Figure 3.4, the three vectors $\mathbf{e}_1$, $\mathbf{v}$, and $\mathbf{w}$ lie in the same plane through the origin. Any one of them is a linear combination of the other two. Hence, the set $\{\mathbf{e}_1, \mathbf{v}, \mathbf{w}\}$ is linearly dependent. On the other hand, the three vectors $\mathbf{e}_2$, $\mathbf{v}$, and $\mathbf{w}$ do not lie in the same plane through the origin. Hence, none of them is a linear combination of the other two, and so the set $\{\mathbf{e}_2, \mathbf{v}, \mathbf{w}\}$ is linearly independent.

Exercises 3.3

1. For each of the following implications identify the hypothesis and the conclusion. Is the implication true? Write down the converse of the implication. Is the converse true? Make a table of values of the variables that illustrate the different possible combinations of truth values for the hypothesis and the conclusion. Determine the corresponding truth values of the implication and the converse.

   a. If $2x + 1 = 5$, then $x = 2$.

   b. If $x = y$, then $|x| = |y|$.

   c. If $\sqrt{x} + 3 = 1$, then $x = 4$.

   d. If $\tan x = 1$, then $x = \frac{\pi}{4}$.

   e. If f is differentiable, then f is continuous.

   f. If today is February 28, then tomorrow is March 1.

   g. If today is February 29, then tomorrow is March 1.

2. Show that $\{(2, 4, 2), (3, 2, 0), (1, -2, 2)\}$ is a linearly independent subset of $\mathbb{R}^3$.

3. Show that $\{(1, 0, 0, 0, 0, 0), (1, 2, 0, 0, 0, 0), (0, 1, 2, 3, 0, 0), (0, 0, 1, 2, 3, 4)\}$ is a linearly independent subset of $\mathbb{R}^6$.

4. Show that $\{(2, 2, 6, 0), (0, -1, 0, 1), (1, 2, 3, 3), (1, -1, 3, -2)\}$ is a linearly dependent subset of $\mathbb{R}^4$. Write one of the vectors in this set as a linear combination of the other three. Show that the remaining three vectors form a linearly independent subset of $\mathbb{R}^4$.

5. Determine whether

$$\left\{ \begin{bmatrix} 1 & 2 & 1 \\ 4 & 0 & 0 \end{bmatrix}, \begin{bmatrix} 0 & 1 & 1 \\ 2 & 0 & 2 \end{bmatrix}, \begin{bmatrix} 1 & 1 & 1 \\ 1 & 1 & 1 \end{bmatrix}, \begin{bmatrix} 1 & 1 & 0 \\ 0 & 0 & 3 \end{bmatrix} \right\}$$

   is a linearly independent subset of $\mathbb{M}(2, 3)$.

6. Show that the polynomials defined by

$$p_0(x) = x^3, \quad p_1(x) = (x - 1)^3, \quad p_2(x) = (x - 2)^3, \quad p_3(x) = (x - 3)^3$$

   form a linearly independent subset of $\mathbb{P}_3$.

7. Determine whether the polynomials defined by

$$p_1(x) = x^2 + 2x + 3, \quad p_2(x) = x^2 + 2x + 1, \quad p_3(x) = x^2 + x + 2$$

   form a linearly independent subset of $\mathbb{P}_2$.

8. Show that $\{\sin, \cos, \exp\}$ is a linearly independent subset of $\mathbb{F}(\mathbb{R})$.

9. Show that the functions defined by $f(x) = e^x$ and $g(x) = e^{-x}$ form a linearly independent subset of $\mathbb{F}(\mathbb{R})$.

10. Show that the functions defined by $f(x) = e^x$, $g(x) = e^{2x}$, and $h(x) = e^{x+2}$ form a linearly dependent subset of $\mathbb{F}(\mathbb{R})$.

11. Determine whether the functions defined by $f_0(x) = e^x$, $f_1(x) = xe^x$, and $f_2(x) = x^2 e^x$ form a linearly independent subset of $\mathbb{F}(\mathbb{R})$.

12. Show that if a finite subset of a vector space contains the zero vector, then the subset is linearly dependent.

13. Suppose $\{v_1, \ldots, v_m, v_{m+1}, \ldots, v_{m+k}\}$ is a linearly independent subset of a vector space V. Show that $\{v_1, \ldots, v_m\}$ is a linearly independent subset of V.

14. Prove that if $\{v, w\}$ is a linearly independent subset of a vector space, then $\{v + w, w\}$ is also linearly independent.

15. Suppose $\{v, w\}$ is a linearly independent subset of a vector space V. If $x \in V$ is not a linear combination of v and w, show that $\{v, w, x\}$ is a linearly independent subset of V.

16. By Exercise 11 of Section 1.5 the set $\{0\}$ containing only the zero vector forms a vector space. What subsets of this vector space are linearly independent?

17. Use Theorem 3.5 to determine which of the following sets are linearly independent. No computation should be necessary.

   a. $\{(-3, 1, 0), (6, 2, 0)\}$

   b. $\{(1, 1, 1), (10, 20, 30), (23, 43, 63)\}$

   c. $\{(1, 0, 0, 0), (1, 2, 0, 0), (1, 2, 3, 0), (1, 2, 3, 4)\}$

   d. $\left\{ \begin{bmatrix} 0 & 0 \\ 0 & 0 \end{bmatrix}, \begin{bmatrix} 1 & 0 \\ 0 & 1 \end{bmatrix}, \begin{bmatrix} 0 & 1 \\ 1 & 0 \end{bmatrix} \right\}$

   e. $\{p_1, p_2, p_3\}$ where $p_1(x) = (x + 1)^3$, $p_2(x) = x^2 + x$, and $p_3(x) = x^3 + 1$.

18. Consider the five vectors depicted in Figure 3.4. For each of the following sets, find a vector that is a linear combination of other vectors in the set or conclude that the set is linearly independent.

   a. $\{e_1, e_2, v\}$ **b.** $\{e_2, e_3, w\}$

   c. $\{e_2, e_3, v, w\}$ **d.** $\{v\}$

   e. $\{v, w\}$ **f.** $\{e_1, e_2 + e_3\}$

   g. $\{e_1, e_1 + w, v, w\}$ **h.** $\{e_1 + e_2, e_1 + e_3, e_2 + e_3\}$

3.4 Basis

The notions of a spanning set and of a linearly independent set are the two main ingredients needed to define the dimension of a vector space. Here are the long-awaited definitions:

Definition 3.6

A finite subset $\{v_1, v_2, \ldots, v_n\}$ of a vector space V is a **basis** for V if and only if the set is linearly independent and spans all of V.

Definition 3.7

A vector space V has **dimension** n if and only if there is a basis for V containing exactly n vectors. In this situation, we say that V is n-**dimensional** and write

$$\dim V = n.$$

In the next section we will prove that if one basis for a vector space V has n elements, then every basis for V will have n elements. Hence, the dimension of V is a well-defined quantity.

Quick Example   *Show that $\{x + 1, x - 1\}$ is a basis for $\mathbb{P}_1$. What is $\dim \mathbb{P}_1$?*

Neither of the two functions in the set is a multiple of the other. Hence, by Theorem 3.5, the set is linearly independent. Any polynomial $ax + b$ can be written as a linear combination

$$ax + b = \tfrac{a+b}{2}(x + 1) + \tfrac{a-b}{2}(x - 1)$$

of the two polynomials in the set. Hence, the set spans $\mathbb{P}_1$. Thus, $\{x + 1, x - 1\}$ satisfies the two requirements to be a basis for $\mathbb{P}_1$. The number of elements in this set determines the dimension $\dim \mathbb{P}_1 = 2$. ■

According to the definitions, we are in the satisfying position of being able to show that $\mathbb{R}^3$ is three-dimensional. It should be apparent that the set $\{(1, 0, 0), (0, 1, 0), (0, 0, 1)\}$ spans $\mathbb{R}^3$, and nearly as apparent that this set is linearly independent. Thus, it is a basis for $\mathbb{R}^3$. The fact that it has three elements leads us to the desired conclusion. In general, $\mathbb{R}^n$ has dimension n since

$$\{(1, 0, 0, \ldots, 0),\ (0, 1, 0, \ldots, 0),\ \ldots,\ (0, 0, \ldots, 0, 1)\}$$

is a basis having n elements. We will refer to this set as the **standard basis** for $\mathbb{R}^n$. We will often let $\mathbf{e}_i$ denote the element of the standard basis that consists of all 0s except for the 1 as the ith coordinate.

Similarly, we can easily check that the powers of x from 0 through n form a linearly independent set that spans $\mathbb{P}_n$. Hence, $\dim \mathbb{P}_n = n + 1$. It is not so easy to find a finite set that spans the vector space $\mathbb{P}$ of all polynomials. In fact, it is impossible. This is one of many important examples of vector spaces that do not have a finite basis. The following definition acknowledges such vector spaces. Although we want to be able to work with such examples, our development of the concept of dimension will be restricted to the cases where we have finite bases. The project on infinite-dimensional vector spaces at the end of the chapter indicates how you can extend the concepts of linear combination, span, linear independence, and basis to infinite-dimensional vector spaces.

Definition 3.8

A vector space is **finite-dimensional** if and only if it contains a finite basis. Otherwise, the vector space is **infinite-dimensional**.

Before considering the idea of dimension any further, let us examine the idea of basis a little closer.

In Section 3.2 we saw that the set $\{(1, -2, 3), (1, 0, 1), (0, 1, -2)\}$ spans all of $\mathbb{R}^3$. Thus, to show it is a basis, we have only to check that it is linearly independent. We begin by assuming that a linear combination of the three vectors is equal to $\mathbf{0}$,

$$r_1(1, -2, 3) + r_2(1, 0, 1) + r_3(0, 1, -2) = \mathbf{0},$$

and solving for the scalars by applying Gaussian elimination to the augmented matrix

$$\begin{bmatrix} 1 & 1 & 0 & 0 \\ -2 & 0 & 1 & 0 \\ 3 & 1 & -2 & 0 \end{bmatrix}.$$

But wait. This is the very same matrix we reduced when we were checking whether the set spanned $\mathbb{R}^3$. There we had arbitrary numbers in the rightmost column, so now just think of $b_1 = b_2 = b_3 = 0$. Looking back at our work in that example, we see without any additional computation that the reduced row-echelon form is

$$\begin{bmatrix} 1 & 0 & 0 & 0 \\ 0 & 1 & 0 & 0 \\ 0 & 0 & 1 & 0 \end{bmatrix}.$$

Only when all the scalars are zero will a linear combination of the three vectors yield the zero vector. Thus, the set is linearly independent.

Crossroads As we will discover, the connection between spanning and linear independence frequently eliminates the need to do two separate computations. In the next section we will develop some theoretical results to take advantage of this situation when it occurs.

Exercises 3.4

1. **a.** Describe a basis for $\mathbb{M}(4, 3)$.

   b. What is dim $\mathbb{M}(4, 3)$?

   c. What is dim $\mathbb{M}(m, n)$?

2. Is it possible to find five vectors in $\mathbb{R}^2$ such that any two of them form a basis for $\mathbb{R}^2$? Does $\mathbb{R}^2$ have a unique basis?

3. Show that $\{(1, 2, 3), (0, 1, 2), (0, 0, 1)\}$ is a basis for $\mathbb{R}^3$.

4. Determine whether $\{(1, 0, 1, 0), (1, 1, 0, 0), (1, 1, 1, 1), (0, 0, 0, 1)\}$ is a basis for $\mathbb{R}^4$.

5. Show that $\left\{ \begin{bmatrix} 2 & 1 \\ 1 & 4 \end{bmatrix}, \begin{bmatrix} 1 & 1 \\ 3 & 0 \end{bmatrix}, \begin{bmatrix} 1 & -2 \\ 1 & 1 \end{bmatrix}, \begin{bmatrix} 3 & 2 \\ 0 & 1 \end{bmatrix} \right\}$ is a basis for $\mathbb{M}(2, 2)$.

6. Determine whether $\left\{ \begin{bmatrix} 1 & 1 \\ 0 & 0 \end{bmatrix}, \begin{bmatrix} 1 & 0 \\ 1 & 0 \end{bmatrix}, \begin{bmatrix} 1 & 0 \\ 0 & 1 \end{bmatrix}, \begin{bmatrix} 1 & 1 \\ 1 & 1 \end{bmatrix} \right\}$ is a basis for $\mathbb{M}(2, 2)$.

7. Show that the polynomials p_1, p_2, and p_3 defined by

$$p_1(x) = x^2 + 1, \qquad p_2(x) = 2x^2 + x - 1, \qquad p_3(x) = x^2 + x$$

form a basis for $\mathbb{P}_2$.

8. Verify that the powers 1, x, x^2, ..., x^n of x define functions that form a basis for $\mathbb{P}_n$.

9. **a.** Show that $S = \{p \in \mathbb{P}_3 \mid p(2) = 0\}$ is a subspace of $\mathbb{P}_3$.

b. Show that $\{x - 2, x(x - 2), x^2(x - 2)\}$ is a basis for S.

c. Show that $\{x - 2, (x - 2)^2, (x - 2)^3\}$ is a basis for S.

d. What is dim S?

10. It is known that every solution to the differential equation $y'' = -y$ is a function of the form $a \sin x + b \cos x$ for some constants a and b. (See the article "An Elementary Approach to $y'' = -y$" by J. L. Brenner in the April 1988 issue of the *American Mathematical Monthly* for a five-line proof of this result.) Show that $\{\sin, \cos\}$ is a basis for the subspace of $\mathbb{F}(\mathbb{R})$ consisting of all solutions to this differential equation.

11. By Exercise 11 of Section 1.5 the set $\{\mathbf{0}\}$ containing only the zero vector forms a vector space. What is the dimension of this vector space?

12. Prove that if $\{\mathbf{v}, \mathbf{w}\}$ is a basis for a vector space, then $\{\mathbf{v} + \mathbf{w}, \mathbf{w}\}$ is also a basis for the vector space.

13. **a.** Prove that if $\{\mathbf{v}, \mathbf{w}\}$ is a basis for a vector space, then $\{2\mathbf{v}, \mathbf{w}\}$ is a basis for the vector space.

b. Suppose $\{\mathbf{v}, \mathbf{w}\}$ is a basis for a vector space. For what scalars a and b will $\{a\mathbf{v}, b\mathbf{w}\}$ be a basis for the vector space? Prove your claim.

c. Generalize your claim in part b to bases with more than two elements.

14. Consider a modification of the definition of vector space in which the set of rational numbers

$$\mathbb{Q} = \left\{ \frac{p}{q} \mid p \text{ and } q \text{ are integers and } q \neq 0 \right\}$$

replaces the set of real numbers $\mathbb{R}$ as scalars. Show that

$$S = \{a + b\sqrt{2} \mid a \in \mathbb{Q} \text{ and } b \in \mathbb{Q}\}$$

is one of these $\mathbb{Q}$-vector spaces. Use the fact that $\sqrt{2}$ is irrational to show that $\{1, \sqrt{2}\}$ is a basis for S as a $\mathbb{Q}$-vector space.

3.5 Dimension

It is now time to face up to a serious flaw in our definition of the dimension of a vector space. As was mentioned in the previous section, a vector space could conceivably have two (or more) different dimensions. After all, we have seen several different bases even for such a common vector space as $\mathbb{R}^3$. Was it just a coincidence that they all had three elements? Your geometric intuition should lead you to reject the possibility that the dimension of $\mathbb{R}^3$ could be other than three, or that a plane could be of dimension other than two. What is at stake here is whether the definition does in fact agree with our intuition. If it doesn't, we would hope to find a better definition rather than have to make a drastic change in our intuition.

The basic fact we need is formulated in the following theorem.

Theorem 3.9   Comparison Theorem

Suppose V is a vector space. If $\{\mathbf{w}_1, \mathbf{w}_2, \ldots, \mathbf{w}_n\}$ spans V and $\{\mathbf{v}_1, \mathbf{v}_2, \ldots, \mathbf{v}_m\}$ is a linearly independent subset of V, then $n \geq m$.

Mathematical Strategy Session   The direct approach to proving this theorem would be to start with the hypotheses that one set spans the vector space and that the other set is linearly independent, spell out what this means in terms of equations and linear combinations, and somehow conclude that $n \geq m$.

This proof seems to be easier to follow if we take a roundabout approach. We will assume that the conclusion is false and derive a contradiction to one of the hypotheses. Thus, if the hypotheses are both true, the conclusion cannot be false. The only alternative is for the conclusion to be true. (This is an appeal to what is known as the **law of the excluded middle**.)

This strategy for proving a theorem is known as a **proof by contradiction** or an **indirect proof**. Although it is occasionally a useful technique, direct proofs are generally more straightforward and thus easier to comprehend.

Proof   Let us assume that the conclusion is false. That is, suppose $n < m$. Set this aside for a short time. Soon we will see that it leads us into trouble with the linear independence of $\{\mathbf{v}_1, \mathbf{v}_2, \ldots, \mathbf{v}_m\}$.

Since $\{\mathbf{w}_1, \mathbf{w}_2, \ldots, \mathbf{w}_n\}$ spans V, we can write the vectors $\mathbf{v}_1, \mathbf{v}_2, \ldots, \mathbf{v}_m$ as linear combinations as follows:

$$a_{11}\mathbf{w}_1 + a_{12}\mathbf{w}_2 + \cdots + a_{1n}\mathbf{w}_n = \mathbf{v}_1$$
$$a_{21}\mathbf{w}_1 + a_{22}\mathbf{w}_2 + \cdots + a_{2n}\mathbf{w}_n = \mathbf{v}_2$$
$$\vdots \qquad \vdots \qquad\qquad \vdots \qquad \vdots$$
$$a_{m1}\mathbf{w}_1 + a_{m2}\mathbf{w}_2 + \cdots + a_{mn}\mathbf{w}_n = \mathbf{v}_m$$

The double subscripting has been chosen so that a_{ij} is the coefficient of $\mathbf{w}_j$ in the expression for $\mathbf{v}_i$.

Because of the assumption that $n < m$, the system

$$a_{11}x_1 + a_{21}x_2 + \cdots + a_{m1}x_m = 0$$
$$a_{12}x_1 + a_{22}x_2 + \cdots + a_{m2}x_m = 0$$
$$\vdots \qquad \vdots \qquad\qquad \vdots \qquad \vdots$$
$$a_{1n}x_1 + a_{2n}x_2 + \cdots + a_{mn}x_m = 0$$

has more unknowns than equations. By the Fundamental Theorem of Homogeneous Systems, the system has nontrivial solutions. Let $(c_1, c_2, \ldots, c_m)$ be such a solution. That is,

$$a_{11}c_1 + a_{21}c_2 + \cdots + a_{m1}c_m = 0$$
$$a_{12}c_1 + a_{22}c_2 + \cdots + a_{m2}c_m = 0$$
$$\vdots \qquad \vdots \qquad\qquad \vdots \qquad \vdots$$
$$a_{1n}c_1 + a_{2n}c_2 + \cdots + a_{mn}c_m = 0$$

and not all the numbers $c_1, c_2, \ldots, c_m$ are equal to zero.

Now the expressions for $\mathbf{v}_1, \mathbf{v}_2, \ldots, \mathbf{v}_m$ and some massive rearranging give

$$
\begin{aligned}
c_1\mathbf{v}_1 + c_2\mathbf{v}_2 + \cdots + c_m\mathbf{v}_m &= c_1(a_{11}\mathbf{w}_1 + a_{12}\mathbf{w}_2 + \cdots + a_{1n}\mathbf{w}_n) \\
&\quad + c_2(a_{21}\mathbf{w}_1 + a_{22}\mathbf{w}_2 + \cdots + a_{2n}\mathbf{w}_n) + \cdots \\
&\quad + c_m(a_{m1}\mathbf{w}_1 + a_{m2}\mathbf{w}_2 + \cdots + a_{mn}\mathbf{w}_n) \\
&= (c_1a_{11} + c_2a_{21} + \cdots + c_ma_{m1})\mathbf{w}_1 \\
&\quad + (c_1a_{12} + c_2a_{22} + \cdots + c_ma_{m2})\mathbf{w}_2 + \cdots \\
&\quad + (c_1a_{1n} + c_2a_{2n} + \cdots + c_ma_{mn})\mathbf{w}_n \\
&= 0\mathbf{w}_1 + 0\mathbf{w}_2 + \cdots + 0\mathbf{w}_n \\
&= \mathbf{0}.
\end{aligned}
$$

Thus, we have a nontrivial linear combination of $\mathbf{v}_1, \mathbf{v}_2, \ldots, \mathbf{v}_n$ that is equal to $\mathbf{0}$. This contradicts the hypothesis that $\{\mathbf{v}_1, \mathbf{v}_2, \ldots, \mathbf{v}_n\}$ is linearly independent.

We must conclude that the number m of unknowns in the homogeneous system does not exceed the number n of equations. That is, $n \geq m$. ∎

The fact that a vector space can have only one dimension is an easy consequence of the Comparison Theorem.

Theorem 3.10

If $\{\mathbf{v}_1, \mathbf{v}_2, \ldots, \mathbf{v}_m\}$ and $\{\mathbf{w}_1, \mathbf{w}_2, \ldots, \mathbf{w}_n\}$ are bases for a vector space V, then $m = n$.

Proof   Since $\{\mathbf{w}_1, \mathbf{w}_2, \ldots, \mathbf{w}_n\}$ spans V and $\{\mathbf{v}_1, \mathbf{v}_2, \ldots, \mathbf{v}_m\}$ is linearly independent (bases have both properties), the Comparison Theorem gives the result that $n \geq m$. But we also have that $\{\mathbf{v}_1, \mathbf{v}_2, \ldots, \mathbf{v}_m\}$ spans V and $\{\mathbf{w}_1, \mathbf{w}_2, \ldots, \mathbf{w}_n\}$ is linearly independent. Again by the Comparison Theorem, $m \geq n$. The two inequalities together yield the desired equality, $m = n$. ∎

The next two theorems are very useful in working with bases for vector spaces. The first provides a way to whittle redundant vectors from a spanning set for a vector space until a linearly independent spanning set (a basis) is obtained.

Theorem 3.11 Contraction Theorem

Suppose $\{\mathbf{v}_1, \mathbf{v}_2, \ldots, \mathbf{v}_n\}$ spans a vector space V. Then some subset of $\{\mathbf{v}_1, \mathbf{v}_2, \ldots, \mathbf{v}_n\}$ is a basis for V.

Proof If $\{\mathbf{v}_1, \mathbf{v}_2, \ldots, \mathbf{v}_n\}$ is a linearly independent set, then we have a basis to begin with. Otherwise, $\{\mathbf{v}_1, \mathbf{v}_2, \ldots, \mathbf{v}_n\}$ is linearly dependent, and we know by Theorem 3.5 that one of the vectors can be written as a linear combination of the other vectors in the set. That is, for some i with $1 \leq i \leq n$, we have

$$\mathbf{v}_i = r_1\mathbf{v}_1 + r_2\mathbf{v}_2 + \cdots + r_{i-1}\mathbf{v}_{i-1} + r_{i+1}\mathbf{v}_{i+1} + \cdots + r_n\mathbf{v}_n.$$

Now it is easy to see that the subset $\{\mathbf{v}_1, \mathbf{v}_2, \ldots, \mathbf{v}_{i-1}, \mathbf{v}_{i+1}, \ldots, \mathbf{v}_n\}$ still spans V. Indeed, any $\mathbf{v} \in V$ can be written as a linear combination $\mathbf{v} = a_1\mathbf{v}_1 + a_2\mathbf{v}_2 + \cdots + a_i\mathbf{v}_i + \cdots + a_n\mathbf{v}_n$, and by substituting the preceding expression for $\mathbf{v}_i$, we obtain $\mathbf{v}$ as a linear combination without using $\mathbf{v}_i$:

$$\begin{aligned}
\mathbf{v} = \; & a_1\mathbf{v}_1 + a_2\mathbf{v}_2 + \cdots + a_{i-1}\mathbf{v}_{i-1} \\
& + a_i(r_1\mathbf{v}_1 + r_2\mathbf{v}_2 + \cdots + r_{i-1}\mathbf{v}_{i-1} + r_{i+1}\mathbf{v}_{i+1} + \cdots + r_n\mathbf{v}_n) \\
& + a_{i+1}\mathbf{v}_{i+1} + \cdots + a_n\mathbf{v}_n \\
= \; & (a_1 + a_ir_1)\mathbf{v}_1 + \cdots + (a_{i-1} + a_ir_{i-1})\mathbf{v}_{i-1} \\
& + (a_{i+1} + a_ir_{i+1})\mathbf{v}_{i+1} + \cdots + (a_n + a_ir_n)\mathbf{v}_n.
\end{aligned}$$

Now if $\{\mathbf{v}_1, \mathbf{v}_2, \ldots, \mathbf{v}_{i-1}, \mathbf{v}_{i+1}, \ldots, \mathbf{v}_n\}$ is linearly independent, we have a basis. Otherwise, we can find another vector that is a linear combination of the remaining $n - 2$ vectors and argue as before that when this vector is eliminated, the remaining set still spans V.

Each time we repeat this operation, we obtain a spanning set for V with one less vector. We must eventually find one of these subsets that is linearly independent. For ultimately we would find that the empty set spans V. In this case V is the trivial vector space $\{\mathbf{0}\}$, and by our convention, $\varnothing$ is linearly independent. ∎

Quick Example *Prove that a vector space is finite-dimensional if and only if it has a finite spanning set.*

If a vector space is finite-dimensional, then it has a basis with a finite number of elements. This finite set will span the vector space.

Conversely, suppose a vector space has a finite spanning set. By the Contraction Theorem, some subset of this set is a basis for the vector space. Since this basis is a finite set, the vector space is finite-dimensional. ∎

Crossroads Proofs of theorems often contain information not given explicitly in the statement of the theorem. This is especially true in proofs that describe the steps in an algorithm, as is the case for the Contraction Theorem. Its proof provides a recipe for constructing a basis from a spanning set. The following example is a typical application of this process. This method of contracting a spanning set down to a basis will reappear in our work with matrices in Chapter 6.

Quick Example   *Let S be the subspace of $\mathbb{R}^3$ spanned by*

$$\{(1, 2, -1), (3, 1, 2), (2, -1, 3), (1, -3, 4)\}.$$

Use the proof of the Contraction Theorem as a recipe for finding a subset of $\{(1, 2, -1), (3, 1, 2), (2, -1, 3), (1, -3, 4)\}$ that is a basis for S.

We begin by determining whether one of the vectors is a linear combination of other vectors in the set. We apply the usual methods to solve the equation

$$r_1(1, 2, -1) + r_2(3, 1, 2) + r_3(2, -1, 3) + r_4(1, -3, 4) = (0, 0, 0).$$

We reduce the coefficient matrix of the resulting homogeneous system and find that r_3 and r_4 are free variables:

$$\begin{bmatrix} 1 & 3 & 2 & 1 \\ 2 & 1 & -1 & -3 \\ -1 & 2 & 3 & 4 \end{bmatrix} \longrightarrow \begin{bmatrix} 1 & 3 & 2 & 1 \\ 0 & -5 & -5 & -5 \\ 0 & 5 & 5 & 5 \end{bmatrix} \longrightarrow \begin{bmatrix} 1 & 0 & -1 & -2 \\ 0 & 1 & 1 & 1 \\ 0 & 0 & 0 & 0 \end{bmatrix}.$$

We can solve for $(1, -3, 4)$ by setting $r_3 = 0$ and $r_4 = 1$ and determining the values $r_1 = 2$ and $r_2 = -1$ for the leading variables. This shows that $(1, -3, 4) = -2(1, 2, -1) - (-1)(3, 1, 2) - 0(2, -1, 3)$ is a linear combination of the other vectors in the set. Hence, $(1, -3, 4)$ can be eliminated from the set, and the remaining three vectors will still span S. Similarly, we can solve for $(2, -1, 3)$ by setting $r_3 = 1$ and $r_4 = 0$ and determining the values $r_1 = 1$, $r_2 = -1$. This shows that $(2, -1, 3) = -1(1, 2, -1) - (-1)(3, 1, 2)$ is a linear combination of the remaining two vectors. Hence, it too can be eliminated from the set. Since neither of these two remaining vectors, $(1, 2, -1)$ and $(3, 1, 2)$, is a multiple of the other, they form a linearly independent set. Since this set still spans the subspace S, it is therefore a basis for S.   ∎

♚ **Mathematical Strategy Session**   If you compare the preceding example with your solutions to Exercises 15, 16, and 17, you will discover a way to automate the selection of the basis vectors from the spanning set $\{v_1, \ldots, v_n\}$. The equation $r_1 v_1 + \cdots + r_n v_n = 0$ produces a homogeneous system of linear equations with unknowns $r_1, \ldots, r_n$. When you put the coefficient matrix in reduced row-echelon form, you will find that some of the unknowns are leading variables and some are free variables. The vectors corresponding to the free variables can be written as linear combinations of other vectors in the set. Once these are eliminated, only leading variables will remain. This makes it easy to see that the vectors corresponding to leading variables form a linearly independent set. Thus, the vectors corresponding to the leading variables form a basis that is a subset of the spanning set.

The companion to the Contraction Theorem treats the opposite situation and describes how to expand a linearly independent set of vectors until it contains enough vectors to span the vector space. As was the case in the proof of the Contraction Theorem, the proof of the Expansion Theorem works with one vector at a time. The verification that a vector can be adjoined in such a way that the resulting set is linearly independent relies on a preliminary result stated next. This lemma is often as useful as the theorem itself.

Lemma 3.12   Expansion Lemma

Suppose $\{v_1, \ldots, v_n\}$ is a linearly independent subset of a vector space V. Consider a vector $v_{n+1} \in V$. If $v_{n+1} \notin \text{span}\{v_1, \ldots, v_n\}$, then $\{v_1, \ldots, v_n, v_{n+1}\}$ is linearly independent.

Proof   Here is an outline of the proof. You should not have a great deal of difficulty writing down a complete proof. Doing so will demonstrate your mastery of the concept of linear independence.

To show $\{v_1, \ldots, v_n, v_{n+1}\}$ is linearly independent, start as usual by assuming that

$$r_1 v_1 + \cdots + r_n v_n + r_{n+1} v_{n+1} = \mathbf{0}.$$

First show that $r_{n+1} = 0$; if not, you can easily write v_{n+1} as a linear combination of $v_1, \ldots, v_n$, contrary to the hypothesis. Now the equation reduces to

$$r_1 v_1 + \cdots + r_n v_n = \mathbf{0}.$$

Say the magic words and derive $r_1 = 0, \ldots, r_n = 0$. Since all $n + 1$ coefficients have been shown to be equal to zero, you can conclude that $\{v_1, \ldots, v_n, v_{n+1}\}$ is linearly independent. ∎

The Expansion Theorem itself has one crucial hypothesis not present in the Contraction Theorem or in the Expansion Lemma. Watch how the assumption that the vector space is finite-dimensional ensures that we will eventually obtain the desired spanning set.

Theorem 3.13   Expansion Theorem

Suppose $\{v_1, \ldots, v_m\}$ is a linearly independent subset of a finite-dimensional vector space V. Then we can adjoin vectors $v_{m+1}, \ldots, v_{m+k}$ so that $\{v_1, \ldots, v_m, v_{m+1}, \ldots, v_{m+k}\}$ is a basis for V. ∎

Proof   If $\{v_1, \ldots, v_m\}$ spans V, then we have a basis to begin with. In this case $k = 0$, and the list of vectors we need to adjoin is empty. Otherwise, there is some vector $v_{m+1} \in V$ that is not a linear combination of $\{v_1, \ldots, v_m\}$. By the Expansion Lemma, $\{v_1, \ldots, v_m, v_{m+1}\}$ is linearly independent.

Now if $\{\mathbf{v}_1, \ldots, \mathbf{v}_m, \mathbf{v}_{m+1}\}$ spans V, we have a basis. Otherwise, we can again adjoin another vector to the set and still maintain the property that it is linearly independent.

Each time we repeat this process, we obtain a linearly independent set with one additional vector. By the Comparison Theorem, the number of elements in a linearly independent set cannot exceed the dimension of V. Hence, by the time the set has been expanded to contain dim V elements, we are guaranteed that we will not be able to produce a larger linearly independent subset. At this point, the set must span V. Hence, it will be a basis for V.   ■

Again we have a proof that provides a recipe for constructing a basis for a vector space.

Quick Example   *Use the proof of the Expansion Theorem to find a basis for $\mathbb{R}^4$ that contains the vectors $(1, 1, 1, 1)$ and $(0, 0, 1, 1)$.*

To keep things simple, let us try to adjoin the standard basis vectors for $\mathbb{R}^4$. You can easily check that $(1, 0, 0, 0)$ is not a linear combination of these two vectors. Hence, $\{(1, 1, 1, 1), (0, 0, 1, 1), (1, 0, 0, 0)\}$ is linearly independent. We cannot use $(0, 1, 0, 0)$, since it is a linear combination of the three vectors we already have. However, $(0, 0, 1, 0)$ is not a linear combination of these three vectors. Hence, we can adjoin it to obtain the linearly independent set

$$\{(1, 1, 1, 1), (0, 0, 1, 1), (1, 0, 0, 0), (0, 0, 1, 0)\}.$$

Now that we have $4 = \dim \mathbb{R}^4$ elements, the argument in the last paragraph of the proof of the Expansion Theorem shows that this set must be a basis for $\mathbb{R}^4$.   ■

You have come a long way in your study of vector spaces and have gained much experience in dealing with mathematical concepts from an abstract point of view. This section contains an especially heavy dose of theoretical mathematics. Following through the arguments of such mathematical proofs requires very meticulous reading with numerous calls to check out earlier results and to work out the details of some points that are left to the reader. Among the rewards for this careful reading are a deeper understanding of the concepts involved, acquisition of techniques that are useful in other situations, and a glimpse at the beauty of mathematics that is rivaled only by classical music and great literature.

Here are two theorems you will be asked to prove in the exercises at the end of this section. They represent a plateau in the development of the theory of vector spaces that began with the list of axioms introduced in Chapter 1.

The first of these theorems is geometrically appealing. It guarantees, for example, that there are no four-dimensional subspaces of $\mathbb{R}^3$, only the zero subspace $\{\mathbf{0}\}$, lines, planes, and $\mathbb{R}^3$ itself. It is interesting to compare this theorem with similar results in other algebraic structures. For example, there are groups (systems with one operation that satisfies several basic algebraic conditions) that can be generated by two elements yet contain subgroups that cannot be generated by any finite subset.

Theorem 3.14

A subspace S of a finite-dimensional vector space V is finite-dimensional, and $\dim S \leq \dim V$.

The final theorem of this section is a great labor-saving result. If you happen to know the dimension of your vector space, you can cut in half the work needed to verify that a set with the appropriate number of elements is a basis. Check either that the set spans the vector space or that the set is linearly independent, and the other condition will automatically hold. Often linear independence is easier to check since it usually amounts to solving a system of linear equations that is homogeneous.

Theorem 3.15

Suppose V is an n-dimensional vector space. If $\{v_1, \ldots, v_n\}$ spans V, then it is also linearly independent. If $\{v_1, \ldots, v_n\}$ is linearly independent, then it also spans V.

Quick Example *Show that the set $\{(2, 4, -2), (3, 2, 0), (1, -2, -2)\}$ is a basis for $\mathbb{R}^3$.*

By Exercise 2a of Section 3.2, the set spans $\mathbb{R}^3$. We also know that $\dim \mathbb{R}^3 = 3$. Since there are three elements in this set, we conclude by Theorem 3.15 that the set is a basis for $\mathbb{R}^3$. ∎

Quick Example *Show that the polynomials defined by $p_0(x) = x^3$, $p_1(x) = (x - 1)^3$, $p_2(x) = (x - 2)^3$, and $p_3(x) = (x - 3)^3$ form a basis for $\mathbb{P}_3$.*

By Exercise 6 of Section 3.3, the set is linearly independent. Since it has $\dim \mathbb{P}_3 = 4$ elements, by Theorem 3.15 the set is a basis for $\mathbb{P}_3$. ∎

Exercises 3.5

1. Fill in the steps to this outline of a classic indirect proof that $\sqrt{2}$ is irrational.
   a. Assume $\sqrt{2}$ is rational. Set up notation for an expression for $\sqrt{2}$ as a ratio of integers that have been reduced to lowest terms.
   b. Square both sides of the equation and multiply both sides by the denominator of the fraction.
   c. Observe that the square of the numerator is even. Explain why the numerator itself must be even.
   d. Write the numerator as 2 times another integer.
   e. Show that the square of the denominator is even; hence, the denominator itself must be even.
   f. Notice the contradiction to the reduction of the fraction to lowest terms.

2.   **a.** Adapt the proof in Exercise 1 to show that $\sqrt{3}$ is irrational.

   b. Determine whether $\sqrt{2} + \sqrt{3}$ is rational.

   c. Although π and e are known to be irrational, it is unknown whether $\pi + e$ is rational. Can you make any progress toward resolving this mystery?

3.   Fill in the steps to this outline of Euclid's beautiful indirect proof that there are an infinite number of primes. (Recall that a positive integer is a **prime** if and only if there are exactly two positive integers, the number itself and 1, that divide evenly into it.)

   a. Suppose there are only a finite number of primes. Set up notation for the list of all primes.

   b. Multiply all these primes together. Add 1 to the product.

   c. Explain why none of the primes in the list divides evenly into this number.

   d. Conclude that the number has a prime factor that is not on the list.

   e. State the contradiction clearly and wrap up the proof.

4.   **a.** What does the Comparison Theorem say about the number of elements in a set that spans $\mathbb{P}_7$?

   b. What does the Comparison Theorem say about the number of elements in a linearly independent subset of $\mathbb{P}_7$?

5.   **a.** What does the Comparison Theorem say about the number of elements in a set that spans $\mathbb{M}(4, 3)$?

   b. What does the Comparison Theorem say about the number of elements in a linearly independent subset of $\mathbb{M}(4, 3)$?

6.   Suppose S is a subset of $\mathbb{P}_4$.

   a. If S has four elements, is it possible for S to be linearly independent? Is it possible for S to span $\mathbb{P}_4$?

   b. If S has five elements, is it possible for S to be linearly independent? Is it possible for S to span $\mathbb{P}_4$?

   c. If S has six elements, is it possible for S to be linearly independent? Is it possible for S to span $\mathbb{P}_4$?

7.   Illustrate the proof of the Comparison Theorem by showing that if $\{\mathbf{v}_1, \ldots, \mathbf{v}_m\}$ is a linearly independent subset of $\mathbb{R}^4$, then $m \le 4$. (Suggestion: Compare with the spanning set $\{\mathbf{w}_1, \mathbf{w}_2, \mathbf{w}_3, \mathbf{w}_4\} = \{(1, 0, 0, 0), (0, 1, 0, 0), (0, 0, 1, 0), (0, 0, 0, 1)\}$.)

8.   Illustrate the proof of the Comparison Theorem by showing that if $\{\mathbf{w}_1, \ldots, \mathbf{w}_n\}$ spans $\mathbb{R}^4$, then $4 \le n$. (Suggestion: Compare with the linearly independent set $\{\mathbf{v}_1, \mathbf{v}_2, \mathbf{v}_3, \mathbf{v}_4\} = \{(1, 0, 0, 0), (0, 1, 0, 0), (0, 0, 1, 0), (0, 0, 0, 1)\}$.)

9.   Suppose V is a vector space. Suppose $\{\mathbf{w}_1, \ldots, \mathbf{w}_n\}$ spans V, $\{\mathbf{v}_1, \ldots, \mathbf{v}_m\}$ is a linearly independent subset of V, and $\{\mathbf{u}_1, \ldots, \mathbf{u}_k\}$ is a basis for V. Use the Comparison Theorem to prove that $m \le k \le n$.

10.   Fill in the details to give a complete, well-organized proof of the Expansion Lemma.

11.   Give an alternative proof of the Expansion Lemma based on Theorem 3.5.

12.   Prove Theorem 3.14.

13. Prove Theorem 3.15.

14. Suppose V is an infinite-dimensional vector space. For any integer $n \geq 0$, show that V contains a subspace of dimension n.

15. By Exercise 3 of Section 3.2, the set

$$\{(3, 1, 1), (1, 5, 3), (-2, 2, 1), (5, 4, 3), (1, 1, 1)\}$$

spans $\mathbb{R}^3$. Use the proof of the Contraction Theorem as a recipe for finding a subset of this set that is a basis for $\mathbb{R}^3$.

16. Use the proof of the Contraction Theorem as a recipe for finding a subset of

$$\{(1, 2, 1, 2), (3, 1, 2, 0), (1, -1, -1, -1), (1, 0, 2, -1)\}$$

that is a basis for the subspace of $\mathbb{R}^4$ spanned by this set.

17. Use the proof of the Contraction Theorem as a recipe for finding a subset of

$$\{x^2 + 2x, x^3 + 3x, x^3 + x^2, x + 1, x^2 + 3x + 1, x^3 + 2x + 1\}$$

that is a basis for a subspace of $\mathbb{P}_3$. Is your subset a basis for $\mathbb{P}_3$?

18. Suppose $\{\mathbf{u}_1, \ldots, \mathbf{u}_{n+1}\}$ spans the n-dimensional vector space V. Suppose $\mathbf{u}_1 + \cdots + \mathbf{u}_{n+1} = \mathbf{0}$. Show that if any vector $\mathbf{u}_i$ is deleted, then the resulting set $\{\mathbf{u}_1, \ldots, \mathbf{u}_{i-1}, \mathbf{u}_{i+1}, \ldots, \mathbf{u}_{n+1}\}$ is a basis for V.

19. Use the proof of the Expansion Theorem as a recipe for finding a basis for $\mathbb{R}^3$ that contains $\{(2, 3, 1), (4, -2, 1)\}$.

20. Use the proof of the Expansion Theorem as a recipe for finding a basis for $\mathbb{P}_3$ that contains $\{x + 3, x^2 + x\}$.

21. Suppose S is a subspace of a finite-dimensional vector space V. Prove that if $\dim S = \dim V$, then $S = V$.

3.6 Coordinates   (cf p.68 first)

You may be wondering what good it does to have a basis for a vector space, let alone several bases. In our mercenary society, everyone wants to know the bottom line—what is it worth?

Well, we have already seen that a basis is key to determining the dimension of a vector space. In this section we will see that a basis also makes it possible to introduce coordinates in arbitrary vector spaces that work very much like the familiar coordinates for vectors in Euclidean spaces.

We begin with an example of how this works in $\mathbb{R}^2$. You can easily check that

$$E = \left\{ \begin{bmatrix} 1 \\ 0 \end{bmatrix}, \begin{bmatrix} 0 \\ 1 \end{bmatrix} \right\} \quad \text{and} \quad B = \left\{ \begin{bmatrix} 2 \\ 1 \end{bmatrix}, \begin{bmatrix} 1 \\ -1 \end{bmatrix} \right\}$$

are bases for $\mathbb{R}^2$. We can write any vector in $\mathbb{R}^2$ as a linear combination of elements in either basis. For example,

$$\begin{bmatrix} 6 \\ 9 \end{bmatrix} = 6 \begin{bmatrix} 1 \\ 0 \end{bmatrix} + 9 \begin{bmatrix} 0 \\ 1 \end{bmatrix}$$

for the basis E. Here the coefficients are precisely the entries of the vector itself. With the basis B,

$$\begin{bmatrix} 6 \\ 9 \end{bmatrix} = 5 \begin{bmatrix} 2 \\ 1 \end{bmatrix} - 4 \begin{bmatrix} 1 \\ -1 \end{bmatrix}.$$

The coefficients are not as directly related to the entries of the original vector as they were with the basis E. Nevertheless, we can use the coefficients 5 and -4 as entries of an element of $\mathbb{R}^2$ that corresponds to the original vector.

Here is the definition that sets up the notation we need. For this definition, the elements of the basis must be listed in some specific order. Although we will continue to write a basis as a set of vectors, in this context we will assume that elements have been designated first, second, third, and so forth.

Definition 3.16

Let $B = \{\mathbf{u}_1, \ldots, \mathbf{u}_n\}$ be an ordered basis for a vector space V. For any $\mathbf{v} \in V$, write $\mathbf{v} = r_1\mathbf{u}_1 + \cdots + r_n\mathbf{u}_n$. The **coordinate vector** for $\mathbf{v}$ with respect to B is the element of $\mathbb{R}^n$ denoted $[\mathbf{v}]_B$ and defined by

$$[\mathbf{v}]_B = \begin{bmatrix} r_1 \\ \vdots \\ r_n \end{bmatrix}.$$

The linear combinations in the preceding example can be expressed very succinctly with this new notation:

$$\begin{bmatrix} 6 \\ 9 \end{bmatrix}_E = \begin{bmatrix} 6 \\ 9 \end{bmatrix} \quad \text{and} \quad \begin{bmatrix} 6 \\ 9 \end{bmatrix}_B = \begin{bmatrix} 5 \\ -4 \end{bmatrix}.$$

Coordinates of vectors with respect to the standard basis are especially easy to interpret geometrically. This is because the basis vectors $\mathbf{e}_1$ and $\mathbf{e}_2$ lie so nicely on the coordinate axes and we are familiar with plotting points in the Cartesian coordinate system. With a little adjustment we can adapt the same geometric ideas to coordinates with respect to the basis B. Simply draw the axes in the directions of the basis vectors $\mathbf{x} = \begin{bmatrix} 2 \\ 1 \end{bmatrix}$ and $\mathbf{y} = \begin{bmatrix} 1 \\ -1 \end{bmatrix}$. Mark the scales along these axes in terms of multiples of the vectors $\mathbf{x}$ and $\mathbf{y}$. Figure 3.5 illustrates the vector $\mathbf{v} = \begin{bmatrix} 6 \\ 9 \end{bmatrix}$ in this new coordinate system.

Whereas determining a coordinate vector usually involves solving a system of linear equations, it is very simple to find a vector with a given set of coordinates. For example, if we know that $\begin{bmatrix} x \\ y \end{bmatrix}_B = \begin{bmatrix} 8 \\ 3 \end{bmatrix}$, then

$$\begin{bmatrix} x \\ y \end{bmatrix} = 8 \begin{bmatrix} 2 \\ 1 \end{bmatrix} + 3 \begin{bmatrix} 1 \\ -1 \end{bmatrix} = \begin{bmatrix} 19 \\ 5 \end{bmatrix}.$$

We have been relying on the fact that a basis spans a vector space to know that we can always find coordinates for any vector. The following theorem shows how linear independence guarantees that, with respect to a given basis, only one coordinate vector is possible for any vector in a vector space.

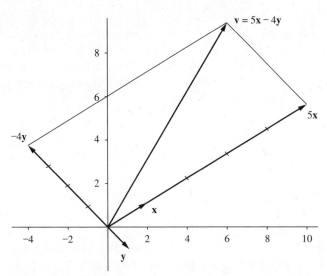

Figure 3.5 The coordinates of $\mathbf{v} = \begin{bmatrix} 6 \\ 9 \end{bmatrix}$ are $[\mathbf{v}]_B = \begin{bmatrix} 5 \\ -4 \end{bmatrix}$.

Theorem 3.17

Suppose $B = \{\mathbf{u}_1, \ldots, \mathbf{u}_n\}$ is an ordered basis for a vector space V. For any $\mathbf{v} \in V$ there is a unique list of scalars $r_1, \ldots, r_n$ such that $\mathbf{v} = r_1\mathbf{u}_1 + \cdots + r_n\mathbf{u}_n$.

Proof The basis B spans V. Thus, any vector $\mathbf{v} \in V$ can be written as a linear combination $\mathbf{v} = r_1\mathbf{u}_1 + \cdots + r_n\mathbf{u}_n$. This gives the existence of the desired list of scalars $r_1, \ldots, r_n$. Next we use the linear independence of the basis B to show that the list is unique. Suppose we have $\mathbf{v} = r_1\mathbf{u}_1 + \cdots + r_n\mathbf{u}_n$ and $\mathbf{v} = s_1\mathbf{u}_1 + \cdots + s_n\mathbf{u}_n$. Then

$$\mathbf{0} = \mathbf{v} - \mathbf{v}$$
$$= (r_1\mathbf{u}_1 + \cdots + r_n\mathbf{u}_n) - (s_1\mathbf{u}_1 + \cdots + s_n\mathbf{u}_n)$$
$$= (r_1 - s_1)\mathbf{u}_1 + \cdots + (r_n - s_n)\mathbf{u}_n.$$

Since B is linearly independent, we must have $r_1 - s_1 = 0, \ldots, r_n - s_n = 0$. That is, $r_1 = s_1, \ldots, r_n = s_n$, as desired. ■

If the basis is especially simple, you can sometimes determine coordinate vectors by inspection. Here is an example.

Quick Example *Consider the ordered basis* $B = \left\{ \begin{bmatrix} 1 \\ 0 \\ 0 \end{bmatrix}, \begin{bmatrix} 1 \\ 1 \\ 0 \end{bmatrix}, \begin{bmatrix} 1 \\ 1 \\ 1 \end{bmatrix} \right\}$, *for* $\mathbb{R}^3$. *De-*

termine the coordinate vector $\begin{bmatrix} 3 \\ -8 \\ 7 \end{bmatrix}_B$.

Only the third vector in B will contribute to the third coordinate in any linear combination of elements in B. Thus, its coefficient must be 7. Unfortunately, this also contributes 7 to the second coordinate, where we want -8. We can remedy this by adding -15 times the second vector in B. The first coordinate can be adjusted from the -8 we have so far to the desired 3 by adding 11 times the first vector in B. Thus, we have determined the linear combination

$$\begin{bmatrix} 3 \\ -8 \\ 7 \end{bmatrix} = 11 \begin{bmatrix} 1 \\ 0 \\ 0 \end{bmatrix} - 15 \begin{bmatrix} 1 \\ 1 \\ 0 \end{bmatrix} + 7 \begin{bmatrix} 1 \\ 1 \\ 1 \end{bmatrix}.$$

This gives $\begin{bmatrix} 3 \\ -8 \\ 7 \end{bmatrix}_B = \begin{bmatrix} 11 \\ -15 \\ 7 \end{bmatrix}$. ∎

In more complicated cases, you will need to solve a system of linear equations to find a coordinate vector. This is still quite straightforward.

Quick Example   *By Exercise 7 of Section 3.4,*

$$B = \{x^2 + 1, \; 2x^2 + x - 1, \; x^2 + x\}$$

is a basis for $\mathbb{P}_2$. *Find* $[x^2 - x + 4]_B$.

We write

$$x^2 - x + 4 = r_1(x^2 + 1) + r_2(2x^2 + x - 1) + r_3(x^2 + x)$$

and solve the resulting system:

$$\begin{aligned} r_1 + 2r_2 + r_3 &= 1 \\ r_2 + r_3 &= -1 \\ r_1 - r_2 &= 4 \end{aligned}$$

$$\begin{bmatrix} 1 & 2 & 1 & 1 \\ 0 & 1 & 1 & -1 \\ 1 & -1 & 0 & 4 \end{bmatrix} \rightarrow \begin{bmatrix} 1 & 2 & 1 & 1 \\ 0 & 1 & 1 & -1 \\ 0 & -3 & -1 & 3 \end{bmatrix} \rightarrow \begin{bmatrix} 1 & 2 & 1 & 1 \\ 0 & 1 & 1 & -1 \\ 0 & 0 & 2 & 0 \end{bmatrix}$$

$$\rightarrow \begin{bmatrix} 1 & 2 & 0 & 1 \\ 0 & 1 & 0 & -1 \\ 0 & 0 & 1 & 0 \end{bmatrix} \rightarrow \begin{bmatrix} 1 & 0 & 0 & 3 \\ 0 & 1 & 0 & -1 \\ 0 & 0 & 1 & 0 \end{bmatrix}$$

So $[x^2 - x + 4]_B = \begin{bmatrix} 3 \\ -1 \\ 0 \end{bmatrix}$. ∎

Coordinate vectors are especially useful in situations where we do not have a standard system of coordinates. For example, it is known that the solution set S of the differential equation $y'' - y = 0$ is a two-dimensional subspace of $\mathbb{F}(\mathbb{R})$. By Exercise 9 of Section 3.3, the set $B = \{e^x, e^{-x}\}$ of two solutions is linearly independent. Hence, by Theorem 3.15, B is a basis. We can use this basis to introduce coordinates for the solution space S and make it seem just like $\mathbb{R}^2$.

Quick Example   *The hyperbolic sine function is a solution to the differential equation $y'' - y = 0$. Write the coordinate vector of* $\sinh$ *with respect to* $B = \{e^x, e^{-x}\}$.

Recall that $\sinh x = \dfrac{e^x - e^{-x}}{2}$. Thus, $[\sinh]_B = \begin{bmatrix} \frac{1}{2} \\ -\frac{1}{2} \end{bmatrix}$. ∎

Exercises 3.6

1. Consider the ordered basis $B = \left\{ \begin{bmatrix} 1 \\ 2 \\ 3 \end{bmatrix}, \begin{bmatrix} 1 \\ 2 \\ 0 \end{bmatrix}, \begin{bmatrix} 1 \\ 0 \\ 0 \end{bmatrix} \right\}$ for $\mathbb{R}^3$. Find the following coordinate vectors.

a. $\begin{bmatrix} 1 \\ 2 \\ 3 \end{bmatrix}_B$ **b.** $\begin{bmatrix} 7 \\ 2 \\ -6 \end{bmatrix}_B$ **c.** $\begin{bmatrix} -1 \\ 3 \\ 0 \end{bmatrix}_B$

2. Consider the ordered basis $B = \left\{ \begin{bmatrix} 2 \\ 0 \\ 3 \end{bmatrix}, \begin{bmatrix} 1 \\ 1 \\ 1 \end{bmatrix}, \begin{bmatrix} 0 \\ 1 \\ -1 \end{bmatrix} \right\}$ for $\mathbb{R}^3$. Find the following coordinate vectors.

a. $\begin{bmatrix} 2 \\ 0 \\ 3 \end{bmatrix}_B$ **b.** $\begin{bmatrix} 4 \\ 0 \\ 6 \end{bmatrix}_B$ **c.** $\begin{bmatrix} 0 \\ 0 \\ 0 \end{bmatrix}_B$

d. $\begin{bmatrix} 5 \\ 6 \\ 2 \end{bmatrix}_B$ **e.** $\begin{bmatrix} 2 \\ 1 \\ 5 \end{bmatrix}_B$ **f.** $\left[\begin{bmatrix} 5 \\ 6 \\ 2 \end{bmatrix} + \begin{bmatrix} 2 \\ 1 \\ 5 \end{bmatrix} \right]_B$

g. $\begin{bmatrix} 50 \\ 60 \\ 20 \end{bmatrix}_B$ **h.** $\begin{bmatrix} 52 \\ 61 \\ 25 \end{bmatrix}_B$ **i.** $\left[\begin{bmatrix} 5 \\ 6 \\ 2 \end{bmatrix}_B \right]_B$

3. Consider the ordered basis $B = \{(0, 0, 0, 1), (0, 0, 1, 1), (0, 1, 1, 1), (1, 1, 1, 1)\}$ for $\mathbb{R}^4$. Find the following coordinate vectors.

a. $[(0, 3, 0, 5)]_B$ **b.** $[(8, -1, 2, 3)]_B$ **c.** $[(a_1, a_2, a_3, a_4)]_B$

4. Consider the subspace of $\mathbb{M}(2, 2)$ with basis $B = \left\{ \begin{bmatrix} 1 & 0 \\ 0 & 0 \end{bmatrix}, \begin{bmatrix} 0 & 0 \\ 0 & 1 \end{bmatrix}, \begin{bmatrix} 0 & 1 \\ 1 & 0 \end{bmatrix} \right\}$. Find the coordinate vectors with respect to B for the following matrices in this subspace.

a. $\begin{bmatrix} -3 & 0 \\ 0 & 4 \end{bmatrix}$ **b.** $\begin{bmatrix} 1 & 2 \\ 2 & 1 \end{bmatrix}$ **c.** $\begin{bmatrix} 1 & 5 \\ 5 & 9 \end{bmatrix}$

5. Use the polynomials defined by

$$p_1(x) = x^2 + 1, \qquad p_2(x) = x^2 + x + 2, \qquad p_3(x) = 3x - 1$$

to form a basis $B = \{p_1, p_2, p_3\}$ for $\mathbb{P}_2$. Compute the following coordinate vectors.

a. $[x^2 + x + 2]_B$ **b.** $[x^2]_B$ **c.** $[x^2 + 3x]_B$
d. $[x^2 - 3x + 6]_B$ **e.** $[(x + 1)^2]_B$ **f.** $[(p_3(x))^2]_B$

6. Consider the basis $B = \left\{ \begin{bmatrix} 1 & 2 \\ -2 & 0 \end{bmatrix}, \begin{bmatrix} 3 & 1 \\ 5 & 1 \end{bmatrix}, \begin{bmatrix} 4 & 2 \\ 0 & -1 \end{bmatrix}, \begin{bmatrix} 3 & 0 \\ 1 & -5 \end{bmatrix} \right\}$

for $\mathbb{M}(2, 2)$. Find the matrix A with coordinate vector $[A]_B = \begin{bmatrix} 3 \\ 1 \\ -1 \\ 1 \end{bmatrix}$.

7. Notice that the function defined by $f(x) = \sin(x + \frac{\pi}{3})$ is a solution to the differential equation $y'' = -y$. Find the coordinate vector of f with respect to the basis $\{\sin, \cos\}$ for the solution space of this differential equation. (See Exercise 10 of Section 3.4.)

8. **a.** Show that coordinate vectors are compatible with the vector space operation of addition. That is, if $B = \{\mathbf{u}_1, \ldots, \mathbf{u}_n\}$ is an ordered basis for the vector space V, and $\mathbf{v}$ and $\mathbf{w}$ are elements of V with

$$[\mathbf{v}]_B = \begin{bmatrix} r_1 \\ r_2 \\ \vdots \\ r_n \end{bmatrix} \quad \text{and} \quad [\mathbf{w}]_B = \begin{bmatrix} s_1 \\ s_2 \\ \vdots \\ s_n \end{bmatrix},$$

then

$$[\mathbf{v} + \mathbf{w}]_B = \begin{bmatrix} r_1 \\ r_2 \\ \vdots \\ r_n \end{bmatrix} + \begin{bmatrix} s_1 \\ s_2 \\ \vdots \\ s_n \end{bmatrix} = [\mathbf{v}]_B + [\mathbf{w}]_B.$$

b. Show that coordinate vectors are compatible with the vector space operation of scalar multiplication. That is, if we also have $r \in \mathbb{R}$, then

$$[r\mathbf{v}]_B = r \begin{bmatrix} r_1 \\ r_2 \\ \vdots \\ r_n \end{bmatrix} = r[\mathbf{v}]_B.$$

9. Suppose B is an ordered basis for $\mathbb{R}^2$. Suppose $\begin{bmatrix} 3 \\ 2 \end{bmatrix}_B = \begin{bmatrix} 1 \\ 1 \end{bmatrix}$ and $\begin{bmatrix} -1 \\ 4 \end{bmatrix}_B = \begin{bmatrix} 2 \\ 1 \end{bmatrix}$. Determine the two vectors in the basis B.

Project: Infinite-Dimensional Vector Spaces

In this chapter the concepts of linear combination, span, linear independence, basis, dimension, and coordinates have been restricted to finite sets of vectors. This is adequate for the development of the theory of dimension for finite-dimensional vector spaces. However, we have seen that not every vector space is finite-dimensional. In particular, spaces of functions that are fundamental to the study of differential equations and functional analysis are typically infinite-dimensional.

Many of the results of this chapter can be extended to apply to infinite-dimensional vector spaces. The material in this section points out some of the modifications to make as you follow the development of the concept of dimension in the context of infinite-dimensional vector spaces.

1. Suppose S is an arbitrary subset of a vector space V. Define a linear combination of the elements of S to be any sum obtained by choosing a finite number of elements of S, multiplying them by scalars, and adding the resulting vectors. If S is finite, show that this reduces to Definition 3.1. Try out your definition where S is an infinite subset of $\mathbb{R}^2$ or $\mathbb{R}^3$. What functions in $\mathbb{C}(\mathbb{R})$ are linear combinations of the functions defined by $1, x, x^2, x^3, \ldots$? Find a function in $\mathbb{C}(\mathbb{R})$ that is not a linear combination of these functions.

Suppose S and T are subsets of a vector space V. Formulate and prove a relation between linear combinations of the vectors in $S \cup T$ and linear combinations of the vectors in S and in T. Can you extend your result to unions of any finite number of sets? Can you extend to unions of an arbitrary family of subsets of V?

Keep in mind that addition in a vector space is a binary operation. That is, we always combine exactly two vectors to form a sum. The concept of linear combination extends the notion of addition (combined with scalar multiplication) by allowing sums of more than two vectors. However, even when the set of vectors is infinite, only finite sums can be considered. Recall from your calculus course that the concept of convergence is necessary to define infinite series of real numbers as a limit of finite sums. In the purely algebraic context of vector spaces, we have no comparable notion of convergence that allows us to talk about infinite sums.

2. Restate Definition 3.2 to define the span of an arbitrary subset of a vector space. Extend Theorem 3.3 to this more general situation.

Let $\mathbb{R}^\infty = \{(x_1, x_2, \ldots \;) \mid x_1 \in \mathbb{R}, \; x_2 \in \mathbb{R}, \; \ldots \}$ be the vector space of infinite sequences of real numbers. Use the operations of addition and scalar multiplication as defined in Exercise 10 of Section 1.7. Let $\mathbf{e}_1 = (1, 0, 0, \ldots \;)$, $\mathbf{e}_2 = (0, 1, 0, \ldots \;)$, $\ldots$ be analogs of the standard basis elements in Euclidean spaces. Give a simple description of the subspace of $\mathbb{R}^\infty$ spanned by $E = \{\mathbf{e}_1, \mathbf{e}_2, \ldots \}$.

Suppose S and T are subsets of a vector space V. How are span S, span T, span$(S \cup T)$, and span$(S \cap T)$ related? If $S \subseteq T$, what is the relation between span S and span T?

3. Extend Definition 3.4 of linear independence to arbitrary subsets of a vector space. Since a finite sum cannot deal with all the vectors in an infinite set, you will have to consider arbitrary linear combinations (rather than a single sum) equal to zero as the hypothesis of your condition. State and prove an extension to Theorem 3.5.

Show that the nonnegative integer powers of x form a linearly independent subset of $\mathbb{C}(\mathbb{R})$. Show that $E = \{\mathbf{e}_1, \mathbf{e}_2, \ldots \}$ as defined previously is a linearly independent subset of $\mathbb{R}^\infty$. What other sequences can be adjoined to E so that the resulting set is still linearly independent?

4. Combine the concepts of span and linear independence to give a definition of a basis for an arbitrary vector space. Where do you encounter problems in trying to find bases for $\mathbb{P}$, $\mathbb{R}^\infty$, and $\mathbb{C}(\mathbb{R})$? Is it any easier to find a basis for the subspace of all bounded sequences in $\mathbb{R}^\infty$ or for the even smaller subspace of all sequences in $\mathbb{R}^\infty$ that converge to zero?

5. Extending the results of Section 3.5 requires some background in dealing with cardinalities of infinite sets. As you pursue this additional reading, be sure to make note

of Zorn's Lemma (now guess what is yellow and equivalent to the Axiom of Choice). This is a standard tool for proving that every vector space has a basis.

6. Definition 3.16 and Theorem 3.17, concerning coordinate vectors, extend fairly easily to bases with an arbitrary number of elements. The main difference involves introducing an indexed basis to replace the ordered basis. If you write the basis $B = \{\mathbf{u}_\alpha \mid \alpha \in A\}$, where A is an appropriate indexing set, then the coordinates can be thought of as a collection of vectors indexed by A. Such an assignment of a vector to each element of A is nothing more than a function $f : A \to V$.

Project: Linear Codes

Information processed with electronic equipment is usually encoded as a digital pattern. Whether the information is a picture from space, a sound signal on a compact disk, or a block of text in a word processor, it will be converted to a sequence of numbers for transmission, manipulation, or storage. Later the numbers will be reinterpreted in the original form.

The digital patterns are usually divided into blocks of numbers that can be conveniently represented and manipulated as vectors. For detection of errors, a certain amount of redundancy is encoded along with the original information. A certain subspace may be designated as containing vectors of legitimate blocks. If a vector is not in the subspace, we know it contains an error. We may even be able to correct the error by deciding which correct sequence most likely produced the corrupted vector.

The vectors in this application have coordinates restricted to a finite set, most frequently a set with just two elements, $\{0, 1\}$. This leads naturally to a modification of the definition of vector spaces we have been dealing with. Rather than using all real numbers as scalars, we take scalars from the finite set of values of the coordinates. Some preliminary work is needed to define arithmetic on such a finite set.

1. Consider the integers modulo a prime number p. Think of this as the set $\mathbb{Z}_p = \{0, 1, \ldots, p - 1\}$ of remainders when integers are divided by p. Check that addition, subtraction, and multiplication are valid in $\mathbb{Z}_p$ provided that all results are reduced modulo p (that is, to the remainder obtained when the ordinary result is divided by p). Show that each integer other than 0 has a multiplicative inverse modulo p. Use this to introduce division into $\mathbb{Z}_p$.

A system such as $\mathbb{Z}_p$ in which the four operations of arithmetic work basically as they do for the real numbers is known as a **field**. The system of rational numbers, the system of complex numbers, and, of course, the real numbers are common examples of fields. Virtually all the results of elementary linear algebra hold if we consider vector spaces defined with an arbitrary field as the set of scalars.

2. Develop the theory of vector spaces with scalars taken from $\mathbb{Z}_p$. This involves checking through the first three chapters to make sure all properties of real numbers used in the proofs have analogous properties in $\mathbb{Z}_p$. You may wish to restrict consideration to the family $\mathbb{Z}_2^n$ of ordered lists of n elements from $\mathbb{Z}_2 = \{0, 1\}$.

3. During the short history of electronic computers, mathematicians, engineers, and computer scientists have developed numerous coding schemes for detecting and correcting errors. This work has developed into an area of research known as algebraic coding theory. The results make ingenious use of vector spaces and matrices with components in $\mathbb{Z}_p$. Some introductory sources of material are:

Elwin R. Berlekamp, *Algebraic Coding Theory,* New York: McGraw-Hill, 1968.

Raymond Hill, *A First Course in Coding Theory,* New York: Oxford University Press, 1988.

W. Wesley Peterson and E. J. Weldon, Jr., *Error Correcting Codes*, Cambridge, MA: MIT Press, 1972.

Bart Rice and Carrol Wilde, "Error Correcting Codes 1," UMAP Module 346, *UMAP Journal,* vol. 1, no. 3, 1980, pp. 91–116.

Consult your library for additional references on coding theory.

Chapter 3 Summary

Chapter 3 develops the concept of the dimension of a vector space. Every step builds on the properties that follow from the vector space axioms. Addition and scalar multiplication are the ingredients in forming linear combinations. The concept of linear combination is used to define span and linear independence. These two concepts combine to give the idea of a basis for a vector space. A basis allows us to introduce coordinates in an abstract vector space and recognize it as very similar to the Euclidean prototype.

Computations

The role of systems of linear equations
Determine whether one vector is a linear combination of others
Determine the span of a set of vectors
Determine whether a set is linearly independent

Coordinate vectors
Compute coordinate vectors with respect to an ordered basis
Compute a vector given its coordinate vector

Theory

Steps toward the definition of dimension
Linear combination
Span
Linear independence
Basis
Dimension

Comparison Theorem
Indirect proof
Use of the Fundamental Theorem of Homogeneous Systems

Subspaces of a finite-dimensional space are finite-dimensional

Uniqueness of coordinate vectors

Applications

Contraction Theorem for finding a basis within a spanning set

Expansion Theorem for enlarging a linearly independent set to a basis

Coordinate vectors to identify arbitrary vector spaces with the Euclidean space of the same dimension.

Review Exercises

1. **a.** Show that $\{(1, 2, 1), (2, 1, 5), (1, -4, 7)\}$ is a linearly dependent subset of $\mathbb{R}^3$.

    b. Write the set spanned by $\{(1, 2, 1), (2, 1, 5), (1, -4, 7)\}$ in the standard form of a line or plane in $\mathbb{R}^3$.

2. Suppose $f \in \mathbb{P}_3$ is defined by $f(x) = 5x^3 + 4x - 2$. Consider the four polynomials defined by

    $$p_1(x) = x^3,$$
    $$p_2(x) = x^3 + x^2,$$
    $$p_3(x) = x^3 + x^2 + x,$$
    $$p_4(x) = x^3 + x^2 + x + 1.$$

    a. Show that $B = \{p_1, p_2, p_3, p_4\}$ is a linearly independent subset of $\mathbb{P}_3$.

    b. With no further computation, explain why B is a basis for $\mathbb{P}_3$.

    c. Write f as a linear combination of p_1, p_2, p_3, and p_4.

    d. Write down the coordinate vector $[f]_B$.

3. **a.** Show that $B = \left\{ \begin{bmatrix} 0 & 0 \\ 0 & 1 \end{bmatrix}, \begin{bmatrix} 0 & 0 \\ 1 & 2 \end{bmatrix}, \begin{bmatrix} 0 & 1 \\ 2 & 3 \end{bmatrix}, \begin{bmatrix} 1 & 2 \\ 3 & 4 \end{bmatrix} \right\}$ is a linearly independent subset of $\mathbb{M}(2, 2)$.

    b. With no further computation, explain why B is a basis for $\mathbb{M}(2, 2)$.

    c. Write $\begin{bmatrix} 2 & 3 \\ 0 & -1 \end{bmatrix}$ as a linear combination of the elements of B.

    d. Find the coordinate vector of $\begin{bmatrix} 2 & 3 \\ 0 & -1 \end{bmatrix}$ with respect to B.

4. Show that the functions defined by $f(x) = \sin x$, $g(x) = \sin 2x$, and $h(x) = \sin 3x$ form a linearly independent subset of $\mathbb{F}(\mathbb{R})$.

5. Show that the functions defined by $f(x) = \sin x$, $g(x) = \sin 2x$, and $h(x) = \sin x \cos x$ form a linearly dependent subset of $\mathbb{F}(\mathbb{R})$.

6. Suppose the vectors v_1, v_2, v_3, and v_4 satisfy

    $$8v_1 - 3v_2 + 4v_3 + 2v_4 = 0.$$

    Use this equation to write v_4 as a linear combination of the other three vectors. Is the set $\{v_1, v_2, v_3, v_4\}$ linearly independent?

7. **a.** Suppose $\{v, w\}$ spans a vector space V. Show that $\{v + w, v - w\}$ spans V.

b. Suppose $\{\mathbf{v}, \mathbf{w}\}$ is a linearly independent subset of a vector space V. Show that $\{\mathbf{v} + \mathbf{w}, \mathbf{v} - \mathbf{w}\}$ is a linearly independent subset of V.

8. Find a basis for the subspace $S = \{p \in \mathbb{P}_3 \mid p(5) = 0\}$.

9. **a.** Is it possible for a set of five vectors to span $\mathbb{M}(2, 3)$?

b. Must any set of five vectors span $\mathbb{M}(2, 3)$?

c. Is it possible for five vectors to form a linearly independent subset of $\mathbb{M}(2, 3)$?

d. Must any five vectors form a linearly independent subset of $\mathbb{M}(2, 3)$?

e. Is it possible for a set of seven vectors to span $\mathbb{M}(2, 3)$?

f. Must any set of seven vectors span $\mathbb{M}(2, 3)$?

g. Is it possible for seven vectors to form a linearly independent subset of $\mathbb{M}(2, 3)$?

h. Must any seven vectors form a linearly independent subset of $\mathbb{M}(2, 3)$?

i. If a set of six vectors spans $\mathbb{M}(2, 3)$, is it possible for the set to be linearly independent?

j. If a set of six vectors spans $\mathbb{M}(2, 3)$, must the set be linearly independent?

k. If six vectors form a linearly independent subset of $\mathbb{M}(2, 3)$, is it possible for the set to span $\mathbb{M}(2, 3)$?

l. If six vectors form a linearly independent subset of $\mathbb{M}(2, 3)$, must the set span $\mathbb{M}(2, 3)$?

10. Use the proof of the Contraction Theorem as a recipe for finding a subset of

$$\{(1, -1, 0, 2), (-1, 1, 0, -2), (1, 1, 1, 0), (0, 2, 1, -2)\}$$

that is a basis for the subspace of $\mathbb{R}^4$ spanned by this set.

11. Consider the polynomials in $\mathbb{P}_2$ defined by $p_1(x) = x^2 + 2x + 3$, $p_2(x) = x^2 + 2x$, and $p_3(x) = x^2$.

a. Show that the set $\{p_1, p_2, p_3\}$ is linearly independent.

b. Without further computation, explain how you know that $\{p_1, p_2, p_3\}$ is a basis for $\mathbb{P}_2$.

12. Consider the ordered basis $B = \{(1, 1, 1), (-1, -1, 0), (-1, 0, -1)\}$ for $\mathbb{R}^3$. Find the following coordinate vectors.

a. $[(-1, -1, 0)]_B$ **b.** $[(0, 0, 1)]_B$ **c.** $[(1, 0, 0)]_B$

d. $[(0, 0, 0)]_B$ **e.** $[(8, -2, 7)]_B$ **f.** $[(a, b, c)]_B$

13. Consider the ordered basis $B = \{\ln \sin x, \ln \cos x, 1\}$ of function in $\mathbb{F}((0, \frac{\pi}{2}))$. Find the following coordinate vectors.

a. $[\ln \sin 2x]_B$ **b.** $[\ln \tan x]_B$

c. $[\ln \sec x]_B$ **d.** $[\ln(1 + \tan^2 x)]_B$

14. Suppose $B = \{\mathbf{u}_1, \mathbf{u}_2, \mathbf{u}_3, \mathbf{u}_4\}$ is an ordered basis for a vector space V. Suppose $\mathbf{v} \in V$ has coordinate vector

$$[\mathbf{v}]_B = \begin{bmatrix} 2 \\ -1 \\ 0 \\ 4 \end{bmatrix}.$$

   a. Write $\mathbf{v}$ as a linear combination of $\mathbf{u}_1$, $\mathbf{u}_2$, $\mathbf{u}_3$, and $\mathbf{u}_4$.

   b. Write $\frac{1}{2}\mathbf{v}$ as a linear combination of $\mathbf{u}_1$, $\mathbf{u}_2$, $\mathbf{u}_3$, and $\mathbf{u}_4$.

   c. Determine $[\frac{1}{2}\mathbf{v}]_B$

15. Suppose $\{\mathbf{v}_1, \ldots, \mathbf{v}_m\}$ and $\{\mathbf{w}_1, \ldots, \mathbf{w}_n\}$ are two finite subsets of a vector space V. Consider the subspaces $S = \mathrm{span}\{\mathbf{v}_1, \ldots, \mathbf{v}_m\}$ and $T = \mathrm{span}\{\mathbf{w}_1, \ldots, \mathbf{w}_n\}$

   a. Show that the subspace $\mathrm{span}\{\mathbf{v}_1, \ldots, \mathbf{v}_m, \mathbf{w}_1, \ldots, \mathbf{w}_n\}$ contains the union $S \cup T$.

   b. Show that $\mathrm{span}\{\mathbf{v}_1, \ldots, \mathbf{v}_m, \mathbf{w}_1, \ldots, \mathbf{w}_n\}$ is the smallest subspace that contains $S \cup T$. That is, if W is any subspace containing $S \cup T$, then

$$\mathrm{span}\{\mathbf{v}_1, \ldots, \mathbf{v}_m, \mathbf{w}_1, \ldots, \mathbf{w}_n\} \subseteq W.$$

16. Suppose S and T are finite-dimensional subspaces of a vector space V. By Exercise 19 of Section 1.8 we know that $S \cap T$ is a subspace of V. Of course, $S \cap T$ is also a subspace of S and a subspace of T.

   a. Give an example in which $\{\mathbf{v}_1, \ldots, \mathbf{v}_m\}$ is a basis for S and $\{\mathbf{w}_1, \ldots, \mathbf{w}_n\}$ is a basis for T, but $\{\mathbf{v}_1, \ldots, \mathbf{v}_m\} \cap \{\mathbf{w}_1, \ldots, \mathbf{w}_n\}$ is not a basis for $S \cap T$.

   b. Show that there are bases $\{\mathbf{v}_1, \ldots, \mathbf{v}_m\}$ for S and $\{\mathbf{w}_1, \ldots, \mathbf{w}_n\}$ for T such that $\{\mathbf{v}_1, \ldots, \mathbf{v}_m\} \cap \{\mathbf{w}_1, \ldots, \mathbf{w}_n\}$ is a basis for $S \cap T$. (Suggestion: Start with a basis for $S \cap T$ and apply the Expansion Theorem to get a basis for S and again to get a basis for T.)

17. Suppose S and T are finite-dimensional subspaces of a vector space V. Suppose $\{\mathbf{x}_1, \ldots, \mathbf{x}_k, \mathbf{v}_{k+1}, \ldots, \mathbf{v}_m\}$ is a basis for S and $\{\mathbf{x}_1, \ldots, \mathbf{x}_k, \mathbf{w}_{k+1}, \ldots, \mathbf{w}_n\}$ is a basis for T chosen as in the previous exercise so that $\{\mathbf{x}_1, \ldots, \mathbf{x}_k\}$ is a basis for $S \cap T$.

   a. Prove that $\{\mathbf{x}_1, \ldots, \mathbf{x}_k, \mathbf{v}_{k+1}, \ldots, \mathbf{v}_m, \mathbf{w}_{k+1}, \ldots, \mathbf{w}_n\}$ is linearly independent. (Suggestion: Start off by assuming a linear combination of the vectors is equal to zero. Use this equation to find a vector that is both in S and in T. Use Theorem 3.17 to derive the conclusion that all the coefficients must be zero.)

   b. Prove that $\dim(\mathrm{span}(S \cup T)) = \dim S + \dim T - \dim(S \cap T)$.

18. Prove the following version of the Contraction Theorem: Suppose $\{\mathbf{v}_1, \ldots, \mathbf{v}_m\}$ is a linearly independent subset of a vector space V. Suppose the set $\{\mathbf{v}_1, \ldots, \mathbf{v}_m, \mathbf{v}_{m+1}, \ldots, \mathbf{v}_{m+k}\}$ spans V. Then some subset of $\{\mathbf{v}_1, \ldots, \mathbf{v}_m, \mathbf{v}_{m+1}, \ldots, \mathbf{v}_{m+k}\}$ that contains $\{\mathbf{v}_1, \ldots, \mathbf{v}_m\}$ is a basis for V.

19. Prove the following version of the Expansion Theorem: Suppose V is a subspace of a finite-dimensional vector space. If $\{\mathbf{v}_1, \ldots, \mathbf{v}_m\}$ is a linearly independent subset of V, then we can adjoin vectors $\mathbf{v}_{m+1}, \ldots, \mathbf{v}_{m+k}$ so that $\{\mathbf{v}_1, \ldots, \mathbf{v}_m, \mathbf{v}_{m+1}, \ldots, \mathbf{v}_{m+k}\}$ is a basis for V.

Inner Product Spaces

I*n the first three chapters we considered vector spaces in their purest form. You may have noticed, however, that in certain examples of vector spaces it is possible to perform operations besides addition and scalar multiplication. For example, in $\mathbb{R}^2$ and $\mathbb{R}^3$ there are notions of distance and angle. In the vector space $\mathbb{F}(X)$ of real-valued functions on any set X it is possible to multiply two functions, and in $\mathbb{F}(\mathbb{R})$ there is also the notion of composition of functions. Furthermore, in certain subspaces of $\mathbb{F}(\mathbb{R})$ powerful analytical operations such as differentiation and integration can be performed.*

The main topic of this chapter is a new kind of multiplication. This product structure proves to be useful in many vector spaces. We will investigate how it leads to quantitative notions of length, distance, and angle, and how it lets us generalize these ideas beyond their ordinary geometric meanings.

4.1 Inner Products and Norms

This section introduces a new kind of multiplication on vector spaces and develops some of the properties. It also provides two examples whose common properties have been abstracted in the general definition. Thus, our work in this section will bring out the underlying features common to these two examples. Also keep in mind the efficiency of proving results in the general setting; once we have verified that the examples satisfy the defining conditions, we instantly have a large body of results at our disposal.

Here is the definition of the new operation.

Definition 4.1

An **inner product** on a vector space V is an operation that combines any two vectors $\mathbf{v} \in V$ and $\mathbf{w} \in V$ to produce a real number denoted $\langle \mathbf{v}, \mathbf{w} \rangle$. This operation must satisfy the following four axioms for all $\mathbf{v}, \mathbf{w}, \mathbf{x} \in V$ and $r \in \mathbb{R}$.

1. $\langle \mathbf{v}, \mathbf{v} \rangle \geq 0$; and if $\langle \mathbf{v}, \mathbf{v} \rangle = 0$, then $\mathbf{v} = \mathbf{0}$ (positive-definite law)
2. $\langle \mathbf{v}, \mathbf{w} \rangle = \langle \mathbf{w}, \mathbf{v} \rangle$ (commutative law)
3. $\langle r\mathbf{v}, \mathbf{w} \rangle = r\langle \mathbf{v}, \mathbf{w} \rangle$ (homogeneous property)
4. $\langle \mathbf{v} + \mathbf{w}, \mathbf{x} \rangle = \langle \mathbf{v}, \mathbf{x} \rangle + \langle \mathbf{w}, \mathbf{x} \rangle$ (distributive law)

A vector space together with an inner product is an **inner product space**.

Crossroads   You have seen the form of the axioms for an inner product in other contexts. For example, if you replace the inner product multiplication of vectors with ordinary multiplication of real numbers, you will recognize these properties as familiar laws of real-number algebra. Make a special note of one peculiar feature of the inner product. Even though both factors are vectors (unlike scalar multiplication, which combines a real number with a vector), the result of the inner product is a real number rather than another vector.

Your friends will be mightily impressed when you mention that you are studying inner product spaces. Remember, however, that the term merely refers to a set with three algebraic operations: the familiar addition and scalar multiplication associated with any vector space, and the newly introduced operation of the inner product.

While the defining properties of inner product are still fresh, we want to state a few basic consequences.

Theorem 4.2

Suppose V is an inner product space. For all $\mathbf{v}, \mathbf{w}, \mathbf{x} \in V$ and $r \in \mathbb{R}$, the following identities hold.

a. $\langle \mathbf{v}, \mathbf{0} \rangle = \langle \mathbf{0}, \mathbf{v} \rangle = 0$
b. $\langle \mathbf{v}, r\mathbf{w} \rangle = r\langle \mathbf{v}, \mathbf{w} \rangle$
c. $\langle \mathbf{v}, \mathbf{w} + \mathbf{x} \rangle = \langle \mathbf{v}, \mathbf{w} \rangle + \langle \mathbf{v}, \mathbf{x} \rangle$

Proof   Here is a proof of part a based on the fact that $0\mathbf{0} = \mathbf{0}$ in any vector space:

$$\langle \mathbf{v}, \mathbf{0} \rangle = \langle \mathbf{0}, \mathbf{v} \rangle$$
$$= \langle 0\mathbf{0}, \mathbf{v} \rangle$$
$$= 0\langle \mathbf{0}, \mathbf{v} \rangle$$
$$= 0.$$

The remaining results are straightforward consequences of the four defining properties of inner product. You are asked to write down the proofs as Exercise 6 at the end of this section.   ∎

The concept of inner product is a generalization of two families of examples. One is a multiplication of vectors in Euclidean space $\mathbb{R}^n$; the other is a multiplication in certain function spaces. We will examine these examples in some detail in later sections of this chapter. Our immediate goal is to define the product structure for each of the vector spaces and to verify that it satisfies the four axioms required to make it an inner product. You may be familiar with the first example from a physics or multivariate calculus course.

Definition 4.3

The **dot product** on a Euclidean space $\mathbb{R}^n$ is an operation that combines two vectors $\mathbf{v} = (v_1, \ldots, v_n)$ and $\mathbf{w} = (w_1, \ldots, w_n)$ to produce the real number denoted $\mathbf{v} \cdot \mathbf{w}$ and defined by

$$\mathbf{v} \cdot \mathbf{w} = v_1 w_1 + \cdots + v_n w_n.$$

The dot product of two vectors is especially easy to compute. Here is an example of how it works with two vectors in $\mathbb{R}^4$.

Quick Example   *Compute the dot product of* $(2, -1, 5, 0)$ *and* $(3, -2, -3, 1)$.

$$(2, -1, 5, 0) \cdot (3, -2, -3, 1) = 2 \cdot 3 + (-1) \cdot (-2) + 5 \cdot (-3) + 0 \cdot 1$$
$$= -7.   \blacksquare$$

We can now run through the verification that $\langle \mathbf{v}, \mathbf{w} \rangle = \mathbf{v} \cdot \mathbf{w}$ defines an inner product on $\mathbb{R}^n$. Let $\mathbf{v} = (v_1, \ldots, v_n)$, $\mathbf{w} = (w_1, \ldots, w_n)$, and $\mathbf{x} = (x_1, \ldots, x_n)$ be arbitrary elements of $\mathbb{R}^n$ and let r be an arbitrary real number.

1.
$$\mathbf{v} \cdot \mathbf{v} = v_1^2 + \cdots + v_n^2$$
$$\geq 0$$

since the square of any real number is nonnegative and the sum of nonnegative real numbers is nonnegative. Also, the only way the sum of nonnegative real numbers can equal zero is if all the terms are zero. In the case of $\mathbf{v} \cdot \mathbf{v} = v_1^2 + \cdots + v_n^2 = 0$, this implies that $v_1 = \cdots = v_n = 0$; that is, $\mathbf{v} = (0, \ldots, 0) = \mathbf{0}$.

2.
$$\mathbf{v} \cdot \mathbf{w} = v_1 w_1 + \cdots + v_n w_n$$
$$= w_1 v_1 + \cdots + w_n v_n$$
$$= \mathbf{w} \cdot \mathbf{v}.$$

3.
$$(r\mathbf{v}) \cdot \mathbf{w} = (rv_1)w_1 + \cdots + (rv_n)w_n$$
$$= r(v_1 w_1) + \cdots + r(v_n w_n)$$
$$= r(v_1 w_1 + \cdots + v_n w_n)$$
$$= r(\mathbf{v} \cdot \mathbf{w}).$$

4.
$$(\mathbf{v} + \mathbf{w}) \cdot \mathbf{x} = (v_1 + w_1, \ldots, v_n + w_n) \cdot (x_1, \ldots, x_n)$$
$$= (v_1 + w_1)x_1 + \cdots + (v_n + w_n)x_n$$
$$= (v_1 x_1 + w_1 x_1) + \cdots + (v_n x_n + w_n x_n)$$
$$= (v_1 x_1 + \cdots + v_n x_n) + (w_1 x_1 + \cdots + w_n x_n)$$
$$= \mathbf{v} \cdot \mathbf{x} + \mathbf{w} \cdot \mathbf{x}.$$

Unless stated to the contrary, whenever we deal with $\mathbb{R}^n$ as an inner product space, we will assume that the inner product is in fact the dot product.

The second family of examples relies on the concept of the definite integral. For our purposes we will consider a closed, bounded interval $[a, b]$ with $a < b$ and define an inner product on the vector space

$$\mathbb{C}([a, b]) = \{f : [a, b] \to \mathbb{R} \mid f \text{ is continuous}\}.$$

Although the inner product can be defined on a larger space of functions, we will be using results from calculus concerning various properties of the definite integral. The safest course is for us to assume these results for continuous functions as discussed in your calculus course. This will be adequate for our purposes; you may want to investigate possible extensions to larger spaces of functions.

Definition 4.4

The **standard inner product** on $\mathbb{C}([a, b])$ is defined by

$$\langle f, g \rangle = \int_a^b f(x)g(x) \, dx$$

for any $f, g \in \mathbb{C}([a, b])$.

To evaluate an inner product on $\mathbb{C}([a, b])$, you may need to review your techniques of integration. Repeated integration by parts is the key technique for evaluating the inner product in the following example.

Quick Example   *Consider the functions defined by $f(x) = x^2$ and $g(x) = e^x$. Compute the standard inner product of f and g as elements of $\mathbb{C}([0, 1])$.*

$$\langle f, g \rangle = \int_0^1 x^2 e^x \, dx$$

$$= x^2 e^x \Big|_0^1 - \int_0^1 2x e^x \, dx$$

$$= e - \left(2x e^x \Big|_0^1 - \int_0^1 2 e^x \, dx \right)$$

$$= e - 2e + 2 e^x \Big|_0^1$$

$$= e - 2. \quad \blacksquare$$

Notice the importance of the domain of the functions. If we substitute another interval for $[0, 1]$, these same formulas define different functions in a different function space, leading to a different value for the inner product.

You are asked to verify in Exercise 19 at the end of this section that this standard inner product on $\mathbb{C}([a, b])$ does indeed satisfy the four axioms of Definition 4.1. Unless stated to the contrary, whenever we deal with $\mathbb{C}([a, b])$ as an inner product space, we will assume the inner product is in fact the standard inner product.

Quick Example   *Verify that* $\langle p, q \rangle = a_2 b_2 + a_1 b_1 + a_0 b_0$ *for* $p(x) = a_2 x^2 + a_1 x + a_0$ *and* $q(x) = b_2 x^2 + b_1 x + b_0$ *defines an inner product on* $\mathbb{P}_2$.

Let p, q, and r be three polynomials in $\mathbb{P}_2$. Let $p(x) = a_2 x^2 + a_1 x + a_0$, $q(x) = b_2 x^2 + b_1 x + b_0$, and $r(x) = c_2 x^2 + c_1 x + c_0$. Also, let s be a real number. Now $\langle p, p \rangle = a_2^2 + a_1^2 + a_0^2 \geq 0$, and this sum of squares is equal to zero only if $a_2 = a_1 = a_0 = 0$. Thus, if $\langle p, p \rangle = 0$, then $p(x) = 0x^2 + 0x + 0 = 0$, so p is the zero polynomial.

The commutative law is a straightforward verification:

$$\langle p, q \rangle = a_2 b_2 + a_1 b_1 + a_0 b_0$$

$$= b_2 a_2 + b_1 a_1 + b_0 a_0$$

$$= \langle q, p \rangle.$$

Since sp is defined by $(sp)(x) = s(a_2 x^2 + a_1 x + a_0) = (sa_2)x^2 + (sa_1)x + (sa_0)$, we have

$$\langle sp, q \rangle = (sa_2) b_2 + (sa_1) b_1 + (sa_0) b_0$$

$$= s(a_2 b_2 + a_1 b_1 + a_0 b_0)$$

$$= s\langle p, q \rangle.$$

Finally, since $p + q$ is defined by $(p + q)(x) = p(x) + q(x) = (a_2 x^2 + a_1 x + a_0) + (b_2 x^2 + b_1 x + b_0) = (a_2 + b_2)x^2 + (a_1 + b_1)x + (a_0 + b_0)$, we have

$$\langle p + q, r \rangle = (a_2 + b_2) c_2 + (a_1 + b_1) c_1 + (a_0 + b_0) c_0$$

$$= (a_2 c_2 + a_1 c_1 + a_0 c_0) + (b_2 c_2 + b_1 c_1 + b_0 c_0)$$

$$= \langle p, r \rangle + \langle q, r \rangle.$$

Crossroads   Many other examples of inner products are important in both theoretical and applied fields ranging from functional analysis and numerical analysis to physics and economics. The preceding example and Exercises 14, 15, 17, 18, and 20 at the end of this section introduce you to a variety of different inner products. The project on orthogonal polynomials at the end of this chapter mentions several modifications of the standard inner product on function spaces.

The remaining sections of this chapter will explore the general properties of inner product spaces and examine how these properties apply to our two principal families of examples. We will see that the dot product on $\mathbb{R}^n$ provides an algebraic tool to quantify geometric concepts such as magnitude and distance. This will guide us in exploring a way to measure the distance between two functions. The key to these ideas is introduced in the following definition.

Definition 4.5

The **norm** of a vector $\mathbf{v}$ in an inner product space V is denoted $\|\mathbf{v}\|$ and is defined by $\|\mathbf{v}\| = \sqrt{\langle \mathbf{v}, \mathbf{v} \rangle}$.

It is reassuring to note that the norm of a vector $\mathbf{v} = (v_1, \ldots, v_n)$ in $\mathbb{R}^n$ is the familiar Euclidean length of the arrow that represents the vector. That is,

$$\|\mathbf{v}\| = \sqrt{\langle \mathbf{v}, \mathbf{v} \rangle} = \sqrt{v_1^2 + \cdots + v_n^2}$$

is the distance from the origin to the point $(v_1, \ldots, v_n)$. It is on the basis of this observation that the norm in any inner product space generalizes the Euclidean concept of length. In particular, this gives a quantitative way to measure the magnitude of a function $f \in C([a, b])$, namely,

$$\|f\| = \sqrt{\langle f, f \rangle} = \sqrt{\int_a^b \left(f(x) \right)^2 dx}.$$

Exercises 4.1

1. **a.** Translate the four inner product axioms into analogous statements about the real numbers using ordinary multiplication. Which ones make sense? Which are true?

    b. Translate the four inner product axioms into analogous statements about scalar multiplication. Which ones make sense? Which ones are true?

2. Evaluate the dot products.

    a. $(3, -2, 4) \cdot (2, 1, -4)$

    b. $(1, -1, 1, -1) \cdot (1, 1, 1, 1)$

    c. $\left(\frac{1}{\sqrt{3}}, \frac{1}{\sqrt{3}}, \frac{1}{\sqrt{3}} \right) \cdot \left(\frac{1}{\sqrt{3}}, \frac{1}{\sqrt{3}}, \frac{1}{\sqrt{3}} \right)$

    d. $(2, 5, -3, 4, -1) \cdot (0, 0, 0, 0, 0)$

3. Use the norm defined in terms of the dot product on Euclidean space to compute the following.

   a. $\|(1, 2, 3)\|$

   b. $\left\|\left(\frac{1}{\sqrt{3}}, \frac{1}{\sqrt{3}}, \frac{1}{\sqrt{3}}\right)\right\|$

   c. $\left\|\left(\frac{1}{100}, \frac{1}{100}, \frac{1}{100}, \frac{1}{100}, \frac{1}{100}\right)\right\|$

   d. $\left\|\left(\frac{1}{100}, \frac{1}{100}, \frac{1}{100}, \frac{1}{100}, \frac{1}{100}, \frac{1}{100}, \frac{1}{100}, \frac{1}{100}, \frac{1}{100}, \frac{1}{100}\right)\right\|$

4. Evaluate the standard inner product of each of the following pairs of functions.

   a. $f(x) = x$, $g(x) = x^2$ in $\mathbb{C}([0, 1])$

   b. $f(x) = x$, $g(x) = x^2$ in $\mathbb{C}([-1, 1])$

   c. $f(x) = x^2$, $g(x) = \sin x$ in $\mathbb{C}([0, 2\pi])$

   d. $f(x) = 1$, $g(x) = \cos x$ in $\mathbb{C}([-\pi, \pi])$

   e. $f(x) = \sin x$, $g(x) = \cos x$ in $\mathbb{C}([-\pi, \pi])$

   f. $f(x) = \sin(2x)$, $g(x) = \cos(3x)$ in $\mathbb{C}([-\pi, \pi])$

5. **a.** Use the norm defined in terms of the standard inner product on $\mathbb{C}([-\pi, \pi])$ to compute $\|\sin\|$, $\|\cos\|$, and $\|\mathbf{1}\|$, where $\mathbf{1}$ denotes the constant function with value 1.

   b. Use the norm defined in terms of the standard inner product on $\mathbb{C}([0, \pi])$ to compute $\|\sin\|$, $\|\cos\|$, and $\|\mathbf{1}\|$.

6. Prove parts b and c of Theorem 4.2.

7. Give an alternate proof that $\langle \mathbf{v}, \mathbf{0} \rangle = 0$ using the distributive law for inner products and the fact that $\mathbf{0} + \mathbf{0} = \mathbf{0}$.

8. Suppose $\mathbf{v}$ and $\mathbf{w}$ are vectors in an inner product space.

   a. Prove that $\langle -\mathbf{v}, \mathbf{w} \rangle = \langle \mathbf{v}, -\mathbf{w} \rangle = -\langle \mathbf{v}, \mathbf{w} \rangle$.

   b. Prove that $\langle -\mathbf{v}, -\mathbf{w} \rangle = \langle \mathbf{v}, \mathbf{w} \rangle$.

9. State some identities that you suspect hold for inner products and subtraction. Try to prove your conjectures or provide counterexamples.

10. Suppose $\mathbf{v}$ and $\mathbf{w}$ are vectors in an inner product space. Prove that
$$\langle \mathbf{v} + \mathbf{w}, \mathbf{v} + \mathbf{w} \rangle = \|\mathbf{v}\|^2 + 2\langle \mathbf{v}, \mathbf{w} \rangle + \|\mathbf{w}\|^2.$$

11. Suppose $\mathbf{v}$ and $\mathbf{w}$ are vectors in an inner product space. Prove that
$$\langle \mathbf{v} + \mathbf{w}, \mathbf{v} - \mathbf{w} \rangle = \|\mathbf{v}\|^2 - \|\mathbf{w}\|^2.$$

12. Suppose $\mathbf{v}$, $\mathbf{w}$, $\mathbf{x}$, and $\mathbf{y}$ are vectors in an inner product space and $a, b, c, d \in \mathbb{R}$. Prove that
$$\langle a\mathbf{v} + b\mathbf{w},\ c\mathbf{x} + d\mathbf{y} \rangle = ac\langle \mathbf{v}, \mathbf{x} \rangle + ad\langle \mathbf{v}, \mathbf{y} \rangle + bc\langle \mathbf{w}, \mathbf{x} \rangle + bd\langle \mathbf{w}, \mathbf{y} \rangle.$$

13. You are undoubtedly familiar with the summation notation from your calculus course. This notation has several advantages over the alternative of writing out several typical terms and using three dots. First, it is more concise. Second, it does not mistakenly imply that one or more terms must actually occur. As we encounter longer expressions in our work with linear algebra,

we will adopt the summation notation when appropriate. For example, we can rewrite the definition of the dot product on $\mathbb{R}^n$ as

$$\mathbf{v} \cdot \mathbf{w} = \sum_{i=1}^{n} v_i w_i.$$

To refresh yourself in the use of this notation, rewrite the verification that the dot product is an inner product on $\mathbb{R}^n$ using the summation notation.

14. Verify that $\langle (v_1, v_2), (w_1, w_2) \rangle = 2v_1 w_1 + 3v_2 w_2$ defines an inner product on $\mathbb{R}^2$.

15. Verify that $\langle (v_1, v_2), (w_1, w_2) \rangle = 2v_1 w_1 - v_1 w_2 - v_2 w_1 + 5v_2 w_2$ defines an inner product on $\mathbb{R}^2$.

16. Show that $\langle (v_1, v_2), (w_1, w_2) \rangle = v_1 w_1 + v_1 w_2 + v_2 w_1 + v_2 w_2$ does not define an inner product on $\mathbb{R}^2$. Give an explicit counterexample to one of the axioms.

17. **a.** Verify that $\langle p, q \rangle = p(-1)q(-1) + p(0)q(0) + p(1)q(1)$ defines an inner product on $\mathbb{P}_2$.

   b. Find polynomials $p, q \in \mathbb{P}_2$ for which the value of the inner product defined in part a differs from the value of the standard inner product of p and q as functions in $\mathbb{C}([-1, 1])$.

   c. Why does the formula in part a not define an inner product on $\mathbb{P}_3$?

18. Verify that $\left\langle \begin{bmatrix} a & b \\ c & d \end{bmatrix}, \begin{bmatrix} a' & b' \\ c' & d' \end{bmatrix} \right\rangle = aa' + bb' + cc' + dd'$ defines an inner product on $\mathbb{M}(2, 2)$.

19. Get out your calculus textbook and look up the results needed to verify that the standard inner product on $\mathbb{C}([a, b])$ is in fact an inner product. Why is it necessary to insist that $a < b$?

20. **a.** Verify that $\langle f, g \rangle = \int_0^1 x f(x)g(x)\,dx$ defines an inner product on $\mathbb{C}([0, 1])$.

   b. Show that $\langle f, g \rangle = \int_{-1}^1 x f(x)g(x)\,dx$ does not define an inner product on $\mathbb{C}([-1, 1])$. Give an explicit counterexample to one of the axioms.

21. Suppose $\mathbf{v}$ and $\mathbf{w}$ are vectors in an inner product space. Prove that $\|\mathbf{v} - \mathbf{w}\|^2 = \|\mathbf{v}\|^2 - 2\langle \mathbf{v}, \mathbf{w} \rangle + \|\mathbf{w}\|^2$.

22. Prove that the norm on an inner product space V satisfies the **parallelogram law:** for any $\mathbf{v}, \mathbf{w} \in V$,

$$\|\mathbf{v} + \mathbf{w}\|^2 + \|\mathbf{v} - \mathbf{w}\|^2 = 2\|\mathbf{v}\|^2 + 2\|\mathbf{w}\|^2.$$

23. Prove that the norm on an inner product space V satisfies the **polarization identity:** for any $\mathbf{v}, \mathbf{w} \in V$,

$$\|\mathbf{v} + \mathbf{w}\|^2 - \|\mathbf{v} - \mathbf{w}\|^2 = 4\langle \mathbf{v}, \mathbf{w} \rangle.$$

24. It is possible to take an axiomatic approach to norms as we did with inner products. In this approach we would define a norm on a vector space V to be a function that assigns to each vector $\mathbf{v} \in V$ a real number, denoted $\|\mathbf{v}\|$, in such a way as to satisfy the following three properties.

a. $\|\mathbf{v}\| \geq 0$; and if $\|\mathbf{v}\| = 0$, then $\mathbf{v} = \mathbf{0}$

b. $\|r\mathbf{v}\| = |r|\,\|\mathbf{v}\|$

c. $\|\mathbf{v} + \mathbf{w}\| \leq \|\mathbf{v}\| + \|\mathbf{w}\|$

You will see in Theorem 4.9 that we can use any inner product to define a norm. Show that the function defined on $\mathbb{R}^3$ by

$$\|(v_1, v_2, v_3)\| = \max\{|v_1|, |v_2|, |v_3|\}$$

satisfies the three properties listed above. Give a specific counterexample to show that this function does not satisfy the parallelogram law of Exercise 22. Conclude that this function cannot be derived from any inner product on $\mathbb{R}^3$ and hence, that this approach to norms is broader than the approach of defining the norm in terms of an inner product.

4.2 Geometry in Euclidean Spaces

The vector space axioms lead to qualitative concepts of magnitude and direction. The additional structure of an inner product is the tool needed to quantify these concepts. In this section we will use the dot product to derive formulas for the concepts of size, distance, angle, and projection in Euclidean space.

As we noted at the end of the previous section, the norm, defined in terms of the dot product, immediately gives a numeric measure of the magnitude of a vector $\mathbf{v} = (v_1, \dots, v_n) \in \mathbb{R}^n$. That is,

$$\|\mathbf{v}\| = \sqrt{\mathbf{v} \cdot \mathbf{v}} = \sqrt{v_1^2 + \cdots + v_n^2}$$

gives the length of the arrow from the origin to the point $(v_1, \dots, v_n)$ that we use to represent the vector $\mathbf{v}$.

It is also very easy to obtain the distance between two vectors $\mathbf{v} = (v_1, \dots, v_n)$ and $\mathbf{w} = (w_1, \dots, w_n)$ in $\mathbb{R}^n$ as the norm of the difference:

$$\|\mathbf{v} - \mathbf{w}\| = \sqrt{(\mathbf{v} - \mathbf{w}) \cdot (\mathbf{v} - \mathbf{w})} = \sqrt{(v_1 - w_1)^2 + \cdots + (v_n - w_n)^2}$$

You undoubtedly recognize this as a generalization of the distance formulas in $\mathbb{R}^2$ and $\mathbb{R}^3$ that are derived from the Pythagorean Theorem (see Exercise 13 at the end of this section).

The following theorem shows how the dot product allows us to measure the angle between two vectors in Euclidean space. Notice that any two vectors lie in a plane through the origin, and in this plane the concept of angle and the results of trigonometry are well established.

Theorem 4.6

The angle θ between two nonzero vectors $\mathbf{v}$ and $\mathbf{w}$ in $\mathbb{R}^n$ satisfies

$$\cos \theta = \frac{\mathbf{v} \cdot \mathbf{w}}{\|\mathbf{v}\|\,\|\mathbf{w}\|}.$$

Proof   First notice that since $\mathbf{v} \neq \mathbf{0}$ and $\mathbf{w} \neq \mathbf{0}$, we have $\|\mathbf{v}\| \neq 0$ and $\|\mathbf{w}\| \neq 0$; so we are not in danger of dividing by zero in the formula for $\cos\theta$.

The key to the proof is to apply the law of cosines to the triangle in $\mathbb{R}^n$ with vertices $\mathbf{v}$, $\mathbf{w}$, and $\mathbf{0}$. In this triangle, θ is the angle between the two sides of length $\|\mathbf{v}\|$ and $\|\mathbf{w}\|$. As we saw in Section 1.9, the arrow representing $\mathbf{v} - \mathbf{w}$ is a parallel translation of the arrow from $\mathbf{w}$ to $\mathbf{v}$. Thus, the side opposite the angle θ has length $\|\mathbf{v} - \mathbf{w}\|$. See Figure 4.1.

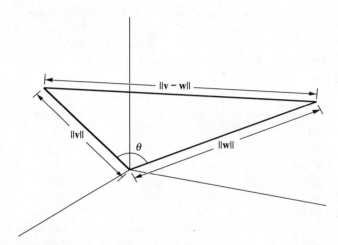

Figure 4.1   Apply the law of cosines in $\mathbb{R}^n$.

The law of cosines gives the equation

$$\|\mathbf{v} - \mathbf{w}\|^2 = \|\mathbf{v}\|^2 + \|\mathbf{w}\|^2 - 2\|\mathbf{v}\|\,\|\mathbf{w}\|\cos\theta.$$

So

$$\|\mathbf{v}\|\,\|\mathbf{w}\|\cos\theta = \tfrac{1}{2}\big(\|\mathbf{v}\|^2 + \|\mathbf{w}\|^2 - \|\mathbf{v} - \mathbf{w}\|^2\big).$$

From the expansion $\|\mathbf{v} - \mathbf{w}\|^2 = \|\mathbf{v}\|^2 - 2\mathbf{v} \cdot \mathbf{w} + \|\mathbf{w}\|^2$, obtained in Exercise 21 of Section 4.1, we also have

$$\mathbf{v} \cdot \mathbf{w} = \tfrac{1}{2}\big(\|\mathbf{v}\|^2 + \|\mathbf{w}\|^2 - \|\mathbf{v} - \mathbf{w}\|^2\big).$$

Hence, $\|\mathbf{v}\|\,\|\mathbf{w}\|\cos\theta = \mathbf{v} \cdot \mathbf{w}$. Dividing both sides by the nonzero quantity $\|\mathbf{v}\|\,\|\mathbf{w}\|$ gives us the desired result.   ∎

Crossroads   For any angle θ, we know that $|\cos\theta| \leq 1$. Thus, the formula of Theorem 4.6 gives $\dfrac{\mathbf{v} \cdot \mathbf{w}}{\|\mathbf{v}\|\,\|\mathbf{w}\|} \leq 1$. We can rewrite this in the form $|\mathbf{v} \cdot \mathbf{w}| \leq \|\mathbf{v}\|\,\|\mathbf{w}\|$, which is valid for all vectors $\mathbf{v}, \mathbf{w} \in \mathbb{R}^n$ (even the zero vectors). In the next section we will prove the Cauchy-Schwarz inequality as a generalization of this result to arbitrary inner product spaces.

An immediate consequence of Theorem 4.6 is a formula for the angle between two nonzero vectors **v** and **w** in $\mathbb{R}^n$:

$$\theta = \arccos \frac{\mathbf{v} \cdot \mathbf{w}}{\|\mathbf{v}\| \|\mathbf{w}\|}.$$

Although it is interesting to have an algebraic/trigonometric formula for measuring angles in higher-dimensional spaces, we will be most interested in the cases where the vectors form a right angle. This occurs if and only if θ is an odd multiple of $\frac{\pi}{2}$. That is, **v** and **w** are perpendicular if and only if

$$\cos \theta = \frac{\mathbf{v} \cdot \mathbf{w}}{\|\mathbf{v}\| \|\mathbf{w}\|} = 0.$$

This fraction is zero if and only if the numerator is zero. Thus, **v** and **w** are perpendicular if and only if $\mathbf{v} \cdot \mathbf{w} = 0$.

If one or both of the vectors are zero, we know that the dot product will automatically equal zero. This motivates the following definition of a term to encompass perpendicularity for nonzero vectors, as well as the possibility that one or both of the vectors are zero.

Definition 4.7

Two vectors **v** and **w** in $\mathbb{R}^n$ are **orthogonal** if and only if $\mathbf{v} \cdot \mathbf{w} = 0$.

This definition effectively translates the geometric concept of orthogonality into an algebraic equation. Here is an illustration of the use of this interplay between geometry and algebra to describe the set of vectors orthogonal to a given vector in $\mathbb{R}^4$. After you work Exercise 9 at the end of this section, you will not be surprised that the resulting set is a subspace of the vector space.

Quick Example   *Find a basis for the subspace S of all vectors in $\mathbb{R}^4$ orthogonal to* $(1, -3, 0, 4)$.

We are looking for the solution set to the equation $(1, -3, 0, 4) \cdot (v_1, v_2, v_3, v_4) = 0$. This is a single homogeneous equation $v_1 - 3v_2 + 0v_3 + 4v_4 = 0$ in four unknowns. Letting $v_2 = r$, $v_3 = s$, and $v_4 = t$ be arbitrary real numbers, we have that $v_1 = 3r - 4t$. Hence, a typical element of S can be written

$$
\begin{aligned}
(v_1, v_2, v_3, v_4) &= (3r - 4t, r, s, t) \\
&= r(3, 1, 0, 0) + s(0, 0, 1, 0) + t(-4, 0, 0, 1).
\end{aligned}
$$

Thus, the set $\{(3, 1, 0, 0), (0, 0, 1, 0), (-4, 0, 0, 1)\}$ spans S. By looking at the last three coordinates, we see that the set is linearly independent. Hence, it is a basis for this subspace. ∎

In Section 1.9 we defined a plane in terms of two direction vectors. The concept of orthogonality as expressed by the dot product gives us another way to write equations for planes in $\mathbb{R}^3$. For a plane described by an equation $ax + by + cz = d$, this alternative formulation shows that (a, b, c) is a **normal** vector to the plane. That is, (a, b, c) is orthogonal to every direction vector for the plane. Here is an example of how this works.

Quick Example   *Show that $(3, 1, -2)$ is a normal vector to the plane defined by the equation $3x + y - 2z = 4$. Use this to write the plane in the standard form of a plane in $\mathbb{R}^3$.*

Begin by rewriting the equation in the form of a dot product of two vectors in $\mathbb{R}^3$ that is equal to zero. The first step is to incorporate the constant 4 with one of the variables, say z. The equivalent equation

$$3x + y - 2(z + 2) = 0$$

can now be rewritten in the desired form

$$(3, 1, -2) \cdot \big((x, y, z) - (0, 0, -2)\big) = 0.$$

The geometric interpretation of this equation is that the plane is the set of all vectors (x, y, z) such that $(x, y, z) - (0, 0, -2)$ is orthogonal to $(3, 1, -2)$. Thus, we see that $(3, 1, -2)$ is a normal vector to the plane. We also see that the plane passes through $(0, 0, -2)$. For direction vectors, simply write down two vectors such as $(1, -3, 0)$ and $(0, 2, 1)$ that are orthogonal to $(3, 1, -2)$. Thus, the plane is

$$\{r(1, -3, 0) + s(0, 2, 1) + (0, 0, -2) \mid r, s \in \mathbb{R}\}. \quad \blacksquare$$

The formulas for orthogonal, or perpendicular, projection of one vector onto another vector are also fairly simple to derive. Suppose **u** and **v** are vectors in $\mathbb{R}^n$ with $\mathbf{u} \neq \mathbf{0}$. We want to decompose **v** as the sum of two vectors, with one in the same direction as **u** and the other orthogonal to **u**. The component of **v** that is in the same direction as **u** will, of course, be a multiple of **u**; that is, it will equal $c\mathbf{u}$ for some scalar c. With this notation, $\mathbf{v} - c\mathbf{u}$ is a vector that we can add to $c\mathbf{u}$ to obtain the original vector **v**. See Figure 4.2. This second vector must be orthogonal to **u**; that is,

$$(\mathbf{v} - c\mathbf{u}) \cdot \mathbf{u} = 0.$$

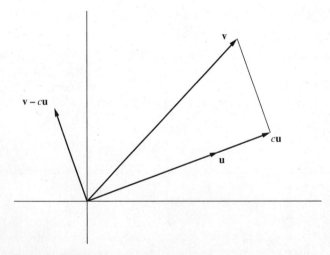

Figure 4.2   Orthogonal projection of **v** onto **u**.

This equation is equivalent to $\mathbf{v} \cdot \mathbf{u} - c(\mathbf{u} \cdot \mathbf{u}) = 0$, or (since $\mathbf{u} \cdot \mathbf{u} \neq 0$)

$$c = \frac{\mathbf{v} \cdot \mathbf{u}}{\mathbf{u} \cdot \mathbf{u}}.$$

In summary, then, with this value of c we can write

$$\mathbf{v} = c\mathbf{u} + (\mathbf{v} - c\mathbf{u}),$$

where $c\mathbf{u}$ is parallel to $\mathbf{u}$ and the second term, $\mathbf{v} - c\mathbf{u}$, is orthogonal to $\mathbf{u}$.

Quick Example   *Compute the orthogonal projection of $(1, 2, 3)$ onto $(1, 0, 1)$. Write $(1, 2, 3)$ as the sum of a vector parallel to $(1, 0, 1)$ and a vector orthogonal to $(1, 0, 1)$.*

We compute $\dfrac{(1, 2, 3) \cdot (1, 0, 1)}{(1, 0, 1) \cdot (1, 0, 1)} = \frac{4}{2} = 2$. Thus, $2(1, 0, 1) = (2, 0, 2)$ is the orthogonal projection of $(1, 2, 3)$ onto $(1, 0, 1)$. Now

$$(1, 2, 3) = 2(1, 0, 1) + \big((1, 2, 3) - 2(1, 0, 1)\big) = (2, 0, 2) + (-1, 2, 1)$$

is a decomposition of $(1, 2, 3)$ into a vector $(2, 0, 2)$ parallel to $(1, 0, 1)$ and a vector $(-1, 2, 1)$ orthogonal to $(1, 0, 1)$.

Exercises 4.2

1. Compute to the nearest minute of arc the angles between the following pairs of vectors. (Recall that there are 60 minutes in a degree.)

   a. $(1, 5)$ and $(-2, 3)$

   b. $(-3, 0, 4)$ and $(1, 1, -1)$

   c. $(2, 1, 2, 1, 1)$ and $(1, 3, 0, -1, -1)$

2. Compute to the nearest minute of arc the angle between

   a. a main diagonal and an adjacent edge of a cube;

   b. a main diagonal and an adjacent diagonal of a face of a cube;

   c. a diagonal of a face and an adjacent edge of a cube.

3. Generalize the previous exercise to four-dimensional cubes in $\mathbb{R}^4$.

4. A model of a methane molecule depicts a carbon atom at the center of a regular tetrahedron bonded to hydrogen atoms at the four vertices of the tetrahedron. What is the angle between the bonds to the hydrogen atoms?

5. Find the set of all vectors in $\mathbb{R}^2$ that are orthogonal to $(2, 3)$. Write the set in the standard form of a line through the origin.

6. Find the set of all vectors in $\mathbb{R}^3$ that are orthogonal to $(2, 3, 5)$. Write the set in the standard form of a plane through the origin.

7. Find the set of all vectors in $\mathbb{R}^3$ that are orthogonal to both $(-1, 0, 2)$ and $(3, 1, -2)$. Write the set in the standard form of a line through the origin.

8. Find the set of all vectors in $\mathbb{R}^4$ that are orthogonal to both $(1, 1, 2, 1)$ and $(2, 2, 1, 2)$. Write the set in the standard form of a plane through the origin.

9. For a given vector $\mathbf{v} \in \mathbb{R}^n$ show that the set of all vectors orthogonal to $\mathbf{v}$ is a subspace of $\mathbb{R}^n$.

10. Determine a vector orthogonal to the plane defined by the equation

$$-x + 5y - 3z = 4.$$

Write this plane in terms of two direction vectors and a point in the plane.

11. Find normal vectors to the planes defined by the equations

$$2x + y + 3z = -2 \quad \text{and} \quad x - 5y + z = 3.$$

Show that these vectors are orthogonal. What geometric conclusion can you derive about the two planes?

12. Find normal vectors to the planes defined by the equations

$$x + y + 2z = 1 \quad \text{and} \quad x - 2y + 3z = 2.$$

Use these normal vectors to approximate to the nearest degree the angle of intersection of the two planes.

13. **a.** Hundreds of proofs of the Pythagorean Theorem have been recorded. Look one up or try to discover one yourself. For instance, the altitude from the right angle to the hypotenuse gives two triangles that are similar to the original right triangle. The equality of the ratios of the lengths of corresponding sides of these triangles leads to the desired conclusion in a few simple algebraic steps.

b. Use the Pythagorean Theorem to derive the formula for the distance between two points (x_1, x_2) and (y_1, y_2) in $\mathbb{R}^2$.

c. Use the Pythagorean Theorem to derive the formula for the distance between two points (x_1, x_2, x_3) and (y_1, y_2, y_3) in $\mathbb{R}^3$. (Suggestion: First find the distance from (x_1, x_2, x_3) to (y_1, y_2, x_3).)

d. Generalize your proof to derive the formula for the distance between two points in $\mathbb{R}^n$.

14. Compute the orthogonal projection of $(5, -1)$ onto $(2, 3)$. Write $(5, -1)$ as the sum of a vector parallel to $(2, 3)$ and a vector orthogonal to $(2, 3)$.

15. Compute the orthogonal projection of $(4, -3, -5)$ onto $(-3, 2, 2)$. Write $(4, -3, -5)$ as the sum of a vector parallel to $(-3, 2, 2)$ and a vector orthogonal to $(-3, 2, 2)$.

16. Compute the orthogonal projection of $(1, 1, 0, 1, 1, 1)$ onto $(2, 1, 1, 1, 1, 1)$. Write $(1, 1, 0, 1, 1, 1)$ as the sum of a vector parallel to $(2, 1, 1, 1, 1, 1)$ and a vector orthogonal to $(2, 1, 1, 1, 1, 1)$.

17. Suppose $\mathbf{u}_1$ and $\mathbf{u}_2$ are vectors in $\mathbb{R}^n$ with neither being a multiple of the other. We want to decompose a given vector $\mathbf{v} \in \mathbb{R}^n$ as the sum of one component $c_1 \mathbf{u}_1 + c_2 \mathbf{u}_2$ in the plane spanned by $\mathbf{u}_1$ and $\mathbf{u}_2$, and a remaining component $\mathbf{v} - (c_1 \mathbf{u}_1 + c_2 \mathbf{u}_2)$ orthogonal to this plane.

a. Use the conditions that $\mathbf{v} - (c_1 \mathbf{u}_1 + c_2 \mathbf{u}_2)$ is to be orthogonal to $\mathbf{u}_1$ and $\mathbf{u}_2$ to derive equations that c_1 and c_2 must satisfy.

b. Use your results from part a to decompose $(1, 4, -3)$ as the sum of a component in the plane spanned by $(2, -4, -1)$ and $(5, -3, 1)$ and a component orthogonal to this plane.

c. Use your results from part a to decompose $(1, 4, -3)$ as the sum of a component in the plane spanned by $(2, -4, -1)$ and $(3, 1, 2)$ and a component orthogonal to this plane.

d. Notice that since $(5, -3, 1) = (2, -4, -1) + (3, 1, 2)$, the planes in parts b and c are identical. What makes the computations in part c so much easier?

18. Let $\mathbf{u}$ be a given nonzero vector in $\mathbb{R}^n$. Show that any vector $\mathbf{v} \in \mathbb{R}^n$ has a unique decomposition as the sum of a vector in the direction of $\mathbf{u}$ and a vector orthogonal to $\mathbf{u}$. That is, if $\mathbf{v} = \mathbf{v}_1 + \mathbf{v}_2$ and $\mathbf{v} = \mathbf{v}_1' + \mathbf{v}_2'$, where $\mathbf{v}_1$ and $\mathbf{v}_1'$ are multiples of $\mathbf{u}$, and $\mathbf{v}_2$ and $\mathbf{v}_2'$ are orthogonal to $\mathbf{u}$, then $\mathbf{v}_1 = \mathbf{v}_1'$ and $\mathbf{v}_2 = \mathbf{v}_2'$.

4.3 The Cauchy-Schwarz Inequality

In the previous section, we investigated algebraic formulas for computing geometric concepts such as distance and angle in Euclidean spaces. The next step involves transferring this relationship between algebra and geometry to other inner product spaces. We will use the algebraic formulas to introduce geometric concepts in situations where we do not have prior geometric intuition. Our specific goals are to develop measures of distance and angle in function spaces. We will use these tools to quantify how good one function is as an approximation to another. This in turn will enable us to select from a particularly nice subspace the function that is closest to a given function not in the subspace.

Let us begin with a result that is the key to developing the concept of distance in an inner product space.

> ### Theorem 4.8 Cauchy-Schwarz Inequality
>
> If $\mathbf{v}$ and $\mathbf{w}$ are two vectors in an inner product space, then $|\langle \mathbf{v}, \mathbf{w} \rangle| \leq \|\mathbf{v}\| \|\mathbf{w}\|$.

Proof If either $\mathbf{v} = \mathbf{0}$ or $\mathbf{w} = \mathbf{0}$, then we know that both $|\langle \mathbf{v}, \mathbf{w} \rangle|$ and $\|\mathbf{v}\| \|\mathbf{w}\|$ will equal zero. The conclusion follows in this case.

Let us now consider the case where $\mathbf{v} \neq \mathbf{0}$ and $\mathbf{w} \neq \mathbf{0}$. Then $\|\mathbf{v}\| \neq 0$ and $\|\mathbf{w}\| \neq 0$, so we can use $\dfrac{1}{\|\mathbf{v}\|}$ and $\dfrac{1}{\|\mathbf{w}\|}$ as scalars in the linear combination

$$\frac{1}{\|\mathbf{v}\|}\mathbf{v} + \frac{1}{\|\mathbf{w}\|}\mathbf{w}.$$

By Axiom 1 of Definition 4.1, the inner product of any vector with itself is nonnegative. In particular,

$$\left\langle \frac{1}{\|\mathbf{v}\|}\mathbf{v} + \frac{1}{\|\mathbf{w}\|}\mathbf{w}, \frac{1}{\|\mathbf{v}\|}\mathbf{v} + \frac{1}{\|\mathbf{w}\|}\mathbf{w} \right\rangle \geq 0.$$

We expand this inner product, recognizing that $\langle \mathbf{v}, \mathbf{v} \rangle = \|\mathbf{v}\|^2$ and $\langle \mathbf{w}, \mathbf{w} \rangle = \|\mathbf{w}\|^2$, to derive the following inequalities:

$$\frac{1}{\|\mathbf{v}\|^2}\langle \mathbf{v}, \mathbf{v} \rangle + \frac{2}{\|\mathbf{v}\|\,\|\mathbf{w}\|}\langle \mathbf{v}, \mathbf{w} \rangle + \frac{1}{\|\mathbf{w}\|^2}\langle \mathbf{w}, \mathbf{w} \rangle \geq 0$$

$$1 + \frac{2}{\|\mathbf{v}\|\,\|\mathbf{w}\|}\langle \mathbf{v}, \mathbf{w} \rangle + 1 \geq 0$$

$$2 \geq -\frac{2}{\|\mathbf{v}\|\,\|\mathbf{w}\|}\langle \mathbf{v}, \mathbf{w} \rangle$$

$$-\|\mathbf{v}\|\,\|\mathbf{w}\| \leq \langle \mathbf{v}, \mathbf{w} \rangle$$

By a similar derivation, we can start with

$$\left\langle \frac{1}{\|\mathbf{v}\|}\mathbf{v} - \frac{1}{\|\mathbf{w}\|}\mathbf{w}, \; \frac{1}{\|\mathbf{v}\|}\mathbf{v} - \frac{1}{\|\mathbf{w}\|}\mathbf{w} \right\rangle \geq 0$$

to conclude

$$\langle \mathbf{v}, \mathbf{w} \rangle \leq \|\mathbf{v}\|\,\|\mathbf{w}\|.$$

These two inequalities give us an upper bound for the absolute value of $\langle \mathbf{v}, \mathbf{w} \rangle$, namely,

$$|\langle \mathbf{v}, \mathbf{w} \rangle| \leq \|\mathbf{v}\|\,\|\mathbf{w}\|. \quad \blacksquare$$

The Cauchy-Schwarz inequality enables us to do several things. First, it allows us to verify the property of the norm known as the **triangle inequality**. This is property c listed in the following theorem. The first two properties follow more directly from the axioms for inner products. Property a is the **positive-definite condition,** and property b is the **homogeneous** condition.

Theorem 4.9

Suppose V is an inner product space. For all $\mathbf{v}, \mathbf{w} \in V$ and $r \in \mathbb{R}$, the following properties hold.

   a. $\|\mathbf{v}\| \geq 0$; and $\|\mathbf{v}\| = 0$ if and only if $\mathbf{v} = \mathbf{0}$
   b. $\|r\,\mathbf{v}\| = |r|\,\|\mathbf{v}\|$
   c. $\|\mathbf{v} + \mathbf{w}\| \leq \|\mathbf{v}\| + \|\mathbf{w}\|$

Proof Axiom 1 of inner products gives that $\langle \mathbf{v}, \mathbf{v} \rangle \geq 0$. Hence, $\|\mathbf{v}\| = \sqrt{\langle \mathbf{v}, \mathbf{v} \rangle} \geq 0$. Axiom 1 along with part a of Theorem 4.2 gives that $\langle \mathbf{v}, \mathbf{v} \rangle = 0$ if and only if $\mathbf{v} = \mathbf{0}$. Hence, $\|\mathbf{v}\| = \sqrt{\langle \mathbf{v}, \mathbf{v} \rangle} = 0$ if and only if $\mathbf{v} = \mathbf{0}$.

The following string of equalities uses Axiom 2 of inner products and part b of Theorem 4.2 to prove part b of the theorem:

$$\|r\,\mathbf{v}\| = \sqrt{\langle r\,\mathbf{v}, r\,\mathbf{v} \rangle} = \sqrt{r\langle \mathbf{v}, r\,\mathbf{v} \rangle} = \sqrt{r^2\langle \mathbf{v}, \mathbf{v} \rangle}$$
$$= \sqrt{r^2}\,\sqrt{\langle \mathbf{v}, \mathbf{v} \rangle} = |r|\,\|\mathbf{v}\|.$$

Finally, the Cauchy-Schwarz inequality makes short work of part c:

$$\begin{aligned}
\|\mathbf{v} + \mathbf{w}\|^2 &= \langle \mathbf{v} + \mathbf{w}, \ \mathbf{v} + \mathbf{w} \rangle \\
&= \|\mathbf{v}\|^2 + 2\langle \mathbf{v}, \mathbf{w} \rangle + \|\mathbf{w}\|^2 \\
&\leq \|\mathbf{v}\|^2 + 2\|\mathbf{v}\| \|\mathbf{w}\| + \|\mathbf{w}\|^2 \\
&= (\|\mathbf{v}\| + \|\mathbf{w}\|)^2.
\end{aligned}$$

Since $\|\mathbf{v} + \mathbf{w}\|$ and $\|\mathbf{v}\| + \|\mathbf{w}\|$ are both nonnegative, we can take the square root of the first and last expressions, to obtain

$$\|\mathbf{v} + \mathbf{w}\| \leq \|\mathbf{v}\| + \|\mathbf{w}\|. \quad \blacksquare$$

The second use we want to make of the Cauchy-Schwarz inequality is to extend the notion of angle between two vectors to arbitrary inner product spaces. Notice that if $\mathbf{v}$ and $\mathbf{w}$ are nonzero vectors in an inner product space V, then $\|\mathbf{v}\| \neq 0$ and $\|\mathbf{w}\| \neq 0$, so we can divide both sides of the Cauchy-Schwarz inequality by $\|\mathbf{v}\| \|\mathbf{w}\|$ to obtain

$$\frac{|\langle \mathbf{v}, \mathbf{w} \rangle|}{\|\mathbf{v}\| \|\mathbf{w}\|} \leq 1.$$

Hence, there is a value of θ such that

$$\cos \theta = \frac{\langle \mathbf{v}, \mathbf{w} \rangle}{\|\mathbf{v}\| \|\mathbf{w}\|}.$$

Furthermore, if we restrict our choice of θ to the interval $[0, \pi]$, then θ is uniquely determined. This leads to the following definition.

Definition 4.10

The **angle** between two nonzero vectors $\mathbf{v}$ and $\mathbf{w}$ in an inner product space is the value of θ in the interval $[0, \pi]$ that satisfies

$$\cos \theta = \frac{\langle \mathbf{v}, \mathbf{w} \rangle}{\|\mathbf{v}\| \|\mathbf{w}\|}.$$

That is,

$$\theta = \arccos \frac{\langle \mathbf{v}, \mathbf{w} \rangle}{\|\mathbf{v}\| \|\mathbf{w}\|}.$$

Theorem 4.6 shows that this definition is compatible with our geometric concept of angle in Euclidean spaces. It provides a reasonable way to introduce this concept into other inner product spaces such as $\mathbb{C}(\mathbb{R})$, where it would be difficult to use a protractor to measure the angle between two functions.

Quick Example   *Compute the angle between the polynomials defined by $p(x) = x^2 - 3x + 1$ and $q(x) = x + 1$ with respect to the standard inner product on $\mathbb{P}_2$ as a subspace of $\mathbb{C}([-1, 1])$. Compute the angle between p and q with respect to the inner product $\langle p, q \rangle = a_2 b_2 + a_1 b_1 + a_0 b_0$ for $p(x) = a_2 x^2 + a_1 x + a_0$ and $q(x) = b_2 x^2 + b_1 x + b_0$.*

With the standard inner product on $\mathbb{P}_2$ as a subspace of $\mathbb{C}([-1, 1])$, we have

$$\langle p, q \rangle = \int_{-1}^{1} (x^2 - 3x + 1)(x + 1) \, dx$$

$$= \int_{-1}^{1} (x^3 - 2x^2 - 2x + 1) \, dx$$

$$= \left. (\tfrac{1}{4}x^4 - \tfrac{2}{3}x^3 - x^2 + x) \right|_{-1}^{1}$$

$$= \tfrac{2}{3}.$$

Similar computations give $\langle p, p \rangle = \int_{-1}^{1} (x^2 - 3x + 1)^2 \, dx = \frac{146}{15}$ and $\langle q, q \rangle =$
$\int_{-1}^{1} (x + 1)^2 \, dx = \frac{8}{3}$. Thus, $\cos \theta = \dfrac{\langle p, q \rangle}{\|\mathbf{v}\| \|\mathbf{w}\|} = \dfrac{\frac{2}{3}}{\sqrt{\frac{146}{15}} \sqrt{\frac{8}{3}}} = \dfrac{\sqrt{365}}{146}$. Therefore,
with respect to this inner product, the angle between the polynomials p and q is
$\theta = \arccos\left(\frac{\sqrt{365}}{146}\right) \approx 82°\, 28'\, 51''$ to the nearest second of arc.

With the inner product on $\mathbb{P}_2$ defined in terms of the product of the coefficients,
we have $\langle p, q \rangle = 1 \cdot 0 + (-3) \cdot 1 + 1 \cdot 1 = -2$. Similarly, $\langle p, p \rangle = 1^2 + (-3)^2 +$
$1^2 = 11$ and $\langle q, q \rangle = 1^2 + 1^2 = 2$. Thus, $\cos \theta = \dfrac{\langle p, q \rangle}{\|\mathbf{v}\| \|\mathbf{w}\|} = \dfrac{-2}{\sqrt{11} \sqrt{2}} = -\dfrac{\sqrt{22}}{11}$.
Therefore, with respect to this inner product, the angle between the polynomials p
and q is $\theta = \arccos\left(-\frac{\sqrt{22}}{11}\right) \approx 115°\, 14'\, 22''$ to the nearest second of arc. ∎

With our discussion of orthogonality and projection in Euclidean spaces as a guide,
let us mimic these concepts in the abstract setting.

Definition 4.11

Two vectors $\mathbf{v}$ and $\mathbf{w}$ in an inner product space are **orthogonal** if and only if
$\langle \mathbf{v}, \mathbf{w} \rangle = 0$.

Notice that this definition of orthogonality includes the case where one or both of
the vectors is equal to zero, as well as the case where the angle between them is $\frac{\pi}{2}$.

Definition 4.12

Suppose $\mathbf{u}$ is a nonzero vector in an inner product space. For any vector $\mathbf{v}$ in this
space, the **projection** of $\mathbf{v}$ onto $\mathbf{u}$ is denoted $\text{proj}_{\mathbf{u}}(\mathbf{v})$ and is defined by

$$\text{proj}_{\mathbf{u}}(\mathbf{v}) = \frac{\langle \mathbf{v}, \mathbf{u} \rangle}{\langle \mathbf{u}, \mathbf{u} \rangle} \mathbf{u}.$$

With vectors **u** and **v** as in the definition, $\text{proj}_\mathbf{u}(\mathbf{v})$ is clearly a multiple of **u**. Also, $\mathbf{v} - \text{proj}_\mathbf{u}(\mathbf{v})$ is orthogonal to **u** since

$$\langle \mathbf{v} - \text{proj}_\mathbf{u}(\mathbf{v}), \mathbf{u} \rangle = \left\langle \mathbf{v} - \frac{\langle \mathbf{v}, \mathbf{u} \rangle}{\langle \mathbf{u}, \mathbf{u} \rangle} \mathbf{u}, \mathbf{u} \right\rangle$$

$$= \langle \mathbf{v}, \mathbf{u} \rangle - \frac{\langle \mathbf{v}, \mathbf{u} \rangle}{\langle \mathbf{u}, \mathbf{u} \rangle} \langle \mathbf{u}, \mathbf{u} \rangle$$

$$= \langle \mathbf{v}, \mathbf{u} \rangle - \langle \mathbf{v}, \mathbf{u} \rangle$$

$$= 0.$$

Thus, the equation

$$\mathbf{v} = \text{proj}_\mathbf{u}(\mathbf{v}) + \left(\mathbf{v} - \text{proj}_\mathbf{u}(\mathbf{v}) \right)$$

gives a decomposition of any vector **v** into a component $\text{proj}_\mathbf{u}(\mathbf{v})$ parallel to **u** and a component $\mathbf{v} - \text{proj}_\mathbf{u}(\mathbf{v})$ orthogonal to **u**.

Quick Example   *Use the standard inner product on $\mathbb{C}([0, \pi])$ to find the projection of the function* sin *onto the function defined by* $f(x) = x$. *Write* sin *as the sum of a function parallel to f and a function orthogonal to f.*

A computer algebra system (or a little more practice with integration by parts) gives $\int_0^\pi x \sin x \, dx = \pi$ and $\int_0^\pi x^2 \, dx = \frac{1}{3}\pi^3$. This gives

$$\left(\text{proj}_f(\sin)\right)(x) = \frac{\langle \sin, f \rangle}{\langle f, f \rangle} f(x) = \frac{\int_0^\pi x \sin x \, dx}{\int_0^\pi x^2 \, dx} x = \frac{\pi}{\frac{1}{3}\pi^3} x = \frac{3}{\pi^2} x.$$

Thus, we can write the desired decomposition as

$$\sin x = \frac{3}{\pi^2} x + \left(\sin x - \frac{3}{\pi^2} x \right). \quad \blacksquare$$

Exercises 4.3

1. Fill in the details in the proof of the Cauchy-Schwarz inequality to show that

$$\langle \mathbf{v}, \mathbf{w} \rangle \leq \|\mathbf{v}\| \|\mathbf{w}\|$$

for elements **v** and **w** in an inner product space.

2. Suppose **v** and **w** are elements of an inner product space. Show that

$$2|\langle \mathbf{v}, \mathbf{w} \rangle| \leq \|\mathbf{v}\|^2 + \|\mathbf{w}\|^2.$$

3. Suppose **v** and **w** are vectors in an inner product space. Use the triangle inequality to show that $\|\mathbf{v}\| \leq \|\mathbf{v} - \mathbf{w}\| + \|\mathbf{w}\|$ or, equivalently, that $\|\mathbf{v} - \mathbf{w}\| \geq \|\mathbf{v}\| - \|\mathbf{w}\|$. Use the same strategy to derive the inequality $\|\mathbf{v} - \mathbf{w}\| \geq \|\mathbf{w}\| - \|\mathbf{v}\|$. Conclude that

$$\|\mathbf{v} - \mathbf{w}\| \geq \big| \|\mathbf{v}\| - \|\mathbf{w}\| \big|.$$

4. **a.** Show that the polynomials defined by $p(x) = x$ and $q(x) = x^2$ are orthogonal with respect to the inner product on $\mathbb{P}_2$ defined in Exercise 17 of Section 4.1.

   b. Show that these polynomials, when considered as functions in $\mathbb{C}([0, 1])$, are not orthogonal with respect to the standard inner product on $\mathbb{C}([0, 1])$.

5. Use the Cauchy-Schwarz inequality to show that the arithmetic mean of $a_1, \ldots, a_n \in \mathbb{R}$ is less than or equal to the root-mean-square of the numbers; that is, show that

$$\frac{a_1 + \cdots + a_n}{n} \leq \sqrt{\frac{a_1^2 + \cdots + a_n^2}{n}}.$$

6. Suppose that $a_1, \ldots, a_n$ are positive real numbers.

   a. Show that $\left(a_1 + \cdots + a_n\right)\left(\dfrac{1}{a_1} + \cdots + \dfrac{1}{a_n}\right) \geq n^2$.

   b. Conclude that the harmonic mean $\dfrac{n}{a_1^{-1} + \cdots + a_n^{-1}}$ is less than or equal to the arithmetic mean $\dfrac{a_1 + \cdots + a_n}{n}$.

7. **a.** Suppose that $f \in \mathbb{C}([a, b])$. Show that

$$\left(\int_a^b f(x)\, dx\right)^2 \leq (b - a) \int_a^b \left(f(x)\right)^2 dx.$$

   b. Suppose in addition that $f' \in \mathbb{C}([a, b])$ and $f(a) = 0$. Show for any $x \in [a, b]$ that

$$\left(f(x)\right)^2 \leq (x - a) \int_a^b \left(f'(x)\right)^2 dx.$$

   c. Finally, show that such a function satisfies

$$\int_a^b \left(f(x)\right)^2 dx \leq \frac{(b - a)^2}{2} \int_a^b \left(f'(x)\right)^2 dx.$$

8. Suppose θ is the angle between two vectors $\mathbf{v}$ and $\mathbf{w}$ in an inner product space. Use the definition of angle along with the properties of norms and inner products to prove the law of cosines

$$\|\mathbf{v} - \mathbf{w}\|^2 = \|\mathbf{v}\|^2 + \|\mathbf{w}\|^2 - 2\|\mathbf{v}\|\,\|\mathbf{w}\|\cos\theta$$

in this general setting.

9. **a.** Use the standard inner product on $\mathbb{C}([0, 1])$ to find the angle between the functions defined by $f(x) = x$ and $g(x) = x^2$.

   b. Compute the projection of g onto f.

   c. Write g as the sum of a multiple of f and a function orthogonal to f.

10. Consider the functions defined by $f(x) = \frac{1}{2}x$ and $g(x) = \sqrt{x}$ as elements of $\mathbb{C}([0, 4])$ with the standard inner product. Decompose g as the sum of one function parallel to f and another function orthogonal to f.

11. Use the standard inner product on $\mathbb{C}([0, 1])$ to find the angle between a function of the form $f(x) = ax$ and the exponential function. Show that the angle depends only on the sign of the constant a.

12. Use the standard inner product on $\mathbb{C}([-\pi, \pi])$ to compute the angles between pairs of the three functions sin, cos, and the constant function 1 with value 1.

13. Consider the inner product on $\mathbb{R}^2$ defined in Exercise 14 of Section 4.1 by the formula $\langle (v_1, v_2), (w_1, w_2) \rangle = 2v_1w_1 + 3v_2w_2$. Use this inner product to compute the following quantities.

    a. $\|(1, 5)\|$ b. $\|(-2, 3)\|$

    c. Compute the angle between $(1, 5)$ and $(-2, 3)$ to the nearest minute of arc. Compare this with the result of Exercise 1a of Section 4.2.

14. Consider the inner product on $\mathbb{P}_2$ defined in Exercise 17 of Section 4.1 by the formula $\langle p, q \rangle = p(-1)q(-1) + p(0)q(0) + p(1)q(1)$. Use this inner product to compute the following quantities.

    a. $\|x\|$ b. $\|x^2\|$

    c. Compute the angle between the polynomials defined by x and x^2.

4.4 Orthogonality

Of all the possible bases for $\mathbb{R}^3$, the standard basis $\{(1, 0, 0), (0, 1, 0), (0, 0, 1)\}$ is certainly the easiest to use for finding coordinates of vectors in $\mathbb{R}^3$. In this section you will discover that the orthogonality of each pair of vectors in this basis accounts for much of this simplicity (although the 0s and 1s are an additional bonus with this set). We will continue to use the dot product in Euclidean spaces as a prototype to introduce geometric concepts in an arbitrary inner product space.

We begin with a definition that extends the concept of orthogonality between a pair of vectors (defined in terms of the inner product of the two vectors being equal to zero) to orthogonality for a set of vectors.

Definition 4.13

A set S of vectors in an inner product space is an **orthogonal set** if and only if any two distinct vectors in S are orthogonal. That is, if $\mathbf{v}, \mathbf{w} \in S$ and $\mathbf{v} \neq \mathbf{w}$, then $\langle \mathbf{v}, \mathbf{w} \rangle = 0$.

Quick Example Show that $\{(2, 2, -1), (2, -1, 2), (-1, 2, 2)\}$ is an orthogonal set in $\mathbb{R}^3$.

From a set with three elements, there are six ordered pairs of distinct elements: for each of the three possibilities for the first element of the pair, two choices remain for the other element. Because of the commutative law for inner products, we only need to check orthogonality of each pair in one order. This reduces the verification to the following three computations:

$$(2, 2, -1) \cdot (2, -1, 2) = 4 - 2 - 2 = 0;$$
$$(2, 2, -1) \cdot (-1, 2, 2) = -2 + 4 - 2 = 0;$$
$$(2, -1, 2) \cdot (-1, 2, 2) = -2 - 2 + 4 = 0. \quad \blacksquare$$

Quick Example *Show that* $\{\sin, \cos, \mathbf{1}, \mathbf{0}\}$ *is an orthogonal set in* $\mathbb{C}([-\pi, \pi])$, *where* $\mathbf{1}$ *denotes the constant function with value* 1 *and* $\mathbf{0}$ *denotes the constant function with value* 0.

From a set of four elements, you can choose $4 \cdot 3 = 12$ ordered pairs of distinct elements. The commutative law for inner products cuts the work in half. In Exercise 12 of Section 4.3 you should have discovered that

$$\langle \sin, \cos \rangle = 0, \qquad \langle \sin, \mathbf{1} \rangle = 0, \qquad \langle \cos, \mathbf{1} \rangle = 0.$$

By part a of Theorem 4.2, the inner product of $\mathbf{0}$ with any of the other three functions is 0 as well. $\blacksquare$

Suppose we want to write $(2, 3, 5)$ as a linear combination of $(2, 2, -1)$, $(2, -1, 2)$, and $(-1, 2, 2)$. Our standard procedure is to assume we have such a linear combination

$$(2, 3, 5) = a(2, 2, -1) + b(2, -1, 2) + c(-1, 2, 2),$$

set up a system of three equations in three unknowns, and put the augmented matrix of this system through the Gaussian reduction process to obtain values for the coefficients a, b, and c. Here is an easier way to determine a, b, and c using the orthogonality of the set $\{(2, 2, -1), (2, -1, 2), (-1, 2, 2)\}$. Watch what happens when we take the inner product (the dot product in $\mathbb{R}^3$) of both sides of the equation with $(2, 2, -1)$:

$$\langle (2, 3, 5), (2, 2, -1) \rangle = \langle a(2, 2, -1) + b(2, -1, 2) + c(-1, 2, 2), (2, 2, -1) \rangle$$
$$= a\langle (2, 2, -1), (2, 2, -1) \rangle$$
$$+ b\langle (2, -1, 2), (2, 2, -1) \rangle + c\langle (-1, 2, 2), (2, 2, -1) \rangle.$$

Since we already know the last two terms are equal to 0, we do not really even need to write them. After evaluating just two inner products, this equation reduces to $5 = 9a$, giving the value $a = \frac{5}{9}$. Similarly, if you take the inner product of both sides of the original equation with the second vector $(2, -1, 2)$ in the orthogonal set, the resulting equation

$$\langle (2, 3, 5), (2, -1, 2) \rangle = b\langle (2, -1, 2), (2, -1, 2) \rangle$$

easily yields $11 = 9b$, or $b = \frac{11}{9}$. Finally, we can repeat the process with the third vector. The resulting equation

$$\langle (2, 3, 5), (-1, 2, 2) \rangle = c\langle (-1, 2, 2), (-1, 2, 2) \rangle$$

yields $c = \frac{14}{9}$.

Crossroads To justify our assumption that

$$\{(2, 2, -1), (2, -1, 2), (-1, 2, 2)\}$$

spans $\mathbb{R}^3$, we can rely on the following theorem, which shows that linear independence is a consequence of the orthogonality of this set of vectors. Since $\dim \mathbb{R}^3 = 3$, we know by Theorem 3.15 that the set $\{(2, 2, -1), (2, -1, 2), (-1, 2, 2)\}$ is a basis for $\mathbb{R}^3$. Hence, the preceding values for a, b, and c are indeed the coefficients in the linear combination.

The proof of the following theorem is based on the idea used to determine the coefficients in the previous example.

Theorem 4.14

An orthogonal set $\{\mathbf{v}_1, \ldots, \mathbf{v}_n\}$ of nonzero vectors in an inner product space is linearly independent.

Proof Suppose $r_1\mathbf{v}_1 + \cdots + r_n\mathbf{v}_n = \mathbf{0}$. The trick to showing that any one of the coefficients, say r_i, must be zero is to take the inner product of both sides of this equation with the vector $\mathbf{v}_i$. This gives

$$\langle r_1\mathbf{v}_1 + \cdots + r_n\mathbf{v}_n, \mathbf{v}_i \rangle = \langle \mathbf{0}, \mathbf{v}_i \rangle;$$
$$\langle r_1\mathbf{v}_1, \mathbf{v}_i \rangle + \cdots + \langle r_i\mathbf{v}_i, \mathbf{v}_i \rangle + \cdots + \langle r_n\mathbf{v}_n, \mathbf{v}_i \rangle = 0;$$
$$r_1\langle \mathbf{v}_1, \mathbf{v}_i \rangle + \cdots + r_i\langle \mathbf{v}_i, \mathbf{v}_i \rangle + \cdots + r_n\langle \mathbf{v}_n, \mathbf{v}_i \rangle = 0;$$
$$r_1 0 + \cdots + r_i\langle \mathbf{v}_i, \mathbf{v}_i \rangle + \cdots + r_n 0 = 0;$$
$$r_i\langle \mathbf{v}_i, \mathbf{v}_i \rangle = 0.$$

Since $\mathbf{v}_i$ is a nonzero vector, the real number $\langle \mathbf{v}_i, \mathbf{v}_i \rangle$ is not equal to zero; so we must have $r_i = 0$.

This argument works for each of the coefficients $r_1, \ldots, r_n$. We conclude, therefore, that the set satisfies the definition of linear independence. ∎

When we are dealing with nonzero vectors in an inner product space, it is useful to multiply them by the appropriate scalars to standardize their norms to equal 1. This process is called **normalization.** It is easy to check that $\dfrac{1}{\|\mathbf{v}\|}$ is a scalar multiplier we can use to normalize the length of $\mathbf{v}$ to equal 1. Indeed, if $r = \dfrac{1}{\|\mathbf{v}\|}$, then

$$\|r\mathbf{v}\| = r\|\mathbf{v}\| = \frac{1}{\|\mathbf{v}\|}\|\mathbf{v}\| = 1.$$

Definition 4.15

A vector $\mathbf{v}$ in an inner product space is a **unit vector** if and only if $\|\mathbf{v}\| = 1$.

Combining the idea of unit vectors with the concept of an orthogonal set of vectors leads to the following definition.

Definition 4.16

A set S of vectors in an inner product space is **orthonormal** if and only if S is orthogonal and each vector in S is a unit vector. That is, for $\mathbf{v}, \mathbf{w} \in S$,

$$\langle \mathbf{v}, \mathbf{w} \rangle = \begin{cases} 0 & \text{if } \mathbf{v} \neq \mathbf{w}; \\ 1 & \text{if } \mathbf{v} = \mathbf{w}. \end{cases}$$

Quick Example   *Prove that if $E = \{\mathbf{e}_1, \ldots, \mathbf{e}_n\}$ is an orthonormal set of n elements in an inner product space V with dim $V = n$, then E is a basis for V.*

The vectors in E are unit vectors, so they are nonzero. By Theorem 4.14, the set E is linearly independent. Because the number of elements in E is equal to dim V, Theorem 3.15 implies that E is a basis for V.   ∎

With an orthonormal basis $E = \{\mathbf{e}_1, \ldots, \mathbf{e}_n\}$ for an inner product space V, it becomes especially simple to write a vector $\mathbf{v} \in V$ as a linear combination $r_1\mathbf{e}_1 + \cdots + r_n\mathbf{e}_n$. The trick introduced at the beginning of this section no longer involves division:

$$\begin{aligned} \langle \mathbf{v}, \mathbf{e}_i \rangle &= \langle r_1\mathbf{e}_1 + \cdots + r_n\mathbf{e}_n, \ \mathbf{e}_i \rangle \\ &= r_1\langle \mathbf{e}_1, \mathbf{e}_i \rangle + \cdots + r_i\langle \mathbf{e}_i, \mathbf{e}_i \rangle + \cdots + r_n\langle \mathbf{e}_n, \mathbf{e}_i \rangle \\ &= r_1 0 + \cdots + r_i 1 + \cdots + r_n 0 \\ &= r_i. \end{aligned}$$

That is, $\mathbf{v} = \langle \mathbf{v}, \mathbf{e}_1 \rangle \mathbf{e}_1 + \cdots + \langle \mathbf{v}, \mathbf{e}_n \rangle \mathbf{e}_n$, or

$$[\mathbf{v}]_E = \begin{bmatrix} \langle \mathbf{v}, \mathbf{e}_1 \rangle \\ \vdots \\ \langle \mathbf{v}, \mathbf{e}_n \rangle \end{bmatrix}.$$

Working with an orthonormal basis is very much like working with the standard basis in $\mathbb{R}^n$. For example, you are asked to show in Exercise 16 at the end of this section that the inner product reduces to the dot product of the coordinate vectors with respect to the orthonormal basis E; that is, if $\mathbf{w}$ is another element of V, then

$$\langle \mathbf{v}, \mathbf{w} \rangle = [\mathbf{v}]_E \cdot [\mathbf{w}]_E.$$

To exploit these remarkable properties, we need to have an orthonormal basis for V. The following result guarantees that such a basis will exist. In fact, it gives an algorithm for converting an arbitrary basis into an orthonormal basis.

Theorem 4.17 Gram-Schmidt Orthonormalization Process

Suppose $\{\mathbf{u}_1, \ldots, \mathbf{u}_n\}$ is a basis for an inner product space V. Define a sequence of vectors as follows:

$$\mathbf{e}_1 = \frac{1}{\|\mathbf{u}_1\|}\mathbf{u}_1;$$

$$\mathbf{v}_2 = \mathbf{u}_2 - \langle \mathbf{u}_2, \mathbf{e}_1 \rangle \mathbf{e}_1 \qquad \text{and} \qquad \mathbf{e}_2 = \frac{1}{\|\mathbf{v}_2\|}\mathbf{v}_2;$$

$$\mathbf{v}_3 = \mathbf{u}_3 - \big(\langle \mathbf{u}_3, \mathbf{e}_1 \rangle \mathbf{e}_1 + \langle \mathbf{u}_3, \mathbf{e}_2 \rangle \mathbf{e}_2\big) \qquad \text{and} \qquad \mathbf{e}_3 = \frac{1}{\|\mathbf{v}_3\|}\mathbf{v}_3;$$

and in general for $k \leq n$,

$$\mathbf{v}_k = \mathbf{u}_k - \big(\langle \mathbf{u}_k, \mathbf{e}_1 \rangle \mathbf{e}_1 + \cdots + \langle \mathbf{u}_k, \mathbf{e}_{k-1} \rangle \mathbf{e}_{k-1}\big) \qquad \text{and} \qquad \mathbf{e}_k = \frac{1}{\|\mathbf{v}_k\|}\mathbf{v}_k.$$

Then $\{\mathbf{e}_1, \ldots, \mathbf{e}_n\}$ is an orthonormal basis for V.

Crossroads The Gram-Schmidt process converts the original basis $\{\mathbf{u}_1, \ldots, \mathbf{u}_n\}$ into an orthonormal basis $\{\mathbf{e}_1, \ldots, \mathbf{e}_n\}$ one vector at a time. The definition of each $\mathbf{e}_k$ (other than the first) involves the definition of an intermediate vector

$$\mathbf{v}_k = \mathbf{u}_k - \big(\langle \mathbf{u}_k, \mathbf{e}_1 \rangle \mathbf{e}_1 + \cdots + \langle \mathbf{u}_k, \mathbf{e}_{k-1} \rangle \mathbf{e}_{k-1}\big).$$

Notice that the coefficient $\langle \mathbf{u}_k, \mathbf{e}_i \rangle$ of any vector $\mathbf{e}_i$ in this linear combination is precisely the coefficient of $\mathbf{e}_i$ we need to write $\mathbf{u}_k$ as a linear combination of the orthonormal basis $\{\mathbf{e}_1, \ldots, \mathbf{e}_n\}$. The scalar product $\langle \mathbf{u}_k, \mathbf{e}_i \rangle \mathbf{e}_i$ is the projection of $\mathbf{u}_k$ onto $\mathbf{e}_i$ (division by $\langle \mathbf{e}_i, \mathbf{e}_i \rangle$ is not necessary since $\mathbf{e}_i$ is a unit vector). Picture this as the component of $\mathbf{u}_k$ in the direction of the vector $\mathbf{e}_i$; when this component is subtracted from $\mathbf{u}_k$, the remaining vector will be orthogonal to $\mathbf{e}_i$.

Proof Let us follow through the algorithm and, as each of the vectors $\mathbf{e}_k$ is defined, check that it is orthogonal to each of the previously defined unit vectors $\mathbf{e}_1, \ldots, \mathbf{e}_{k-1}$. Actually, it is easier to check that the intermediate vector $\mathbf{v}_k$ is orthogonal to each of the vectors $\mathbf{e}_1, \ldots, \mathbf{e}_{k-1}$. These are straightforward computations:

$\mathbf{v}_2$ is orthogonal to $\mathbf{e}_1$:

$$\begin{aligned}
\langle \mathbf{v}_2, \mathbf{e}_1 \rangle &= \langle \mathbf{u}_2 - \langle \mathbf{u}_2, \mathbf{e}_1 \rangle \mathbf{e}_1, \ \mathbf{e}_1 \rangle \\
&= \langle \mathbf{u}_2, \mathbf{e}_1 \rangle - \langle \mathbf{u}_2, \mathbf{e}_1 \rangle \langle \mathbf{e}_1, \mathbf{e}_1 \rangle \\
&= \langle \mathbf{u}_2, \mathbf{e}_1 \rangle - \langle \mathbf{u}_2, \mathbf{e}_1 \rangle \\
&= 0.
\end{aligned}$$

$\mathbf{v}_3$ is orthogonal to $\mathbf{e}_1$ and $\mathbf{e}_2$:

$$\langle \mathbf{v}_3, \mathbf{e}_1 \rangle = \langle \mathbf{u}_3 - \langle \mathbf{u}_3, \mathbf{e}_1 \rangle \mathbf{e}_1 - \langle \mathbf{u}_3, \mathbf{e}_2 \rangle \mathbf{e}_2,\ \mathbf{e}_1 \rangle$$
$$= \langle \mathbf{u}_3, \mathbf{e}_1 \rangle - \langle \mathbf{u}_3, \mathbf{e}_1 \rangle \langle \mathbf{e}_1, \mathbf{e}_1 \rangle - \langle \mathbf{u}_3, \mathbf{e}_2 \rangle \langle \mathbf{e}_2, \mathbf{e}_1 \rangle$$
$$= \langle \mathbf{u}_3, \mathbf{e}_1 \rangle - \langle \mathbf{u}_3, \mathbf{e}_1 \rangle$$
$$= 0;$$
$$\langle \mathbf{v}_3, \mathbf{e}_2 \rangle = \langle \mathbf{u}_3 - \langle \mathbf{u}_3, \mathbf{e}_1 \rangle \mathbf{e}_1 - \langle \mathbf{u}_3, \mathbf{e}_2 \rangle \mathbf{e}_2,\ \mathbf{e}_2 \rangle$$
$$= \langle \mathbf{u}_3, \mathbf{e}_2 \rangle - \langle \mathbf{u}_3, \mathbf{e}_1 \rangle \langle \mathbf{e}_1, \mathbf{e}_2 \rangle - \langle \mathbf{u}_3, \mathbf{e}_2 \rangle \langle \mathbf{e}_2, \mathbf{e}_2 \rangle$$
$$= \langle \mathbf{u}_3, \mathbf{e}_2 \rangle - \langle \mathbf{u}_3, \mathbf{e}_2 \rangle$$
$$= 0.$$

In general, for $i < k$, $\mathbf{v}_k$ is orthogonal to $\mathbf{e}_i$:

$$\langle \mathbf{v}_k, \mathbf{e}_i \rangle = \langle \mathbf{u}_k - (\langle \mathbf{u}_k, \mathbf{e}_1 \rangle \mathbf{e}_1 + \cdots + \langle \mathbf{u}_k, \mathbf{e}_{k-1} \rangle \mathbf{e}_{k-1}),\ \mathbf{e}_i \rangle$$
$$= \langle \mathbf{u}_k, \mathbf{e}_i \rangle - \langle \mathbf{u}_k, \mathbf{e}_1 \rangle \langle \mathbf{e}_1, \mathbf{e}_i \rangle - \cdots - \langle \mathbf{u}_k, \mathbf{e}_i \rangle \langle \mathbf{e}_i, \mathbf{e}_i \rangle$$
$$\quad - \cdots - \langle \mathbf{u}_k, \mathbf{e}_{k-1} \rangle \langle \mathbf{e}_{k-1}, \mathbf{e}_i \rangle$$
$$= \langle \mathbf{u}_k, \mathbf{e}_i \rangle - \langle \mathbf{u}_k, \mathbf{e}_i \rangle$$
$$= 0.$$

Since $\mathbf{e}_k$ is a scalar multiple of $\mathbf{v}_k$, it too will be orthogonal to each of the vectors $\mathbf{e}_1, \ldots, \mathbf{e}_{k-1}$. (See Exercise 2 at the end of this section.)

One remaining concern is that the scalars used in normalizing the vectors $\mathbf{v}_k$ to produce $\mathbf{e}_k$ may involve division by zero. However, $\mathbf{u}_1$ is a basis vector, so it is nonzero. Hence, $\|\mathbf{u}_1\| \neq 0$. Since $\mathbf{e}_1$ is a scalar multiple of $\mathbf{u}_1$, we see that the second basis vector $\mathbf{u}_2$ is not a scalar multiple of $\mathbf{e}_1$; so $\mathbf{v}_2$ is nonzero. Hence, $\|\mathbf{v}_2\| \neq 0$. Since $\mathbf{e}_2$ is a linear combination of $\mathbf{u}_2$ and $\mathbf{e}_1$ (which is a multiple of $\mathbf{u}_1$), we see that the third basis vector $\mathbf{u}_3$ is not a linear combination of $\mathbf{e}_1$ and $\mathbf{e}_2$; so $\mathbf{v}_3$ is nonzero. Hence, $\|\mathbf{v}_3\| \neq 0$. In general, we can accumulate the facts that $\mathbf{e}_1, \ldots, \mathbf{e}_{k-1}$ are linear combinations of $\mathbf{u}_1, \ldots, \mathbf{u}_{k-1}$. As a result, we see that $\mathbf{u}_k$ is not a linear combination of $\mathbf{e}_1, \ldots, \mathbf{e}_{k-1}$; so $\mathbf{v}_k \neq 0$. Hence, $\|\mathbf{v}_k\| \neq 0$.

Once we have the orthonormal set $\{\mathbf{e}_1, \ldots, \mathbf{e}_n\}$, Theorem 4.14 guarantees that the set is linearly independent. Then, by Theorem 3.15, we know it is a basis for the n-dimensional vector space V. ∎

In practice the Gram-Schmidt process is a simple matter of evaluating formulas. Although the arithmetic of paper-and-pencil computations often becomes quite involved, the process is not inherently difficult. Always keep the geometric concepts in mind. First modify $\mathbf{u}_k$ by subtracting its projection onto the first $k - 1$ orthonormal vectors. The resulting vector $\mathbf{v}_k$ is then orthogonal to each of these vectors. Then stretch or shrink $\mathbf{v}_k$ to produce a unit vector $\mathbf{e}_k$ in this same direction.

Quick Example   *Find an orthonormal basis for $\mathbb{R}^3$ that contains a multiple of the vector $(1, 2, 2)$.*

Let us apply the Gram-Schmidt process to the basis

$$\{(1, 2, 2), (0, 1, 0), (0, 0, 1)\}.$$

First we adjust the length of $(1, 2, 2)$:

$$\mathbf{e}_1 = \frac{1}{\|(1, 2, 2)\|}(1, 2, 2) = (\tfrac{1}{3}, \tfrac{2}{3}, \tfrac{2}{3}).$$

Next we straighten up $(0, 1, 0)$ so the result is orthogonal to $\mathbf{e}_1$. We do this by subtracting the projection of $(0, 1, 0)$ onto $\mathbf{e}_1$:

$$\mathbf{v}_2 = (0, 1, 0) - \big((0, 1, 0) \cdot (\tfrac{1}{3}, \tfrac{2}{3}, \tfrac{2}{3})\big)(\tfrac{1}{3}, \tfrac{2}{3}, \tfrac{2}{3})$$

$$= (0, 1, 0) - (\tfrac{2}{9}, \tfrac{4}{9}, \tfrac{4}{9})$$

$$= (-\tfrac{2}{9}, \tfrac{5}{9}, -\tfrac{4}{9}).$$

Then we adjust the length of $\mathbf{v}_2$. Since any multiple of $\mathbf{v}_2$ will be orthogonal to $\mathbf{e}_1$, we might as well normalize a convenient multiple of $\mathbf{v}_2$ such as $(-2, 5, -4)$:

$$\mathbf{e}_2 = \frac{1}{\|(-2, 5, -4)\|}(-2, 5, -4) = \frac{1}{\sqrt{45}}(-2, 5, -4) = \Big(-\frac{2}{\sqrt{45}}, \frac{5}{\sqrt{45}}, -\frac{4}{\sqrt{45}}\Big).$$

Finally, we straighten up $(0, 0, 1)$ so the result is orthogonal to $\mathbf{e}_1$ and $\mathbf{e}_2$. We do this by subtracting the projections of $(0, 0, 1)$ onto $\mathbf{e}_1$ and $\mathbf{e}_2$:

$$\mathbf{v}_3 = (0, 0, 1) - \big((0, 0, 1) \cdot (\tfrac{1}{3}, \tfrac{2}{3}, \tfrac{2}{3})\big)(\tfrac{1}{3}, \tfrac{2}{3}, \tfrac{2}{3})$$

$$- \big((0, 0, 1) \cdot \big(-\tfrac{2}{\sqrt{45}}, \tfrac{5}{\sqrt{45}}, -\tfrac{4}{\sqrt{45}}\big)\big)\big(-\tfrac{2}{\sqrt{45}}, \tfrac{5}{\sqrt{45}}, -\tfrac{4}{\sqrt{45}}\big)$$

$$= (0, 0, 1) - (\tfrac{2}{9}, \tfrac{4}{9}, \tfrac{4}{9}) - (\tfrac{8}{45}, -\tfrac{20}{45}, \tfrac{16}{45})$$

$$= (-\tfrac{18}{45}, 0, \tfrac{9}{45}).$$

And then we adjust the length of a convenient multiple of $\mathbf{v}_3$:

$$\mathbf{e}_3 = \frac{1}{\|(-2, 0, 1)\|}(-2, 0, 1) = \Big(-\frac{2}{\sqrt{5}}, 0, \frac{1}{\sqrt{5}}\Big).$$

Thus, $\big\{(\tfrac{1}{3}, \tfrac{2}{3}, \tfrac{2}{3}), \big(-\frac{2}{\sqrt{45}}, \frac{5}{\sqrt{45}}, -\frac{4}{\sqrt{45}}\big), \big(-\frac{2}{\sqrt{5}}, 0, \frac{1}{\sqrt{5}}\big)\big\}$ is an orthonormal basis that contains a multiple of $(1, 2, 2)$. ∎

Exercises 4.4

1. The second longest word in this text occurs in this section. What is that word?

2. If $\mathbf{v}$ and $\mathbf{w}$ are orthogonal vectors in an inner product space, show that any scalar multiples of $\mathbf{v}$ and $\mathbf{w}$ are orthogonal.

3. Suppose $\mathbf{v}$ is a nonzero vector in an inner product space. Find all scalars r such that $r\mathbf{v}$ is a unit vector.

4. Verify that $\{(3, 6, -2), (-2, 3, 6), (6, -2, 3)\}$ is an orthogonal subset of $\mathbb{R}^3$.

5. Show that $\big\{(\tfrac{3}{7}, \tfrac{6}{7}, -\tfrac{2}{7}), (-\tfrac{2}{7}, \tfrac{3}{7}, \tfrac{6}{7}), (\tfrac{6}{7}, -\tfrac{2}{7}, \tfrac{3}{7})\big\}$ is an orthonormal subset of $\mathbb{R}^3$. With no further computation, explain why it is a basis for $\mathbb{R}^3$.

6. Apply the Gram-Schmidt process to find an orthonormal basis for the subspace of $\mathbb{R}^4$ spanned by $\{(1, 0, 1, 0), (1, 1, 0, 1), (1, 1, 1, 1)\}$.

7. Find an orthonormal basis for $\mathbb{P}_2$ as a subspace of $\mathbb{C}([-1, 1])$ with the standard inner product.

8. Find an orthonormal basis for $\mathbb{P}_2$ with the inner product defined in Exercise 17 of Section 4.1.

9. Consider the functions defined by $f(x) = 1$ and $g(x) = e^x$ as elements of $\mathbb{C}([0, 1])$. Find an orthonormal basis for the subspace of $\mathbb{C}([0, 1])$ spanned by $\{f, g\}$.

10. Consider the inner product space $\mathbb{C}([-a, a])$, where a is any positive real number. Recall that $f \in \mathbb{C}([-a, a])$ is an **even** function if and only if $f(-x) = f(x)$ for all $x \in [-a, a]$, and f is an **odd** function if and only if $f(-x) = -f(x)$ for all $x \in [-a, a]$. Show that any even function is orthogonal to any odd function in $\mathbb{C}([-a, a])$.

11. Suppose $\mathbf{w}$ is a vector in an inner product space V. Show that the set

$$\mathbf{w}^{\perp} = \{\mathbf{v} \in V \mid \langle \mathbf{v}, \mathbf{w} \rangle = 0\}$$

is a subspace of V.

12. Suppose $\mathbf{w}$ is a nonzero vector in a finite-dimensional inner product space V. With $\mathbf{w}^{\perp}$ defined as in the previous exercise, show that

$$\dim(\mathbf{w}^{\perp}) = \dim V - 1.$$

(Suggestion: Apply the Gram-Schmidt orthonormalization process with $\mathbf{u}_1 = \mathbf{w}$.)

13. Suppose S is a subset of an inner product space V. Show that the set

$$S^{\perp} = \{\mathbf{v} \in V \mid \langle \mathbf{v}, \mathbf{w} \rangle = 0 \text{ for all } \mathbf{w} \in S\}$$

is a subspace of V.

14. Suppose S is a subspace of a finite-dimensional inner product space V. With $S^{\perp}$ defined as in the previous exercise, show that $\dim S + \dim S^{\perp} = \dim V$.

15. Here is a generalization of the Pythagorean Theorem to arbitrary inner product spaces. Use the formula in Exercise 10 of Section 4.1 to give a three-step proof of this result.

If $\mathbf{v}$ and $\mathbf{w}$ are orthogonal vectors in an inner product space, then

$$\|\mathbf{v} + \mathbf{w}\|^2 = \|\mathbf{v}\|^2 + \|\mathbf{w}\|^2.$$

16. Suppose $E = \{\mathbf{e}_1, \dots, \mathbf{e}_n\}$ is an orthonormal basis for an inner product space V. For any vectors $\mathbf{v}, \mathbf{w} \in V$, show that

a. $\langle \mathbf{v}, \mathbf{w} \rangle = [\mathbf{v}]_E \cdot [\mathbf{w}]_E$

b. $\|\mathbf{v}\| = \|[\mathbf{v}]_E\|$

c. $\|\mathbf{v} - \mathbf{w}\| = \|[\mathbf{v}]_E - [\mathbf{w}]_E\|$

17. Suppose $E = \{\mathbf{e}_1, \dots, \mathbf{e}_n\}$ is an orthonormal basis for an inner product space V. For any $\mathbf{v} \in V$, prove Bessel's equality:

$$\sum_{i=1}^{n} \langle \mathbf{v}, \mathbf{e}_i \rangle^2 = \|\mathbf{v}\|^2.$$

18. Suppose the Gram-Schmidt process is applied to a basis $\{\mathbf{u}_1, \ldots, \mathbf{u}_n\}$, where the first k vectors $\mathbf{u}_1, \ldots, \mathbf{u}_k$ are orthonormal. Show that $\mathbf{e}_i = \mathbf{u}_i$ for $i = 1, \ldots, k$.

4.5 Fourier Analysis

In this section we will extend the idea of projecting one vector onto a nonzero vector. This will enable us to determine the vector in a subspace that is the best approximation to a given vector not in the subspace.

Definition 4.18

Suppose $\{\mathbf{e}_1, \ldots, \mathbf{e}_n\}$ is an orthonormal basis for a subspace S of an inner product space V. The **projection** of a vector $\mathbf{v} \in V$ onto the subspace S is denoted $\text{proj}_S(\mathbf{v})$ and is defined by

$$\text{proj}_S(\mathbf{v}) = \langle \mathbf{v}, \mathbf{e}_1 \rangle \mathbf{e}_1 + \cdots + \langle \mathbf{v}, \mathbf{e}_n \rangle \mathbf{e}_n.$$

This definition (see Figure 4.3) merits several comments. First, the definition requires a finite basis for the subspace S but not for the inner product space V. Second, for any $\mathbf{v} \in V$, we always have $\text{proj}_S(\mathbf{v}) \in S$. Third, if $\mathbf{v} \in S$, then $\langle \mathbf{v}, \mathbf{e}_i \rangle$ is the coefficient of $\mathbf{e}_i$ needed to write $\mathbf{v}$ as a linear combination of $\{\mathbf{e}_1, \ldots, \mathbf{e}_n\}$; thus, in this case, $\text{proj}_S(\mathbf{v}) = \mathbf{v}$. Finally, the equation

$$\mathbf{v} = \text{proj}_S(\mathbf{v}) + \left(\mathbf{v} - \text{proj}_S(\mathbf{v})\right)$$

gives a decomposition of an arbitrary element $\mathbf{v} \in V$ as a sum of a vector $\text{proj}_S(\mathbf{v})$ in S and a vector $\mathbf{v} - \text{proj}_S(\mathbf{v})$ that is orthogonal to every element of S. Indeed, for any one of the basis elements $\mathbf{e}_i$,

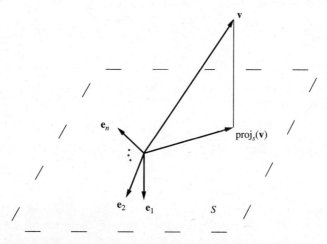

Figure 4.3 The projection of $\mathbf{v}$ onto the subspace S.

$$\langle \mathbf{v} - \mathrm{proj}_S(\mathbf{v}), \mathbf{e}_i \rangle = \langle \mathbf{v} - \langle \mathbf{v}, \mathbf{e}_1 \rangle \mathbf{e}_1 - \cdots - \langle \mathbf{v}, \mathbf{e}_n \rangle \mathbf{e}_n, \mathbf{e}_i \rangle$$

$$= \langle \mathbf{v}, \mathbf{e}_i \rangle - \langle \mathbf{v}, \mathbf{e}_1 \rangle \langle \mathbf{e}_1, \mathbf{e}_i \rangle - \cdots - \langle \mathbf{v}, \mathbf{e}_i \rangle \langle \mathbf{e}_i, \mathbf{e}_i \rangle - \cdots - \langle \mathbf{v}, \mathbf{e}_n \rangle \langle \mathbf{e}_n, \mathbf{e}_i \rangle$$

$$= \langle \mathbf{v}, \mathbf{e}_i \rangle - \langle \mathbf{v}, \mathbf{e}_i \rangle$$

$$= 0.$$

Exercise 1 at the end of this section gives the desired result.

The following theorem justifies our intuition that the orthogonal projection of a vector onto a subspace is the vector in the subspace that is closest to the original vector.

Theorem 4.19   Approximation Theorem

Suppose $\{\mathbf{e}_1, \ldots, \mathbf{e}_n\}$ is an orthonormal basis for a subspace S of an inner product space V. For a given vector $\mathbf{v} \in V$, there is a unique vector $\mathbf{w}$ in S that minimizes $\|\mathbf{v} - \mathbf{w}\|$. This vector is

$$\mathbf{w} = \mathrm{proj}_S(\mathbf{v}) = \sum_{i=1}^{n} \langle \mathbf{v}, \mathbf{e}_i \rangle \mathbf{e}_i.$$

In other words, of all the vectors $\mathbf{w} \in S$, we can make $\|\mathbf{v} - \mathbf{w}\|$ smallest by choosing $\mathbf{w} = \mathrm{proj}_S(\mathbf{v})$. This characterization of the projection shows that $\mathrm{proj}_S(\mathbf{v})$ depends only on the vector $\mathbf{v}$ and the subspace S, and does not depend on the basis $\{\mathbf{e}_1, \ldots, \mathbf{e}_n\}$ used to compute the projection.

Proof   Let $\mathbf{w}$ be an element of S. We can write $\mathbf{v} - \mathbf{w} = \big(\mathbf{v} - \mathrm{proj}_S(\mathbf{v})\big) + \big(\mathrm{proj}_S(\mathbf{v}) - \mathbf{w}\big)$. Since $\mathrm{proj}_S(\mathbf{v})$ and $\mathbf{w}$ are in the subspace S, so is $\mathrm{proj}_S(\mathbf{v}) - \mathbf{w}$. Since $\mathbf{v} - \mathrm{proj}_S(\mathbf{v})$ is orthogonal to every vector in S, it is orthogonal to $\mathrm{proj}_S(\mathbf{v}) - \mathbf{w}$. By the generalization of the Pythagorean Theorem (Exercise 15 of Section 4.4), we have $\|\mathbf{v} - \mathbf{w}\|^2 = \|\mathbf{v} - \mathrm{proj}_S(\mathbf{v})\|^2 + \|\mathrm{proj}_S(\mathbf{v}) - \mathbf{w}\|^2$. The first term in this sum is not affected by our choice of $\mathbf{w}$. Thus, the minimum value of the sum occurs when $\mathbf{w} = \mathrm{proj}_S(\mathbf{v})$ and the second term is equal to zero. Only this choice of $\mathbf{w}$ gives this minimum value for $\|\mathbf{v} - \mathbf{w}\|^2$. This choice is also the unique vector that minimizes the nonnegative quantity $\|\mathbf{v} - \mathbf{w}\|$. ∎

Let us try out the Approximation Theorem in $\mathbb{C}([-\pi, \pi])$. By Exercise 2 at the end of this section, the functions in $\mathbb{C}([-\pi, \pi])$ defined by

$$\frac{1}{\sqrt{2\pi}}, \quad \frac{1}{\sqrt{\pi}} \cos x, \quad \frac{1}{\sqrt{\pi}} \sin x, \quad \frac{1}{\sqrt{\pi}} \cos(2x),$$

$$\frac{1}{\sqrt{\pi}} \sin(2x), \quad \frac{1}{\sqrt{\pi}} \cos(3x), \quad \frac{1}{\sqrt{\pi}} \sin(3x), \ldots$$

form an orthonormal set with respect to the standard inner product on this space. Thus, we can choose any finite number of these functions to generate a subspace from which to select approximations to any function in $\mathbb{C}([-\pi, \pi])$.

Quick Example *Find the linear combination of the five functions defined by*

$$\frac{1}{\sqrt{2\pi}}, \quad \frac{1}{\sqrt{\pi}}\cos x, \quad \frac{1}{\sqrt{\pi}}\sin x, \quad \frac{1}{\sqrt{\pi}}\cos(2x), \quad \frac{1}{\sqrt{\pi}}\sin(2x)$$

that best approximates the function defined by $f(x) = x$.

The five functions form an orthonormal basis for a subspace S of $\mathbb{C}([-\pi, \pi])$. Thus, we can find coefficients of the best approximation

$$(\text{proj}_S(f))(x) = a_0\frac{1}{\sqrt{2\pi}} + a_1\frac{1}{\sqrt{\pi}}\cos x + b_1\frac{1}{\sqrt{\pi}}\sin x$$

$$+ a_2\frac{1}{\sqrt{\pi}}\cos(2x) + b_2\frac{1}{\sqrt{\pi}}\sin(2x)$$

by taking inner products. To begin with,

$$a_0 = \left\langle f, \frac{1}{\sqrt{2\pi}} \right\rangle = \int_{-\pi}^{\pi} x\frac{1}{\sqrt{2\pi}}\,dx = \frac{x^2}{2\sqrt{2\pi}}\Bigg|_{-\pi}^{\pi} = 0.$$

For $k \geq 1$, integration by parts gives

$$a_k = \left\langle f, \frac{1}{\sqrt{\pi}}\cos(kx) \right\rangle$$

$$= \int_{-\pi}^{\pi} x\frac{1}{\sqrt{\pi}}\cos(kx)\,dx$$

$$= \frac{1}{\sqrt{\pi}}x\frac{1}{k}\sin(kx)\Bigg|_{-\pi}^{\pi} - \frac{1}{\sqrt{\pi}}\int_{-\pi}^{\pi}\frac{1}{k}\sin(kx)\,dx$$

$$= 0 + \frac{1}{\sqrt{\pi}}\frac{1}{k^2}\cos(kx)\Bigg|_{-\pi}^{\pi}$$

$$= 0$$

and

$$b_k = \left\langle f, \frac{1}{\sqrt{\pi}}\sin(kx) \right\rangle$$

$$= \int_{-\pi}^{\pi} x\frac{1}{\sqrt{\pi}}\sin(kx)\,dx$$

$$= \frac{1}{\sqrt{\pi}}x\left(-\frac{1}{k}\cos(kx)\right)\Bigg|_{-\pi}^{\pi} - \frac{1}{\sqrt{\pi}}\int_{-\pi}^{\pi}\left(-\frac{1}{k}\cos(kx)\right)dx$$

$$= -\frac{1}{\sqrt{\pi}}\frac{\pi}{k}(-1)^k + \frac{1}{\sqrt{\pi}}\frac{-\pi}{k}(-1)^k + \frac{1}{\sqrt{\pi}}\frac{1}{k^2}\sin(kx)\Bigg|_{-\pi}^{\pi}$$

$$= (-1)^{k+1}\frac{2\sqrt{\pi}}{k}.$$

(Notice that $a_k = 0$ also follows from the general fact stated in Exercise 10 of Section 4.4.) We can now write the best approximation as

$$\left(\text{proj}_S(f)\right)(x) = 0 + 0\cos x + (-1)^2 2\sqrt{\pi}\frac{1}{\sqrt{\pi}}\sin x$$

$$+ 0\cos(2x) + (-1)^3\frac{2\sqrt{\pi}}{2}\frac{1}{\sqrt{\pi}}\sin(2x)$$

$$= 2\sin x - \sin(2x). \quad \blacksquare$$

If we want a better approximation to f, we can enlarge the subspace by adjoining to the orthonormal set trigonometric functions with smaller periods. For example, if we include $\frac{1}{\sqrt{\pi}}\cos(3x)$ and $\frac{1}{\sqrt{\pi}}\sin(3x)$, we can compute coefficients 0 and $\frac{2\sqrt{\pi}}{3}$, respectively. Of course, the coefficients of the original basis functions remain unchanged. Hence, the function defined by

$$2\sin x - \sin(2x) + \tfrac{2}{3}\sin(3x)$$

is the function in the subspace spanned by the first seven of the orthogonal functions that best approximates $f(x) = x$.

These linear combinations are known as the **Fourier approximations** to the given function. With the coefficients a_k and b_k defined as above, the infinite series

$$a_0\frac{1}{\sqrt{2\pi}} + a_1\frac{1}{\sqrt{\pi}}\cos x + b_1\frac{1}{\sqrt{\pi}}\sin x + a_2\frac{1}{\sqrt{\pi}}\cos(2x) + b_2\frac{1}{\sqrt{\pi}}\sin(2x) + \cdots$$

is known as the **Fourier series** of the function f. In Figure 4.4 you can see how the graphs of the Fourier approximations

$$f_1(x) = 2\sin x,$$
$$f_2(x) = 2\sin x - \sin(2x),$$
$$f_3(x) = 2\sin x - \sin(2x) + \tfrac{2}{3}\sin(3x)$$

are indeed getting closer to the graph of f. Even though all the Fourier approximations to $f(x) = x$ will have value 0 at $-\pi$ and at π, the graphs viewed over the entire interval $[-\pi, \pi]$ are getting closer to the graph of f as required to give small values for

$$\|f_k - f\| = \sqrt{\int_{-\pi}^{\pi}\left(f_k(x) - f(x)\right)^2 dx}.$$

The original function f can be viewed as just one period of a periodic function defined on all of $\mathbb{R}$. The Fourier analysis of a function as illustrated in the preceding example is the mathematical basis for decomposing a wave form or other periodic phenomenon into simple harmonic oscillations of various frequencies.

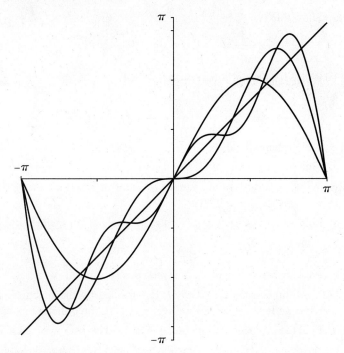

Figure 4.4 Fourier approximations to $f(x) = x$.

Exercises 4.5

1. Show that if **v** is orthogonal to the vectors $\mathbf{v}_1, \ldots, \mathbf{v}_n$ in an inner product space, then **v** is orthogonal to every vector in the subspace

$$S = \mathrm{span}\{\mathbf{v}_1, \ldots, \mathbf{v}_n\}.$$

2. **a.** Show that any two of the functions defined by 1, $\cos(kx)$, and $\sin(kx)$, where k is a positive integer, are orthogonal vectors in $\mathbb{C}([-\pi, \pi])$. (Suggestion: To evaluate the integrals you will encounter, you may want to consult a table of integrals or, better yet, a symbolic algebra software package. If you insist on working through these integrals by hand, you will find the following trigonometric identities helpful:

$$\sin a \sin b = \tfrac{1}{2}(\cos(a - b) - \cos(a + b)),$$

$$\cos a \cos b = \tfrac{1}{2}(\cos(a - b) + \cos(a + b)),$$

$$\sin a \cos b = \tfrac{1}{2}(\sin(a - b) + \sin(a + b)).)$$

b. Show that

$$\left\{ \frac{1}{\sqrt{2\pi}}, \ \frac{1}{\sqrt{\pi}}\cos x, \ \frac{1}{\sqrt{\pi}}\sin x, \ \frac{1}{\sqrt{\pi}}\cos(2x), \ \frac{1}{\sqrt{\pi}}\sin(2x), \ \ldots \right\}$$

is an orthonormal subset of $\mathbb{C}([-\pi, \pi])$.

3.   **a.** Verify that

$$\left\{ \left(\tfrac{1}{\sqrt{7}}, \tfrac{2}{\sqrt{7}}, -\tfrac{1}{\sqrt{7}}, \tfrac{1}{\sqrt{7}}\right), \left(\tfrac{1}{\sqrt{3}}, -\tfrac{1}{\sqrt{3}}, -\tfrac{1}{\sqrt{3}}, 0\right), \left(\tfrac{2}{\sqrt{23}}, -\tfrac{1}{\sqrt{23}}, \tfrac{3}{\sqrt{23}}, \tfrac{3}{\sqrt{23}}\right) \right\}$$

is an orthonormal basis for a subspace S of $\mathbb{R}^4$.

b. Find the projection of $(1, 0, 0, 1)$ onto S.

c. Write $(1, 0, 0, 1)$ as the sum of a vector in S and a vector orthogonal to every element in S.

4.   **a.** Find an orthonormal basis for the subspace of $\mathbb{R}^3$ spanned by

$$\{(2, 1, -2), (1, -1, 1)\}.$$

b. Find the projection of $(1, 2, 3)$ onto this subspace.

5.   Let S be the subspace of $\mathbb{C}([-1, 1])$ spanned by the functions defined by $f(x) = x$ and $g(x) = x^2$. (Warning: $\{f, g\}$ is not an orthonormal basis for S.)

a. Find the projection of the function defined by $h(x) = x^3$ onto S.

b. Write h as the sum of a function in S and a function orthogonal to S.

c. How close is h to S?

6.   Find the first three nonzero terms of the Fourier series for the function $f : [-\pi, \pi] \to \mathbb{R}$ defined by

$$f(x) = \begin{cases} -\pi & \text{if } x \in [-\pi, 0); \\ \pi & \text{if } x \in [0, \pi]. \end{cases}$$

7.   Find the first three nonzero terms of the Fourier series for the function $f \in \mathbb{C}([-\pi, \pi])$ defined by $f(x) = x^2$.

8.   Suppose $\{e_1, \ldots, e_n\}$ is an orthonormal subset of an inner product space V. For any $v \in V$, prove Bessel's inequality:

$$\sum_{i=1}^{n} \langle v, e_i \rangle^2 \leq \|v\|^2.$$

(Suggestion: Show that the summation on the left side of the inequality is equal to $\|\text{proj}_S(v)\|^2$, where $S = \text{span}\{e_1, \ldots, e_n\}$.)

9.   **a.** Identify the graphs of f_1, f_2, and f_3 in Figure 4.4.

b. Compute the distances $\|f_1 - f\|$, $\|f_2 - f\|$, and $\|f_3 - f\|$ to show that the sequence f_1, f_2, f_3 is progressively closer to f.

Project: Continuity

The goals of this project are to introduce the concept of continuity for functions between vector spaces and to examine the continuity of some familiar functions.

Begin by recalling the ingredients used to define continuity for a function $f : \mathbb{R} \to \mathbb{R}$. Continuity of f at a point $x_0 \in \mathbb{R}$ essentially means that we can get $f(x)$ close to $f(x_0)$. The expression $|f(x) - f(x_0)|$ for the distance between $f(x)$ and $f(x_0)$ quantifies this notion of closeness. Letting the positive real number ϵ designate the degree of closeness, we require that $|f(x) - f(x_0)| < \epsilon$. Of course, we do not require $|f(x) - f(x_0)| < \epsilon$ for all real numbers x, but only for those that are close to x_0. Again, we can express the degree of closeness by the distance $|x - x_0|$ between x and x_0. The restriction on x can thus be expressed by the requirement that $|x - x_0| < \delta$ for some positive real number δ.

Putting this together, we say that $f : \mathbb{R} \to \mathbb{R}$ is **continuous at a point** $x_0 \in \mathbb{R}$ if and only if for any $\epsilon > 0$ there is $\delta > 0$ such that if $|x - x_0| < \delta$, then $|f(x) - f(x_0)| < \epsilon$. We can state this fairly elaborate definition more concisely by using the standard symbols $\forall$ for the universal quantifier, $\exists$ for the existential quantifier, and an abbreviation for *such that*. Thus, $f : \mathbb{R} \to \mathbb{R}$ is continuous at x_0 if and only if

$$\forall \epsilon > 0 \; \exists \delta > 0 \text{ s.t. } |x - x_0| < \delta \implies |f(x) - f(x_0)| < \epsilon.$$

Finally, recall that $f : \mathbb{R} \to \mathbb{R}$ is **continuous** if and only if it is continuous at every point $x_0 \in \mathbb{R}$.

1. Now we are ready to make the transition to a function $f : V \to W$ between the vector spaces V and W. The basic ingredients are a measure of how close a vector $\mathbf{v}$ is to another vector $\mathbf{v}_0$ in V and a measure of how close the corresponding vectors $f(\mathbf{v})$ and $f(\mathbf{v}_0)$ are in W. Why not require that V and W have norms defined on them? That is, assume there are functions $\| \; \|_V : V \to \mathbb{R}$ and $\| \; \|_W : W \to \mathbb{R}$ that satisfy the three properties listed in Exercise 24 of Section 4.1. The expressions $\|\mathbf{v} - \mathbf{v}_0\|$ and $\|f(\mathbf{v}) - f(\mathbf{v}_0)\|$ are the obvious candidates for the distances. Everything is now set up for you to state a definition of continuity of $f : V \to W$ at a point $\mathbf{v}_0 \in V$. Then define continuity of f in terms of your definition of pointwise continuity.

Suppose V and W are normed vector spaces. Show that the following functions are continuous:

For a fixed $\mathbf{w}_0 \in W$, the constant function $C_{\mathbf{w}_0} : V \to W$ defined by $C_{\mathbf{w}_0}(\mathbf{v}) = \mathbf{w}_0$;

The identity function $\mathrm{id}_V : V \to V$ defined by $\mathrm{id}_V(\mathbf{v}) = \mathbf{v}$;

For a fixed $\mathbf{v}_0 \in V$ the translation function $\tau_{\mathbf{v}_0} : V \to V$ defined by $\tau_{\mathbf{v}_0}(\mathbf{v}) = \mathbf{v} + \mathbf{v}_0$;

For a fixed $r \in \mathbb{R}$ the scalar multiplication function $\sigma_r : V \to V$ defined by $\sigma_r(\mathbf{v}) = r\mathbf{v}$.

2. In order to consider continuity of a function of two variables, we first need a way to combine two vector spaces into one. For vector spaces V and W, let $V \oplus W$ denote the set of all ordered pairs of an element in V followed by an element of W. That is,

$$V \oplus W = \{(\mathbf{v}, \mathbf{w}) \mid \mathbf{v} \in V \text{ and } \mathbf{w} \in W\}.$$

Define addition and scalar multiplication on $V \oplus W$ by

$$(\mathbf{v}_1, \mathbf{w}_1) + (\mathbf{v}_2, \mathbf{w}_2) = (\mathbf{v}_1 + \mathbf{v}_2, \mathbf{w}_1 + \mathbf{w}_2),$$
$$r(\mathbf{v}_1, \mathbf{w}_1) = (r\mathbf{v}_1, r\mathbf{w}_1)$$

for any $(\mathbf{v}_1, \mathbf{w}_1), (\mathbf{v}_2, \mathbf{w}_2) \in V \oplus W$ and any $r \in \mathbb{R}$. This new vector space is known as the **direct sum** of V and W. As a set, it is probably familiar to you as the Cartesian product of V and W.

Let $\| \ \|_V$ be a norm on V and let $\| \ \|_W$ be a norm on W. Consider the function $\| \ \| : V \oplus W \to \mathbb{R}$ defined by $\|(\mathbf{v}, \mathbf{w})\| = \|\mathbf{v}\|_V + \|\mathbf{w}\|_W$ for all $(\mathbf{v}, \mathbf{w}) \in V \oplus W$. Show that this is a norm on $V \oplus W$ in the sense that it satisfies the three properties of a norm given in Exercise 24 of Section 4.1.

How does $\mathbb{R}^1 \oplus \mathbb{R}^1$ compare with $\mathbb{R}^2$? How do the norms compare? If V and W are finite-dimensional vector spaces, what is the dimension of $V \oplus W$? The preceding definitions generalize easily to the direct sum of any finite number of vector spaces. You may be interested in carrying out the details. Can you generalize to the direct sum of an infinite collection of vector spaces?

3. The pieces are now in place for us to consider the operation of addition in a vector space V as a function $A : V \oplus V \to V$ where $A(\mathbf{v}, \mathbf{w}) = \mathbf{v} + \mathbf{w}$ for any $\mathbf{v}, \mathbf{w} \in V$. Suppose V has a norm. Use the resulting norm on $V \oplus V$ to show that A is continuous. Similarly, consider scalar multiplication to be a function from $\mathbb{R} \oplus V$ to V. Show that this function is continuous. A **topological vector space** is a vector space with a concept of closeness such that addition and scalar multiplication are continuous operations. These results show that a normed vector space is a topological vector space.

We have encountered another function of two variables, the inner product. Suppose V is an inner product space. With the norm on V defined in terms of the inner product, and with the resulting norm on $V \oplus V$, show that the inner product is a continuous function from $V \oplus V$ to $\mathbb{R}$.

4. Once you have covered the concept of linear function in Section 6.1, you can consider the interaction between linearity and continuity. Show, for example, that every linear function $T : \mathbb{R}^n \to \mathbb{R}^m$ is continuous. Is integration a continuous function from $\mathbb{C}([a, b])$ to $\mathbb{R}$? Is differentiation a continuous function from $\mathbb{D}^{(2)}([a, b])$ to $\mathbb{D}^{(1)}([a, b])$? The interplay between linearity and continuity leads directly into functional analysis, one of the richest branches of mathematics.

Project: Orthogonal Polynomials

In this chapter we have seen the advantages of working with bases that are orthonormal. In particular, Fourier analysis gives linear combinations of simple trigonometric functions that best approximate a function in $\mathbb{C}([-\pi, \pi])$. Polynomials are another set of functions that are especially simple to work with. A set of orthogonal polynomials often gives additional insight into differential equations, approximations, and numerical integration. You may want to investigate one of the four families of orthogonal polynomials mentioned in this project. Each of these topics provides an opportunity for you to try your hand at developing a mathematical theory, illustrating the theory with numbers and formulas, and exploring an application of this material to your favorite cognate subject. You are encouraged to explore the mathematical resources in your library. Your professor may be able to provide additional reference material. Some places to start are:

William Boyce and Richard DiPrima, *Elementary Differential Equations,* New York: John Wiley & Sons, 1992.

David Kincaid and Ward Cheney, *Numerical Analysis,* Pacific Grove, CA: Brooks/Cole, 1991.

Shepley Ross, *Introduction to Ordinary Differential Equations,* New York: John Wiley & Sons, 1989.

1. Legendre polynomials can be defined recursively by

$$P_0(x) = 1, \qquad P_1(x) = x, \qquad P_{n+1}(x) = \frac{(2n+1)xP_n(x) - nP_{n-1}(x)}{n+1} \qquad \text{for } n \geq 1.$$

The same polynomials are obtained from the formula of Rodrigues:

$$P_n(x) = \frac{1}{2^n n!} \frac{d^n}{dx^n} (x^2 - 1)^n.$$

These polynomials are solutions to the Legendre differential equation

$$(1 - x^2)y'' - 2xy' + n(n+1)y = 0.$$

The Legendre polynomials are orthogonal functions in $C([-1, 1])$ with the standard inner product. The coefficients

$$c_k = \frac{1}{\|P_k\|} \left\langle f, \frac{P_k}{\|P_k\|} \right\rangle = \frac{2k+1}{2} \int_{-1}^{1} f(x)P_k(x)\,dx$$

will give the polynomial $p = \sum_{k=0}^{n} c_k P_k$ in $\mathbb{P}_n$ that best approximates the continuous function f.

2. Laguerre polynomials can be defined recursively by

$$L_0(x) = 1, \qquad L_1(x) = -x + 1,$$

$$L_{n+1}(x) = (2n + 1 - x)L_n(x) - n^2 L_{n-1}(x) \qquad \text{for} \quad n \geq 1.$$

Rodrigues' formula is

$$L_n(x) = e^x \frac{d^n}{dx^n} (x^n e^{-x}).$$

These polynomials are solutions to the Laguerre differential equation

$$xy'' + (1 - x)y' + ny = 0.$$

Multiply both sides of this equation by e^{-x} and rewrite it in the form $(xe^{-x}y')' + ne^{-x}y = 0$ to see how the exponential function enters the picture.

Consider the subspace of $C([0, \infty))$ of functions f for which the improper integral $\int_0^\infty (f(x))^2 e^{-x}\,dx$ converges. The Laguerre polynomials are orthogonal functions in this subspace with respect to the inner product defined by $\langle f, g \rangle = \int_0^\infty f(x)g(x)e^{-x}\,dx$. Thus, Laguerre polynomials give best approximations to functions defined on $[0, \infty)$ with the measure of closeness weighted by the negative exponential factor.

3. Hermite polynomials can be defined recursively by

$$H_0(x) = 1, \qquad H_1(x) = 2x, \qquad H_{n+1}(x) = 2xH_n(x) - 2nH_{n-1}(x) \qquad \text{for } n \geq 1.$$

Rodrigues' formula is

$$H_n(x) = (-1)^n e^{x^2} \frac{d^n}{dx^n} (e^{-x^2}).$$

These polynomials satisfy the Hermite differential equation

$$y'' - 2xy' + 2ny = 0.$$

Consider the subspace of $\mathbb{C}(\mathbb{R})$ of functions f for which the improper integral $\int_{-\infty}^{\infty} \left(f(x)\right)^2 e^{-x^2} dx$ converges. The Hermite polynomials are orthogonal functions in this subspace with respect to the inner product defined by $\langle f, g \rangle = \int_{-\infty}^{\infty} f(x)g(x)e^{-x^2} dx$.

4. Chebyshev polynomials can be defined recursively by

$$T_0(x) = 1, \qquad T_1(x) = x, \qquad T_{n+1}(x) = 2xT_n(x) - T_{n-1}(x) \quad \text{for } n \geq 1.$$

For $x \in [-1, 1]$ they can also be defined by the formula

$$T_n(x) = \cos(n \arccos x).$$

These polynomials satisfy the Chebyshev differential equation

$$(1 - x^2)y'' - xy' + n^2 y = 0.$$

Consider the subspace of $\mathbb{C}((-1, 1))$ of functions for which the improper integral $\int_{-1}^{1} \frac{\left(f(x)\right)^2}{\sqrt{1 - x^2}} dx$ converges. The Chebyshev polynomials are orthogonal in this subspace with respect to the inner product defined by $\langle f, g \rangle = \int_{-1}^{1} \frac{f(x)g(x)}{\sqrt{1 - x^2}} dx$. Chebyshev polynomials are important in numerical analysis in trying to find a polynomial that agrees with a given function at certain points and has minimum deviation from the function at any point in an interval. If the points are chosen to be zeros of T_n, one of the factors in the error formula for the interpolation will be minimized.

Chapter 4 Summary

Chapter 4 enriches the operations available on certain vector spaces with a new kind of multiplication. The inner product allows us to quantify geometric concepts of length, distance, angle, and projection. Other vector spaces with inner products become visible through the miracle of geometry.

Computations

Evaluation of dot product on $\mathbb{R}^n$

Evaluation of the standard inner product on $\mathbb{C}([a, b])$

Length, distance, angle

Orthogonal projection
Projection onto a nonzero vector
Projection onto a subspace spanned by an orthonormal set

Decomposition of a vector into orthogonal components
Projection onto a subspace as the best approximation to a vector

Normalizing a vector

Gram-Schmidt orthonormalization process

Theory

Axioms for an inner product
Basic properties derived from the axioms
Verification that given examples are inner products
Dot product on $\mathbb{R}^n$
Standard inner product on $\mathbb{C}([a, b])$
Nonstandard examples

Norm defined in terms of an inner product

Cauchy-Schwarz inequality
Used to define angle in an inner product space
Used to prove the triangle inequality

Orthogonality
Definition and geometric interpretation
Linear independence of a set of nonzero orthogonal vectors

Orthonormal basis and coefficients $\langle v, e_i \rangle$
Components of coordinate vectors
Coefficients in the Gram-Schmidt formulas
Coefficients in the orthogonal projection
Coefficients in a best approximation

Applications

Use of the Cauchy-Schwarz inequality to prove other inequalities

Angles and distance in Euclidean spaces

Planes in $\mathbb{R}^3$ represented by a normal vector

Fourier analysis for approximating a function

Review Exercises

1. **a.** Verify that $\langle (v_1, v_2, v_3), (w_1, w_2, w_3) \rangle = 4v_1w_1 + 2v_2w_2 + v_3w_3$ defines an inner product on $\mathbb{R}^3$.

    b. Find the angle between $(1, -2, 3)$ and $(5, 1, 1)$ with respect to this inner product.

    c. Find the angle between $(1, -2, 3)$ and $(5, 1, 1)$ with respect to the standard inner product on $\mathbb{R}^3$.

2. a. Verify that $\langle p, q \rangle = p(0)q(0) + p'(0)q'(0) + p''(0)q''(0)$ defines an inner product on $\mathbb{P}_2$.

  b. Consider the polynomial p defined by $p(x) = x^2 - 3x + 2$. Compute the norm of p with respect to this inner product.

  c. Compute the norm of p with respect to the standard inner product on $\mathbb{C}([-1, 1])$.

  d. Compute the norm of p with respect to the standard inner product on $\mathbb{C}([0, 1])$.

  e. Compute the norm of p with respect to the inner product on $\mathbb{P}_2$ defined by $\langle p, q \rangle = p(-1)q(-1) + p(0)q(0) + p(1)q(1)$.

3. Suppose a regular tetrahedron (a geometrical figure whose four faces are equilateral triangles) is sitting on a flat table.

  a. What is the angle of elevation of an edge between one of the vertices on the table to the vertex at the top of the tetrahedron?

  b. What is the angle of elevation along one of the faces from the center of an edge on the table to the vertex at the top of the tetrahedron?

4. Imagine the unit hypercube sitting in $\mathbb{R}^4$ with one corner at the origin and with edges along the positive axes. Find the angle formed by the main diagonal from the origin to $(1, 1, 1, 1)$ and a face diagonal from the origin to $(1, 1, 1, 0)$.

5. Suppose a, b, c, and d are nonzero real numbers. Use the Cauchy-Schwarz inequality to prove that

$$(a^2 + b^2 + c^2 + d^2)\left(\frac{1}{a^2} + \frac{1}{b^2} + \frac{1}{c^2} + \frac{1}{d^2}\right) \geq 16.$$

6. a. Find the orthogonal projection of $(2, 0, -1, 3)$ onto $(1, 1, 0, 1)$.

  b. Write $(2, 0, -1, 3)$ as the sum of a vector parallel to $(1, 1, 0, 1)$ and another vector orthogonal to $(1, 1, 0, 1)$.

7. Consider the functions f and g in $\mathbb{C}([-1, 1])$ defined by $f(x) = x^2$ and $g(x) = 3x - 1$.

  a. Find the orthogonal projection of f onto g.

  b. Write f as the sum of a function parallel to g and another function orthogonal to g.

  c. Find the orthogonal projection of g onto f.

  d. Write g as the sum of a function parallel to f and another function orthogonal to f.

8. Let $f : [0, 1] \rightarrow \mathbb{R}$ be a continuous function. Apply the Cauchy-Schwarz inequality to f and the constant function with value 1 to derive the inequality

$$\left(\int_0^1 f(x)\, dx\right)^2 \leq \int_0^1 (f(x))^2\, dx.$$

9. Suppose $\{e_1, e_2, e_3\}$ is an orthonormal subset of an arbitrary inner product space V. (Do not assume this is the standard basis for $\mathbb{R}^3$ or even that V is a Euclidean space.) Let

$$\mathbf{v} = 4\mathbf{e}_1 + 3\mathbf{e}_2 + \mathbf{e}_3 \quad \text{and} \quad \mathbf{w} = \mathbf{e}_1 - 2\mathbf{e}_2.$$

   a. Compute the inner product $\langle \mathbf{v}, \mathbf{e}_2 \rangle$.

   b. Compute the inner product $\langle \mathbf{v}, \mathbf{w} \rangle$.

10. **a.** Verify that $\{1, \sqrt{12}(x - \tfrac{1}{2})\}$ is an orthonormal set in the space $\mathbb{C}([0, 1])$ with the standard inner product.

   b. Let $S = \mathrm{span}\{1, \sqrt{12}(x - \tfrac{1}{2})\}$. Determine the projection of x^2 onto the subspace S.

11. **a.** Verify that $\{(\tfrac{4}{5}, \tfrac{2}{5}, \tfrac{2}{5}, \tfrac{1}{5}), (-\tfrac{1}{5}, \tfrac{2}{5}, -\tfrac{2}{5}, \tfrac{4}{5})\}$ is an orthonormal subset of $\mathbb{R}^4$.

   b. Let $S = \mathrm{span}\{(\tfrac{4}{5}, \tfrac{2}{5}, \tfrac{2}{5}, \tfrac{1}{5}), (-\tfrac{1}{5}, \tfrac{2}{5}, -\tfrac{2}{5}, \tfrac{4}{5})\}$ Compute the projection of $(5, 10, 0, -5)$ onto S.

   c. What vector in S is closest to $(5, 10, 0, -5)$?

12. Let S be the subspace of $\mathbb{R}^4$ spanned by the set

$$B = \left\{ \left(\tfrac{1}{\sqrt{2}}, 0, \tfrac{1}{\sqrt{2}}, 0\right), \left(\tfrac{-1}{\sqrt{6}}, \tfrac{2}{\sqrt{6}}, \tfrac{1}{\sqrt{6}}, 0\right) \right\}$$

   a. Verify that B is an orthonormal subset of $\mathbb{R}^4$.

   b. What is the dimension of S?

   c. Find the orthogonal projection of $(0, 1, 2, 3)$ onto S.

   d. What is the distance from $(0, 1, 2, 3)$ to S?

   e. Apply the Gram-Schmidt orthonormalization process to modify $\mathbf{u}_3 = (0, 1, 2, 3)$ to obtain a vector $\mathbf{e}_3$ that can be adjoined to B to yield an orthonormal set of three vectors.

13. **a.** Show that $B = \{(2, 1, 1), (-1, 1, 1), (0, 2, -2)\}$ is an orthogonal subset of $\mathbb{R}^3$.

   b. With no further computation, explain how you know that B is a basis for $\mathbb{R}^3$.

   c. Show how to take advantage of the orthogonality of this set to find the coefficient a so that $(5, 7, 2) = a(2, 1, 1) + b(-1, 1, 1) + c(0, 2, -2)$.

   d. Give the coordinate vector $[(5, 7, 2)]_B$.

14. Consider the subset $\{p_1, p_2\}$ of $\mathbb{C}([0, 1])$ where $p_1(x) = 1$ and $p_2(x) = 2\sqrt{3}x - \sqrt{3}$.

   a. Show that $\{p_1, p_2\}$ is an orthonormal subset of $\mathbb{C}([0, 1])$.

   b. Find the projection $\mathrm{proj}_S(q)$ of the polynomial defined by $q(x) = x^2$ onto the space S spanned by $\{p_1, p_2\}$.

   c. Use the result of part b to write a nonzero polynomial that is orthogonal to both p_1 and p_2.

   d. Use the result of part c to produce a polynomial p_3 such that $\{p_1, p_2, p_3\}$ is an orthonormal set.

15. Compute the constant term and the coefficients of $\sin x$, $\cos x$, $\sin 2x$, and $\cos 2x$ in the Fourier approximation of the function $f(x) = x^2 - x$ considered as an element of $\mathbb{C}([-\pi, \pi])$.

Chapter 5

Matrices

In the previous chapter we saw how an inner product on a vector space enables us to develop geometric structures on the space. In particular, we introduced length, angle, and orthogonal projection in Euclidean spaces and certain function spaces. The goal of this chapter is to introduce a multiplicative structure on spaces of matrices. This operation is somewhat curious in that it often combines matrices from different spaces to produce a matrix in yet another space. Matrix multiplication obeys most of the familiar properties of real-number algebra. However, we need to be on the alert for few properties that do not hold.

The results of matrix algebra along with intuitive ideas of probability lead to the mathematical modeling technique of Markov chains. We will be able to derive some nontrivial results about the long-term behavior of evolving systems.

5.1 Matrix Algebra

The definition of matrix multiplication is somewhat more complicated than you might expect. The ultimate justification for this definition will be revealed in Chapter 6 when we use matrices to represent certain functions between vector spaces. In the meantime, we will find numerous applications for this new operation on matrices. For example, matrix multiplication gives a concise notation for representing a system of linear equations. In general, this definition has built-in capabilities for computing linear combinations: it uses addition and scalar multiplication, the essential operations on vector spaces.

Let us begin with a simple application to motivate the definition of multiplication of matrices.

Suppose a delivery service is studying the different possible routes for shipping packages. The following tables give the number of routes they can use from the small towns of Albion, Homer, and Parma to the regional distribution centers, Jackson and Battle Creek, and from these regional distribution centers to the national distribution centers, Chicago, Detroit, Indianapolis, and Cleveland.

	To Jackson	To Battle Creek
From Albion	3	2
From Homer	2	0
From Parma	1	1

	To Chicago	To Detroit	To Indianapolis	To Cleveland
From Jackson	2	1	1	3
From Battle Creek	4	2	1	1

Shipments from Albion to Chicago can be sent to Jackson (3 ways) and forwarded to Chicago (2 ways) in $3 \cdot 2 = 6$ different ways, or they can be sent to Battle Creek (2 ways) and on to Chicago (4 ways) in $2 \cdot 4 = 8$ different ways. Thus, the number of shipping routes from Albion to Chicago is

$$3 \cdot 2 + 2 \cdot 4 = 14.$$

The number of routes from any of the small towns to any of the national distribution centers can be computed as a similar sum of products. You can easily verify the entries in the following table.

	To Chicago	To Detroit	To Indianapolis	To Cleveland
From Albion	14	7	5	11
From Homer	4	2	2	6
From Parma	6	3	2	4

Each of the rows of the matrix $\begin{bmatrix} 3 & 2 \\ 2 & 0 \\ 1 & 1 \end{bmatrix}$ from the first table has been combined with each of the columns of the matrix $\begin{bmatrix} 2 & 1 & 1 & 3 \\ 4 & 2 & 1 & 1 \end{bmatrix}$ from the second table in a dot-product type of multiplication to yield one entry of the resultant matrix

$$\begin{bmatrix} 14 & 7 & 5 & 11 \\ 4 & 2 & 2 & 6 \\ 6 & 3 & 2 & 4 \end{bmatrix}.$$

This example leads to the official definition of this new operation for matrices.

Definition 5.1

The **matrix product** of an $m \times n$ matrix $A = [a_{ij}]$ with an $n \times p$ matrix $B = [b_{jk}]$ is the $m \times p$ matrix, denoted AB, whose ik-entry is the dot product

$$\sum_{j=1}^{n} a_{ij}b_{jk}$$

of the ith row of A with the kth column of B.

The pattern of adding the products of corresponding entries across the rows of the first factor and down the columns of the second factor soon becomes automatic. Pay particular attention to the sizes of the matrices being multiplied. The number of columns of the first factor must equal the number of rows of the second factor. The product will then be a matrix with the same number of rows as the first factor and the same number of columns as the second factor.

Quick Example   *Determine the sizes of the products*

$$\begin{bmatrix} 3 & -1 & 1 \\ 0 & 2 & 4 \\ 5 & 1 & -1 \\ -2 & 0 & 0 \end{bmatrix} \begin{bmatrix} 2 & 1 \\ 0 & 1 \\ 1 & 1 \end{bmatrix} \quad \text{and} \quad \begin{bmatrix} 2 & 0 & 1 \\ 1 & 1 & 1 \end{bmatrix} \begin{bmatrix} 3 & -1 \\ 0 & 2 \end{bmatrix}.$$

Compute the products.

In the first example the left factor is a 4×3 matrix, and the right factor is a 3×2 matrix. Thus, the product is a 4×2 matrix. It is computed as follows:

$$\begin{bmatrix} 3 & -1 & 1 \\ 0 & 2 & 4 \\ 5 & 1 & -1 \\ -2 & 0 & 0 \end{bmatrix} \begin{bmatrix} 2 & 1 \\ 0 & 1 \\ 1 & 1 \end{bmatrix} = \begin{bmatrix} 3 \cdot 2 + (-1) \cdot 0 + 1 \cdot 1 & 3 \cdot 1 + (-1) \cdot 1 + 1 \cdot 1 \\ 0 \cdot 2 + 2 \cdot 0 + 4 \cdot 1 & 0 \cdot 1 + 2 \cdot 1 + 4 \cdot 1 \\ 5 \cdot 2 + 1 \cdot 0 + (-1) \cdot 1 & 5 \cdot 1 + 1 \cdot 1 + (-1) \cdot 1 \\ -2 \cdot 2 + 0 \cdot 0 + 0 \cdot 1 & -2 \cdot 1 + 0 \cdot 1 + 0 \cdot 1 \end{bmatrix}$$

$$= \begin{bmatrix} 7 & 3 \\ 4 & 6 \\ 9 & 5 \\ -4 & -2 \end{bmatrix}.$$

In the second example, the left factor has three columns but the second factor has only two rows. Thus, this product is undefined.   ■

Notice that any row of a matrix A multiplied by a matrix B gives the corresponding row of the product AB. Similarly, A times any column of B gives the corresponding column of AB.

Quick Example   *Compute the second row of the following product. Compute the third column of this product.*

$$\begin{bmatrix} 7 & 1 & 2 \\ 3 & -2 & 1 \\ -1 & 2 & 0 \end{bmatrix} \begin{bmatrix} 4 & 2 & -1 \\ 2 & 5 & 2 \\ 3 & 1 & 3 \end{bmatrix}$$

The second row of the product involves only the second row of the first factor:

$$[3 \ -2 \quad 1] \begin{bmatrix} 4 & 2 & -1 \\ 2 & 5 & 2 \\ 3 & 1 & 3 \end{bmatrix}$$

$$= [3 \cdot 4 + (-2)2 + 1 \cdot 3 \quad 3 \cdot 2 + (-2)5 + 1 \cdot 1 \quad 3(-1) + (-2)2 + 1 \cdot 3]$$

$$= [11 \ -3 \ -4]$$

The third column of the product involves only the third column of the second factor:

$$\begin{bmatrix} 7 & 1 & 2 \\ 3 & -2 & 1 \\ -1 & 2 & 0 \end{bmatrix} \begin{bmatrix} -1 \\ 2 \\ 3 \end{bmatrix} = \begin{bmatrix} 7 \cdot (-1) + 1 \cdot 2 + 2 \cdot 3 \\ 3 \cdot (-1) + (-2) \cdot 2 + 1 \cdot 3 \\ -1 \cdot (-1) + 2 \cdot 2 + 0 \cdot 3 \end{bmatrix} = \begin{bmatrix} 1 \\ -4 \\ 5 \end{bmatrix}. \quad \blacksquare$$

Crossroads The definition of matrix multiplication in terms of sums of products of entries of the factors opens a door to our work with linear equations. In fact, a linear system such as

$$2x + 3y - z = 8$$
$$5x \qquad + 2z = 4$$

can be reformulated as a single matrix equation

$$\begin{bmatrix} 2 & 3 & -1 \\ 5 & 0 & 2 \end{bmatrix} \begin{bmatrix} x \\ y \\ z \end{bmatrix} = \begin{bmatrix} 8 \\ 4 \end{bmatrix}.$$

As we will quickly discover, this door swings both ways. The properties of matrix multiplication will yield results about linear systems, and our knowledge of linear systems will be crucial in establishing some of the basic facts about matrix multiplication.

The spaces $\mathbb{M}(n, n)$ are especially compatible with matrix multiplication since such a space is closed under this operation. This calls for a definition.

Definition 5.2

A **square** matrix is a matrix in which the number of rows equals the number of columns.

The operation of matrix multiplication satisfies several pleasant properties listed in the upcoming theorem. One of the properties refers to the special square matrices defined as follows.

Definition 5.3

For any positive integer n, the $n \times n$ **identity matrix** is the matrix

$$I_n = \begin{bmatrix} 1 & 0 & \cdots & 0 \\ 0 & 1 & \cdots & 0 \\ \vdots & \vdots & & \vdots \\ 0 & 0 & \cdots & 1 \end{bmatrix},$$

whose entries are 1s along the diagonal running from upper left to lower right and 0s elsewhere. If the size of the matrix is understood, such a matrix is denoted simply as I.

By part e of the following theorem, we see that these matrices are the multiplicative identities for matrix multiplication in the same sense as the real number 1 is the multiplicative identity for real-number multiplication.

Theorem 5.4

Suppose $A, A' \in \mathbb{M}(m, n)$, $B, B' \in \mathbb{M}(n, p)$, $C \in \mathbb{M}(p, q)$, and $r \in \mathbb{R}$. Then

   a. $(AB)C = A(BC)$ (associative law of matrix multiplication)

   b. $(A + A')B = AB + A'B$ (right distributive law of matrix multiplication)

   c. $A(B + B') = AB + AB'$ (left distributive law of matrix multiplication)

   d. $(rA)B = r(AB) = A(rB)$ (associative law of scalar and matrix multiplication)

   e. $AI_n = A = I_mA$ (multiplicative identity property)

Proof Let us look at the computations that prove property b. You are asked to prove the other properties as exercises. The notation involving double subscripts is perhaps the greatest stumbling block in proving these basic results about matrix multiplication.

Let $A = [a_{ij}]$, $A' = [a'_{ij}]$, and $B = [b_{jk}]$, where i runs from 1 to m, j runs from 1 to n, and k runs from 1 to p. Then

$$
\begin{aligned}
(A + A')B &= \big([a_{ij}] + [a'_{ij}]\big)[b_{jk}] & \text{\textit{substitution;}} \\[2mm]
&= [a_{ij} + a'_{ij}][b_{jk}] & \text{\textit{definition of matrix addition;}} \\[2mm]
&= \left[\sum_{j=1}^{n}(a_{ij} + a'_{ij})b_{jk}\right] & \text{\textit{definition of matrix multiplication;}} \\[2mm]
&= \left[\sum_{j=1}^{n}(a_{ij}b_{jk} + a'_{ij}b_{jk})\right] & \text{\textit{distributive law of real numbers;}}
\end{aligned}
$$

$$= \left[\sum_{j=1}^{n} a_{ij}b_{jk} + \sum_{j=1}^{n} a'_{ij}b_{jk} \right] \qquad \text{associative and commutative}$$
$$\text{properties of real numbers;}$$

$$= \left[\sum_{j=1}^{n} a_{ij}b_{jk} \right] + \left[\sum_{j=1}^{n} a'_{ij}b_{jk} \right] \qquad \text{definition of matrix addition;}$$

$$= [a_{ij}][b_{jk}] + [a'_{ij}][b_{jk}] \qquad \text{definition of matrix multiplication;}$$

$$= AB + A'B \qquad \text{substitution.} \quad \blacksquare$$

The commutative law is conspicuously absent from the list of properties in Theorem 5.4. The fact of the matter is that the matrix products AB and BA are in general *not* equal. First of all, only when the number of rows of A equals the number of columns of B and the number of rows of B equals the number of columns of A, will the products AB and BA both be defined. Furthermore, even for square matrices it is easy to give examples where the commutative law fails. For example,

$$\begin{bmatrix} 1 & 0 \\ 0 & 0 \end{bmatrix} \begin{bmatrix} 0 & 1 \\ 0 & 0 \end{bmatrix} = \begin{bmatrix} 0 & 1 \\ 0 & 0 \end{bmatrix} \quad \text{whereas} \quad \begin{bmatrix} 0 & 1 \\ 0 & 0 \end{bmatrix} \begin{bmatrix} 1 & 0 \\ 0 & 0 \end{bmatrix} = \begin{bmatrix} 0 & 0 \\ 0 & 0 \end{bmatrix}.$$

Incidentally, this example shows that it is possible for the product of two nonzero matrices to equal the zero matrix.

Situations in which two matrices give equal products in either order of multiplication will be noteworthy events. Such matrices are said to commute. For example, consider the matrices obtained by repeatedly multiplying a square matrix A by itself:

$$A^0 = I, \quad A^1 = A, \quad A^2 = AA, \quad A^3 = AAA, \quad \ldots.$$

These powers of a square matrix commute. Indeed, both the products $A^m A^n$ and $A^n A^m$ are equal to A^{m+n}.

The final theorem in this section describes how matrix multiplication interacts with elementary row operations. The ideas are simple enough. However, we need to introduce enough notation to describe the matrices involved and to specify the row operation in use. This theorem will play an important role in the proof of Theorem 7.7.

Theorem 5.5

Suppose A is an $m \times n$ matrix and B is an $n \times q$ matrix. Let A' be the result of applying an elementary row operation to A. Then $A'B$ is the result of applying the same elementary row operation to AB.

Proof   Here is what happens when we interchange two rows. In Exercise 15 at the end of this section, you can try your hand at proving the results for the other two types of row operations.

Other than the two rows that are interchanged, the rows of A are identical to the rows of A'. Hence, the corresponding rows of AB and $A'B$ will be identical. We need to check the rows i and j involved in the interchange. Suppose $i < j$. The easiest way to see what is going on is to write out the product $A'B$:

$$A'B = \begin{bmatrix} a_{11} & \cdots & a_{1n} \\ & \vdots & \\ a_{j1} & \cdots & a_{jn} \\ & \vdots & \\ a_{i1} & \cdots & a_{in} \\ & \vdots & \\ a_{m1} & \cdots & a_{mn} \end{bmatrix} \begin{bmatrix} b_{11} & \cdots & b_{1q} \\ \vdots & & \vdots \\ b_{n1} & \cdots & b_{nq} \end{bmatrix} = \begin{bmatrix} \sum_{k=1}^{n} a_{1k}b_{k1} & \cdots & \sum_{k=1}^{n} a_{1k}b_{kq} \\ \vdots & & \vdots \\ \sum_{k=1}^{n} a_{jk}b_{k1} & \cdots & \sum_{k=1}^{n} a_{jk}b_{kq} \\ \vdots & & \vdots \\ \sum_{k=1}^{n} a_{ik}b_{k1} & \cdots & \sum_{k=1}^{n} a_{ik}b_{kq} \\ \vdots & & \vdots \\ \sum_{k=1}^{n} a_{mk}b_{k1} & \cdots & \sum_{k=1}^{n} a_{mk}b_{kq} \end{bmatrix}.$$

This looks impressive, but each entry involves nothing more than going across a row of A' and down a column of B and adding up the products. We recognize this as the product AB with rows i and j interchanged. ∎

Exercises 5.1

1. Suppose four computers can communicate through the links of a network. The existing links allow the direct transmission of data only from the first computer to the third, from the second computer to the first, from the third computer to the second and fourth, and from the fourth computer to the first.

   a. Set up a 4×4 matrix whose entries are 0s and 1s to model the direct links in this network.

   b. Multiply your matrix by itself. Interpret the entries in the product in terms of communication through the network.

   c. Multiply together three factors of your matrix.

   d. Add the matrices obtained in parts a, b, and c. What conclusion can you draw about the number of direct transmissions necessary before any computer can send a message to any other computer?

2. Consider the matrices

$$A = \begin{bmatrix} 2 & 1 & -1 \\ 1 & 3 & 0 \\ 5 & -4 & 1 \end{bmatrix}, \qquad B = \begin{bmatrix} 1 & 3 & 4 \\ -2 & 1 & 0 \end{bmatrix}, \qquad C = \begin{bmatrix} 0 & 2 \\ -1 & -2 \\ 3 & 0 \end{bmatrix}.$$

   In each of the following, perform the indicated operations or explain why the operation is undefined.

   a. AC
   b. CA
   c. BC
   d. CB
   e. $BA + B$
   f. $B(A + I_3)$
   g. $(BC)^2$
   h. B^2C^2

3. a. Find a 3×3 matrix A and 3×1 matrices $\mathbf{x}$ and $\mathbf{b}$ so that the linear system

$$2x_1 - x_2 - 3x_3 = 4$$
$$x_1 + x_2 + x_3 = -2$$
$$x_1 + 2x_2 + 3x_3 = 5$$

   can be represented by the equation $A\mathbf{x} = \mathbf{b}$.

b. With your matrix A and the matrix $C = \begin{bmatrix} 1 & -3 & 2 \\ -2 & 9 & -5 \\ 1 & -5 & 3 \end{bmatrix}$, compute the products AC and CA.

c. Compute the product $C\mathbf{b}$.

d. Use the results of parts a and b to show that $\mathbf{x} = C\mathbf{b}$ is the unique solution to the linear system.

4.   a. If A is an $m \times n$ matrix, what sizes must the zero matrices be in the equation $A0 = 0$?

b. Show that with **0** matrices of the proper sizes, the equation does indeed hold.

c. Give a precise statement and proof of the matrix equation $0A = 0$.

5. Find matrices $A, B \in \mathbb{M}(2, 2)$ that have no entries equal to zero, but such that $AB = \mathbf{0}$.

6.   a. Find four matrices $A \in \mathbb{M}(2, 2)$ with $A^2 = I$.

b. Determine all matrices $A \in \mathbb{M}(2, 2)$ with $A^2 = I$.

7.   a. Let $A = \begin{bmatrix} 0 & 3 & -4 \\ 0 & 0 & 2 \\ 0 & 0 & 0 \end{bmatrix}$. Compute A^2 and A^3.

b. Formulate and prove a theorem about 3×3 matrices based on the outcome of part a.

c. State a generalization of your theorem for $n \times n$ matrices.

8. Let $A = \begin{bmatrix} 1 & 2 & 1 \\ 0 & -2 & 3 \\ 4 & 1 & 2 \end{bmatrix}$.

a. Compute A^2 and A^3.

b. Show that $A^3 - A^2 - 11A - 25I = \mathbf{0}$.

9.   a. Use the Comparison Theorem of Section 3.5 to show that for any 2×2 matrix A, some nontrivial linear combination of A^4, A^3, A^2, A, and I is equal to $\mathbf{0}$.

b. State and prove a generalization of this result for $n \times n$ matrices.

c. Compare the result of part a with Review Exercise 4 at the end of this chapter.

d. Look up the Cayley-Hamilton Theorem in an advanced book on linear algebra.

10. Prove Theorem 5.4, part a.

11. Prove Theorem 5.4, part c.

12. Prove Theorem 5.4, part d.

13. Prove Theorem 5.4, part e.

14.   a. Suppose A is a 4×3 matrix. Find a 3×1 matrix $\mathbf{b}$ such that $A\mathbf{b}$ is the first column of A.

b. Suppose A is an $m \times n$ matrix. Find an $n \times 1$ matrix $\mathbf{b}$ such that $A\mathbf{b}$ is the jth column of A.

c. How can the ith row of a matrix be obtained in terms of matrix multiplication?

15. Suppose A is an $m \times n$ matrix and B is an $n \times q$ matrix.

a. Let A' be the result of multiplying row i of A by a constant c. Prove that $A'B$ is the result of multiplying row i of AB by c.

b. Let A' be the result of adding a constant c times row i to row j of A. Prove that $A'B$ is the result of adding c times row i of AB to row j of AB.

16. Prove that the solution set of a homogeneous system of m linear equations in n unknowns is a subspace of $\mathbb{R}^n$. (Suggestion: Let $A \in \mathbb{M}(m, n)$ be the

coefficient matrix, and view $\mathbf{x} = \begin{bmatrix} x_1 \\ \vdots \\ x_n \end{bmatrix} \in \mathbb{R}^n$ as an element of $\mathbb{M}(n, 1)$. Work

with the matrix equation $A\mathbf{x} = \mathbf{0}$.)

17. Suppose $A \in \mathbb{M}(m, n)$ and $\mathbf{b} \in \mathbb{M}(m, 1)$. Use properties of matrix multiplication and the identification of $\mathbb{R}^n$ with $\mathbb{M}(n, 1)$ to prove the following facts about systems of linear equations.

a. If $\mathbf{v}_0 \in \mathbb{R}^n$ is any solution to the linear system $A\mathbf{x} = \mathbf{b}$ and if $\mathbf{v} \in \mathbb{R}^n$ is any solution to the homogeneous system $A\mathbf{x} = \mathbf{0}$, then $\mathbf{v}_0 + \mathbf{v}$ is a solution to the system $A\mathbf{x} = \mathbf{b}$.

b. If $\mathbf{v}_0 \in \mathbb{R}^n$ and $\mathbf{v}_1 \in \mathbb{R}^n$ are any two solutions of the linear system $A\mathbf{x} = \mathbf{b}$, then the difference $\mathbf{v}_0 - \mathbf{v}_1$ is a solution to the homogeneous system $A\mathbf{x} = \mathbf{0}$.

c. If $\mathbf{v}_0 \in \mathbb{R}^n$ is any solution to the linear system $A\mathbf{x} = \mathbf{b}$ and S is the solution space of the homogeneous system $A\mathbf{x} = \mathbf{0}$, then the set that is naturally enough denoted $\mathbf{v}_0 + S$ and defined by

$$\mathbf{v}_0 + S = \{\mathbf{v}_0 + \mathbf{v} \mid \mathbf{v} \in S\}$$

is the solution space of the original system.

5.2 Inverses

In the previous section we noted that a linear system such as

$$\begin{aligned} x_1 + x_2 + x_3 &= 5 \\ -x_1 + 3x_2 + 2x_3 &= -2 \\ 2x_1 + x_2 + x_3 &= 1 \end{aligned}$$

can be written in terms of a single matrix equation

$$A\mathbf{x} = \mathbf{b},$$

where

$$A = \begin{bmatrix} 1 & 1 & 1 \\ -1 & 3 & 2 \\ 2 & 1 & 1 \end{bmatrix}, \qquad \mathbf{x} = \begin{bmatrix} x_1 \\ x_2 \\ x_3 \end{bmatrix}, \qquad \mathbf{b} = \begin{bmatrix} 5 \\ -2 \\ 1 \end{bmatrix}.$$

The goal of this section is to develop the notion of an inverse A^{-1} for the matrix A with respect to the operation of matrix multiplication. This will enable us to solve for $\mathbf{x}$ much as if we were dealing with an equation involving real numbers:

$$A\mathbf{x} = \mathbf{b}$$
$$A^{-1}(A\mathbf{x}) = A^{-1}\mathbf{b}$$
$$(A^{-1}A)\mathbf{x} = A^{-1}\mathbf{b}$$
$$I\mathbf{x} = A^{-1}\mathbf{b}$$
$$\mathbf{x} = A^{-1}\mathbf{b}$$

Definition 5.6

Let $A \in \mathbb{M}(m, n)$. The matrix $B \in \mathbb{M}(n, m)$ is a multiplicative **inverse** of A if and only if $AB = I_m$ and $BA = I_n$. If the first equation holds, we say that B is a **right inverse** of A. If the second holds, we say that B is a **left inverse** of A. If A has an inverse, we say that A is **invertible** or **nonsingular**.

Quick Example   *Verify that the matrix $B = \begin{bmatrix} 3 & -5 \\ -1 & 2 \end{bmatrix}$ is an inverse of the matrix $A = \begin{bmatrix} 2 & 5 \\ 1 & 3 \end{bmatrix}$.*

We simply compute the products AB and BA:

$$\begin{bmatrix} 2 & 5 \\ 1 & 3 \end{bmatrix}\begin{bmatrix} 3 & -5 \\ -1 & 2 \end{bmatrix} = \begin{bmatrix} 1 & 0 \\ 0 & 1 \end{bmatrix} \quad \text{and} \quad \begin{bmatrix} 3 & -5 \\ -1 & 2 \end{bmatrix}\begin{bmatrix} 2 & 5 \\ 1 & 3 \end{bmatrix} = \begin{bmatrix} 1 & 0 \\ 0 & 1 \end{bmatrix}.$$

Since both products give the identity matrix, B is an inverse of A.   ∎

The concepts of one-sided inverses are useful for working with matrices of arbitrary sizes. However, a matrix must be a square matrix if it has a left inverse and a right inverse. Be sure to look at Exercise 9 at the end of this section to see how to prove this important fact. Exercises 10 and 11 give an alternative proof of this result.

The following theorem states that a matrix has only one possible inverse. This justifies our reference to *the* inverse of a matrix, as well as the notation A^{-1} for the multiplicative inverse of A.

Theorem 5.7

An invertible matrix has a unique inverse.

Proof   The proof of this theorem is as concise as its statement. Suppose the matrices B and C are inverses of the matrix A. Then

$$B = BI = B(AC) = (BA)C = IC = C. \quad \blacksquare$$

Here are two theorems whose proofs are straightforward verifications of the definition of multiplicative inverse.

Theorem 5.8

If a matrix A is invertible, then A^{-1} is also invertible. In this case, $(A^{-1})^{-1} = A$.

Proof   Simply observe that the defining conditions for A^{-1} to be the inverse of A are $AA^{-1} = I$ and $A^{-1}A = I$, and that the defining conditions for A to be the inverse of A^{-1} are $A^{-1}A = I$ and $AA^{-1} = I$. $\blacksquare$

Theorem 5.9

Suppose $A, B \in \mathbb{M}(n, n)$. If A and B are invertible, then so is AB. In this case, $(AB)^{-1} = B^{-1}A^{-1}$.

You are asked to prove this theorem as Exercise 6 at the end of this section. Since matrix multiplication is not commutative, be sure to notice the reversal of the factors in the formula for $(AB)^{-1}$.

The time has finally come for us to face up to the problem of finding the inverse of a given square matrix. The method is best illustrated by means of an example. To find the inverse of

$$A = \begin{bmatrix} 1 & 1 & 1 \\ -1 & 3 & 2 \\ 2 & 1 & 1 \end{bmatrix},$$

let us begin by finding a 3×3 matrix C so that $AC = I$. The matrix C will at least be a right inverse of A. Notice that the columns of the product AC are determined by multiplying A by the respective columns of C. Thus, if we let C_1, C_2, and C_3 denote the three columns of C, we see that we need to solve the three systems

$$AC_1 = \begin{bmatrix} 1 \\ 0 \\ 0 \end{bmatrix}, \qquad AC_2 = \begin{bmatrix} 0 \\ 1 \\ 0 \end{bmatrix}, \qquad \text{and} \quad AC_3 = \begin{bmatrix} 0 \\ 0 \\ 1 \end{bmatrix}.$$

We now proceed with the row reduction of the augmented matrices:

$$\begin{bmatrix} 1 & 1 & 1 & 1 \\ -1 & 3 & 2 & 0 \\ 2 & 1 & 1 & 0 \end{bmatrix} \rightarrow \begin{bmatrix} 1 & 1 & 1 & 1 \\ 0 & 4 & 3 & 1 \\ 0 & -1 & -1 & -2 \end{bmatrix} \rightarrow \begin{bmatrix} 1 & 1 & 1 & 1 \\ 0 & 1 & 1 & 2 \\ 0 & 4 & 3 & 1 \end{bmatrix}$$

$$\rightarrow \begin{bmatrix} 1 & 1 & 1 & 1 \\ 0 & 1 & 1 & 2 \\ 0 & 0 & -1 & -7 \end{bmatrix} \rightarrow \begin{bmatrix} 1 & 1 & 0 & -6 \\ 0 & 1 & 0 & -5 \\ 0 & 0 & 1 & 7 \end{bmatrix} \rightarrow \begin{bmatrix} 1 & 0 & 0 & -1 \\ 0 & 1 & 0 & -5 \\ 0 & 0 & 1 & 7 \end{bmatrix},$$

$$\begin{bmatrix} 1 & 1 & 1 & 0 \\ -1 & 3 & 2 & 1 \\ 2 & 1 & 1 & 0 \end{bmatrix} \rightarrow \begin{bmatrix} 1 & 1 & 1 & 0 \\ 0 & 4 & 3 & 1 \\ 0 & -1 & -1 & 0 \end{bmatrix} \rightarrow \begin{bmatrix} 1 & 1 & 1 & 0 \\ 0 & 1 & 1 & 0 \\ 0 & 4 & 3 & 1 \end{bmatrix}$$

$$\rightarrow \begin{bmatrix} 1 & 1 & 1 & 0 \\ 0 & 1 & 1 & 0 \\ 0 & 0 & -1 & 1 \end{bmatrix} \rightarrow \begin{bmatrix} 1 & 1 & 0 & 1 \\ 0 & 1 & 0 & 1 \\ 0 & 0 & 1 & -1 \end{bmatrix} \rightarrow \begin{bmatrix} 1 & 0 & 0 & 0 \\ 0 & 1 & 0 & 1 \\ 0 & 0 & 1 & -1 \end{bmatrix},$$

$$\begin{bmatrix} 1 & 1 & 1 & 0 \\ -1 & 3 & 2 & 0 \\ 2 & 1 & 1 & 1 \end{bmatrix} \rightarrow \begin{bmatrix} 1 & 1 & 1 & 0 \\ 0 & 4 & 3 & 0 \\ 0 & -1 & -1 & 1 \end{bmatrix} \rightarrow \begin{bmatrix} 1 & 1 & 1 & 0 \\ 0 & 1 & 1 & -1 \\ 0 & 4 & 3 & 0 \end{bmatrix}$$

$$\rightarrow \begin{bmatrix} 1 & 1 & 1 & 0 \\ 0 & 1 & 1 & -1 \\ 0 & 0 & -1 & 4 \end{bmatrix} \rightarrow \begin{bmatrix} 1 & 1 & 0 & 4 \\ 0 & 1 & 0 & 3 \\ 0 & 0 & 1 & -4 \end{bmatrix} \rightarrow \begin{bmatrix} 1 & 0 & 0 & 1 \\ 0 & 1 & 0 & 3 \\ 0 & 0 & 1 & -4 \end{bmatrix}.$$

Notice that the coefficient matrix A controls the row operations that are performed. So why not combine all three systems by augmenting A by the three columns of C and reducing the single 3×6 matrix $[A \mid C]$? The reduction goes as follows:

$$\begin{bmatrix} 1 & 1 & 1 & 1 & 0 & 0 \\ -1 & 3 & 2 & 0 & 1 & 0 \\ 2 & 1 & 1 & 0 & 0 & 1 \end{bmatrix} \rightarrow \begin{bmatrix} 1 & 1 & 1 & 1 & 0 & 0 \\ 0 & 4 & 3 & 1 & 1 & 0 \\ 0 & -1 & -1 & -2 & 0 & 1 \end{bmatrix}$$

$$\rightarrow \begin{bmatrix} 1 & 1 & 1 & 1 & 0 & 0 \\ 0 & 1 & 1 & 2 & 0 & -1 \\ 0 & 4 & 3 & 1 & 1 & 0 \end{bmatrix} \rightarrow \begin{bmatrix} 1 & 1 & 1 & 1 & 0 & 0 \\ 0 & 1 & 1 & 2 & 0 & -1 \\ 0 & 0 & -1 & -7 & 1 & 4 \end{bmatrix}$$

$$\rightarrow \begin{bmatrix} 1 & 1 & 0 & -6 & 1 & 4 \\ 0 & 1 & 0 & -5 & 1 & 3 \\ 0 & 0 & 1 & 7 & -1 & -4 \end{bmatrix} \rightarrow \begin{bmatrix} 1 & 0 & 0 & -1 & 0 & 1 \\ 0 & 1 & 0 & -5 & 1 & 3 \\ 0 & 0 & 1 & 7 & -1 & -4 \end{bmatrix}$$

Thus, we find that $C = \begin{bmatrix} -1 & 0 & 1 \\ -5 & 1 & 3 \\ 7 & -1 & -4 \end{bmatrix}$ is a matrix that satisfies $AC = I$.

Since we cannot rely on matrix multiplication to be commutative, it is somewhat of a surprise to compute CA and find that we also have

$$\begin{bmatrix} -1 & 0 & 1 \\ -5 & 1 & 3 \\ 7 & -1 & -4 \end{bmatrix} \begin{bmatrix} 1 & 1 & 1 \\ -1 & 3 & 2 \\ 2 & 1 & 1 \end{bmatrix} = \begin{bmatrix} 1 & 0 & 0 \\ 0 & 1 & 0 \\ 0 & 0 & 1 \end{bmatrix}.$$

Hence, C is the inverse of A.

Mathematical Strategy Session   The remainder of this section is devoted to showing that the preceding observation was not just a lucky coincidence: a right inverse of a square matrix will always be a left inverse. This

amazing result justifies the following recipe for finding the inverse of a square
matrix A:

1. Augment A by the identity matrix of the same size.
2. Perform Gaussian elimination to put $[A \mid I]$ in reduced row-echelon
   form $[I \mid C]$.
3. Read off the matrix C as the right inverse of A, which will also be the
   left inverse of A.

Really, the only thing that can go wrong is that in the reduction process
the leading 1s may not appear on the left half of the augmented matrix. The
following discussion confirms this as a signal that the matrix A does not have
an inverse.

We begin by considering an important quantity associated with any matrix. This
number measures how close a matrix comes to being invertible.

Definition 5.10

The **rank** of a matrix A, denoted rank A, is the number of leading 1s in the
reduced row-echelon form of the matrix.

Thus, to find the rank of a matrix, we reduce the matrix to row-echelon form and count the
number of leading 1s. Since each leading 1 appears in a different row and in a different
column, the rank can be computed by counting the number of rows that have leading 1s
or by counting the number of columns that have leading 1s.

Quick Example   *Find the ranks of*

$$A = \begin{bmatrix} 1 & 3 & 1 & 1 & -2 \\ 2 & 6 & 2 & 5 & 1 \\ 0 & 0 & 0 & 3 & 2 \\ -3 & -9 & -3 & 1 & 4 \end{bmatrix},$$

the $m \times n$ zero matrix $\mathbf{0}$, but the $n \times n$ identity matrix I_n.

The reduced row-echelon form of A is

$$\begin{bmatrix} 1 & 3 & 1 & 0 & 0 \\ 0 & 0 & 0 & 1 & 0 \\ 0 & 0 & 0 & 0 & 1 \\ 0 & 0 & 0 & 0 & 0 \end{bmatrix}.$$

There are three leading 1s. Thus, rank $A = 3$.
   Any zero matrix is already in reduced row-echelon form. It has no leading 1s.
Thus, rank $\mathbf{0} = 0$.
   The $n \times n$ identity matrix is already in reduced row-echelon form. There are n
leading 1s. Thus, rank $I_n = n$.   ∎

By Theorem 2.5 the reduced row-echelon form of a matrix is unique. Hence, the rank of a matrix depends only on the matrix and not the sequence of row operations used to reduce it. The following paragraph gives an alternative proof of this result. It uses only results we have already established about the solution space of a system of linear equations.

Suppose $A \in \mathbb{M}(m, n)$ has rank r. By Exercise 16 of Section 5.1, the solution set S of the homogeneous system $A\mathbf{x} = \mathbf{0}$ is a subspace of $\mathbb{R}^n$. A basis for this subspace can be obtained by the usual process of introducing arbitrary values for the free variables as determined by the reduced form of A. We get one basis element for each column of the reduced matrix that does not contain a leading 1. Of course, rank A is equal to the number of columns that do contain leading 1s. Therefore, $\dim S + \operatorname{rank} A = n$, the number of columns of A. Hence, we can write rank $A = n - \dim S$ in terms of quantities that do not depend on the reduction process.

Our work so far in finding a right inverse of a matrix is summarized in the following theorem.

Theorem 5.11

An $n \times n$ matrix A has a right inverse C if and only if rank $A = n$. In this case, the right inverse is unique.

Proof   Suppose rank $A = n$. Then A can be reduced to the $n \times n$ identity matrix I, the only $n \times n$ matrix in reduced row-echelon form with n leading 1s. Apply the row operations that reduce A to I to reduce $[A \mid I]$ to $[I \mid C]$ for some $n \times n$ matrix C. For $j = 1, \ldots, n$, let C_j denote the jth column of C and let $\mathbf{e}_j$ denote the jth column of I. Then C_j is the unique solution to the equation $A\mathbf{x} = \mathbf{e}_j$. Hence, C is the unique matrix with $AC = I$.

Now suppose rank $A \neq n$. We want to conclude that there is no matrix C such that $AC = I$. Our hypothesis implies that the reduced form of A will contain one or more rows that are entirely zeros. Apply the row operations that transform A to reduced row-echelon form to reduce the $n \times 2n$ matrix $[A \mid I]$. This will create a row that has n zeros followed by at least one nonzero entry. Indeed, a row with n zeros on the right would mean that rank $I < n$. If a row begins with n zeros and has a nonzero entry in column $n + j$, then the system $AC_j = \mathbf{e}_j$ has no solutions. Hence, there is no matrix C such that $AC = I$.   ∎

Next we drop the condition that C is only a right inverse.

Theorem 5.12

An $n \times n$ matrix A has an inverse C if and only if rank $A = n$.

Proof   Suppose rank $A = n$. From the proof of the previous theorem, we know that we can reduce $[A \mid I]$ to $[I \mid C]$ where $AC = I$. For each row operation used in this reduction, find the row operation that is its inverse. Apply these inverse operations in reverse order to reduce $[C \mid I]$ to $[I \mid A]$. Thus, we see that A is a right inverse of C; that is, $CA = I$. We conclude from these two equations that C is the inverse of A.

Conversely, suppose A has an inverse C. Then C is certainly a right inverse of A. By the previous theorem, rank $A = n$. ■

Now the pieces are ready to fall into place. In spite of the fact that matrix multiplication is not commutative, if a square matrix A has a right inverse C, then C will also be a left inverse of A. This is certainly an amazing result.

Theorem 5.13

If A and C are $n \times n$ matrices with $AC = I$, then $CA = I$.

Proof   The equation $AC = I$ means that C is a right inverse of A. By Theorem 5.11, rank $A = n$. Thus, by Theorem 5.12, A has an inverse. This inverse is in particular a right inverse of A. By the uniqueness statement of Theorem 5.11, this inverse must be C. Hence, C is a two-sided inverse, and so we also have $CA = I$. ■

Exercises 5.2

1. Determine the ranks of the following matrices.

a.
$$\begin{bmatrix} 1 & 4 & 0 & 3 \\ 2 & 8 & 1 & 6 \\ 0 & 1 & 1 & 1 \end{bmatrix}$$

b.
$$\begin{bmatrix} 4 & 6 \\ 6 & 9 \\ 0 & 0 \\ 2 & 3 \end{bmatrix}$$

c.
$$\begin{bmatrix} 1 & 1 & 1 & 1 \\ 1 & 1 & 1 & 1 \\ 1 & 1 & 1 & 1 \\ 1 & 1 & 1 & 1 \end{bmatrix}$$

d.
$$\begin{bmatrix} 16 & 3 & 2 & 13 \\ 5 & 10 & 11 & 8 \\ 9 & 6 & 7 & 12 \\ 4 & 15 & 14 & 1 \end{bmatrix}$$

2. Determine the ranks of:

a.
$$\begin{bmatrix} 7 & 9 & -8 & 4 \\ 0 & 1 & 5 & 9 \\ 0 & 0 & 3 & 7 \\ 0 & 0 & 0 & -2 \end{bmatrix}$$

b.
$$\begin{bmatrix} 7 & 9 & -8 & 4 \\ 0 & 0 & 5 & 9 \\ 0 & 0 & 3 & 7 \\ 0 & 0 & 0 & -2 \end{bmatrix}$$

c.
$$\begin{bmatrix} 7 & 9 & -8 & 4 \\ 0 & 0 & 5 & 9 \\ 0 & 0 & 3 & 7 \\ 0 & 0 & 0 & 0 \end{bmatrix}$$

d. Determine the rank of an arbitrary 4×4 matrix of the form

$$\begin{bmatrix} a_{11} & a_{12} & a_{13} & a_{14} \\ 0 & a_{22} & a_{23} & a_{24} \\ 0 & 0 & a_{33} & a_{34} \\ 0 & 0 & 0 & a_{44} \end{bmatrix}$$

in terms of the entries a_{11}, a_{22}, a_{33}, and a_{44}.

e. Generalize the result of part d to $n \times n$ matrices in which the only nonzero entries are in the upper right triangular region of the matrix.

3. **a.** Determine the relations between the rank of the matrix

$$A = \begin{bmatrix} 2 & -6 & 4 \\ 4 & 1 & 3 \\ 3 & 4 & a \end{bmatrix}$$

and the value of the entry a.

b. If a is a randomly chosen real number, what is the probability that the matrix A will be invertible?

4. Use the algorithm developed in this section to find the inverses of the following matrices (or to conclude that the inverse does not exist).

a. $\begin{bmatrix} 3 & 4 \\ 7 & 9 \end{bmatrix}$ **b.** $\begin{bmatrix} 1 & 2 & 1 \\ 0 & -2 & 3 \\ 4 & 1 & 2 \end{bmatrix}$

c. $\begin{bmatrix} -1 & -3 & -3 \\ 1 & 2 & -1 \\ 2 & 3 & -6 \end{bmatrix}$ **d.** $\begin{bmatrix} -1 & -3 & -3 \\ 1 & 2 & -1 \\ 2 & 3 & -3 \end{bmatrix}$

e. $\begin{bmatrix} 1 & -\frac{1}{2} & 0 \\ -\frac{1}{2} & 1 & -\frac{1}{2} \\ 0 & -\frac{1}{2} & 1 \end{bmatrix}$ **f.** $\begin{bmatrix} 1 & 8 & 2 \\ -2 & 3 & -1 \\ 3 & -1 & 2 \end{bmatrix}$

5. Suppose A is a nonsingular square matrix and r is a nonzero real number. Show that rA is nonsingular and that

$$(rA)^{-1} = \frac{1}{r}A^{-1}.$$

6. Prove Theorem 5.9.

7. Formulate and prove a version of Theorem 5.9 for matrices that have right inverses. Do not assume the matrices are square.

8. **a.** Show that $\begin{bmatrix} 2 \\ -1 \end{bmatrix}$ is a right inverse of the matrix $\begin{bmatrix} 1 & 1 \end{bmatrix}$.

b. Show that $\begin{bmatrix} 2 \\ -1 \end{bmatrix}$ is not a left inverse of the matrix $\begin{bmatrix} 1 & 1 \end{bmatrix}$.

c. Show that $\begin{bmatrix} 1 & 1 \end{bmatrix}$ does not have a left inverse.

9. Suppose $A \in \mathbb{M}(m, n)$. Let $\{e_1, \ldots, e_n\}$ be the standard basis for $\mathbb{R}^n$. Prove the following results.

a. If A has a right inverse C, then $\{Ae_1, \ldots, Ae_n\}$ spans $\mathbb{R}^m$. (Suggestion: For any $v \in \mathbb{R}^m$, notice that $Cv \in \mathbb{R}^n$).

b. If A has a right inverse, then $m \le n$.

c. If A has a left inverse C', then $\{Ae_1, \ldots, Ae_n\}$ is a linearly independent subset of $\mathbb{R}^m$.

d. If A has a left inverse, then $m \ge n$.

e. If A is invertible, then A is a square matrix.

10. Suppose $A \in \mathbb{M}(m, n)$. Use the ideas of the proof of Theorem 5.11 to show there is a matrix $C \in \mathbb{M}(n, m)$ such that $AC = I$ if and only if rank $A = m$. Conclude that if A has a right inverse, then $m \le n$.

11. Suppose $A \in \mathbb{M}(m, n)$ has a left inverse C' and a right inverse C.

   a. Determine the sizes of C' and C.

   b. Apply the conclusion of the previous exercise to A and to C' to show that A must be a square matrix.

12. Show that $\begin{bmatrix} a & b \\ c & d \end{bmatrix}$ has an inverse if and only if $ad - bc \ne 0$. (Suggestion: Consider separately the cases where $a = 0$ and where $a \ne 0$.)

13. Suppose a square matrix A satisfies the equation $A^3 - A^2 - 11A - 25I = 0$.

   a. Find a formula for A^{-1} as a linear combination of powers of A.

   b. The 3×3 matrix of Exercise 8 of Section 5.1 satisfies this equation. Use your formula to compute the inverse of this matrix. Compare this result with your work in part b of Exercise 4 above.

5.3 Markov Chains

The interaction between two branches of mathematics invariably leads to the enrichment of both. In this section and the next we will investigate a topic from the theory of probability that makes nontrivial use of our work with matrix algebra. The discussion will occasionally require some basic facts about probability. You may be interested in investigating the formal definitions and logical development of this intriguing subject. The material in these two sections, however, will be based on an intuitive understanding of these results. Often the idea behind a proof will merely be outlined or illustrated by an example.

Let us begin with a demographic model of a simple migration pattern. This model could easily be enhanced to describe more sophisticated situations or adapted to other kinds of migrations: traffic flow, transfer of goods in an economy, diffusion of liquids and gases, and chemical reactions.

Suppose the population of a closed society is composed of a rural segment (initially 60%) and an urban segment (initially 40%). Suppose that each year $\frac{2}{10}$ of the people on farms move to cities, with the remaining $\frac{8}{10}$ staying on farms, and $\frac{1}{10}$ of the people in cities move to farms, with the remaining $\frac{9}{10}$ staying in cities. We would like to investigate the long-term behavior of the population under this model. It appears that there is a net migration to the cities. Does this mean that the rural population will eventually be depleted? Will the population stabilize to some equilibrium distribution where the urban migration is balanced by the rural migration?

Let us begin by considering the distribution of the population at the end of the first year. On farms there will be $\frac{8}{10}$ of the initial 60% rural population plus $\frac{1}{10}$ of the initial 40% urban population. That is, the fraction of the population that is rural will be

$$\frac{8}{10} \frac{6}{10} + \frac{1}{10} \frac{4}{10} = .52$$

after one year. Similarly,

$$\frac{2}{10}\frac{6}{10} + \frac{9}{10}\frac{4}{10} = .48$$

of the population will be in the cities after one year.

You undoubtedly noticed that these computations are precisely the kind of sums of products involved in matrix multiplication. From the table of migration data

	From farm	From city
To farm	$\frac{8}{10}$	$\frac{1}{10}$
To city	$\frac{2}{10}$	$\frac{9}{10}$

we can read off the matrix

$$P = \begin{bmatrix} .8 & .1 \\ .2 & .9 \end{bmatrix}.$$

Let us write the initial population distribution as a vector $ebdv_0 = \begin{bmatrix} .6 \\ .4 \end{bmatrix}$. The matrix multiplication

$$P\mathbf{v}_0 = \begin{bmatrix} .8 & .1 \\ .2 & .9 \end{bmatrix}\begin{bmatrix} .6 \\ .4 \end{bmatrix} = \begin{bmatrix} .52 \\ .48 \end{bmatrix} = \mathbf{v}_1$$

gives a vector $\mathbf{v}_1$ whose entries describe the two segments of the population after one year. You should check that the population distribution vector after the second year can be computed by a similar multiplication:

$$P\mathbf{v}_1 = \begin{bmatrix} .8 & .1 \\ .2 & .9 \end{bmatrix}\begin{bmatrix} .52 \\ .48 \end{bmatrix} = \begin{bmatrix} .464 \\ .536 \end{bmatrix} = \mathbf{v}_2.$$

Notice that the associative law of matrix multiplication gives

$$\mathbf{v}_2 = P\mathbf{v}_1 = P(P\mathbf{v}_0) = (PP)\mathbf{v}_0 = P^2\mathbf{v}_0.$$

In general, we can calculate the population vector $\mathbf{v}_n$ for the end of year n in terms of the population vector $\mathbf{v}_{n-1}$ for the end of the previous year by a simple matrix multiplication $\mathbf{v}_n = P\mathbf{v}_{n-1}$. If we trace the computation of the population vectors back to the initial distribution $\mathbf{v}_0$, we can derive an alternative formula for $\mathbf{v}_n$:

$$\begin{aligned} \mathbf{v}_n = P\mathbf{v}_{n-1} &= P(P\mathbf{v}_{n-2}) = P^2\mathbf{v}_{n-2} \\ &= P^2(P\mathbf{v}_{n-3}) = P^3\mathbf{v}_{n-3} \\ &= \quad \cdots \quad = P^{n-1}\mathbf{v}_1 \\ &= P^{n-1}(P\mathbf{v}_0) = P^n\mathbf{v}_0. \end{aligned}$$

The following table shows the results (rounded to three decimal places) of computing the distribution vectors for the first twenty years. It appears that after a decade or so, the population stabilizes, with approximately one-third of the population rural and two-thirds urban.

Year	Rural	Urban
0	.600	.400
1	.520	.480
2	.464	.536
3	.425	.575
4	.397	.603
5	.378	.622
6	.365	.635
7	.355	.645
8	.349	.651
9	.344	.656
10	.341	.659
11	.339	.661
12	.337	.663
13	.336	.664
14	.335	.665
15	.335	.665
16	.334	.666
17	.334	.666
18	.334	.666
19	.334	.666
20	.334	.666

We have been specifying the population distributions in terms of the percentages of people who live in the rural and urban areas. We can equally well interpret these percentages as the probabilities that an individual randomly chosen from the population will live on a farm or in a city. With this change in perspective, the following definition captures the essential features of this example.

Definition 5.14

A **Markov chain** consists of

1. A list of a finite number r of **states**,
2. A **transition matrix** $P = [p_{ij}] \in \mathbb{M}(r, r)$, and
3. An **initial distribution** vector $\mathbf{v}_0 \in \mathbb{R}^r$.

The system begins in exactly one of the r states. The ith entry of $\mathbf{v}_0$ indicates the probability that the system begins in state i. The system moves from state to state in discrete steps. After each step the system is in exactly one of the r states. The probability of taking a step into the ith state depends only on the state the system was in at the end of the previous step. The probability of moving from state j to state i in one step is p_{ij}.

Thus, our demographic model is a Markov chain with two states, rural and urban, with transition matrix $\begin{bmatrix} .8 & .1 \\ .2 & .9 \end{bmatrix}$, and with initial distribution vector $\begin{bmatrix} .6 \\ .4 \end{bmatrix}$.

Quick Example *A community access television station has four sponsors: a clothing store, a hardware store, and two grocery stores. The station starts the day with an ad for one of the grocery stores (chosen by the flip of a coin). At the end of each program it runs an ad for one of the sponsors. Although the station never runs an ad for the same type of store twice in a row, all other sequences of ads are equally likely. Set up a Markov chain to model this situation.*

There are four states, corresponding to the ads for the four stores. Let us choose labels such as C, H, G_1, and G_2 to indicate the type of store associated with each state. With the four states in the order listed, the transition matrix is

$$\begin{bmatrix} 0 & \frac{1}{3} & \frac{1}{2} & \frac{1}{2} \\ \frac{1}{3} & 0 & \frac{1}{2} & \frac{1}{2} \\ \frac{1}{3} & \frac{1}{3} & 0 & 0 \\ \frac{1}{3} & \frac{1}{3} & 0 & 0 \end{bmatrix}.$$

The initial distribution vector is $\begin{bmatrix} 0 \\ 0 \\ \frac{1}{2} \\ \frac{1}{2} \end{bmatrix}$. ∎

Since each entry p_{ij} of the transition matrix P of a Markov chain represents a probability, we have $0 \le p_{ij} \le 1$. Since the jth column of P contains probabilities for all possible states the system can move to in one step from state j, these entries must add up to 1. That is, for any $j = 1, \ldots, r$,

$$\sum_{i=1}^{r} p_{ij} = 1.$$

Similarly, the entries of the initial distribution vector $\mathbf{v}_0$ all lie in the interval $[0, 1]$ and add up to 1.

In the demographic example we observed that the entries of the nth power of the transition matrix give the probabilities of transitions between the two states after n years. This should convince you of the reasonableness of the following theorem.

Theorem 5.15

Suppose a Markov chain with r states has transition matrix P and initial distribution vector $\mathbf{v}_0$. The probability of moving from the jth state to the ith state in n steps is the ij-entry of P^n. The probability that the system is in the ith state after n steps is the ith entry of the probability distribution vector $P^n \mathbf{v}_0$.

The demographic model suggests that we should look for an equilibrium vector $\mathbf{s}$ whose entries specify a probability distribution that will not change as the system evolves. That is, $\mathbf{s}$ should satisfy the equation $P\mathbf{s} = \mathbf{s}$. We would expect the probability distribution vector $\mathbf{v}_n = P^n \mathbf{v}_0$ to converge to $\mathbf{s}$ for any initial distribution vector $\mathbf{v}_0$. In particular, if we choose the ith standard basis element $\mathbf{e}_i$ of $\mathbb{R}^r$ to be the initial distribution vector, then $\mathbf{v}_n = P^n\mathbf{e}_i$ is the ith column of P^n. Hence, we should also expect all the columns of P^n to converge to $\mathbf{s}$.

There is one mild condition we need to impose upon a Markov chain before we are assured of this convergence to an equilibrium distribution.

Definition 5.16

A Markov chain is **regular** if and only if some power of the transition matrix has only positive entries.

The condition that some power P^n of the transition matrix has only positive entries means that after exactly n steps, it is possible to enter any state from any starting state.

Theorem 5.17   Fundamental Theorem of Regular Markov Chains

Suppose P is the transition matrix for a regular Markov chain. Then there is a unique vector $\mathbf{s}$ whose components add up to 1 and that satisfies $P\mathbf{s} = \mathbf{s}$. The entries of the columns of P^n converge to the corresponding entries of $\mathbf{s}$ as n increases to infinity.

A straightforward way to find the equilibrium distribution vector $\mathbf{s}$ is to solve the equation $P\mathbf{s} = \mathbf{s}$. If we use the $r \times r$ identity matrix I to write the right side of this equation as $I\mathbf{s}$, we can transform this equation to $P\mathbf{s} - I\mathbf{s} = \mathbf{0}$. By the distributive law for matrix multiplication, we can rewrite this one more time as a homogeneous system $(P - I)\mathbf{s} = \mathbf{0}$. The Fundamental Theorem of Regular Markov Chains states that this homogeneous system will have nontrivial solutions and that there will be a unique solution whose components add up to 1.

Quick Example   *In the demographic model developed in this section, determine the equilibrium vector and the limiting matrix for P^n as n increases to infinity.*

The equilibrium vector $\mathbf{s} = \begin{bmatrix} s_1 \\ s_2 \end{bmatrix}$ satisfies the equation

$$\begin{bmatrix} .8 & .1 \\ .2 & .9 \end{bmatrix}\begin{bmatrix} s_1 \\ s_2 \end{bmatrix} = \begin{bmatrix} s_1 \\ s_2 \end{bmatrix}.$$

Along with the condition that the entries of $\mathbf{s}$ add up to 1, this is equivalent to solving the system

$$-.2s_1 + .1s_2 = 0$$
$$.2s_1 - .1s_2 = 0$$
$$s_1 + s_2 = 1$$

A few quick row operations,

$$\begin{bmatrix} -.2 & .1 & 0 \\ .2 & -.1 & 0 \\ 1 & 1 & 1 \end{bmatrix} \rightarrow \begin{bmatrix} 1 & 1 & 1 \\ -2 & 1 & 0 \\ 0 & 0 & 0 \end{bmatrix} \rightarrow \begin{bmatrix} 1 & 1 & 1 \\ 0 & 1 & \frac{2}{3} \\ 0 & 0 & 0 \end{bmatrix} \rightarrow \begin{bmatrix} 1 & 0 & \frac{1}{3} \\ 0 & 1 & \frac{2}{3} \\ 0 & 0 & 0 \end{bmatrix},$$

give $\mathbf{s} = \begin{bmatrix} s_1 \\ s_2 \end{bmatrix} = \begin{bmatrix} \frac{1}{3} \\ \frac{2}{3} \end{bmatrix}$. Hence, P^n converges to $\begin{bmatrix} \frac{1}{3} & \frac{1}{3} \\ \frac{2}{3} & \frac{2}{3} \end{bmatrix}$. This result is consistent

with our earlier calculations.   ■

Exercises 5.3

1. Suppose the demographic model studied in this section had an initial distribution with all of the people on farms and none of the people in cities. Using the transition matrix of the model in this section, compute the distribution vector after one, two, and three years. Compare your results with the results presented in the original example. What do you predict will be the equilibrium distribution? Approximately how long will it take for the population to stabilize near its equilibrium distribution?

2. Set up a Markov chain to model the population in which 40% of the rural people move to cities each year and 30% of the urban people move to farms. Compare the predictions of your model with the results of the example in this section. Notice that both models have the same difference between the percentages of urban and rural migration rates.

3. Consider the Markov chain developed in this section for the sequence of ads played at the community access station.

    a. Compute the square of the transition matrix.

    b. What is the probability that an ad for the clothing store will repeat after two programs?

    c. What is the probability that an ad for the first grocery store will be played two programs after an ad for the second grocery store?

    d. Show that this is a regular Markov chain.

    e. Use the Fundamental Theorem of Regular Markov Chains to find the equilibrium vector.

    f. Which sponsor's ad is played most often in the long run?

4. A **bit** of information stored in a computer can have two values, represented by 0 and 1. Suppose that every time a bit is read, there is a very small probability p that its value will change. Use the fact (see Exercise 19 of Section 7.1 for a verification and Exercise 5 of Section 8.3 for a derivation) that

    $$\begin{bmatrix} 1-p & p \\ p & 1-p \end{bmatrix}^n = \begin{bmatrix} \dfrac{1+(1-2p)^n}{2} & \dfrac{1-(1-2p)^n}{2} \\ \dfrac{1-(1-2p)^n}{2} & \dfrac{1+(1-2p)^n}{2} \end{bmatrix}$$

    to investigate the long-term behavior of the status of the bit. Suppose a manufacturer wants to ensure that after the bit has been read 1000 times, the probability that its value will differ from the original value is less than .00001. How large can p be?

5. Show that a Markov chain with two states and transition matrix $\begin{bmatrix} 0 & 1 \\ 1 & 0 \end{bmatrix}$ is not regular. Describe the long-term behavior of this system.

6. Suppose a field mouse travels among four feeding sites. One of the sites is central, and the mouse can travel to any of the other three from the central site. There is also a trail between two of the three remote sites. Each day the mouse leaves its current site and travels to one of the neighboring sites along a path chosen at random, with all available paths being equally likely.

    a. Set up a Markov chain to model this situation. You will need to specify the states and the transition matrix. You have artistic license to propose a reasonable initial distribution vector.

    b. If the mouse is at the central site on one day, what are the probabilities that it is at each of the four sites three days later?

    c. Show that this Markov chain is regular.

    d. Use the Fundamental Theorem of Regular Markov Chains to find the equilibrium vector.

    e. How many days must pass before the knowledge of the original location of the mouse is essentially useless in predicting its current location?

7. An economist has created a new index that categorizes the daily movements of the stock market as *up sharply, up, steady, down,* or *down sharply.* Suppose she determines that from one day to the next, the index either remains in the same category or moves to an adjacent category, with all possibilities being equally likely to occur.

    a. Formulate this model as a Markov chain.

    b. Use the Fundamental Theorem of Regular Markov Chains to determine the steady-state distribution of the index.

    c. In the long run, how often would you expect the index to be up sharply?

8. Formulate the physiological example in Section 2.4 in terms of a Markov chain. What role does the Fundamental Theorem of Regular Markov Chains play in this example? Compare the results of this theorem with those obtained in Section 2.4.

9. Formulate the mutation example in Exercise 7 of Section 2.4 in terms of a Markov chain. Interpret the Fundamental Theorem of Regular Markov Chains for this model. Compare these conclusions with the solution obtained by the methods of Section 2.4.

5.4 Absorbing Markov Chains

In the previous section we saw how matrix multiplication and the relation between matrices and linear systems can be exploited to obtain nontrivial results about situations modeled by Markov chains. The principal result pertained to regular Markov chains, systems in which it is always possible to go from one state to any other state. In this section we want to deal with Markov chains on the other end of the spectrum, systems with states that trap the process from ever moving to another state. We will see that the inverse of a square matrix plays a crucial role in describing the behavior of these systems.

Definition 5.18

A state of a Markov chain is an **absorbing state** if and only if it is impossible to leave that state. An **absorbing Markov chain** is one that has at least one absorbing state and is such that from every state it is possible to reach an absorbing state, possibly after several steps.

The random walk on a line is a classical example of an absorbing Markov chain. Suppose a particle moves among the positions 1, 2, 3, 4, and 5 arranged along a line. From positions 2, 3, and 4 the particle moves one position to the left with probability $\frac{1}{2}$ and moves one position to the right with probability $\frac{1}{2}$. Positions 1 and 5 are absorbing: once the particle reaches either of these positions, it remains in that position forever. With the states listed in the same order as the five positions on the line, the transition matrix is

$$
P = \begin{bmatrix}
1 & \frac{1}{2} & 0 & 0 & 0 \\
0 & 0 & \frac{1}{2} & 0 & 0 \\
0 & \frac{1}{2} & 0 & \frac{1}{2} & 0 \\
0 & 0 & \frac{1}{2} & 0 & 0 \\
0 & 0 & 0 & \frac{1}{2} & 1
\end{bmatrix}.
$$

Notice that the absorbing states can be identified as corresponding to columns of P that contain a single nonzero entry (which is necessarily equal to 1).

Theorem 5.19

In an absorbing Markov chain, the probability that the process will eventually reach an absorbing state is 1.

Sketch of proof   There is an integer n such that from any starting state it is possible to reach an absorbing state in n steps. Thus, there is some number $p < 1$ such that from any starting state, the probability of not being absorbed after n steps is less than p. Now the probability of not being absorbed after $2n$ steps is less than p^2. In general, the probability of not being absorbed after kn steps is less than p^k. Since $0 \leq p < 1$, this quantity decreases to 0 as k goes to infinity.   ■

This theorem settles the long-run fate of an absorbing Markov chain. The interesting questions that remain concern the behavior of the system prior to its inevitable absorption. For example, what is the average number of steps a system will take before entering an absorbing state? In particular, what is the expected number of times the system will be in each of the nonabsorbing states during this transient period?

The general results can be illustrated by means of the random walk example introduced at the beginning of this section. For convenience, rearrange the order of the states so the absorbing states 1 and 5 are listed before the nonabsorbing states 2, 3, and 4. The transition matrix now breaks into four rectangular blocks

$$P = \begin{bmatrix} 1 & 0 & \frac{1}{2} & 0 & 0 \\ 0 & 1 & 0 & 0 & \frac{1}{2} \\ \hline 0 & 0 & 0 & \frac{1}{2} & 0 \\ 0 & 0 & \frac{1}{2} & 0 & \frac{1}{2} \\ 0 & 0 & 0 & \frac{1}{2} & 0 \end{bmatrix} = \begin{bmatrix} I & R \\ \hline \mathbf{0} & Q \end{bmatrix}.$$

Notice that P^2 also has this form:

$$P^2 = \begin{bmatrix} 1 & 0 & \frac{1}{2} & 0 & 0 \\ 0 & 1 & 0 & 0 & \frac{1}{2} \\ \hline 0 & 0 & 0 & \frac{1}{2} & 0 \\ 0 & 0 & \frac{1}{2} & 0 & \frac{1}{2} \\ 0 & 0 & 0 & \frac{1}{2} & 0 \end{bmatrix} \begin{bmatrix} 1 & 0 & \frac{1}{2} & 0 & 0 \\ 0 & 1 & 0 & 0 & \frac{1}{2} \\ \hline 0 & 0 & 0 & \frac{1}{2} & 0 \\ 0 & 0 & \frac{1}{2} & 0 & \frac{1}{2} \\ 0 & 0 & 0 & \frac{1}{2} & 0 \end{bmatrix}$$

$$= \begin{bmatrix} 1 & 0 & \frac{1}{2} & \frac{1}{4} & 0 \\ 0 & 1 & 0 & \frac{1}{4} & \frac{1}{2} \\ \hline 0 & 0 & \frac{1}{4} & 0 & \frac{1}{4} \\ 0 & 0 & 0 & \frac{1}{2} & 0 \\ 0 & 0 & \frac{1}{4} & 0 & \frac{1}{4} \end{bmatrix} = \begin{bmatrix} I & R + RQ \\ \hline \mathbf{0} & Q^2 \end{bmatrix}.$$

In general,

$$P^n = \begin{bmatrix} I & R + RQ + \cdots + RQ^{n-1} \\ \hline \mathbf{0} & Q^n \end{bmatrix}.$$

The entries of Q^n in the lower right block of P^n provide answers to the questions posed above. To see what is going on, let us concentrate on position 2, the first non-absorbing state. This position corresponds to the third row and the third column of the powers of P. It corresponds to the first row and first column of the powers of Q. Suppose the particle starts in this state and that we want to determine the expected number of visits to this state before the particle is absorbed. We begin by counting one visit for starting in this state. This corresponds to the 1 in the 11-entry of

$$Q^0 = I = \begin{bmatrix} 1 & 0 & 0 \\ 0 & 1 & 0 \\ 0 & 0 & 1 \end{bmatrix}.$$

The particle can never visit this position after one step. This corresponds to the 0 in the 11-entry of

$$Q^1 = Q = \begin{bmatrix} 0 & \frac{1}{2} & 0 \\ \frac{1}{2} & 0 & \frac{1}{2} \\ 0 & \frac{1}{2} & 0 \end{bmatrix}.$$

The $\frac{1}{4}$ as the 11-entry of

$$Q^2 = \begin{bmatrix} \frac{1}{4} & 0 & \frac{1}{4} \\ 0 & \frac{1}{2} & 0 \\ \frac{1}{4} & 0 & \frac{1}{4} \end{bmatrix}.$$

indicates that in one-fourth of the times we observe this random walk, we should expect a visit to this position after two steps. In general, the 11-entry of Q^n gives the frequency of a visit to position 2 after n steps. If we add this series of numbers, we will obtain the expected number of visits to position 2. This is the 11-entry of the sum of the powers of Q:

$$I + Q + Q^2 + Q^3 + \cdots .$$

By the same reasoning, if the system starts in the jth nonabsorbing state, the expected number of visits to the ith nonabsorbing state is the ij-entry of

$$I + Q + Q^2 + Q^3 + \cdots .$$

In this situation, the series of matrices is known to converge in the sense that the entries of the partial sums each converge. Furthermore, this geometric series of matrices converges to $(I-Q)^{-1}$ in strict analogy with the convergence of the geometric series of real numbers

$$1 + r + r^2 + r^3 + \cdots$$

to $(1 - r)^{-1}$ provided that $|r| < 1$. In our example,

$$I - Q = \begin{bmatrix} 1 & -\frac{1}{2} & 0 \\ -\frac{1}{2} & 1 & -\frac{1}{2} \\ 0 & -\frac{1}{2} & 1 \end{bmatrix} \quad \text{and} \quad (I - Q)^{-1} = \begin{bmatrix} \frac{3}{2} & 1 & \frac{1}{2} \\ 1 & 2 & 1 \\ \frac{1}{2} & 1 & \frac{3}{2} \end{bmatrix}.$$

(See Exercise 4e of Section 5.2.) Thus, we can read from the 11-entry that if the system starts in position 2, we can expect 1.5 visits to position 2. Reading down the column, we also see the expected number of visits to position 3 is 1 and the expected number of visits to position 4 is $\frac{1}{2}$. This accounts for all the nonabsorbing states. Thus, if the system starts in position 2, the expected number of steps until the particle is absorbed is $\frac{3}{2} + 1 + \frac{1}{2} = 3$.

The matrix $(I - Q)^{-1}$ is known as the **fundamental matrix** of an absorbing Markov chain. The random walk example illustrates the general results summarized in the concluding theorem of this section.

Theorem 5.20 Fundamental Theorem of Absorbing Markov Chains

Suppose Q is the portion of the transition matrix of an absorbing Markov chain corresponding to the nonabsorbing states. Suppose the system starts in the jth nonabsorbing state. The expected number of visits to the ith nonabsorbing state is the ij-entry of the fundamental matrix $(I - Q)^{-1}$. The sum of the entries in the jth column of this matrix is the expected number of steps until the system reaches an absorbing state.

Quick Example   *The school board wants to model the progress of students at the local high school. Data from recent years show that 2% of the students in the ninth* *grade drop out during the year, along with 3% in the tenth grade, 4% in the eleventh grade, and 5% in the twelfth grade. In each grade, 3% of the students will repeat a grade, and 1% will skip a grade. What is the average number of years a student will attend high school?*

Let us consider a Markov chain with five states, one for out-of-school (drop out or graduation) and four for the different grade levels. The transition matrix is

$$P = \begin{bmatrix} 1 & .02 & .03 & .04 & .97 \\ 0 & .03 & 0 & 0 & 0 \\ 0 & .94 & .03 & 0 & 0 \\ 0 & .01 & .93 & .03 & 0 \\ 0 & 0 & .01 & .93 & .03 \end{bmatrix}.$$

The fundamental matrix is

$$\begin{bmatrix} .97 & 0 & 0 & 0 \\ -.94 & .97 & 0 & 0 \\ -.01 & -.93 & .97 & 0 \\ 0 & -.01 & -.93 & .97 \end{bmatrix}^{-1} = \begin{bmatrix} 1.031 & 0 & 0 & 0 \\ 0.999 & 1.031 & 0 & 0 \\ 0.968 & 0.988 & 1.031 & 0 \\ 0.939 & 0.958 & 0.988 & 1.031 \end{bmatrix}.$$

to three decimal places. Thus, the students who enter ninth grade will spend an average of 1.031 years in that grade, 0.999 years in tenth grade, 0.968 years in eleventh grade, and 0.939 years in twelfth grade. This is a total of 3.937 years.

Exercises 5.4

1. If the random walk discussed in this section begins in position 3 (the second nonabsorbing state), what is the expected number of steps until the particle is absorbed? What if the particle starts in position 4? Explain why the relative sizes of these values are reasonable in relation to the expected number of steps until absorption when the particle begins in position 2.

2. Write a computer program to simulate the random walk discussed in this section. Have the program repeat the walk a large number of times and print out the average number of visits to the nonabsorbing states as well as the average number of steps until absorption. Compare the results of your program with the theoretical results obtained in this section.

3. Modify the random walk example discussed in this section so that the probability of a step to the right is $\frac{2}{3}$ and the probability of a step to the left is $\frac{1}{3}$. Compute the fundamental matrix and read off the information it gives about the transient behavior of the system.

4. Set up an absorbing Markov chain to model the progress of students through four years of college. Make reasonable estimates for the probabilities of advancing to the next year of study, of taking a year off, and of dropping out. Based on your model, what is the average number of years a student attends college?

5. Many characteristics of plants and animals are determined genetically. Suppose the height of a variety of corn is determined by a gene that comes in two forms,

T and *t*. Each plant has a pair of these genes and can thus be pure dominant type *TT*, pure recessive type *tt*, or hybrid *Tt*. Each seed obtains one of the two genes from one plant and the second from another plant. In each generation, a geneticist randomly selects a pair of plants from all possible offspring. These are used to breed the plants for the next generation.

a. Set up a Markov chain to model this experiment. The states will be the pairs of genotypes of the parents. Since it does not matter which parent plant contributed which gene to the offspring, you need only consider six states: (*TT, TT*), (*TT, Tt*), (*TT, tt*), (*Tt, Tt*), (*Tt, tt*), and (*tt, tt*).

b. Show that this is an absorbing Markov chain.

c. Interpret the results of the Fundamental Theorem of Absorbing Markov Chains for this model.

6. Suppose a particle moves among the positions 1, 2, 3, 4, 5, and 6 along a line. Suppose states 1 and 6 are absorbing; otherwise, each step consists of the particle moving one position to the left with probability .4, one step to the right with probability .5, or remaining fixed with probability .1.

a. If the particle starts in position 3, what are the expected numbers of times the particle will visit the four nonabsorbing positions?

b. Show that starting position 3 results in the longest expected time for the particle to be absorbed.

c. If the particle starts in position 3, what is the probability it will be absorbed in position 1? (Suggestion: Notice that the only way the particle can reach position 1 is through position 2. Each time the particle visits position 2, it has a 40% chance of being absorbed into position 1 at the next step.)

Project: Series of Matrices

The discussion in Section 5.4 illustrates how infinite series of matrices arise naturally in the analysis of absorbing Markov chains. One technique for solving a system of differential equations involves evaluating an infinite series of matrices derived from the coefficients of the equations. This project lays the foundation for such applications. You will be using properties familiar from your work with sequences and series of real numbers to derive similar results for matrices. This is an excellent opportunity for you to review the logical development of convergence results for real numbers. In many cases, the proofs you find in your calculus book for real numbers will serve as models for deriving the analogous properties in this new context.

1. Begin by defining the concept of a sequence of matrices. Define the convergence of a sequence of matrices to a limiting matrix. You can't go wrong by using the convergence of the corresponding entries of the matrices. Illustrate your definitions with an example or two.

State and prove some basic results about limits of sums and limits of products. Consider both scalar multiplication and matrix multiplication. Be sure to accompany your formulas with explicit statements of the convergence results you use for your hypotheses and your conclusions.

Dealing with limits of inverses is slightly tricky. Let A_k denote the kth term in a sequence of $n \times n$ matrices. Under the hypotheses that all the matrices in the sequence are invertible and that $\lim_{k \to \infty} A_k = A$ exists and is also invertible, it is not hard to prove that

$$\lim_{k \to \infty} A_k^{-1} = A^{-1}.$$

After you have covered the material on determinants in Section 7.4, you might want to improve this result so that it does not require the assumption that the matrices in the sequence are invertible.

Define the sum of an infinite series of matrices as the limit of the sequence of partial sums. Try a few examples, perhaps one involving the sum of powers of a 2×2 matrix.

2. In the transition matrix of an absorbing Markov chain, let Q denote the portion that indicates the probabilities of moving among the nonabsorbing states. As in the proof of Theorem 5.19, argue that there is an integer n and a number p with $0 \leq p < 1$ such that each entry of Q^{nk} is less than p^k. Show that the series

$$\sum_{k=0}^{\infty} Q^k$$

converges by relating the entries of appropriate groups of terms to the terms of the convergent geometric series

$$\sum_{k=0}^{\infty} p^k.$$

Also show that

$$\lim_{k \to \infty} Q^{k+1} = \mathbf{0}.$$

Verify the formula

$$(I + Q + Q^2 + \cdots + Q^k)(I - Q) = I - Q^{k+1}.$$

Put all these pieces together to conclude that $\sum_{k=0}^{\infty} Q^k$ is the inverse of $I - Q$.

For an invertible matrix P, let $D = P^{-1}QP$. For any nonnegative integer k, show that $Q^k = PD^kP^{-1}$ and that

$$(I + Q + Q^2 + \cdots + Q^k) = P(I + D + D^2 + \cdots + D^k)P^{-1}.$$

Conclude that

$$\sum_{k=0}^{\infty} Q^k = P\left(\sum_{k=0}^{\infty} D^k\right)P^{-1}.$$

With $D = \begin{bmatrix} 0.8 & 0 \\ 0 & 0.5 \end{bmatrix}$, evaluate $\sum_{k=0}^{\infty} D^k$. Let $P = \begin{bmatrix} 2.5 & -1.25 \\ 2.0 & 3.0 \end{bmatrix}$ and

$$Q = PDP^{-1} = \begin{bmatrix} 2.5 & -1.25 \\ 2.0 & 3.0 \end{bmatrix}\begin{bmatrix} 0.8 & 0 \\ 0 & 0.5 \end{bmatrix}\begin{bmatrix} 0.3 & -0.125 \\ -0.2 & 0.25 \end{bmatrix} = \begin{bmatrix} 0.725 & 0.09375 \\ 0.18 & 0.575 \end{bmatrix}.$$

Use the preceding formula to compute $\sum_{k=0}^{\infty} Q^k$. Write a computer program to add powers of Q to approximate the value of this series. Compare these two results with $(I - Q)^{-1}$.

3. Let A and B be $n \times n$ matrices. Let α denote the maximum of the absolute values of the entries of A, and let β denote the maximum of the absolute values of the entries of B. Show that the maximum of the absolute values of the entries of AB is less than or equal to $n\alpha\beta$. Apply this result repeatedly to conclude that the maximum of the absolute values of the entries of A^k is $n^{k-1}\alpha^k$. Now show that the infinite series

$$\sum_{k=0}^{\infty} \frac{1}{k!} A^k$$

converges. By analogy with the power series representation of the exponential function of a real variable, define e^A as the sum of this infinite series of matrices.

Let Q and D be the 2×2 matrices from the previous example. Compute e^D. Use the relation $Q = PDP^{-1}$ to compute e^Q.

The law of exponentials $e^{x+y} = e^x e^y$ for real numbers x and y can be derived from the power series representations as follows:

$$
\begin{aligned}
e^{x+y} &= \sum_{k=0}^{\infty} \frac{1}{k!} (x + y)^k \\
&= \sum_{k=0}^{\infty} \frac{1}{k!} \sum_{j=0}^{k} \frac{k!}{(k-j)!\,j!} x^{k-j} y^j \\
&= \sum_{k=0}^{\infty} \sum_{j=0}^{k} \frac{1}{(k-j)!\,j!} x^{k-j} y^j \\
&= \sum_{j=0}^{\infty} \sum_{k=j}^{\infty} \frac{1}{(k-j)!\,j!} x^{k-j} y^j \\
&= \sum_{j=0}^{\infty} \left(\sum_{k=j}^{\infty} \frac{1}{(k-j)!} x^{k-j} \right) \frac{1}{j!} y^j \\
&= \left(\sum_{k=j}^{\infty} \frac{1}{(k-j)!} x^{k-j} \right) \sum_{j=0}^{\infty} \frac{1}{j!} y^j \\
&= \left(\sum_{k=0}^{\infty} \frac{1}{k!} x^k \right) \left(\sum_{j=0}^{\infty} \frac{1}{j!} y^j \right) \\
&= e^x e^y.
\end{aligned}
$$

Notice the role of the Binomial Theorem in the expansion of $(x + y)^k$. The tricky step where the order of summation is interchanged is legitimate since the series converges absolutely.

Suppose the two $n \times n$ matrices A and B commute; that is, suppose $AB = BA$. Show that powers of $A + B$ can be expanded by the binomial theorem:

$$(A + B)^k = \sum_{j=0}^{k} \frac{k!}{(k-j)!\,j!} A^{k-j} B^j.$$

Now rewrite the proof of the law of exponentials $e^{A+B} = e^A e^B$ for the matrices A and B. Again, the tricky step is the interchange of the order of summation. You will need to show that each of the entries of the matrix series is an absolutely convergent series of real numbers.

Guess what the inverse of e^A might be. In the course of verifying your conjecture, conclude that e^A is in fact invertible.

4. Again let A be an $n \times n$ matrix. Give a definition of $\dfrac{d}{dt}e^{tA}$ in terms of a limit. Use your definition to derive the formula

$$\frac{d}{dt}e^{tA} = Ae^{tA}.$$

Project: Linear Models

Applications of mathematics frequently involve relations among a large number of variables. Often the variables are related in a linear fashion, such as we have seen between unknowns in a system of linear equations and the constants on the right side of the equations. We have seen how matrix multiplication simplifies the notation by representing such relations as a single equation. In a Markov chain, matrix multiplication again gives a linear relation between the frequency distribution among the possible states from one step to the next.

In general and in the simplest form, a linear model is simply an $m \times n$ matrix A that describes the relation $A\mathbf{x} = \mathbf{y}$ between n components of an input vector $\mathbf{x} \in \mathbb{R}^n$ and m components of an output vector $\mathbf{y} \in \mathbb{R}^m$. This relation can also be expressed in the terminology of Chapter 3 by saying that $\mathbf{y}$ is a linear combination of the columns of A where the coefficients of the linear combination are the components of $\mathbf{x}$.

Here are some typical cases where these ideas lead to an understanding of the workings of the economy, results concerning the pricing of commodities and stock options, and consequences of various strategies for resource management.

1. Suppose an industry uses n different raw materials to produce m different items. Let x_j denote the unit price of raw material j, and let a_{ij} denote the amount of raw material j consumed in producing item i. Then $\mathbf{x} = [x_j]$ is a vector with n components and $A = [a_{ij}]$ is an $m \times n$ matrix. The product $A\mathbf{x}$ is a vector in $\mathbb{R}^m$, and component i of this vector is the cost of materials for producing item i.

The mathematical formula expressing the cost of the items in terms of the prices of the raw materials is simple enough. However, if you try to locate some data for such a manufacturing process, you will undoubtedly encounter problems determining exactly what should be considered a raw material as well as the unit prices for these raw materials.

The situation may be reversed. That is, you may be given the allowable costs of the products and have to determine prices you are willing to pay for the raw materials.

Suppose the products of the industry are actually the same as the raw materials. This gives rise to the economic models of Wassily Leontief, winner of the Nobel Prize in economics.

Some references on these and other economic models are:

David Gale, *The Theory of Linear Economic Models*, New York: McGraw-Hill, 1960.

James Quirk and Rubin Saposnik, *Introduction to General Equilibrium Theory and Welfare Economics,* New York: McGraw-Hill, 1968.

2. Suppose a_{ij} denotes the payoff of holding one share of asset j of a list of n possible assets if the economy is in state i of m possible states. The portfolio of an investor can be described as a vector **x**, with component j being the number of shares of asset j. A negative component of the portfolio vector indicates that the investor has taken a short position in that asset. The product $A\mathbf{x}$ describes the income that portfolio **x** will yield in all the various states. This setup is developed in the article "The Arbitrage Principle in Financial Economics" by Hal Varian, appearing in the *Journal of Economic Perspectives* (Fall 1987, vol. 1, no. 2, pp. 55–72). Some reasonable assumptions about the nature of a market economy lead to results on pricing of securities. In particular, two different stock option schemes lead to identical prices.

3. Prior to the synthesis of vitamin A, shark livers were a major source of this vitamin, taken as a supplement by millions of children in the United States. Unfortunately, information on the reproductive rates of sharks was virtually nonexistent. Biologists are now gathering such data, using models to predict the populations of various species, and finding that continued harvesting would have led to extinction of the sharks. This same type of analysis can be used to manage forests for lumber and paper products, fisheries as a source of food and a source of employment, and other such renewable resources.

The first step in setting up a model for a certain species is to classify the population into groups with common properties important for reproduction. Age grouping and sex are the typical factors in this classification. For each group, the mortality rate must be estimated, as well as the probabilities that an individual will change to another group or produce an offspring in another group during a given period of time. Here again, matrix multiplication provides a compact notation for a large array of relations.

Once the effects of harvesting are entered in the model, we can predict the fate of the population. Then various policies can be compared. Measures can be taken to ensure the perpetuation of the species, or ignored with known consequences. Consult UMAP Module 207, "Management of a Buffalo Herd" by Philip Tuchinsky (*UMAP Journal,* 1981, vol. 2, no. 1, pp 75–120), for further details on this type of model, some data for experimentation, and additional references.

Chapter 5 Summary

Chapter 5 deals with a product structure on spaces of matrices. This matrix multiplication obeys many properties familiar from other types of multiplication. The failure of the commutative law is one notable exception. Matrix algebra provides a convenient notation for systems of linear equations, Markov chains, and applications to transportation and communication networks.

Computations

Multiplication
Definition and notation
Identity matrix
Properties and counterexamples
Effects of performing row operations on the left factor of a product

Inverses
   One-sided and two-sided
   Augment by *I* and reduce

Rank of a matrix

Equilibrium vector for a regular Markov chain

Fundamental matrix of an absorbing Markov chain

Theory

Inverses
   Uniqueness
   $(A^{-1})^{-1} = A$ and $(AB)^{-1} = B^{-1}A^{-1}$
   A one-sided inverse of a square matrix is its inverse

Fundamental Theorem of Regular Markov Chains

Fundamental Theorem of Absorbing Markov Chains

Applications

Transportation and communication models

$A\mathbf{x} = \mathbf{b}$ represents a system of linear equations
   Solution sets and subspaces
   Solution $\mathbf{x} = A^{-1}\mathbf{b}$ if *A* is invertible

Markov chains
   States
   Transition matrix
   Initial distribution
   Models with regular Markov chains
   Models with absorbing Markov chains

Review Exercises

1.   Consider the matrix product

$$\begin{bmatrix} 8 & 9 & 3 \\ 2 & 7 & -1 \\ 0 & 3 & 6 \\ -4 & 5 & 8 \\ 6 & 3 & -1 \end{bmatrix} \begin{bmatrix} -3 & 0 & 2 & 9 \\ 5 & 11 & 1 & 17 \\ 7 & -7 & 3 & 8 \end{bmatrix}.$$

   a. The product matrix will be of what size?

   b. Compute the entry in the second row, third column of this product.

2.   Suppose *A* and *B* are $n \times n$ matrices.

   a. Prove that $(A + B)^2 = A^2 + AB + BA + B^2$.

    b. Prove that if $AB = BA$, then $(A + B)^2 = A^2 + 2AB + B^2$.

    c. Prove that if $(A + B)^2 = A^2 + 2AB + B^2$, then $AB = BA$.

3. A matrix obtained by applying a single elementary row operation to the identity matrix is an **elementary matrix.** Suppose A is an $m \times n$ matrix.

    a. Suppose E is an elementary matrix obtained by interchanging two rows of the $m \times m$ identity matrix. Show that EA is the matrix obtained by interchanging the corresponding rows of A.

    b. Suppose E is an elementary matrix obtained by multiplying a row of the $m \times m$ identity matrix by a nonzero constant. Show that EA is the matrix obtained by multiplying the corresponding row of A by the same constant.

    c. Suppose E is an elementary matrix obtained by adding a multiple of one row of the $m \times m$ identity matrix to another row. Show that EA is the matrix obtained by performing the corresponding row operation to A.

    d. Use the results of parts a, b, and c together with the associative law of matrix multiplication to give an alternative proof of Theorem 5.5.

4. Let $A = \begin{bmatrix} 3 & 0 & -4 \\ -2 & -1 & -1 \\ 1 & -2 & 0 \end{bmatrix}$.

    a. Compute A^2.

    b. Compute A^3.

    c. Show that $A^3 - 2A^2 - A + 26I = \mathbf{0}$.

    d. Use the formula in part b to write A^{-1} as a linear combination of A^2, A, and I.

5. For any value of θ show that $\begin{bmatrix} \cos\theta & -\sin\theta \\ \sin\theta & \cos\theta \end{bmatrix}^{-1} = \begin{bmatrix} \cos\theta & \sin\theta \\ -\sin\theta & \cos\theta \end{bmatrix}$.

6. For any values of θ and φ show that

$$\begin{bmatrix} \cos\theta & -\sin\theta \\ \sin\theta & \cos\theta \end{bmatrix} \begin{bmatrix} \cos\varphi & -\sin\varphi \\ \sin\varphi & \cos\varphi \end{bmatrix} = \begin{bmatrix} \cos(\theta + \varphi) & -\sin(\theta + \varphi) \\ \sin(\theta + \varphi) & \cos(\theta + \varphi) \end{bmatrix}.$$

7. Find the inverses of the following matrices:

    a. $\begin{bmatrix} 1 & 2 & 0 \\ 3 & 7 & 1 \\ 0 & -1 & 0 \end{bmatrix}$
    b. $\begin{bmatrix} 1 & 1 & 0 \\ 0 & 2 & 4 \\ -1 & 1 & 5 \end{bmatrix}$

8. Find the inverses of the following matrices:

    a. $\begin{bmatrix} 3 & 4 \\ 2 & 3 \end{bmatrix}$
    b. $\begin{bmatrix} 1 & 2 \\ 3 & 7 \end{bmatrix}$
    c. $\begin{bmatrix} -2 & 5 \\ 3 & -8 \end{bmatrix}$

    d. Based on the results of your computations, make a conjecture about the inverse of an arbitrary 2×2 matrix $A = \begin{bmatrix} a & b \\ c & d \end{bmatrix}$.

    e. Test your conjecture on a few other 2×2 matrices.

    f. What role does the expression $ad - bc$ play in a formula for the inverse of A?

9. Determine the ranks of the following matrices.

a. $\begin{bmatrix} 2 & 3 & 5 \\ 1 & 1 & 2 \\ 1 & 2 & 3 \end{bmatrix}$

b. $\begin{bmatrix} 8 & 1 & 6 \\ 3 & 5 & 7 \\ 4 & 9 & 2 \end{bmatrix}$

c. $\begin{bmatrix} 1 & 2 & 3 & 4 \\ 5 & 6 & 7 & 8 \\ 9 & 10 & 11 & 12 \end{bmatrix}$

d. $\begin{bmatrix} 1 & 2 & 3 & 4 \\ 5 & 6 & 7 & 8 \\ 9 & 10 & 11 & 12 \\ 13 & 14 & 15 & 16 \end{bmatrix}$

10. Suppose A and B are $n \times n$ matrices.

a. Prove that if rank A = rank B = n, then rank AB = n.

b. Prove that if rank AB = n, then rank A = rank B = n.

11. Suppose a Markov chain with two states has transition matrix $P = \begin{bmatrix} .2 & .5 \\ .8 & .5 \end{bmatrix}$.

a. If the system starts in the first state (corresponding to the first row and first column of P), what is the probability that it is in this state after a single step?

b. What is the probability that it is in the first state after two steps?

c. Determine the equilibrium distribution vector **s**.

12. In an analysis of the sequence of letters of the alphabet in a passage of text, 60% of the vowels were followed by a consonant, and 80% of the consonants were followed by a vowel.

a. What are the states of Markov chain for modeling this analysis?

b. What is the transition matrix?

c. What is the probability that the second letter after a consonant is another consonant?

d. What proportion of the letters in the passage would you expect to be consonants?

13. A particle moves among the four quadrants of the xy-plane. It starts in the second quadrant. The first quadrant is an absorbing state. Otherwise, the particle moves counterclockwise to the next quadrant with probability $\frac{3}{4}$ and clockwise to the next quadrant with probability $\frac{1}{4}$.

a. What is the transition matrix for a Markov chain that models this system?

b. What is the probability that the particle will move from the third quadrant to the first quadrant in two steps?

c. Compute the fundamental matrix of this Markov chain.

d. What is the expected number of times the particle will return to the starting state?

e. What is the expected number of steps until the particle is absorbed?

14. You have just discovered a new superconducting material. It consists of microscopic regions that exhibit a phase change at the temperature of liquid nitrogen into crystals of type X, Y, or Z. When a new batch of the material is first cooled, all regions are of type X. From then on, whenever the material is cooled, each region crystallizes into a type other than what it was during the previous cooling, the two alternatives being equally likely. Formulate a Markov chain to model this remarkable phenomenon.

a. What are the states of this Markov chain?

b. What is the transition matrix of this Markov chain?

c. What is the initial distribution vector?

d. Is this a regular Markov chain? Explain.

e. Is this an absorbing Markov chain? Explain.

f. Show that $\mathbf{s} = \begin{bmatrix} \frac{1}{3} \\ \frac{1}{3} \\ \frac{1}{3} \end{bmatrix}$ is the equilibrium vector for this Markov chain.

Linearity

The previous chapters of this text have been devoted to the study of the mathematical objects known as vector spaces. This chapter introduces the study of relations between pairs of vector spaces. This step is analogous to the transition from the arithmetic of the systems of numbers (integers, rational, real, and complex) to the algebraic operations and geometric transformations that relate these number systems. In your study of calculus you experienced a similar shift of emphasis from considering functions as mathematical objects toward using the various operations of limits, differentiation, and integration to establish a web of interconnections among familiar functions.

6.1 Linear Functions

Let us begin this section by listing some familiar functions between vector spaces. We are particularly interested in the interaction of the function with the vector space operations of addition and scalar multiplication. Notice the variety of topics that are related under this unifying concept.

Crossroads Recall from Section 1.7 that a function $f : X \to Y$ is a well-defined rule that assigns to each element x of the domain X an element $f(x)$ of the range Y. Here are some simple operations on common vector spaces.

Consider $m : \mathbb{R} \to \mathbb{R}$ defined by $m(x) = 5x$. Notice that

$$m(x + y) = 5(x + y) = 5x + 5y = m(x) + m(y), \quad \text{and}$$
$$m(rx) = 5(rx) = r(5x) = rm(x).$$

Consider $D : \mathbb{D}(\mathbb{R}) \to \mathbb{F}(\mathbb{R})$ defined by $D(f) = f'$. Recall the basic rules of differentiation

$$D(f + g) = (f + g)' = f' + g' = D(f) + D(g), \quad \text{and}$$
$$D(rf) = (rf)' = r(f') = rD(f).$$

Consider $I : \mathbb{C}([0, 1]) \to \mathbb{R}$ defined by $I(f) = \int_0^1 f(x)\, dx$. Recall the basic rules of integration

$$I(f + g) = \int_0^1 \big(f(x) + g(x)\big)\,dx = \int_0^1 f(x)\, dx + \int_0^1 g(x)\, dx = I(f) + I(g), \quad \text{and}$$

$$I(rf) = \int_0^1 rf(x)\, dx = r\int_0^1 f(x)\, dx = rI(f).$$

Of course, not all functions are so compatible with the two vector space operations. Consider $s : \mathbb{R} \to \mathbb{R}$ defined by $s(x) = x^2$. In general,

$$s(x + y) = (x + y)^2 \neq x^2 + y^2 = s(x) + s(y), \quad \text{and}$$
$$s(rx) = (rx)^2 = r^2 x^2 \neq r(x^2) = rs(x).$$

These examples lead us to the conditions we should require of functions that are compatible with the vector space operations of addition and scalar multiplication.

Definition 6.1

A function $T : V \to W$ from a vector space V to a vector space W is **linear** if and only if for all $\mathbf{v}, \mathbf{w} \in V$ and $r \in \mathbb{R}$, we have

$$T(\mathbf{v} + \mathbf{w}) = T(\mathbf{v}) + T(\mathbf{w}) \quad \text{and} \quad T(r\mathbf{v}) = rT(\mathbf{v}).$$

The two conditions for a function to be linear are known as **additivity** and **homogeneity**. Specifying that a function is linear presupposes that its domain and range are vector spaces. For the sake of brevity, this assumption will often be suppressed. A linear function is also called a **linear map** or a **linear transformation**. A linear function whose domain is the same as its range is frequently referred to as a **linear operator**.

Of the examples defined above, the functions m, D, and I are linear, whereas the function s is not linear.

Another instance of linear functions arises from matrix multiplication. For example, let

$$A = \begin{bmatrix} 3 & 0 & 1 \\ 0 & 1 & -1 \end{bmatrix}.$$

Define $T : \mathbb{R}^3 \rightarrow \mathbb{R}^2$ by $T(\mathbf{v}) = A\mathbf{v}$, where we identify $\mathbf{v} \in \mathbb{R}^3$ with the corresponding element of $\mathbb{M}(3, 1)$ and $A\mathbf{v} \in \mathbb{M}(2, 1)$ with the corresponding element of $\mathbb{R}^2$. Rather than trying to plow through the verification that T is linear from its formula

$$T\left(\begin{bmatrix} v_1 \\ v_2 \\ v_3 \end{bmatrix}\right) = \begin{bmatrix} 3 & 0 & 1 \\ 0 & 1 & -1 \end{bmatrix}\begin{bmatrix} v_1 \\ v_2 \\ v_3 \end{bmatrix} = \begin{bmatrix} 3v_1 + v_3 \\ v_2 - v_3 \end{bmatrix}$$

in terms of coordinates, let us reap some of the fruits of the work we did in establishing the basic properties of matrix multiplication. Theorem 5.4 makes it very easy to verify that T satisfies the two defining properties of linearity. In fact, we can just as easily handle the general case of a linear map defined in terms of matrix multiplication. We will use the Greek letter μ, corresponding to the Roman letter m, to denote this multiplication (or μultiplication) by a matrix. For an arbitrary matrix $A \in \mathbb{M}(m, n)$ define $\mu_A : \mathbb{R}^n \rightarrow \mathbb{R}^m$ by $\mu_A(\mathbf{v}) = A\mathbf{v}$. To show that μ_A is linear, let $\mathbf{v}, \mathbf{w} \in \mathbb{R}^n$ and $r \in \mathbb{R}$. Then

$$\mu_A(\mathbf{v} + \mathbf{w}) = A(\mathbf{v} + \mathbf{w}) = A\mathbf{v} + A\mathbf{w} = \mu_A(\mathbf{v}) + \mu_A(\mathbf{w}), \quad \text{and}$$

$$\mu_A(r\mathbf{v}) = A(r\mathbf{v}) = r(A\mathbf{v}) = r\mu_A(\mathbf{v}).$$

Here again we have identified vectors in Euclidean spaces $\mathbb{R}^n$ and $\mathbb{R}^m$ with matrices in $\mathbb{M}(n, 1)$ and $\mathbb{M}(m, 1)$, respectively. We will make frequent use of this identification throughout the remainder of the text. We will also use the notation μ_A for the linear function as defined above.

Our work with coordinate vectors provides two important examples of linear maps. Let $B = \{\mathbf{u}_1, \ldots, \mathbf{u}_n\}$ be an ordered basis for a vector space V. Define $C_B : V \rightarrow \mathbb{R}^n$ by $C_B(\mathbf{v}) = [\mathbf{v}]_B$. Exercise 8 of Section 3.6 is precisely the verification that C_B is linear. We can reverse this coordinate vector function by using the coordinates of an element

$$\begin{bmatrix} r_1 \\ \vdots \\ r_n \end{bmatrix} \in \mathbb{R}^n$$

as coefficients in a linear combination of the vectors in any set $B = \{\mathbf{v}_1, \ldots, \mathbf{v}_n\}$. That is, define $L_B : \mathbb{R}^n \rightarrow V$ by

$$L_B\left(\begin{bmatrix} r_1 \\ \vdots \\ r_n \end{bmatrix}\right) = r_1\mathbf{v}_1 + \cdots + r_n\mathbf{v}_n.$$

Quick Example   *Show that L_B as defined above is linear.*

Begin by setting up the notation needed for this straightforward verification of the definition of linearity. Let

$$\begin{bmatrix} r_1 \\ \vdots \\ r_n \end{bmatrix}, \begin{bmatrix} s_1 \\ \vdots \\ s_n \end{bmatrix} \in \mathbb{R}^n \quad \text{and} \quad c \in \mathbb{R}.$$

Then

$$L_B\left(\begin{bmatrix} r_1 \\ \vdots \\ r_n \end{bmatrix} + \begin{bmatrix} s_1 \\ \vdots \\ s_n \end{bmatrix}\right) = L_B\left(\begin{bmatrix} r_1 + s_1 \\ \vdots \\ r_n + s_n \end{bmatrix}\right)$$

$$= (r_1 + s_1)\mathbf{v}_1 + \cdots + (r_n + s_n)\mathbf{v}_n$$

$$= (r_1\mathbf{v}_1 + \cdots + r_n\mathbf{v}_n) + (s_1\mathbf{v}_1 + \cdots + s_n\mathbf{v}_n)$$

$$= L_B\left(\begin{bmatrix} r_1 \\ \vdots \\ r_n \end{bmatrix}\right) + L_B\left(\begin{bmatrix} s_1 \\ \vdots \\ s_n \end{bmatrix}\right),$$

and

$$L_B\left(c\begin{bmatrix} r_1 \\ \vdots \\ r_n \end{bmatrix}\right) = L_B\left(\begin{bmatrix} cr_1 \\ \vdots \\ cr_n \end{bmatrix}\right)$$

$$= cr_1\mathbf{v}_1 + \cdots + cr_n\mathbf{v}_n$$

$$= c(r_1\mathbf{v}_1 + \cdots + r_n\mathbf{v}_n)$$

$$= cL_B\left(\begin{bmatrix} r_1 \\ \vdots \\ r_n \end{bmatrix}\right). \qquad \blacksquare$$

The following theorem lists three basic properties of linear maps.

Theorem 6.2

Suppose $T : V \to W$ is linear. Then

a. $T(\mathbf{0}_V) = \mathbf{0}_W$, where $\mathbf{0}_V$ is the additive identity for V and $\mathbf{0}_W$ is the additive identity for W.

b. $T(-\mathbf{v}) = -T(\mathbf{v})$ for any $\mathbf{v} \in V$.

c. $T(r_1\mathbf{v}_1 + \cdots + r_n\mathbf{v}_n) = r_1 T(\mathbf{v}_1) + \cdots + r_n T(\mathbf{v}_n)$ for any $\mathbf{v}_1, \ldots, \mathbf{v}_n \in V$ and $r_1, \ldots, r_n \in \mathbb{R}$.

Proof   The proof of the first result uses Theorem 1.4:

$$T(\mathbf{0}_V) = T(0\mathbf{0}_V) = 0T(\mathbf{0}_V) = \mathbf{0}_W.$$

The proof of the second result uses Theorem 1.5, part d:

$$T(-\mathbf{v}) = T((-1)\mathbf{v}) = (-1)T(\mathbf{v}) = -T(\mathbf{v}).$$

The proof of the third result uses the additivity condition $n - 1$ times and the homogeneity condition n times:

$$T(r_1\mathbf{v}_1 + \cdots + r_n\mathbf{v}_n) = T(r_1\mathbf{v}_1) + \cdots + T(r_n\mathbf{v}_n)$$

$$= r_1 T(\mathbf{v}_1) + \cdots + r_n T(\mathbf{v}_n). \qquad \blacksquare$$

The final theorem of this section states that a linear map is completely determined by its values on a spanning set. This result is extremely helpful in verifying formulas for linear maps. If we have a spanning set for the domain of the linear map, we can often check the formula for the elements of this set and use the theorem to extend the validity of the formula to the entire domain.

Theorem 6.3

Suppose $\{v_1, \ldots, v_n\}$ spans a vector space V. Suppose $T : V \to W$ and $T' : V \to W$ are linear maps such that $T(v_i) = T'(v_i)$ for $i = 1, \ldots, n$. Then $T = T'$.

Proof   In order to conclude that the two functions T and T' are equal, we need to show that their values $T(v)$ and $T'(v)$ are equal for any vector $v \in V$. So let v be an arbitrary vector in V. Since $\{v_1, \ldots, v_n\}$ spans V, we can write $v = r_1 v_1 + \cdots + r_n v_n$ for real numbers $r_1, \ldots, r_n$. Now

$$
\begin{aligned}
T(v) &= T(r_1 v_1 + \cdots + r_n v_n) \\
&= r_1 T(v_1) + \cdots + r_n T(v_n) \\
&= r_1 T'(v_1) + \cdots + r_n T'(v_n) \\
&= T'(r_1 v_1 + \cdots + r_n v_n) \\
&= T'(v). \quad \blacksquare
\end{aligned}
$$

Here is a typical application of this theorem.

Quick Example   *Suppose the linear map $T : \mathbb{R}^3 \to \mathbb{R}^4$ satisfies*

$$T(0, 0, 1) = \mathbf{0}, \qquad T(0, 1, 1) = \mathbf{0}, \qquad T(1, 1, 1) = \mathbf{0}.$$

Show that $T(v) = \mathbf{0}$ for all vectors $v \in \mathbb{R}^3$.

By Exercise 10 at the end of this section, the zero function $Z : \mathbb{R}^3 \to \mathbb{R}^4$ defined by $Z(v) = \mathbf{0}$ for all $v \in \mathbb{R}^3$ is linear. It is easy to verify that $\{(0, 0, 1), (0, 1, 1), (1, 1, 1)\}$ is a spanning set for $\mathbb{R}^3$. Since T and Z assign the same values to the elements of this spanning set, Theorem 6.3 says that these functions are equal. That is, $T(v) = Z(v) = \mathbf{0}$ for all $v \in \mathbb{R}^3$. $\quad \blacksquare$

Exercises 6.1

1. Show that $P : \mathbb{R}^3 \to \mathbb{R}^3$ defined by $P\left(\begin{bmatrix} x \\ y \\ z \end{bmatrix} \right) = \begin{bmatrix} x \\ y \\ 0 \end{bmatrix}$ is linear. Find a matrix A such that $P\left(\begin{bmatrix} x \\ y \\ z \end{bmatrix} \right) = A \begin{bmatrix} x \\ y \\ z \end{bmatrix}$.

2. Show that the function $f : \mathbb{R} \to \mathbb{R}$ defined by $f(x) = 5x + 3$ is not linear.

3. Suppose v_0 is a vector in an inner product space V. Show that $T : V \to \mathbb{R}$ defined by $T(v) = \langle v, v_0 \rangle$ is linear.

4. Suppose x_0 is a real number. Show that the evaluation map $T : \mathbb{F}(\mathbb{R}) \to \mathbb{R}$ defined by $T(f) = f(x_0)$ is linear.

5. Show that $T : \mathbb{P}_2 \to \mathbb{P}_3$ defined by $T(ax^2 + bx + c) = a(x-1)^2 + b(x-1) + c$ is linear.

6. Show that $T : \mathbb{D}^{(2)}(\mathbb{R}) \to \mathbb{F}(\mathbb{R})$ defined by $T(f) = f'' + 3f' - 2f$ is linear.

7. Show that $T : \mathbb{P}_2 \to \mathbb{P}_4$ defined by $T(p(x)) = x^3 p'(x)$ is linear.

8. Consider the transpose operator $T : \mathbb{M}(2,2) \to \mathbb{M}(2,2)$ defined by $T\left(\begin{bmatrix} a & b \\ c & d \end{bmatrix} \right) = \begin{bmatrix} a & c \\ b & d \end{bmatrix}$. Show that T is linear.

9. Consider the identity function $\mathrm{id}_V : V \to V$ defined on a vector space V by $\mathrm{id}_V(v) = v$ for all $v \in V$. Show that id_V is linear.

10. Consider the zero function $Z : V \to W$ defined on a vector space V by $Z(v) = 0_W$ for all $v \in V$. Show that Z is linear.

11. For $a \in \mathbb{R}$ consider the scalar multiplication function $\sigma_a : V \to V$ defined on a vector space V by $\sigma_a(v) = av$. Show that σ_a is linear.

12. For an element v_0 of a vector space V, consider the translation function $\tau_{v_0} : V \to V$ defined by $\tau_{v_0}(v) = v + v_0$. Under what conditions on v_0 will τ_{v_0} be linear?

13. a. Show that $S = \{f \in \mathbb{C}([0, \infty)) \mid f \text{ is bounded}\}$ is a subspace of $\mathbb{C}([0, \infty))$.

    b. For any $f \in S$ and any $x > 0$, show that the improper integral $\int_0^\infty e^{-xt} f(t)\, dt$ converges.

    c. The **Laplace transform** of a function $f \in S$ is a function $\mathcal{L}(f) : (0, \infty) \to \mathbb{R}$ defined by $(\mathcal{L}(f))(x) = \int_0^\infty e^{-xt} f(t)\, dt$. Show that $\mathcal{L} : S \to \mathbb{F}((0, \infty))$ is linear.

14. Give an alternative proof of part a of Theorem 6.2 based on the fact that $0 + 0 = 0$.

15. Suppose $T : V \to W$ is linear.

    a. Show that $\{v \in V \mid T(v) = 0\}$ is a subspace of the domain V.

    b. Show that $\{w \in W \mid w = T(v) \text{ for some } v \in V\}$ is a subspace of the range W.

16. Provide a justification for each link in the chain of equalities between $T(v)$ and $T'(v)$ in the proof of Theorem 6.3.

17. Suppose $T : \mathbb{R}^3 \to \mathbb{P}_3$ is linear and that

$$T\left(\begin{bmatrix} 1 \\ 1 \\ 1 \end{bmatrix} \right) = x^3 + 2x, \quad T\left(\begin{bmatrix} 0 \\ 1 \\ 1 \end{bmatrix} \right) = x^2 + x + 1, \quad T\left(\begin{bmatrix} 0 \\ 0 \\ 1 \end{bmatrix} \right) = 2x^3 - x^2.$$

a. Compute $T\left(\begin{bmatrix} 5 \\ 3 \\ -1 \end{bmatrix} \right)$. b. Compute $T\left(\begin{bmatrix} a \\ b \\ c \end{bmatrix} \right)$.

18. Suppose the linear map $L : \mathbb{R}^2 \to \mathbb{R}^2$ satisfies $L(1, 1) = (3, 3)$ and $L(1, -1) = (3, -3)$. Show that $L(\mathbf{v}) = 3\mathbf{v}$ for all $\mathbf{v} \in \mathbb{R}^2$.

19. Suppose the linear map $L : \mathbb{M}(2, 2) \to \mathbb{M}(2, 2)$ satisfies

$$L\left(\begin{bmatrix} 1 & 0 \\ 0 & 0 \end{bmatrix}\right) = \begin{bmatrix} 1 & 0 \\ 0 & 0 \end{bmatrix}, \qquad L\left(\begin{bmatrix} 0 & 1 \\ 0 & 0 \end{bmatrix}\right) = \begin{bmatrix} 0 & 0 \\ 1 & 0 \end{bmatrix},$$

$$L\left(\begin{bmatrix} 0 & 0 \\ 1 & 0 \end{bmatrix}\right) = \begin{bmatrix} 0 & 1 \\ 0 & 0 \end{bmatrix}, \qquad L\left(\begin{bmatrix} 0 & 0 \\ 0 & 1 \end{bmatrix}\right) = \begin{bmatrix} 0 & 0 \\ 0 & 1 \end{bmatrix}.$$

Show that $L\left(\begin{bmatrix} a & b \\ c & d \end{bmatrix}\right) = \begin{bmatrix} a & c \\ b & d \end{bmatrix}$ for all matrices $\begin{bmatrix} a & b \\ c & d \end{bmatrix} \in \mathbb{M}(2, 2)$.

20. Suppose the linear map $T : \mathbb{P}_2 \to \mathbb{P}_2$ satisfies $T(x^2) = 2x$, $T(x) = 1$, and $T(1) = 0$. Show that T is the differentiation operator on $\mathbb{P}_2$.

21. Suppose $T : \mathbb{R}^2 \to \mathbb{R}^2$ satisfies $T\left(\begin{bmatrix} 1 \\ 0 \end{bmatrix}\right) = \begin{bmatrix} -2 \\ 3 \end{bmatrix}$, $T\left(\begin{bmatrix} 0 \\ 1 \end{bmatrix}\right) = \begin{bmatrix} 5 \\ 1 \end{bmatrix}$, and $T\left(\begin{bmatrix} 1 \\ 1 \end{bmatrix}\right) = \begin{bmatrix} 3 \\ 2 \end{bmatrix}$. Show that T cannot be linear.

22. Suppose $T : \mathbb{R}^2 \to \mathbb{P}_3$ satisfies $T\left(\begin{bmatrix} 1 \\ 4 \end{bmatrix}\right) = x$, $T\left(\begin{bmatrix} 4 \\ 1 \end{bmatrix}\right) = x^2$, and $T\left(\begin{bmatrix} 1 \\ 1 \end{bmatrix}\right) = x^2 + x$. Show that T cannot be linear.

23. Suppose a function $T : V \to W$ from a vector space V to a vector space W satisfies the property that $T(\mathbf{v} + \mathbf{w}) = T(\mathbf{v}) + T(\mathbf{w})$ for all $\mathbf{v}, \mathbf{w} \in V$. Let $\mathbf{v}$ be an arbitrary element of V.

    a. Show that $T(2\mathbf{v}) = 2T(\mathbf{v})$.

    b. Show that $T(r\mathbf{v}) = rT(\mathbf{v})$ for any positive integer r.

    c. Show that $T(0\mathbf{v}) = 0T(\mathbf{v})$.

    d. Show that $T(r\mathbf{v}) = rT(\mathbf{v})$ for any integer r.

    e. Show that $T(r\mathbf{v}) = rT(\mathbf{v})$ for any rational number r.

    f. Speculate as to problems in extending the homogeneous property to all real numbers.

6.2 Compositions and Inverses

In your study of algebra and calculus you undoubtedly encountered compositions and inverses of functions. In this section, we will review these two operations for obtaining new functions from old ones and see how they work for linear functions between vector spaces.

The concept of the inverse of a function is motivated by the problem of solving an equation. This can range from something extremely simple, such as

$$2x + 1 = 7,$$

to an equation that requires a bit of work,

$$2x^3 - 5x^2 + 5x = 3,$$

or even

$$e^x - x^2 = 5,$$

where an approximate solution is the best we can hope to obtain. In each case the value of a function is set equal to a real number, and we want to know if there are any values of the independent variable that satisfy the equation. That is, we want to know if the real number on the right side of the equation is in the image of the function. In the first example, the image of the function defined by $f(x) = 2x+1$ is the set of all real numbers. From this we know that the equation

$$f(x) = 2x + 1 = c$$

has a solution for any value of the constant c. Furthermore, a little algebra or a glance at the graph of this function shows that the only way two real numbers x_1 and x_2 will give the same value for $f(x_1)$ and $f(x_2)$ is when x_1 is equal to x_2. Thus, once a constant c is specified, the equation $f(x) = 2x + 1 = c$ has a unique solution. Exercise 3 at the end of this section asks you to show that the functions involved in the other two examples above behave similarly.

The two concepts hinted at in these examples are made explicit in the following definition.

Definition 6.4

Suppose $f : X \to Y$ is a function from the set X to the set Y. The function f is **one-to-one** if and only if for any $x_1, x_2 \in X$,

$$f(x_1) = f(x_2) \quad \text{implies} \quad x_1 = x_2.$$

The function f is **onto** if and only if for every $y \in Y$ there is $x \in X$ with $f(x) = y$.

This definition is worthy of a few brief comments. First, notice that the condition for a function to be one-to-one involves an implication:

$$f(x_1) = f(x_2) \implies x_1 = x_2.$$

It is occasionally helpful to work with the logically equivalent contrapositive implication

$$x_1 \neq x_2 \implies f(x_1) \neq f(x_2).$$

Second, keep in mind the distinction between the image of a function and the range of the function. For a function $f : X \to Y$, the **image** is the subset

$$\{y \in Y \mid f(x) = y \text{ for some } x \in X\}$$

of the range Y. The function f is onto if and only if these two sets are equal; that is, the image is all of the range. Any function can be modified so that it becomes onto simply by replacing the range with the image. This demonstrates the importance of specifying the range as part of the definition of any function. The same rule with different ranges results in different functions. For example, $\sin : \mathbb{R} \to [-1, 1]$ is onto whereas $\sin : \mathbb{R} \to \mathbb{R}$ is not onto.

Here is an example to illustrate these definitions in a typical example of a function between vector spaces.

Quick Example   *Consider the linear function* $T : \mathbb{R}^2 \to \mathbb{R}^2$ *defined by*

$$T\left(\begin{bmatrix} x \\ y \end{bmatrix}\right) = \begin{bmatrix} 2x + y \\ -3x \end{bmatrix}.$$

Show that T is one-to-one and onto.

To verify that T is one-to-one, we assume that vectors $\begin{bmatrix} x_1 \\ y_1 \end{bmatrix}$ and $\begin{bmatrix} x_2 \\ y_2 \end{bmatrix}$ of the domain have equal images under the function T. The assumption that

$$T\left(\begin{bmatrix} x_1 \\ y_1 \end{bmatrix}\right) = T\left(\begin{bmatrix} x_2 \\ y_2 \end{bmatrix}\right)$$

translates into the vector equation

$$\begin{bmatrix} 2x_1 + y_1 \\ -3x_1 \end{bmatrix} = \begin{bmatrix} 2x_2 + y_2 \\ -3x_2 \end{bmatrix}.$$

The equalities of the corresponding components of these vectors in turn give the two equations

$$2x_1 + y_1 = 2x_2 + y_2 \quad \text{and} \quad -3x_1 = -3x_2.$$

From the second equation it follows that $x_1 = x_2$. From this and the first equation it follows that $y_1 = y_2$. Hence, we have the equality $\begin{bmatrix} x_1 \\ y_1 \end{bmatrix} = \begin{bmatrix} x_2 \\ y_2 \end{bmatrix}$ required to show that T is one-to-one.

To verify that T is onto, we consider an arbitrary element $\begin{bmatrix} a \\ b \end{bmatrix}$ of the range $\mathbb{R}^2$. Our job is to demonstrate that we can always solve the equation

$$T\left(\begin{bmatrix} x \\ y \end{bmatrix}\right) = \begin{bmatrix} 2x + y \\ -3x \end{bmatrix} = \begin{bmatrix} a \\ b \end{bmatrix}.$$

The Gauss-Jordan reduction algorithm is guaranteed to solve the resulting system of linear equations; however, in this simple case a little algebra is all it takes to obtain the explicit solution $x = -\frac{1}{3}b$ and $y = a + \frac{2}{3}b$. ∎

The formulas for x and y in the previous example give us more than we bargained for. They provide a recipe for undoing the effect of the function T. In other words, we have obtained a function that assigns to the image $T\left(\begin{bmatrix} x \\ y \end{bmatrix}\right)$ the original point $\begin{bmatrix} x \\ y \end{bmatrix}$. The following definition reveals the complete story.

Definition 6.5

Consider two functions $f : X \to Y$ and $g : Y \to Z$. The **composition** of f followed by g is the function $g \circ f : X \to Z$ defined by $(g \circ f)(x) = g(f(x))$. The **identity function** on any set X is the function $\mathrm{id}_X : X \to X$ defined by

$\mathrm{id}_X(x) = x$. A function $g : Y \rightarrow X$ is an **inverse** of the function $f : X \rightarrow Y$ if and only if

$$g \circ f = \mathrm{id}_X \quad \text{and} \quad f \circ g = \mathrm{id}_Y.$$

This definition also is worthy of a few comments. First, keep in mind that even in cases where $f \circ g$ and $g \circ f$ are both defined, they will not in general be equal. For example, if $f(x) = \sin x$ and $g(x) = x^2$, no self-respecting mathematics student would confuse $(f \circ g)(x) = \sin x^2$ with $(g \circ f)(x) = \sin^2 x = (\sin x)^2$. Second, concerning inverses, if a function $f : X \rightarrow Y$ has an inverse function, then it has only one inverse function. You are asked to prove this result in Exercise 11 at the end of this section. The inverse of f is often denoted f^{-1}. You need to rely on the context of the discussion to distinguish this notation for the inverse with respect to the operation of composing functions from the notation of the ordinary reciprocal (the inverse with respect to the operation of multiplication). Finally, the condition for g to be an inverse of f is that the two identities hold:

$$g(f(x)) = x \text{ for all } x \in X \quad \text{and} \quad f(g(y)) = y \text{ for all } y \in Y.$$

The example preceding Definition 6.5 hinted at a relation between the concepts of one-to-one and onto on one hand and the existence of an inverse function on the other hand. The following theorem shows that this relation was no accident.

Theorem 6.6

A function $f : X \rightarrow Y$ has an inverse if and only if f is one-to-one and onto.

Proof   Suppose $g : Y \rightarrow X$ is an inverse of f. To show that f is one-to-one, we need to start with the hypothesis that $f(x_1) = f(x_2)$ for arbitrary elements $x_1, x_2 \in X$. To derive the conclusion that $x_1 = x_2$, we apply the function g to both sides of the given equation and use the fact that $g \circ f = \mathrm{id}_X$:

$$x_1 = g(f(x_1)) = g(f(x_2)) = x_2.$$

To show f is onto, let y denote an arbitrary element of Y. Then $g(y)$ is an element of X such that $f(g(y)) = y$.

Conversely, suppose f is one-to-one and onto. Define a function $g : Y \rightarrow X$ that assigns to each element $y \in Y$ some element $x \in X$ with $f(x) = y$. The assumption that f is onto ensures that we can always locate such an element. (In fact, because f is also one-to-one, there will be exactly one element of X that f maps to a given $y \in Y$.) Now $f(g(y)) = f(x) = y$. Also, for any $x \in X$, $g(f(x))$ is an element of X such that $f(g(f(x))) = f(x)$. Since f is one-to-one, this latter equation implies $g(f(x)) = x$.   ∎

The next two theorems apply the concepts of composition and inverse to linear functions. The proof of the first theorem is a good warm-up exercise for your next exam. See Exercise 15 at the end of this section.

Theorem 6.7

Suppose $S : U \to V$ and $T : V \to W$ are linear maps. Then the composition $T \circ S : U \to W$ is linear.

The proof of the second theorem is rather endearing in the way the properties of inverses and linearity eagerly fall into place to give the desired result.

Theorem 6.8

Suppose the linear function $T : V \to W$ has an inverse function. Then $T^{-1} : W \to V$ is linear.

Proof   Let $\mathbf{w}_1, \mathbf{w}_2 \in W$ and $r \in \mathbb{R}$. Let $\mathbf{v}_1 = T^{-1}(\mathbf{w}_1)$ and $\mathbf{v}_2 = T^{-1}(\mathbf{w}_2)$. Then

$$T(\mathbf{v}_1) = T(T^{-1}(\mathbf{w}_1)) = \mathbf{w}_1 \quad \text{and} \quad T(\mathbf{v}_2) = T(T^{-1}(\mathbf{w}_2)) = \mathbf{w}_2.$$

So

$$
\begin{aligned}
T^{-1}(\mathbf{w}_1 + \mathbf{w}_2) &= T^{-1}\big(T(\mathbf{v}_1) + T(\mathbf{v}_2)\big) && \text{substitution;} \\
&= T^{-1}\big(T(\mathbf{v}_1 + \mathbf{v}_2)\big) && \text{T is linear;} \\
&= \mathbf{v}_1 + \mathbf{v}_2 && \text{$T^{-1} \circ T = \mathrm{id}_V$;} \\
&= T^{-1}(\mathbf{w}_1) + T^{-1}(\mathbf{w}_2) && \text{substitution.}
\end{aligned}
$$

Also,

$$
\begin{aligned}
T^{-1}(r\mathbf{w}_1) &= T^{-1}\big(rT(\mathbf{v}_1)\big) && \text{substitution;} \\
&= T^{-1}\big(T(r\mathbf{v}_1)\big) && \text{T is linear;} \\
&= r\mathbf{v}_1 && \text{$T^{-1} \circ T = \mathrm{id}_V$;} \\
&= rT^{-1}(\mathbf{w}_1) && \text{substitution.} \quad \blacksquare
\end{aligned}
$$

The final theorem of this section is a companion to Theorem 6.3 at the end of the previous section. It shows how easy it is to define a linear map if we are given a basis for the domain.

Theorem 6.9

Suppose $B = \{\mathbf{v}_1, \ldots, \mathbf{v}_n\}$ is a basis for a vector space V. Then for any elements $\mathbf{w}_1, \ldots, \mathbf{w}_n$ of a vector space W, there is a unique linear map $T : V \to W$ such that $T(\mathbf{v}_i) = \mathbf{w}_i$ for $i = 1, \ldots, n$.

Proof   Define $T : V \to W$ by $T(\mathbf{v}) = r_1\mathbf{w}_1 + \cdots + r_n\mathbf{w}_n$ where $[\mathbf{v}]_B = \begin{bmatrix} r_1 \\ \vdots \\ r_n \end{bmatrix}$ is the coordinate vector of $\mathbf{v}$ with respect to the basis B. If we let $B' = \{\mathbf{w}_1, \ldots, \mathbf{w}_n\}$, then we can write $T(\mathbf{v}) = L_{B'}([\mathbf{v}]_B)$, where $L_{B'}$ is the linear combination function defined in Section 6.1. Since $L_{B'}$ and the coordinate vector function are linear, Theorem 6.7 gives that the composition T is linear. Of course,

$$T(\mathbf{v}_i) = L_{B'}([\mathbf{v}_i]_B) = L_{B'}(\mathbf{e}_i) = 0\mathbf{w}_1 + \cdots + 1\mathbf{w}_i + \cdots + 0\mathbf{w}_n = \mathbf{w}_i.$$

By Theorem 6.3, there is only one linear map with these values on the elements of the spanning set B. ∎

Exercises 6.2

1. For each of the following functions, either show the function is one-to-one by verifying the logical implication given in the definition, or show that the implication does not hold by finding two points with the same image.

   a. $f : \mathbb{R} \to \mathbb{R}$ defined by $f(x) = \frac{1}{3}x - 2$.

   b. $p : \mathbb{R} \to \mathbb{R}$ defined by $p(x) = x^2 - 3x + 2$.

   c. $s : \mathbb{R} \to \mathbb{R}$ defined by $s(x) = (e^x - e^{-x})/2$.

   d. $W : \mathbb{R} \to \{(x, y) \in \mathbb{R}^2 \mid x^2 + y^2 = 1\}$ defined by $W(t) = (\cos t, \sin t)$.

   e. $L : \mathbb{R}^3 \to \mathbb{R}^3$ defined by $L\left(\begin{bmatrix} x \\ y \\ z \end{bmatrix}\right) = \begin{bmatrix} 2x + y - z \\ -x + 2z \\ x + y + z \end{bmatrix}$.

2. For each of the following functions, either show the function is onto by choosing an arbitrary element of the range and finding an element of the domain that the function maps to the chosen element, or show the function is not onto by finding an element of the range that is not in the image of the function.

   a. $f : \mathbb{R} \to \mathbb{R}$ defined by $f(x) = \frac{1}{3}x - 2$.

   b. $p : \mathbb{R} \to \mathbb{R}$ defined by $p(x) = x^2 - 3x + 2$.

   c. $s : \mathbb{R} \to \mathbb{R}$ defined by $s(x) = (e^x - e^{-x})/2$.

   d. $W : \mathbb{R} \to \{(x, y) \in \mathbb{R}^2 \mid x^2 + y^2 = 1\}$ defined by $W(t) = (\cos t, \sin t)$.

   e. $L : \mathbb{R}^3 \to \mathbb{R}^3$ defined by $L\left(\begin{bmatrix} x \\ y \\ z \end{bmatrix}\right) = \begin{bmatrix} 2x + y - z \\ -x + 2z \\ x + y + z \end{bmatrix}$.

3. The three functions in this exercise come from the examples given at the beginning of this section. In each case, use the derivative of the function to show that the function is increasing and hence is one-to-one. Use the behavior of the function for real numbers that are arbitrarily large in absolute value (positive as well as negative) to show that the function is onto.

   a. $f : \mathbb{R} \to \mathbb{R}$ defined by $f(x) = 2x + 1$.

   b. $g : \mathbb{R} \to \mathbb{R}$ defined by $g(x) = 2x^3 - 5x^2 + 5x$.

   c. $h : \mathbb{R} \to \mathbb{R}$ defined by $h(x) = e^x - x^2$.

4. **a.** Let $D : \mathbb{P}_3 \to \mathbb{P}_3$ denote the differentiation operation. Is D one-to-one? Is it onto?

b. Let $D : \mathbb{P} \to \mathbb{P}$ denote the differentiation operation. Is D one-to-one? Is it onto?

c. Let $D : \mathbb{D}(\mathbb{R}) \to \mathbb{F}(\mathbb{R})$ denote the differentiation operation. Is D one-to-one? Is it onto?

5. Let A denote the antidifferentiation operation that assigns to an integrable function $f : \mathbb{R} \to \mathbb{R}$ the function $A(f) : \mathbb{R} \to \mathbb{R}$ defined by

$$(A(f))(x) = \int_0^x f(t)\, dt.$$

a. Is $A : \mathbb{P} \to \mathbb{P}$ one-to-one? Is it onto?

b. Is $A : \mathbb{C}(\mathbb{R}) \to \mathbb{D}(\mathbb{R})$ one-to-one? Is it onto?

c. Let D denote the differentiation operation as in the previous exercise. For any $f \in \mathbb{D}(\mathbb{R})$ show that $(D \circ A)(f) = f$. Does $(A \circ D)(f) = f$?

6. Explain why the exponential function $\exp : \mathbb{R} \to \mathbb{R}$ does not have an inverse. What simple modification will remedy this problem?

7. Explain why the sine function $\sin : \mathbb{R} \to \mathbb{R}$ does not have an inverse. What are the traditional modifications made to this function to create a function that does have an inverse?

8. For each of the following functions, find a formula for the inverse of the function or explain the difficulties you encounter.

a. $f : \mathbb{R} \to \mathbb{R}$ defined by $f(x) = 2x + 1$.

b. $p : \mathbb{R} \to \mathbb{R}$ defined by $p(x) = x^2 - 3x + 2$.

c. $s : \mathbb{R} \to \mathbb{R}$ defined by $s(x) = (e^x - e^{-x})/2$.

d. $W : \mathbb{R} \to \{(x, y) \in \mathbb{R}^2 \mid x^2 + y^2 = 1\}$ defined by $W(t) = (\cos t, \sin t)$.

e. $L : \mathbb{R}^3 \to \mathbb{R}^3$ defined by $L\left(\begin{bmatrix} x \\ y \\ z \end{bmatrix}\right) = \begin{bmatrix} 2x + y - z \\ -x + 2z \\ x + y + z \end{bmatrix}$.

f. $g : \mathbb{R} \to \mathbb{R}$ defined by $g(x) = 2x^3 - 5x^2 + 5x$.

g. $h : \mathbb{R} \to \mathbb{R}$ defined by $h(x) = e^x - x^2$.

9. An analysis of the proof of Theorem 6.6 reveals that the theorem actually breaks into two independent statements. Give separate proofs of the following results.

a. A function $f : X \to Y$ is onto if and only if there is a function $g : Y \to X$ such that $f \circ g = \text{id}_Y$.

b. Suppose X is a nonempty set. A function $f : X \to Y$ is one-to-one if and only if there is a function $g : Y \to X$ such that $g \circ f = \text{id}_X$.

10. Show that composition of functions is associative. That is, if $f : W \to X$, $g : X \to Y$, and $h : Y \to Z$, then $h \circ (g \circ f) = (h \circ g) \circ f$.

11. Prove that if a function $f : X \to Y$ has an inverse, the inverse is unique. Model your proof on the proof of Theorem 5.7.

12. Prove that the composition of one-to-one functions is one-to-one.

13. Prove that the composition of onto functions is onto.

14. Suppose the functions $f : X \to Y$ and $g : Y \to Z$ have inverses. Show that $g \circ f$ has an inverse and $(g \circ f)^{-1} = f^{-1} \circ g^{-1}$.

15. Prove Theorem 6.7. (Suggestion: Write out the hypothesis in terms of the defining equations for the linearity of S and T. Write out the two equations you need to prove to establish that $T \circ S$ is linear. In each of these equations, start with one side and use the given equations to string together equalities until you reach the other side.)

16. For an element $\mathbf{v}_0$ of a vector space V, consider the translation function $\tau_{\mathbf{v}_0} : V \to V$ defined by $\tau_{\mathbf{v}_0}(\mathbf{v}) = \mathbf{v} + \mathbf{v}_0$. Show that $\tau_{\mathbf{v}_0}$ is invertible. Show that $\tau_{\mathbf{v}_0}^{-1}$ is also a translation function.

17. For $a \in \mathbb{R}$ consider the scalar multiplication function $\sigma_a : V \to V$ defined on a vector space V by $\sigma_a(\mathbf{v}) = a\mathbf{v}$. For what values of a does σ_a have an inverse function? For such a value, show that σ_a^{-1} is also a scalar multiplication function.

18. Give a geometric description of a one-to-one function from a square onto a circle.

19. Give a geometric description of a one-to-one function from the region of the plane bounded by a square onto the disk bounded by a circle.

20. Find a formula for the function from the x-axis to the unit circle centered at the origin defined as follows: Draw a line through a point on the x-axis and the point $(0, 1)$ on the unit circle. The other point of intersection between this line and the circle is assigned to the point on the x-axis. Show this function is one-to-one. What is the image?

21. **a.** Describe a one-to-one function from the natural numbers onto the even natural numbers.

   b. Describe a one-to-one function from the natural numbers onto the integers.

   c. Describe a one-to-one function from the natural numbers onto the rational numbers.

   d. Describe a one-to-one function from the interval $[0, 1)$ onto the interval $[0, 1]$.

   e. Describe a function from $\mathbb{R}$ onto $\mathbb{R}^2$. Is it possible for such a function to be one-to-one?

22. Consider the function whose domain is the set of finite strings of letters of the English alphabet. The function replaces each letter in a string with the next letter in the alphabet and replaces Z with A.

   a. Apply this rule to find out what is behind the omniscient computer aboard the spaceship in Stanley Kubrick's *2001: A Space Odyssey*.

   b. Show that this function is one-to-one.

   c. Show that this function is onto.

   d. Describe the inverse of this function.

6.3 Matrix of a Linear Function

Section 6.1 hinted at a relation between linear functions and matrix multiplication. The first theorem of this section shows how to write a linear function between Euclidean spaces in terms of matrix multiplication. Then, with the help of coordinate vectors, we will do the same for linear functions between any finite-dimensional vector spaces.

We begin by considering two examples that motivate the general development. In each instance, notice that the columns of the matrix are the vectors the function assigns to the standard basis vectors in $\mathbb{R}^2$.

Quick Example   *Suppose $T : \mathbb{R}^2 \to \mathbb{R}^2$ is a linear map with*

$$T\left(\begin{bmatrix} 1 \\ 0 \end{bmatrix}\right) = \begin{bmatrix} 9 \\ 4 \end{bmatrix} \quad \text{and} \quad T\left(\begin{bmatrix} 0 \\ 1 \end{bmatrix}\right) = \begin{bmatrix} 2 \\ -5 \end{bmatrix}.$$

Compute $T\left(\begin{bmatrix} 5 \\ 6 \end{bmatrix}\right)$. Give a formula for $T\left(\begin{bmatrix} a \\ b \end{bmatrix}\right)$ in terms of matrix multiplication.

$$T\left(\begin{bmatrix} 5 \\ 6 \end{bmatrix}\right) = T\left(5\begin{bmatrix} 1 \\ 0 \end{bmatrix} + 6\begin{bmatrix} 0 \\ 1 \end{bmatrix}\right)$$

$$= 5T\left(\begin{bmatrix} 1 \\ 0 \end{bmatrix}\right) + 6T\left(\begin{bmatrix} 0 \\ 1 \end{bmatrix}\right)$$

$$= 5\begin{bmatrix} 9 \\ 4 \end{bmatrix} + 6\begin{bmatrix} 2 \\ -5 \end{bmatrix}$$

$$= \begin{bmatrix} 57 \\ -10 \end{bmatrix}.$$

In general,

$$T\left(\begin{bmatrix} a \\ b \end{bmatrix}\right) = T\left(a\begin{bmatrix} 1 \\ 0 \end{bmatrix} + b\begin{bmatrix} 0 \\ 1 \end{bmatrix}\right)$$

$$= aT\left(\begin{bmatrix} 1 \\ 0 \end{bmatrix}\right) + bT\left(\begin{bmatrix} 0 \\ 1 \end{bmatrix}\right)$$

$$= a\begin{bmatrix} 9 \\ 4 \end{bmatrix} + b\begin{bmatrix} 2 \\ -5 \end{bmatrix}$$

$$= \begin{bmatrix} 9a + 2b \\ 4a - 5b \end{bmatrix}.$$

Thus, we see that we can write

$$T\left(\begin{bmatrix} a \\ b \end{bmatrix}\right) = \begin{bmatrix} 9 & 2 \\ 4 & -5 \end{bmatrix}\begin{bmatrix} a \\ b \end{bmatrix}$$

in terms of matrix multiplication. ∎

The second example has a more geometric flavor.

Quick Example   *Let $R_\theta : \mathbb{R}^2 \to \mathbb{R}^2$ be the function that rotates each point in $\mathbb{R}^2$ counterclockwise about the origin through an angle of θ. Compute $R_\theta\left(\begin{bmatrix} 1 \\ 0 \end{bmatrix}\right)$ and $R_\theta\left(\begin{bmatrix} 0 \\ 1 \end{bmatrix}\right)$. Give a geometric argument that R_θ is linear. Find a formula for $R_\theta\left(\begin{bmatrix} a \\ b \end{bmatrix}\right)$ in terms of matrix multiplication.*

A little trigonometry shows that

$$R_\theta\left(\begin{bmatrix} 1 \\ 0 \end{bmatrix}\right) = \begin{bmatrix} \cos\theta \\ \sin\theta \end{bmatrix} \quad \text{and} \quad R_\theta\left(\begin{bmatrix} 0 \\ 1 \end{bmatrix}\right) = \begin{bmatrix} -\sin\theta \\ \cos\theta \end{bmatrix}.$$

Since R_θ rotates parallelograms into parallelograms and preserves the lengths of vectors, we can see from Figure 6.1 that R_θ is linear.

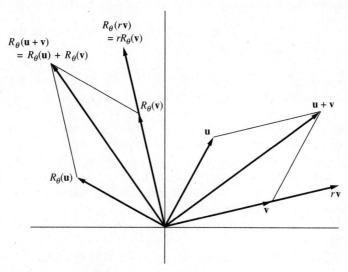

Figure 6.1   Rotation R_θ is a linear function.

We can put this information together now to find a formula for $R_\theta\left(\begin{bmatrix} a \\ b \end{bmatrix}\right)$ for any $\begin{bmatrix} a \\ b \end{bmatrix} \in \mathbb{R}^2$.

$$\begin{aligned}
R_\theta\left(\begin{bmatrix} a \\ b \end{bmatrix}\right) &= R_\theta\left(a\begin{bmatrix} 1 \\ 0 \end{bmatrix} + b\begin{bmatrix} 0 \\ 1 \end{bmatrix}\right) \\
&= aR_\theta\left(\begin{bmatrix} 1 \\ 0 \end{bmatrix}\right) + bR_\theta\left(\begin{bmatrix} 0 \\ 1 \end{bmatrix}\right) \\
&= a\begin{bmatrix} \cos\theta \\ \sin\theta \end{bmatrix} + b\begin{bmatrix} -\sin\theta \\ \cos\theta \end{bmatrix} \\
&= \begin{bmatrix} a\cos\theta - b\sin\theta \\ a\sin\theta + b\cos\theta \end{bmatrix} \\
&= \begin{bmatrix} \cos\theta & -\sin\theta \\ \sin\theta & \cos\theta \end{bmatrix}\begin{bmatrix} a \\ b \end{bmatrix}.
\end{aligned}$$

The following theorem gives the recipe for cooking up the matrix associated with a linear map between Euclidean spaces. Notice how simple Theorem 6.3 makes the proof.

Theorem 6.10

Suppose $T : \mathbb{R}^n \to \mathbb{R}^m$ is linear. Let $\{e_1, \ldots, e_n\}$ be the standard basis for $\mathbb{R}^n$. There is a unique $m \times n$ matrix A such that $T(v) = Av$ for all $v \in \mathbb{R}^n$. The columns of this matrix are the m-tuples $T(e_1), \ldots, T(e_n)$ in that order.

Proof From the condition that $T(v) = Av$ for all $v \in \mathbb{R}^n$, we see that for $j = 1, \ldots, n$, the jth column of the matrix A must be $Ae_j = T(e_j)$. Thus, this condition determines A uniquely.

Now let A be the matrix whose columns are $T(e_1), \ldots, T(e_n)$ in that order. Since both T and multiplication by A are linear functions, and since $T(e_j) = Ae_j$ for $j = 1, \ldots, n$, Theorem 6.3 gives the result that these two functions are equal. That is, $T(v) = Av$ for all $v \in \mathbb{R}^n$. ∎

This theorem shows that any linear map between Euclidean spaces has a concrete representation in terms of matrix multiplication. What can we do if the range and domain of a linear map are not known to be Euclidean spaces? In general, this is quite a difficult question. However, if the range and domain are finite-dimensional vector spaces, we can use coordinate vectors to adapt the techniques that work for Euclidean spaces.

Here is the setup. Suppose $B = \{u_1, \ldots, u_n\}$ is a basis for a vector space V and $B' = \{u'_1, \ldots, u'_m\}$ is a basis for a vector space V'. Let $[\ ]_B$ and $[\ ]_{B'}$ denote the coordinate vector functions. Suppose $T : V \to V'$ is linear. Recall from Section 6.1 that $\mu_A : \mathbb{R}^n \to \mathbb{R}^m$ denotes the function defined in terms of multiplying a vector in $\mathbb{R}^n$ (considered as an element of $\mathbb{M}(n, 1)$) on the left by the matrix $A \in \mathbb{M}(m, n)$.

Here is an example to help you become acquainted with this fancy notation. By Exercise 5 of Section 6.1, the function $T : \mathbb{P}_2 \to \mathbb{P}_3$ defined by $T(ax^2 + bx + c) = a(x - 1)^2 + b(x - 1) + c$ is linear. Consider bases $B = \{1, x, x^2\}$ for $\mathbb{P}_2$ and $B' = \{1, x, x^2, x^3\}$ for $\mathbb{P}_3$. We want a matrix A such that for any $p \in \mathbb{P}_2$ we have

$$[T(p)]_{B'} = A[p]_B.$$

In light of Theorem 6.10 it is reasonable to expect that such a matrix will exist. All we have done is write the polynomials $p \in \mathbb{P}_2$ and $T(p) \in \mathbb{P}_3$ in terms of their coordinate vectors $[p]_B \in \mathbb{R}^3$ and $[T(p)]_{B'} \in \mathbb{R}^4$ so that matrix multiplication will make sense. Theorem 6.11 will fulfill our expectations.

We will rely extensively on diagrams such as

$$
\begin{array}{ccc}
\mathbb{P}_2 & \xrightarrow{\ T\ } & \mathbb{P}_3 \\
\ [\ ]_B \downarrow & & \downarrow [\ ]_{B'} \\
\mathbb{R}^3 & \xrightarrow{\ \mu_A\ } & \mathbb{R}^4
\end{array}
$$

$$A = \begin{bmatrix} a_{11} & \cdots & a_{1n} \\ & & \\ & & \\ a_{m1} & \cdots & a_{mn} \end{bmatrix}$$

to keep track of all the functions and vector spaces involved. Each arrow points from the domain to the range of the indicated function.

Here is how to determine the entries of the matrix A for this example. As is true of any $m \times n$ matrix, the jth column of A is $A\mathbf{e}_j$, where $\mathbf{e}_j$ is the jth standard basis element of $\mathbb{R}^n$. In our case the three columns of A are $A\mathbf{e}_1$, $A\mathbf{e}_2$, and $A\mathbf{e}_3$. We also know from past experience that

$$[1]_B = \mathbf{e}_1, \qquad [x]_B = \mathbf{e}_2, \qquad [x^2]_B = \mathbf{e}_3.$$

Hence, if the equation $[T(p)]_{B'} = A[p]_B$ is to hold, we must have

$$A\mathbf{e}_1 = A[1]_B = [T(1)]_{B'} = [1]_{B'} = \begin{bmatrix} 1 \\ 0 \\ 0 \\ 0 \end{bmatrix},$$

$$A\mathbf{e}_2 = A[x]_B = [T(x)]_{B'} = [x - 1]_{B'} = \begin{bmatrix} -1 \\ 1 \\ 0 \\ 0 \end{bmatrix},$$

$$A\mathbf{e}_3 = A[x^2]_B = [T(x^2)]_{B'} = [(x - 1)^2]_{B'} = \begin{bmatrix} 1 \\ -2 \\ 1 \\ 0 \end{bmatrix}.$$

Putting these columns together gives the matrix

$$A = \begin{bmatrix} 1 & -1 & 1 \\ 0 & 1 & -2 \\ 0 & 0 & 1 \\ 0 & 0 & 0 \end{bmatrix}.$$

Here is the theorem we have been waiting for. It guarantees that the matrix created as in the preceding example will indeed do the job.

Theorem 6.11

Suppose $B = \{\mathbf{u}_1, \ldots, \mathbf{u}_n\}$ is a basis for a vector space V and $B' = \{\mathbf{u}'_1, \ldots, \mathbf{u}'_m\}$ is a basis for a vector space V'. Suppose $T : V \to V'$ is linear. There is a unique $m \times n$ matrix A such that $[T(\mathbf{v})]_{B'} = A[\mathbf{v}]_B$ for all $\mathbf{v} \in V$. The columns of this matrix are the m-tuples $[T(\mathbf{u}_1)]_{B'}, \ldots, [T(\mathbf{u}_n)]_{B'}$ in that order.

This theorem can be summarized by the following diagram. If the matrix A is chosen as described in the theorem, then the composition of the functions across the top and down the right is equal to the composition of the functions down the left and across the bottom.

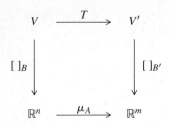

With the matrix A as specified, we can take any element $\mathbf{v} \in V$ and go across the top of the diagram to obtain $T(\mathbf{v}) \in V'$ and then down the right side to obtain $[T(\mathbf{v})]_{B'} \in \mathbb{R}^m$; or we can first go down the left to reach $[\mathbf{v}]_B \in \mathbb{R}^n$ and then across the bottom to obtain $A[\mathbf{v}]_B \in \mathbb{R}^m$. The conclusion of the theorem is that the two expressions denote the same vector in $\mathbb{R}^m$.

Proof   The identity $[T(\mathbf{v})]_{B'} = A[\mathbf{v}]_B$ implies that the jth column of A is $A\mathbf{e}_j = A[\mathbf{u}_j]_B = [T(\mathbf{u}_j)]_{B'}$. Thus, the condition determines A uniquely.

Now let A be the matrix whose columns are $[T(\mathbf{u}_1)]_{B'}, \ldots, [T(\mathbf{u}_n)]_{B'}$ in that order. This means that $[T(\mathbf{u}_j)]_{B'} = A\mathbf{e}_j = A[\mathbf{u}_j]_B$ for $j = 1, \ldots, n$. That is, the two functions have equal values for each element of the basis B of V. Notice that $[T(\mathbf{v})]_{B'}$ is the image of $\mathbf{v}$ under a composition of two linear functions, T and the coordinate vector function $[\ ]_{B'}$. By Theorem 6.7 this composition is linear. Similarly, $A[\mathbf{v}]_B$ is the result of applying to $\mathbf{v}$ the composition of two linear functions, the coordinate vector function $[\ ]_B$ and multiplication by A. Again by Theorem 6.7 this composition is linear. Theorem 6.3 gives the result that these two compositions are equal. That is, $[T(\mathbf{v})]_{B'} = A[\mathbf{v}]_B$ for all $\mathbf{v} \in V$. ∎

Definition 6.12

With the notation as in the previous theorem, A is the **matrix of T relative to the bases B and B'**.

♟ **Mathematical Strategy Session**   The statement of Theorem 6.11 gives a recipe for cooking up the matrix A of a linear transformation $T : V \to V'$ with respect to bases B and B': the jth column of A is $[T(\mathbf{u}_j)]_{B'}$. You may prefer not to rely on your memory to get the details straight in this eminently forgettable formula. As an alternative, you can apply the three simple principles used to prove this theorem:

1. The jth column of A is $A\mathbf{e}_j$.
2. $\mathbf{e}_j$ is the coordinate vector $[\mathbf{u}_j]_B$.
3. The diagram for A to be the matrix of T gives the equation $[T(\mathbf{v})]_{B'} = A[\mathbf{v}]_B$.

Then put these principles together for a foolproof derivation of the columns of A:

$$A\mathbf{e}_j = A[\mathbf{u}_j]_B = [T(\mathbf{u}_j)]_{B'}.$$

This conceptual approach was used in the example prior to the statement of Theorem 6.11. The few extra steps are a small price to pay for the confidence that you are using the right formula.

Exercises 6.3

1. Suppose $T : \mathbb{R}^2 \to \mathbb{R}^3$ is linear and that

$$T\left(\begin{bmatrix} 1 \\ 0 \end{bmatrix}\right) = \begin{bmatrix} 2 \\ 3 \\ -1 \end{bmatrix} \quad \text{and} \quad T\left(\begin{bmatrix} 0 \\ 1 \end{bmatrix}\right) = \begin{bmatrix} -5 \\ 1 \\ 1 \end{bmatrix}.$$

   a. Compute $T\left(\begin{bmatrix} 4 \\ 7 \end{bmatrix}\right)$.

   b. Compute $T\left(\begin{bmatrix} a \\ b \end{bmatrix}\right)$.

   c. Find a matrix A such that $T\left(\begin{bmatrix} a \\ b \end{bmatrix}\right) = A\begin{bmatrix} a \\ b \end{bmatrix}$ for all $\begin{bmatrix} a \\ b \end{bmatrix} \in \mathbb{R}^2$.

2. Find the matrix of the function $T : \mathbb{R}^3 \to \mathbb{R}^2$ defined by

$$T\left(\begin{bmatrix} x \\ y \\ z \end{bmatrix}\right) = \begin{bmatrix} 3x + y - 2z \\ x - y \end{bmatrix}.$$

3. Find the matrix of the function $T : \mathbb{R}^3 \to \mathbb{R}^4$ defined by

$$T\left(\begin{bmatrix} x \\ y \\ z \end{bmatrix}\right) = \begin{bmatrix} -y + z \\ x + 4y + 2z \\ 0 \\ -x - 3z \end{bmatrix}.$$

4. The letter "R" is drawn in the first quadrant so that its left side extends 1 unit along the y-axis and its base extends 1 unit along the x-axis. Each of the following diagrams shows the image of the letter under a linear transformation of $\mathbb{R}^2$ onto itself. Find the matrices of these linear transformations.

a.

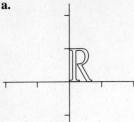

b.

c.

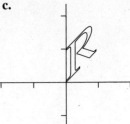

d.

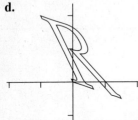

e.

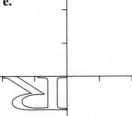

f.

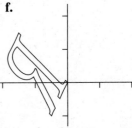

5. Suppose $T : \mathbb{R}^n \to \mathbb{R}^m$ is defined by $T(\mathbf{v}) = A\mathbf{v}$ for some matrix $A \in M(m, n)$. Prove that A is the matrix of T relative to the standard bases for $\mathbb{R}^n$ and $\mathbb{R}^m$.

6. Consider the function that maps each point of $\mathbb{R}^2$ to its reflection through the x-axis. Show that this function is linear. Find a matrix of this function relative to the standard basis for $\mathbb{R}^2$.

7. Consider the function $R_\varphi : \mathbb{R}^3 \to \mathbb{R}^3$ that rotates each point about the x-axis through an angle of φ.

   a. Give a geometric argument that R_φ is linear.

   b. Find a matrix A_φ such that $R_\varphi(\mathbf{v}) = A_\varphi \mathbf{v}$ for all $\mathbf{v} \in \mathbb{R}^3$.

   c. Set up appropriate notation and derive similar results for rotations about the y-axis and rotations about the z-axis.

8. Consider the linear map $D : \mathbb{P}_3 \to \mathbb{P}_3$ defined by $D(p(x)) = p'(x)$.

   a. Find the matrix of D relative to the basis $B = \{1, x, x^2, x^3\}$ used for both the domain and the range.

   b. Find the matrix of D relative to the basis $B' = \{1, x - 1, (x + 1)^2, (x - 1)^3\}$ used for both the domain and the range.

9. Consider $T : \mathbb{R}^2 \to \mathbb{R}^2$ that maps each point to its reflection through the line

$$\left\{ r \begin{bmatrix} 1 \\ 1 \end{bmatrix} \,\middle|\, r \in \mathbb{R} \right\}.$$

   a. Give a geometric argument that T is linear.

   b. Find the matrix of T relative to the standard basis for $\mathbb{R}^2$ used for both the domain and the range.

    c. Find the matrix of T relative to the basis $B = \left\{ \begin{bmatrix} 1 \\ 1 \end{bmatrix}, \begin{bmatrix} -1 \\ 1 \end{bmatrix} \right\}$ used for both the domain and the range.

10. Consider the linear map $T : \mathbb{P}_2 \rightarrow \mathbb{P}_4$ defined by $T(p(x)) = x^2 p(x)$. Find the matrix of T relative to the bases $B = \{1, x, x^2\}$ for $\mathbb{P}_2$ and $B' = \{1, x, x^2, x^3, x^4\}$ for $\mathbb{P}_4$.

11. Consider the linear map $T : \mathbb{P}_3 \rightarrow \mathbb{P}_3$ defined by $T(p(x)) = p(2x + 1)$. Find the matrix of T relative to the basis $B = \{1, x, x^2, x^3\}$ used for both the domain and the range.

12. Let V be a finite-dimensional vector space. Show that the identity matrix is the matrix of the identity function $\text{id}_V : V \rightarrow V$ relative to a basis B used for both the domain and the range.

13. Let V and W be finite-dimensional vector spaces. Find the matrix of the zero function (defined in Exercise 10 of Section 6.1) relative to bases for V and W.

6.4 Matrices of Compositions and Inverses

The two marvelous theorems of this section demonstrate how matrix multiplication and inverses correspond to composition and inverses of linear maps. To a certain degree, these theorems explain the astounding usefulness of the seemingly peculiar definition of matrix multiplication.

    The results in this section follow quite easily from the work we have done so far; the main difficulty arises in keeping track of all the notation that is involved.

Theorem 6.13

Suppose $B = \{\mathbf{u}_1, \ldots, \mathbf{u}_n\}$ is a basis for a vector space V, $B' = \{\mathbf{u}'_1, \ldots, \mathbf{u}'_m\}$ is a basis for a vector space V', and $B'' = \{\mathbf{u}''_1, \ldots, \mathbf{u}''_l\}$ is a basis for a vector space V''. Suppose A is the matrix of a linear map $T : V \rightarrow V'$ relative to the bases B and B'. Suppose A' is the matrix of a linear map $T' : V' \rightarrow V''$ relative to the bases B' and B''. Then $A'A$ is the matrix of the linear map $T' \circ T : V \rightarrow V''$ relative to the bases B and B''.

Proof   We are given that

$$[T(\mathbf{v})]_{B'} = A[\mathbf{v}]_B \quad \text{for all } \mathbf{v} \in V$$

and that

$$[T'(\mathbf{v}')]_{B''} = A'[\mathbf{v}']_{B'} \quad \text{for all } \mathbf{v}' \in V'.$$

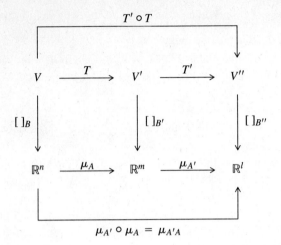

$$\mu_{A'} \circ \mu_A = \mu_{A'A}$$

Thus, for any $\mathbf{v} \in V$ we can use the second of these equations with $\mathbf{v}' = T(\mathbf{v})$ along with the first equation and the associative law of matrix multiplication to obtain

$$[(T' \circ T)(\mathbf{v})]_{B''} = [T'(T(\mathbf{v}))]_{B''}$$
$$= A'[T(\mathbf{v})]_{B'}$$
$$= A'(A[\mathbf{v}]_B)$$
$$= (A'A)[\mathbf{v}]_B$$

as desired.   ∎

Quick Example   *Relative to the bases $\{1, x, x^2\}$ for $\mathbb{P}_2$ and $\{1, x, x^2, x^3\}$ for $\mathbb{P}_3$, we know from work in the previous section that $T : \mathbb{P}_2 \to \mathbb{P}_3$ defined by $T(ax^2+bx+c) = a(x - 1)^2 + b(x - 1) + c$ has matrix*

$$A = \begin{bmatrix} 1 & -1 & 1 \\ 0 & 1 & -2 \\ 0 & 0 & 1 \\ 0 & 0 & 0 \end{bmatrix}$$

and the differentiation operator $D : \mathbb{P}_3 \to \mathbb{P}_3$ has matrix

$$A' = \begin{bmatrix} 0 & 1 & 0 & 0 \\ 0 & 0 & 2 & 0 \\ 0 & 0 & 0 & 3 \\ 0 & 0 & 0 & 0 \end{bmatrix}.$$

Compute the matrix of the composition $D \circ T$ directly and verify that it equals the product $A'A$.

Let A'' be the matrix of the composition $D \circ T$. The columns of A'' are

$$A''e_1 = A''[1]_B = \left[(D \circ T)(1)\right]_{B'} = [0]_{B'} = \begin{bmatrix} 0 \\ 0 \\ 0 \\ 0 \end{bmatrix},$$

$$A''e_2 = A''[x]_B = \left[(D \circ T)(x)\right]_{B'} = \left[D(x-1)\right]_{B'} = [1]_{B'} = \begin{bmatrix} 1 \\ 0 \\ 0 \\ 0 \end{bmatrix},$$

$$A''e_3 = A''[x^2]_B = \left[(D \circ T)(x^2)\right]_{B'} = \left[D\big((x-1)^2\big)\right]_{B'} = [2(x-1)]_{B'} = \begin{bmatrix} -2 \\ 2 \\ 0 \\ 0 \end{bmatrix}.$$

These agree with the columns of the product

$$A'A = \begin{bmatrix} 0 & 1 & 0 & 0 \\ 0 & 0 & 2 & 0 \\ 0 & 0 & 0 & 3 \\ 0 & 0 & 0 & 0 \end{bmatrix} \begin{bmatrix} 1 & -1 & 1 \\ 0 & 1 & -2 \\ 0 & 0 & 1 \\ 0 & 0 & 0 \end{bmatrix} = \begin{bmatrix} 0 & 1 & -2 \\ 0 & 0 & 2 \\ 0 & 0 & 0 \\ 0 & 0 & 0 \end{bmatrix}. \quad \blacksquare$$

Theorem 6.11 gives a recipe for determining the matrix of a linear map relative to bases for the domain and range vector spaces. We can reverse this process and use a matrix $A \in \mathbb{M}(m, n)$ to define a linear map $T : V \rightarrow V'$. If $B = \{u_1, \ldots, u_n\}$ is a basis for V and $B' = \{u'_1, \ldots, u'_m\}$ is a basis for V', simply define $T : V \rightarrow V'$ by

$$T(v) = L_{B'}(A[v]_B)$$

where $L_{B'} : \mathbb{R}^m \rightarrow V'$ is the function defined in Section 6.1 that uses the entries of an element in $\mathbb{R}^m$ as coefficients of a linear combination of the respective elements in B'. Since T is the composition of three linear maps, it is linear. Furthermore, since $L_{B'}$ is the inverse of the coordinate vector function (see Exercise 6 at the end of this section), we have

$$[T(v)]_{B'} = [L_{B'}(A[v]_B)]_{B'} = A[v]_B.$$

Thus, A is the matrix of T relative to the bases B and B'.

This observation will be useful in the proof of the second theorem of this section.

Theorem 6.14

Suppose $B = \{u_1, \ldots, u_n\}$ is a basis for a vector space V and $B' = \{u'_1, \ldots, u'_m\}$ is a basis for a vector space V'. Suppose A is the matrix of a linear map $T : V \rightarrow V'$ relative to the bases B and B'. Then T has an inverse function if and only if A is an invertible matrix. In this case, A^{-1} is the matrix of T^{-1} relative to the bases B' and B.

Proof Suppose first that T has an inverse function T^{-1}. Let $A' \in \mathbb{M}(n, m)$ be the matrix of T^{-1} relative to B' and B. By Theorem 6.13, the product $A'A$ is the matrix of the composition $T^{-1} \circ T$ relative to the basis B. By Exercise 12 of Section 6.3, the identity matrix is also the matrix of $T^{-1} \circ T = \text{id}_V$ relative to B. Since the matrix of a linear map with respect to a given basis is unique, we have that $A'A = I$. A similar consideration of the matrix of $T \circ T'$ gives that $AA' = I$. Hence, A is invertible and $A' = A^{-1}$.

Conversely, suppose A is an invertible matrix. Let $T' : V' \rightarrow V$ be the linear map whose matrix relative to B' and B is A^{-1}. For any $\mathbf{v} \in V$ we have

$$
\begin{aligned}
[T'(T(\mathbf{v}))]_B &= A^{-1}[T(\mathbf{v})]_{B'} \\
&= A^{-1}A[\mathbf{v}]_B \\
&= [\mathbf{v}]_B.
\end{aligned}
$$

Similarly, for any $\mathbf{v}' \in V'$ we have

$$
\begin{aligned}
[T(T'(\mathbf{v}'))]_{B'} &= A[T'(\mathbf{v}')]_B \\
&= AA^{-1}[\mathbf{v}']_{B'} \\
&= [\mathbf{v}']_{B'}.
\end{aligned}
$$

Since the coordinate functions $[\ ]_B$ and $[\ ]_{B'}$ are one-to-one (by Exercise 6 at the end of this section, they have inverses L_B and $L_{B'}$), we conclude that $T'(T(\mathbf{v})) = \mathbf{v}$ for all $\mathbf{v} \in V$ and $T(T'(\mathbf{v}')) = \mathbf{v}'$ for all $\mathbf{v}' \in V'$. ∎

Quick Example *From Exercise 11 of Section 6.3, the matrix of $T : \mathbb{P}_3 \rightarrow \mathbb{P}_3$ defined by $T(p(x)) = p(2x + 1)$ is*

$$
A = \begin{bmatrix} 1 & 1 & 1 & 1 \\ 0 & 2 & 4 & 6 \\ 0 & 0 & 4 & 12 \\ 0 & 0 & 0 & 8 \end{bmatrix}
$$

relative to the basis $B = \{1, x, x^2, x^3\}$. Show that the inverse of T is defined by $T^{-1}(p(x)) = p(\frac{1}{2}(x - 1))$. Find the matrix A' of T^{-1} relative to B and show that this matrix is the inverse of A.

We first verify the two conditions to confirm that the function is the inverse of T:

$$
(T \circ T^{-1})(p(x)) = T(T^{-1}(p(x))) = T(p(\tfrac{1}{2}(x - 1))) = p(2(\tfrac{1}{2}(x - 1)) + 1) = p(x),
$$

$$
(T^{-1} \circ T)(p(x)) = T^{-1}(T(p(x))) = T^{-1}(p(2x + 1)) = p(\tfrac{1}{2}((2x + 1) - 1)) = p(x).
$$

The columns of the matrix A' are

$$A'\mathbf{e}_1 = A'[1]_B = [T^{-1}(1)]_B = [1]_B = \begin{bmatrix} 1 \\ 0 \\ 0 \\ 0 \end{bmatrix},$$

$$A'\mathbf{e}_2 = A'[x]_B = [T^{-1}(x)]_B = [\tfrac{1}{2}(x-1)]_B = \begin{bmatrix} -\tfrac{1}{2} \\ \tfrac{1}{2} \\ 0 \\ 0 \end{bmatrix},$$

$$A'\mathbf{e}_3 = A'[x^2]_B = [T^{-1}(x^2)]_B = [(\tfrac{1}{2}(x-1))^2]_B = \begin{bmatrix} \tfrac{1}{4} \\ -\tfrac{1}{2} \\ \tfrac{1}{4} \\ 0 \end{bmatrix},$$

$$A'\mathbf{e}_4 = A'[x^3]_B = [T^{-1}(x^3)]_B = [(\tfrac{1}{2}(x-1))^3]_B = \begin{bmatrix} -\tfrac{1}{8} \\ \tfrac{3}{8} \\ -\tfrac{3}{8} \\ \tfrac{1}{8} \end{bmatrix}.$$

Thus,

$$A' = \begin{bmatrix} 1 & -\tfrac{1}{2} & \tfrac{1}{4} & -\tfrac{1}{8} \\ 0 & \tfrac{1}{2} & -\tfrac{1}{2} & \tfrac{3}{8} \\ 0 & 0 & \tfrac{1}{4} & -\tfrac{3}{8} \\ 0 & 0 & 0 & \tfrac{1}{8} \end{bmatrix}.$$

We compute that

$$AA' = \begin{bmatrix} 1 & 1 & 1 & 1 \\ 0 & 2 & 4 & 6 \\ 0 & 0 & 4 & 12 \\ 0 & 0 & 0 & 8 \end{bmatrix} \begin{bmatrix} 1 & -\tfrac{1}{2} & \tfrac{1}{4} & -\tfrac{1}{8} \\ 0 & \tfrac{1}{2} & -\tfrac{1}{2} & \tfrac{3}{8} \\ 0 & 0 & \tfrac{1}{4} & -\tfrac{3}{8} \\ 0 & 0 & 0 & \tfrac{1}{8} \end{bmatrix} = \begin{bmatrix} 1 & 0 & 0 & 0 \\ 0 & 1 & 0 & 0 \\ 0 & 0 & 1 & 0 \\ 0 & 0 & 0 & 1 \end{bmatrix}.$$

By Theorem 5.13, $A' = A^{-1}$. ∎

Exercises 6.4

1. Consider the linear map $T : \mathbb{P}_2 \to \mathbb{P}_3$ defined by $T(p(x)) = xp(x)$. Consider the basis $B = \{1, x, x^2\}$ for $\mathbb{P}_2$ and the basis $B' = \{1, x, x^2, x^3\}$ for $\mathbb{P}_3$.
   a. Find the matrix A of T relative to B and B'.
   b. Find the matrix A' of the differentiation operator $D : \mathbb{P}_3 \to \mathbb{P}_2$ relative to B' and B.

    c. Use Theorem 6.11 to compute the matrix of the composition $D \circ T : \mathbb{P}_2 \to \mathbb{P}_2$ relative to the basis B.

    d. Confirm that this is the product $A'A$.

    e. Use Theorem 6.11 to compute the matrix of the composition $T \circ D : \mathbb{P}_3 \to \mathbb{P}_3$ relative to the basis B'.

    f. Confirm that this is the product AA'.

2.   Consider the linear map $L : \mathbb{P}_2 \to \mathbb{P}_2$ defined by $L(p(x)) = 2p(x + 1)$.

    a. Find the matrix A of L relative to the basis $B = \{1, x, x^2\}$.

    b. Show that the inverse of L is defined by $L^{-1}(p(x)) = \frac{1}{2}p(x - 1)$.

    c. Find the matrix A' of L^{-1} relative to B.

    d. Confirm that A' is the inverse of A.

3.   Let $R_\theta : \mathbb{R}^2 \to \mathbb{R}^2$ be the linear function that rotates each point counterclockwise about the origin through an angle of θ.

    a. For any value of θ, give geometric arguments to show that R_θ is one-to-one and onto.

    b. Explain geometrically why $R_\theta \circ R_\varphi = R_{\theta+\varphi}$.

    c. Use the result of part b to show that $R_\theta^{-1} = R_{-\theta}$.

    d. The matrix of a rotation relative to the standard basis for $\mathbb{R}^2$ is given in the second example of Section 6.3. Verify directly that the product of the matrices of R_θ and R_φ is the matrix of $R_{\theta+\varphi}$.

    e. Verify directly that the matrix of $R_{-\theta}$ is the inverse of the matrix of R_θ.

4.   Can we exploit the associativity of composition of functions (see Exercise 10 of Section 6.2) and Theorem 6.13 to prove the associative law of matrix multiplication (part a of Theorem 5.4)?

5.   Let $B = \{v_1, \ldots, v_n\}$ be a set of vectors in a vector space V. In Section 6.1 we verified the linearity of the linear combination function $L_B : \mathbb{R}^n \to V$ defined by

$$L_B\left(\begin{bmatrix} r_1 \\ \vdots \\ r_n \end{bmatrix}\right) = r_1 v_1 + \cdots + r_n v_n.$$

    a. Show that L_B is one-to-one if and only if B is a linearly independent set.

    b. Show that L_B is onto if and only if B spans V.

6.   Suppose $B = \{u_1, \ldots, u_n\}$ is a basis for a vector space V. Let $L_B : \mathbb{R}^n \to V$ be the linear combination function defined in Exercise 5. Show that L_B is the inverse of the coordinate vector function $[\ ]_B$.

7.   Suppose $T : V \to V'$ and $T' : V \to V'$ are linear.

    a. Show that $T + T' : V \to V'$ defined by $(T + T')(v) = T(v) + T'(v)$ is linear.

    b. If V and V' are finite-dimensional, determine the matrix of $T + T'$ in terms of the matrices of T and T'.

8. Suppose $T : V \to V'$ is linear and let $r \in \mathbb{R}$.

    a. Show that $rT : V \to V'$ defined by $(rT)(\mathbf{v}) = r(T(\mathbf{v}))$ is linear.

    b. If V and V' are finite-dimensional, determine the matrix of rT in terms of the matrix of T.

9. In this exercise, let addition and scalar multiplication of linear maps be as defined in the previous two exercises. For any $r \in \mathbb{R}$, let r also denote the scalar multiplication function as defined in Exercise 11 of Section 6.1. Let juxtaposition denote composition, with integer exponents denoting repeated numbers of copies of the same function. Finally, let $D : \mathbb{D}(\mathbb{R}) \to \mathbb{F}(\mathbb{R})$ denote the differentiation operation.

    a. Compute $(D^2 + 3D + 2)(x^2 \sin x)$.

    b. Compute $(D + 1)(D + 2)(x^2 \sin x)$.

    c. For any real numbers r and s, show that $(D+r)(D+s) = D^2+(r+s)D+rs$.

    d. For any linear maps $R : V \to W$, $R' : V \to W$, $S : W \to X$, and $S' : W \to X$, show that $S(R + R') = SR + SR'$ and $(S + S')R = SR + S'R$.

10. Suppose $A \in \mathbb{M}(m, n)$ and $A' \in \mathbb{M}(l, m)$. With μ_A and $\mu_{A'}$ as defined in Section 6.1, show that

    $$\mu_{A'} \circ \mu_A = \mu_{A'A} \quad \text{and} \quad \mu_A^{-1} = \mu_{A^{-1}}.$$

6.5 Change of Basis

We have seen from Exercises 8 and 9 of Section 6.3 that the matrix of a linear map $T : V \to V'$ depends on the bases for the vector spaces V and V'. In this section we want to investigate how changing to different bases will affect the matrix of a linear map. We will be especially interested in linear maps $T : V \to V$, with domain equal to the range. In such a case we will often use the same basis for the domain and range. If $B = \{\mathbf{u}_1, \ldots, \mathbf{u}_n\}$ and $B' = \{\mathbf{u}'_1, \ldots, \mathbf{u}'_n\}$ are bases for V, we will investigate the relation between the matrix of T relative to B and the matrix of T relative to B'. One of our goals in Chapter 8 will be to choose a basis for V such that the matrix of T relative to that basis is especially simple. This will give some insight into the nature of the transformation T.

Let us begin with a theorem that tells how the coordinate vectors change when we switch from one basis to another.

Theorem 6.15

Suppose $B = \{\mathbf{u}_1, \ldots, \mathbf{u}_n\}$ and $B' = \{\mathbf{u}'_1, \ldots, \mathbf{u}'_n\}$ are bases for a vector space V. There is a unique $n \times n$ matrix P such that $[\mathbf{v}]_B = P[\mathbf{v}]_{B'}$ for all $\mathbf{v} \in V$. The columns of this matrix are the n-tuples $[\mathbf{u}'_1]_B, \ldots, [\mathbf{u}'_n]_B$ in that order.

Proof   Let P be the matrix of the identity map $\mathrm{id}_V : V \to V$ relative to the bases B' and B.

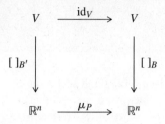

By Theorem 6.11, P is the unique matrix satisfying $[\mathbf{v}]_B = [\mathrm{id}_V(\mathbf{v})]_B = P[\mathbf{v}]_{B'}$ for all $\mathbf{v} \in V$. Furthermore, the jth column of this matrix is $[\mathrm{id}_V(\mathbf{u}'_j)]_B = [\mathbf{u}'_j]_B$, as prescribed. ∎

The conclusion of this theorem can be visualized by means of the triangular diagram:

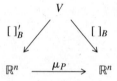

That is, starting with any element $\mathbf{v} \in V$, we can go down the left side of the triangle to obtain $[\mathbf{v}]_{B'}$ and the across the base to reach $P[\mathbf{v}]_{B'}$, or we can reach the same element by mapping $\mathbf{v}$ down the right side of the triangle to $[\mathbf{v}]_B$.

Definition 6.16

The matrix P as described in the previous theorem is the **change-of-basis matrix** for changing from the basis B' to the basis B.

Mathematical Strategy Session   The change-of-basis matrix P is another case where you can memorize that the columns of P are $[\mathbf{u}_j]_B$ and mechanically compute P. Alternatively, you can use the same ideas as for deriving the columns of the matrix of a linear map. Apply the three principles:

1. The jth column of P is $P\mathbf{e}_j$.
2. $\mathbf{e}_j$ is the coordinate vector $[\mathbf{u}'_j]_{B'}$.
3. The diagram for P to be the change-of-basis matrix gives the equation $P[\mathbf{v}]_{B'} = [\mathbf{v}]_B$.

Obtain a foolproof derivation of the columns of P:

$$P\mathbf{e}_j = P[\mathbf{u}_j']_{B'} = [\mathbf{u}_j']_B.$$

Quick Example   *Find the change-of-basis matrix P for changing from the basis*
$B' = \left\{ \begin{bmatrix} 2 \\ 2 \end{bmatrix}, \begin{bmatrix} 4 \\ -1 \end{bmatrix} \right\}$ *to the basis* $B = \left\{ \begin{bmatrix} 2 \\ 0 \end{bmatrix}, \begin{bmatrix} 0 \\ -1 \end{bmatrix} \right\}$ *in* $\mathbb{R}^2$. *Verify directly   that*
$P \begin{bmatrix} 3 \\ -2 \end{bmatrix}_{B'} = \begin{bmatrix} 3 \\ -2 \end{bmatrix}_B.$

The columns of P are $P\mathbf{e}_1 = P \begin{bmatrix} 2 \\ 2 \end{bmatrix}_{B'} = \begin{bmatrix} 2 \\ 2 \end{bmatrix}_B = \begin{bmatrix} 1 \\ -2 \end{bmatrix}$ and $P\mathbf{e}_2 = P \begin{bmatrix} 4 \\ -1 \end{bmatrix}_{B'} = $

$\begin{bmatrix} 4 \\ -1 \end{bmatrix}_B = \begin{bmatrix} 2 \\ 1 \end{bmatrix}$. Thus, $P = \begin{bmatrix} 1 & 2 \\ -2 & 1 \end{bmatrix}$. A little work with a system of linear equa-

tions gives $\begin{bmatrix} 3 \\ -2 \end{bmatrix}_{B'} = \begin{bmatrix} -\frac{1}{2} \\ 1 \end{bmatrix}$. So

$$P \begin{bmatrix} 3 \\ -2 \end{bmatrix}_{B'} = \begin{bmatrix} 1 & 2 \\ -2 & 1 \end{bmatrix} \begin{bmatrix} -\frac{1}{2} \\ 1 \end{bmatrix} = \begin{bmatrix} \frac{3}{2} \\ 2 \end{bmatrix}.$$

This is indeed the coordinate vector $\begin{bmatrix} 3 \\ -2 \end{bmatrix}_B$.   ∎

The preceding example involved a lot of work to obtain a coordinate vector that could have been written down instantly with a minimal amount of mental computation. More useful, however, is the fact that $P^{-1} = \begin{bmatrix} \frac{1}{5} & -\frac{2}{5} \\ \frac{2}{5} & \frac{1}{5} \end{bmatrix}$ exists. So from the equation $P[\mathbf{v}]_{B'} = [\mathbf{v}]_B$, we conclude that $[\mathbf{v}]_{B'} = P^{-1}[\mathbf{v}]_B$. Hence, P^{-1} is the matrix for changing bases from B to B'. Thus, we can obtain the coordinate vector with respect to B' by finding the coordinate vector with respect to B (a much easier computation for this example) and then multiplying this vector by P^{-1}.

Let us pause briefly for an important announcement.

Theorem 6.17

Suppose P is the change-of-basis matrix for changing from a basis B' to a basis B. Then P is invertible, and P^{-1} is the change-of-basis matrix for changing from the basis B to the basis B'.

Exercises 6 and 7 at the end of this section outline two ways to prove this theorem.

We now return to the work in progress. Suppose we also have a linear function $T : \mathbb{R}^2 \to \mathbb{R}^2$ whose matrix relative to the basis B is

$$A = \begin{bmatrix} 1 & -4 \\ 3 & 2 \end{bmatrix}.$$

The functions involved are in the upper square of the following diagram. We would like to determine the matrix A' of T relative to the basis B' in terms of the matrix A and the change-of-basis matrix P. These functions are found in the large outer square of the diagram. These two squares are connected by two copies of the triangular diagrams for P as the change-of-basis matrix.

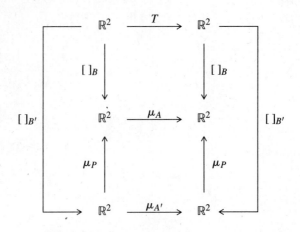

As you trace through this diagram, you will find that we can also put $\mu_{P^{-1}AP}$ across the bottom so that the composition of μ_P followed by μ_A is the same as the composition of $\mu_{P^{-1}AP}$ followed by μ_P. The key is that by Exercise 10 of Section 6.4, we have

$$\mu_P \circ \mu_{P^{-1}AP} = \mu_{P(P^{-1}AP)} = \mu_{AP} = \mu_A \circ \mu_P.$$

Since the matrix of T relative to B' is unique, we have that $A' = P^{-1}AP$.

Quick Example   *Verify for our example that* $\left[T\left(\begin{bmatrix} 3 \\ -2 \end{bmatrix}\right)\right]_{B'} = P^{-1}AP\begin{bmatrix} 3 \\ -2 \end{bmatrix}_{B'}.$

We first compute

$$P^{-1}AP = \begin{bmatrix} \frac{1}{5} & -\frac{2}{5} \\ \frac{2}{5} & \frac{1}{5} \end{bmatrix}\begin{bmatrix} 1 & -4 \\ 3 & 2 \end{bmatrix}\begin{bmatrix} 1 & 2 \\ -2 & 1 \end{bmatrix} = \begin{bmatrix} \frac{1}{5} & -\frac{2}{5} \\ \frac{2}{5} & \frac{1}{5} \end{bmatrix}\begin{bmatrix} 9 & -2 \\ -1 & 8 \end{bmatrix} = \begin{bmatrix} \frac{11}{5} & -\frac{18}{5} \\ \frac{17}{5} & \frac{4}{5} \end{bmatrix}.$$

Then $P^{-1}AP\begin{bmatrix} 3 \\ -2 \end{bmatrix}_{B'} = \begin{bmatrix} \frac{11}{5} & -\frac{18}{5} \\ \frac{17}{5} & \frac{4}{5} \end{bmatrix}\begin{bmatrix} -\frac{1}{2} \\ 1 \end{bmatrix} = \begin{bmatrix} -\frac{47}{10} \\ -\frac{9}{10} \end{bmatrix}$. Now $\left[T\left(\begin{bmatrix} 3 \\ -2 \end{bmatrix}\right)\right]_B =$

$A\begin{bmatrix} 3 \\ -2 \end{bmatrix}_B = \begin{bmatrix} 1 & -4 \\ 3 & 2 \end{bmatrix}\begin{bmatrix} \frac{3}{2} \\ 2 \end{bmatrix} = \begin{bmatrix} -\frac{13}{2} \\ -\frac{17}{2} \end{bmatrix}$. Thus, $T\left(\begin{bmatrix} 3 \\ -2 \end{bmatrix}\right) = -\frac{13}{2}\begin{bmatrix} 2 \\ 0 \end{bmatrix} + \frac{17}{2}\begin{bmatrix} 0 \\ -1 \end{bmatrix} =$

$\begin{bmatrix} -13 \\ -\frac{17}{2} \end{bmatrix}$. It is easy to confirm that $\left[T\left(\begin{bmatrix} 3 \\ -2 \end{bmatrix} \right) \right]_{B'} = \begin{bmatrix} -\frac{47}{10} \\ -\frac{9}{10} \end{bmatrix}$. Indeed, $-\frac{47}{10} \begin{bmatrix} 2 \\ 2 \end{bmatrix} -$

$\frac{9}{10} \begin{bmatrix} 4 \\ -1 \end{bmatrix} = \begin{bmatrix} -13 \\ -\frac{17}{2} \end{bmatrix}$. ∎

Here is the theorem this example has been leading up to.

Theorem 6.18

Suppose $B = \{\mathbf{u}_1, \ldots, \mathbf{u}_n\}$ and $B' = \{\mathbf{u}'_1, \ldots, \mathbf{u}'_n\}$ are bases for a vector space V. Let P be the matrix for changing basis from B' to B. Suppose A is the matrix of a linear map $T : V \to V$ relative to the basis B. Then $P^{-1}AP$ is the matrix of T relative to the basis B'.

Proof   For any $\mathbf{v} \in V$ we know that $[T(\mathbf{v})]_B = A[\mathbf{v}]_B$ and $[\mathbf{v}]_B = P[\mathbf{v}]_{B'}$. From the second equation we derive $[\mathbf{v}]_{B'} = P^{-1}[\mathbf{v}]_B$ and hence, substituting $T(\mathbf{v})$ for $\mathbf{v}$, we have $[T(\mathbf{v})]_{B'} = P^{-1}[T(\mathbf{v})]_B$.

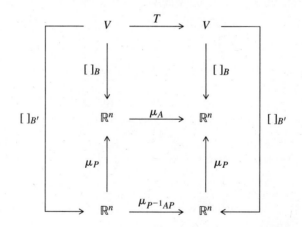

Thus,

$$\begin{aligned}
[T(\mathbf{v})]_{B'} &= P^{-1}[T(\mathbf{v})]_B \\
&= P^{-1}(A[\mathbf{v}]_B) \\
&= (P^{-1}A)[\mathbf{v}]_B \\
&= (P^{-1}A)(P[\mathbf{v}]_{B'}) \\
&= (P^{-1}AP)[\mathbf{v}]_{B'}
\end{aligned}$$

as desired.   ∎

Exercises 6.5

1. Find the change-of-basis matrix for changing from

$$B' = \left\{ \begin{bmatrix} 1 \\ 1 \end{bmatrix}, \begin{bmatrix} -1 \\ 1 \end{bmatrix} \right\} \quad \text{to} \quad B = \left\{ \begin{bmatrix} 1 \\ 2 \end{bmatrix}, \begin{bmatrix} 2 \\ 1 \end{bmatrix} \right\}.$$

2. Find the change-of-basis matrix for changing from

$$B' = \{1 + x, x, 1 - x^2\} \quad \text{to} \quad B = \{x, 1 - x, 1 + x^2\}.$$

3. Show that the matrix for changing from an ordered basis $\{u_1, \ldots, u_n\}$ for $\mathbb{R}^n$ to the standard basis for $\mathbb{R}^n$ consists of the columns $u_1, \ldots, u_n$ in that order.

4. Suppose the matrix of the linear map $T : \mathbb{R}^3 \to \mathbb{R}^3$ relative to

$$B = \left\{ \begin{bmatrix} 1 \\ 0 \\ 1 \end{bmatrix}, \begin{bmatrix} 0 \\ 1 \\ 1 \end{bmatrix}, \begin{bmatrix} 1 \\ 0 \\ 0 \end{bmatrix} \right\}$$

is $A = \begin{bmatrix} 1 & 2 & -1 \\ 2 & 0 & 1 \\ -1 & 1 & 2 \end{bmatrix}$.

a. Find the matrix P for changing from the basis

$$B' = \left\{ \begin{bmatrix} 1 \\ 0 \\ 0 \end{bmatrix}, \begin{bmatrix} 0 \\ 1 \\ 1 \end{bmatrix}, \begin{bmatrix} 0 \\ -1 \\ 1 \end{bmatrix} \right\}$$

to the basis B.

b. Compute the matrix $P^{-1}AP$ of T relative to B'.

c. Use the formula $[T(\mathbf{v})]_B = A[\mathbf{v}]_B$ to compute $T(\mathbf{e}_1)$, $T(\mathbf{e}_2)$, and $T(\mathbf{e}_3)$.

d. Use the formula $[T(\mathbf{v})]_{B'} = P^{-1}AP[\mathbf{v}]_{B'}$ to confirm your values of $T(\mathbf{e}_1)$, $T(\mathbf{e}_2)$, and $T(\mathbf{e}_3)$ from part c.

5. If the matrix of $T : \mathbb{P}_2 \to \mathbb{P}_2$ is $\begin{bmatrix} 1 & -1 & 2 \\ 2 & 1 & 0 \\ -1 & 0 & 3 \end{bmatrix}$ relative to $B = \{1, x - 1, x^2 - x\}$, find the matrix of T relative to $B' = \{1 + x^2, 1 - x^2, x\}$.

6. Suppose $B = \{u_1, \ldots, u_n\}$, $B' = \{u'_1, \ldots, u'_n\}$, and $B'' = \{u''_1, \ldots, u''_n\}$ are bases for a vector space V.

a. Show that the identity matrix I is the matrix for changing bases from B to B.

b. Suppose P is the matrix for changing bases from B to B' and Q is the matrix for changing bases from B' to B''. Show that QP is the matrix for changing bases from B to B''.

c. Suppose in part b that $B'' = B$. Use parts a and b and the uniqueness of the change-of-basis matrix to show that $QP = I$ and $PQ = I$. Conclude that $Q = P^{-1}$ and, in particular, that the change-of-basis matrix P is invertible.

7. Suppose $B = \{\mathbf{u}_1, \ldots, \mathbf{u}_n\}$ and $B' = \{\mathbf{u}'_1, \ldots, \mathbf{u}'_n\}$ are bases for a vector space V. Suppose P is the matrix for changing bases from B' to B.

a. Examine the proof of Theorem 6.15 to confirm that P is the matrix of the identity function id_V relative to the bases B' and B.

b. Explain why id_V is an invertible linear map.

c. Use Theorem 6.14 to conclude that P is invertible.

6.6 Image and Kernel

Now that you have worked with linear maps in terms of their matrices for a while, you are probably convinced that linear maps can be quite complicated. In order to understand some of the important characteristics of a linear map without getting bogged down in the numerical details, we will turn our attention to some subspaces associated with a linear map.

Definition 6.19

Suppose $T : V \to W$ is linear. The **kernel** (or **null space**) of T is denoted $\ker T$ and is defined by

$$\ker T = \{\mathbf{v} \in V \mid T(\mathbf{v}) = \mathbf{0}\}.$$

The **image** of T is denoted $\mathrm{im}\, T$ and is defined by

$$\mathrm{im}\, T = \{\mathbf{w} \in W \mid \mathbf{w} = T(\mathbf{v}) \text{ for some } \mathbf{v} \in V\}.$$

Of course, the image of a linear map is none other than its image as a function. The significance of these definitions lies in the fact, stated in the following theorem, that the kernel and image of a linear map are not merely subsets of the domain and range.

Theorem 6.20

Suppose $T : V \to W$ is linear. Then $\ker T$ is a subspace of V, and $\mathrm{im}\, T$ is a subspace of W.

The proof of this theorem is a straightforward application of the Subspace Theorem. In Exercise 15 of Section 6.1 you were asked to carry out the details of these verifications.

Mathematical Strategy Session Whenever you encounter a new concept, you should examine a few simple examples to help develop your intuition about the concept. Let us look at the kernel and image of some especially simple linear maps.

Consider the projection $P : \mathbb{R}^3 \to \mathbb{R}^3$ onto the *xy*-plane in $\mathbb{R}^3$. The kernel consists of all points that project onto the origin. This is easily seen to be the *z*-axis, a one-dimensional subspace of the domain. The image, of course, is the *xy*-plane, a two-dimensional subspace of the range. In this example, where the domain and range are both $\mathbb{R}^3$, it appears that these two subspaces are somehow complementary.

Compare the previous example with the projection of $\mathbb{R}^3$ onto the *x*-axis again in $\mathbb{R}^3$. Here the kernel is the two-dimensional *yz*-plane, whereas the image is the one-dimensional *x*-axis. Again we see a complementary relation between the kernel and the image.

In general, it is reasonable that a linear map with a relatively large kernel will collapse many vectors to zero, and hence will have a relatively small image. On the other side of the coin, a linear map with a smaller kernel will do a smaller amount of collapsing, and hence have a larger image. These remarks will be made precise in the main theorem of the next section.

Let us examine another linear map, one whose kernel and image are not immediately apparent from geometrical consideration.

Quick Example   *Consider the* 3 × 3 *matrix* $A = \begin{bmatrix} 1 & 1 & 1 \\ 2 & 4 & 5 \\ 1 & 3 & 4 \end{bmatrix}$. *Find bases for the kernel and image of the associated linear map* $\mu_A : \mathbb{R}^3 \to \mathbb{R}^3$.

The problem of obtaining an explicit expression for $\ker \mu_A$ can be instantly reformulated as a problem of solving a homogeneous linear system. Indeed, if we let $\mathbf{v} = \begin{bmatrix} v_1 \\ v_2 \\ v_3 \end{bmatrix}$ denote an arbitrary element of $\mathbb{R}^3$, we have the following string of logically equivalent statements:

$$\begin{aligned} \mathbf{v} \in \ker \mu_A \quad &\Longleftrightarrow \quad \mu_A(\mathbf{v}) = \mathbf{0} \\ &\Longleftrightarrow \quad A\mathbf{v} = \mathbf{0} \\ &\Longleftrightarrow \quad \begin{matrix} v_1 + v_2 + v_3 = 0 \\ 2v_1 + 4v_2 + 5v_3 = 0 \\ v_1 + 3v_2 + 4v_3 = 0 \end{matrix} \end{aligned}$$

We bring out the Gauss-Jordan reduction process to put A in reduced row-echelon form:

$$\begin{bmatrix} 1 & 1 & 1 \\ 2 & 4 & 5 \\ 1 & 3 & 4 \end{bmatrix} \longrightarrow \begin{bmatrix} 1 & 1 & 1 \\ 0 & 2 & 3 \\ 0 & 2 & 3 \end{bmatrix} \longrightarrow \begin{bmatrix} 1 & 1 & 1 \\ 0 & 1 & \frac{3}{2} \\ 0 & 0 & 0 \end{bmatrix} \longrightarrow \begin{bmatrix} 1 & 0 & -\frac{1}{2} \\ 0 & 1 & \frac{3}{2} \\ 0 & 0 & 0 \end{bmatrix}$$

The one column without a leading 1 gives a one-dimensional kernel, namely,

$$\ker \mu_A = \left\{ r \begin{bmatrix} 1 \\ -3 \\ 2 \end{bmatrix} \middle| r \in \mathbb{R} \right\}.$$

In particular, $\left\{ \begin{bmatrix} 1 \\ -3 \\ 2 \end{bmatrix} \right\}$ is a basis for ker μ_A.

Any element in im μ_A can be written

$$\mu_A\left(\begin{bmatrix} v_1 \\ v_2 \\ v_3 \end{bmatrix}\right) = \mu_A(v_1\mathbf{e}_1 + v_2\mathbf{e}_2 + v_3\mathbf{e}_3)$$

$$= v_1\mu_A(\mathbf{e}_1) + v_2\mu_A(\mathbf{e}_2) + v_3\mu_A(\mathbf{e}_3)$$
$$= v_1(A\mathbf{e}_1) + v_2(A\mathbf{e}_2) + v_3(A\mathbf{e}_3)$$
$$= v_1\begin{bmatrix} 1 \\ 2 \\ 1 \end{bmatrix} + v_2\begin{bmatrix} 1 \\ 4 \\ 3 \end{bmatrix} + v_3\begin{bmatrix} 1 \\ 5 \\ 4 \end{bmatrix}.$$

Thus, we see that im μ_A is equal to

$$\text{span}\left\{ \begin{bmatrix} 1 \\ 2 \\ 1 \end{bmatrix}, \begin{bmatrix} 1 \\ 4 \\ 3 \end{bmatrix}, \begin{bmatrix} 1 \\ 5 \\ 4 \end{bmatrix} \right\},$$

the set of all linear combinations of the three columns of A. To find a basis for im μ_A, we apply the result discussed in the Mathematical Strategy Session following the Contraction Theorem. According to that result, a basis consists of the columns of A that correspond to leading variables in the reduced row-echelon form of A. Since

A reduces to $\begin{bmatrix} 1 & 0 & -\frac{1}{2} \\ 0 & 1 & \frac{3}{2} \\ 0 & 0 & 0 \end{bmatrix}$, we know that the set $\left\{ \begin{bmatrix} 1 \\ 2 \\ 1 \end{bmatrix}, \begin{bmatrix} 1 \\ 4 \\ 3 \end{bmatrix} \right\}$ is a basis for

im μ_A. ∎

The preceding example illustrates the general principle that the columns with leading 1s in the reduced form of a matrix indicate the columns of the original matrix that constitute a basis for the image of the associated linear map.

Here are a definition and a theorem to formalize some of our observations.

Definition 6.21

The **column space** of an $m \times n$ matrix A is the subspace of $\mathbb{R}^m$ spanned by the n columns of A considered as elements of $\mathbb{R}^m$.

Theorem 6.22

Suppose A is an $m \times n$ matrix. Then im μ_A is equal to the column space of A.

Proof   The elements of im μ_A can be written

$$\mu_A\left(\begin{bmatrix} v_1 \\ \vdots \\ v_n \end{bmatrix}\right) = \mu_A(v_1\mathbf{e}_1 + \cdots + v_n\mathbf{e}_n)$$

$$= v_1\mu_A(\mathbf{e}_1) + \cdots + v_n\mu_A(\mathbf{e}_n)$$

$$= v_1(A\mathbf{e}_1) + \cdots + v_n(A\mathbf{e}_n).$$

where $\begin{bmatrix} v_1 \\ \vdots \\ v_n \end{bmatrix}$ is an arbitrary element of $\mathbb{R}^n$. Since $A\mathbf{e}_1, \ldots, A\mathbf{e}_n$ are the n columns of A, we see that the elements of im μ_A are precisely the linear combinations of the columns of A.   ■

Whereas the kernel and image are defined for any linear function, the column space comes into play only when there is a matrix involved. On the other hand, it is often useful to consider the column space of a matrix even when the matrix is not associated with a linear map. In this spirit, we can also consider the space spanned by the rows of a matrix.

Definition 6.23

The **row space** of an $m \times n$ matrix A is the subspace of $\mathbb{R}^n$ spanned by the m rows of A considered as elements of $\mathbb{R}^n$.

Theorem 6.24

Row operations do not change the row space of a matrix.

Proof   Let $\mathbf{v}_1, \ldots, \mathbf{v}_m$ denote the rows of an $m \times n$ matrix A. We can denote an arbitrary element of the row space of A by $r_1\mathbf{v}_1 + \cdots + r_m\mathbf{v}_m$, where $r_1, \ldots, r_m \in \mathbb{R}$.

If we interchange rows i and j of A, it is easy enough to interchange the corresponding terms in the linear combination to write

$$r_1\mathbf{v}_1 + \cdots + r_i\mathbf{v}_i + \cdots + r_j\mathbf{v}_j + \cdots + r_m\mathbf{v}_m$$

$$= r_1\mathbf{v}_1 + \cdots + r_j\mathbf{v}_j + \cdots + r_i\mathbf{v}_i + \cdots + r_m\mathbf{v}_m$$

as a linear combination of the rows in the new order.

If we multiply row i by the nonzero constant c, we can write

$$r_1\mathbf{v}_1 + \cdots + r_i\mathbf{v}_i + \cdots + r_m\mathbf{v}_m = r_1\mathbf{v}_1 + \cdots + \frac{r_i}{c}(c\mathbf{v}_i) + \cdots + r_m\mathbf{v}_m$$

as a linear combination of the rows $\mathbf{v}_1, \ldots, c\mathbf{v}_i, \ldots, \mathbf{v}_m$ of the new matrix.

Finally, if we add c times row i to row j, we can write

$$r_1\mathbf{v}_1 + \cdots + r_i\mathbf{v}_i + \cdots + r_j\mathbf{v}_j + \cdots + r_m\mathbf{v}_m$$
$$= r_1\mathbf{v}_1 + \cdots + (r_i - cr_j)\mathbf{v}_i + \cdots + r_j(c\mathbf{v}_i + \mathbf{v}_j) + \cdots + r_m\mathbf{v}_m$$

as a linear combination of the rows $\mathbf{v}_1, \ldots, \mathbf{v}_i, \ldots, c\mathbf{v}_i + \mathbf{v}_j, \ldots, \mathbf{v}_m$ of the new matrix.

Thus, every element of the row space of the matrix A is in the row space of any matrix A' produced by applying a row operation to A. Since A' can be transformed back to A by applying the inverse row operation, the preceding argument demonstrates the reverse containment. Hence, the two matrices have the same row space. ∎

This theorem makes it particularly easy to determine a basis for the row space of a matrix. Simply put the matrix in reduced row-echelon form. The row space of the resulting matrix is the same as the row space of the original matrix. The fact that the columns with leading 1s contain no other nonzero entries makes it easy to see that the rows with leading 1s will be a linearly independent subset of the row space. Since all other rows contain only 0s, it is also clear that these nonzero rows form a basis for the row space of the matrix.

Be sure to note that we read off a basis for the row space of a matrix directly from the reduced form of the matrix. This is in contrast to the situation for the column space, where we must remember to refer to the original matrix to read off a basis.

Quick Example   *Find bases for the row space and the column space of the matrix*

$$A = \begin{bmatrix} 2 & -2 & -1 & -2 & 8 \\ -4 & 4 & -1 & -3 & -8 \\ 1 & -1 & -1 & -2 & 5 \\ 3 & -3 & -2 & 1 & 3 \end{bmatrix}.$$

The reduced row-echelon form of A is

$$\begin{bmatrix} 1 & -1 & 0 & 0 & 3 \\ 0 & 0 & 1 & 0 & 2 \\ 0 & 0 & 0 & 1 & -2 \\ 0 & 0 & 0 & 0 & 0 \end{bmatrix}.$$

Thus, the top three rows in the reduced row-echelon matrix form a basis for the row space of A:

$$\{(1, -1, 0, 0, 3), (0, 0, 1, 0, 2), (0, 0, 0, 1, -2)\}.$$

The first, third, and fourth columns of A form a basis for the column space of A:

$$\left\{ \begin{bmatrix} 2 \\ -4 \\ 1 \\ 3 \end{bmatrix}, \begin{bmatrix} -1 \\ -1 \\ -1 \\ -2 \end{bmatrix}, \begin{bmatrix} -2 \\ -3 \\ -2 \\ 1 \end{bmatrix} \right\}. ∎$$

A few weeks ago, would you have believed that the dimension of the row space of a matrix is always equal to the dimension of its column space? Why, the rows and columns are not even in the same Euclidean space. You now understand, however, how both subspaces have their dimensions determined in different ways by the number of leading 1s in the reduced row-echelon form of the matrix. This simple observation is all that is necessary to prove the following result. Don't let your newly found sophistication blind you to the remarkable nature of results such as this.

Theorem 6.25

The dimension of the row space of any matrix is equal to the dimension of its column space.

Exercises 6.6

1. Find bases for the column space and the row space of each of the matrices

   a. $\begin{bmatrix} 1 & 3 & -1 \\ -1 & 2 & 0 \\ 1 & 8 & -2 \end{bmatrix}$

   b. $\begin{bmatrix} 1 & 3 & 1 & 0 & 1 \\ 2 & 7 & 0 & 5 & 1 \\ -1 & -2 & -3 & 6 & 0 \\ -2 & -5 & -4 & 8 & 3 \end{bmatrix}$

2. Find bases for the kernels and images of the linear maps defined in terms of multiplication by the matrices

   a. $\begin{bmatrix} 1 & 3 & -1 \\ -1 & 2 & 0 \\ 1 & 8 & -2 \end{bmatrix}$

   b. $\begin{bmatrix} 1 & 3 & 1 & 0 & 1 \\ 2 & 7 & 0 & 5 & 1 \\ -1 & -2 & -3 & 6 & 0 \\ -2 & -5 & -4 & 8 & 3 \end{bmatrix}$

   c. In each case compare the dimensions of the kernel, the image, and the domain. Make a conjecture that quantifies the observations at the beginning of this section about the relation among the sizes of these spaces.

3. Find the kernel and image of the identity function $id_V : V \to V$.

4. Find the kernel and image of the zero function $Z : V \to W$ defined by $Z(\mathbf{v}) = \mathbf{0}_W$ for all $\mathbf{v} \in V$.

5. Choose a line through the origin of $\mathbb{R}^2$. Give geometric descriptions of the kernel and the image of the linear function that project each point of $\mathbb{R}^2$ orthogonally onto the chosen line.

6. Find a basis for the kernel of the differentiation map $D : \mathbb{D}(\mathbb{R}) \to \mathbb{F}(\mathbb{R})$.

7. Find a basis for the kernel of the linear map $T : \mathbb{D}(\mathbb{R}) \to \mathbb{F}(\mathbb{R})$ defined by $T(f) = f' - f$.

8. Find a basis for the kernel of the linear map $T : \mathbb{D}^{(2)}(\mathbb{R}) \to \mathbb{F}(\mathbb{R})$ defined by $T(f) = f'' + f$. (Suggestion: See Exercise 10 of Section 3.4.)

9. Find a basis for the kernel of the linear map $E : \mathbb{P}_3 \to \mathbb{R}$ defined by $E(p) = p(2)$. (Suggestion: See Exercise 9 of Section 3.4.)

10. Let a be a real number. Find a basis for the kernel of the evaluation map $E_a : \mathbb{P}_3 \to \mathbb{R}$ defined by $E_a(p) = p(a)$.

6.7 Rank and Nullity

The previous section contains hints of a complementary relation between the kernel and the image of a linear map. This relation can most clearly be expressed in terms of the dimensions of these spaces as defined below.

Definition 6.26

Suppose $T : V \to W$ is linear. If $\ker T$ is a finite-dimensional subspace of V, then the dimension of $\ker T$ is called the **nullity** of T. This is denoted nullity T. If $\operatorname{im} T$ is a finite-dimensional subspace of W, then the dimension of $\operatorname{im} T$ is called the **rank** of T. This is denoted rank T.

Recall that we have previously defined the rank of a matrix A as the number of leading 1s in the reduced row-echelon form of A. In the previous section, we noted that leading 1s in the reduced form of A correspond to columns in A that form a basis for the column space of A. By Theorem 6.22, the column space of A is the same as $\operatorname{im} \mu_A$. Hence, the notion of the rank of a matrix A coincides with the definition of the rank of the corresponding linear map μ_A.

The following theorem is a culmination of much of our work with vector spaces and linear functions.

Theorem 6.27   Dimension Theorem

Suppose $T : V \to W$ is a linear map whose domain V is a finite-dimensional vector space. Then

$$\operatorname{rank} T + \operatorname{nullity} T = \dim V.$$

As we examine various linear maps defined on a finite-dimensional vector space V, the Dimension Theorem says that we can measure the amount $T : V \to W$ collapses V by looking at either $\dim(\ker T)$ or $\dim(\operatorname{im} T)$. Every collapsed dimension of V increases the nullity of T by one and decreases the rank of T by one.

Proof   By Theorem 3.14, any subspace of the finite-dimensional vector space V is also finite-dimensional. Thus, we know there is a basis $\{\mathbf{v}_1, \ldots, \mathbf{v}_m\}$ for $\ker T$. By the Expansion Theorem there are vectors $\mathbf{v}_{m+1}, \ldots, \mathbf{v}_{m+k}$ in V such that

$$\{\mathbf{v}_1, \ldots, \mathbf{v}_m, \mathbf{v}_{m+1}, \ldots, \mathbf{v}_{m+k}\}$$

is a basis for V.

We want to verify the claim that $\{T(\mathbf{v}_{m+1}), \ldots, T(\mathbf{v}_{m+k})\}$ is a basis for $\operatorname{im} T$. To show the set spans $\operatorname{im} T$, choose an arbitrary element $\mathbf{w} \in \operatorname{im} T$. There is $\mathbf{v} \in V$ such that $\mathbf{w} = T(\mathbf{v})$. Also, there are coefficients $r_1, \ldots, r_{m+k} \in \mathbb{R}$ such that $\mathbf{v} = r_1 \mathbf{v}_1 + \cdots + r_{m+k}\mathbf{v}_{m+k}$. Hence,

$$\mathbf{w} = T(\mathbf{v})$$

$$= T(r_1\mathbf{v}_1 + \cdots + r_m\mathbf{v}_m + r_{m+1}\mathbf{v}_{m+1} + \cdots + r_{m+k}\mathbf{v}_{m+k})$$

$$= r_1 T(\mathbf{v}_1) + \cdots + r_m T(\mathbf{v}_m) + r_{m+1} T(\mathbf{v}_{m+1}) + \cdots + r_{m+k} T(\mathbf{v}_{m+k})$$

$$= r_1\mathbf{0} + \cdots + r_m\mathbf{0} + r_{m+1} T(\mathbf{v}_{m+1}) + \cdots + r_{m+k} T(\mathbf{v}_{m+k})$$

$$= r_{m+1} T(\mathbf{v}_{m+1}) + \cdots + r_{m+k} T(\mathbf{v}_{m+k}).$$

Thus, we have verified the condition for $\{T(\mathbf{v}_{m+1}), \ldots, T(\mathbf{v}_{m+k})\}$ to span im T. To show the set is linearly independent, suppose

$$r_{m+1} T(\mathbf{v}_{m+1}) + \cdots + r_{m+k} T(\mathbf{v}_{m+k}) = \mathbf{0}.$$

Then $T(r_{m+1}\mathbf{v}_{m+1} + \cdots + r_{m+k}\mathbf{v}_{m+k}) = \mathbf{0}$. In other words,

$$r_{m+1}\mathbf{v}_{m+1} + \cdots + r_{m+k}\mathbf{v}_{m+k} \in \ker T.$$

Therefore,

$$r_{m+1}\mathbf{v}_{m+1} + \cdots + r_{m+k}\mathbf{v}_{m+k} = r_1\mathbf{v}_1 + \cdots + r_m\mathbf{v}_m$$

for some $r_1, \ldots, r_m \in \mathbb{R}$. This equation can be rewritten as

$$-r_1\mathbf{v}_1 - \cdots - r_m\mathbf{v}_m + r_{m+1}\mathbf{v}_{m+1} + \cdots + r_{m+k}\mathbf{v}_{m+k} = \mathbf{0}.$$

Since the vectors in the linear combination on the left side of this equation form a linearly independent set, all of the coefficients must equal zero. In particular,

$$r_{m+1} = \cdots = r_{m+k} = 0.$$

Thus, we have verified the condition for $\{T(\mathbf{v}_{m+1}), \ldots, T(\mathbf{v}_{m+k})\}$ to be linearly independent. It follows that this set of k vectors is a basis for im T.

We can now read off the required dimensions. We know im T has a basis with k elements; thus, rank $T = \dim(\text{im } T) = k$. We started the proof with a basis for ker T with m elements; thus, nullity $T = \dim(\ker T) = m$. Also, V has a basis with $m + k$ elements. Therefore,

$$\text{rank } T + \text{nullity } T = k + m = \dim V. \quad \blacksquare$$

We can often use the rank and nullity of a linear map $T : V \to W$ to determine basic facts about T. For example, if W is finite-dimensional, then T is onto if and only if rank $T = \dim W$. This is an immediate consequence of Exercise 21 of Section 3.5. The following theorem provides a similar connection between the nullity of T and whether T is one-to-one.

Theorem 6.28

A linear map $T : V \to W$ is one-to-one if and only if ker $T = \{\mathbf{0}\}$.

Proof Suppose T is one-to-one. We want to show that $\mathbf{0}$ is the only element of ker T. So let $\mathbf{v} \in \ker T$. This means, of course, that $T(\mathbf{v}) = \mathbf{0}$. But we also know that $T(\mathbf{0}) = \mathbf{0}$. Since T is one-to-one, we reach the desired conclusion that $\mathbf{v} = \mathbf{0}$.

Conversely, suppose $\ker T = \{0\}$. To show T is one-to-one, we start with $\mathbf{v}_1, \mathbf{v}_2 \in V$ such that $T(\mathbf{v}_1) = T(\mathbf{v}_2)$. Then $\mathbf{0} = T(\mathbf{v}_1) - T(\mathbf{v}_2) = T(\mathbf{v}_1 - \mathbf{v}_2)$. From this we derive the fact that $\mathbf{v}_1 - \mathbf{v}_2 \in \ker T$. The hypothesis gives that $\mathbf{v}_1 - \mathbf{v}_2 = \mathbf{0}$. Hence, $\mathbf{v}_1 = \mathbf{v}_2$, as desired to show T is one-to-one.   ■

Since $\{0\}$ is the only zero-dimensional subspace of a vector space, this theorem can be restated in terms of the nullity of the linear map $T : V \to W$:

T is one-to-one if and only if nullity $T = 0$.

With the tools we have perfected in this section, we can often determine whether or not a linear function is one-to-one or onto simply by knowing the dimensions of the domain and range.

Here are two examples of how to show that certain conditions are impossible with linear maps. The negative conclusions make these ideal candidates for indirect proofs.

Quick Example   *Prove that a linear map $T : \mathbb{R}^3 \to \mathbb{R}^2$ cannot be one-to-one.*

Suppose that T is one-to-one. Then by Theorem 6.28, nullity $T = 0$. The Dimension Theorem gives

$$\operatorname{rank} T = \operatorname{rank} T + 0 = \operatorname{rank} T + \operatorname{nullity} T = \dim \mathbb{R}^3 = 3.$$

Since im T is a subspace of $\mathbb{R}^2$, Theorem 3.14 gives

$$3 = \operatorname{rank} T = \dim(\operatorname{im} T) \leq \dim \mathbb{R}^2 = 2.$$

This contradiction forces us to conclude that the original assumption is false. That is, T cannot be one-to-one.   ■

Quick Example   *Prove that a linear map $T : \mathbb{M}(2, 2) \to \mathbb{P}_4$ cannot be onto.*

Suppose T is onto. Then im $T = \mathbb{P}_4$, so rank $T = \dim(\operatorname{im} T) = \dim \mathbb{P}_4 = 5$. By the Dimension Theorem, rank $T + \operatorname{nullity} T = \dim \mathbb{M}(2, 2)$. In other words, $5 + \operatorname{nullity} T = 4$. Thus, $\dim(\ker T) = \operatorname{nullity} T = -1$. This leads to the impossible situation of a subspace $\ker T$ of dimension -1. From this contradiction, we conclude that T cannot be onto.   ■

And here is an example to show how the Dimension Theorem can lead to a positive result.

Quick Example   *Suppose a linear function $T : \mathbb{P}_4 \to \mathbb{R}^5$ is onto. Prove that T is one-to-one.*

We instantly compute

$$\begin{aligned} \operatorname{nullity} T &= \dim \mathbb{P}_4 - \operatorname{rank} T \\ &= \dim \mathbb{P}_4 - \dim \mathbb{R}^5 \\ &= 0. \end{aligned}$$

Thus, $\ker T = \{0\}$, and by Theorem 6.28, T is one-to-one.   ■

If you want to prove that a linear function $T : V \to W$ is one-to-one and onto, you will usually find it easier to work first at showing that T is one-to-one. This often involves a homogeneous system that will be easier to solve than the nonhomogeneous system involved in showing T is onto. Once you establish that T is one-to-one, you can put nullity $T = 0$ into the equation of the Dimension Theorem and compare

$$\text{rank } T = \dim V - \text{nullity } T = \dim V$$

with $\dim W$ to determine whether T is onto.

Exercises 6.7

1. Is it possible for a linear map to be onto if
    a. the domain is $\mathbb{R}^5$ and the range is $\mathbb{R}^4$?
    b. the domain is $\mathbb{R}^5$ and the range is $\mathbb{P}_4$?
    c. the domain is $\mathbb{R}^5$ and the range is $\mathbb{M}(4, 4)$?
    d. the domain is $\mathbb{R}^5$ and the range is $\mathbb{F}(\mathbb{R})$?

2. Is it possible for a linear map to be one-to-one if
    a. the domain is $\mathbb{R}^5$ and the range is $\mathbb{R}^4$?
    b. the domain is $\mathbb{R}^5$ and the range is $\mathbb{P}_4$?
    c. the domain is $\mathbb{R}^5$ and the range is $\mathbb{M}(4, 4)$?
    d. the domain is $\mathbb{R}^5$ and the range is $\mathbb{F}(\mathbb{R})$?

3. **a.** Explain why there cannot be a linear function $f : \mathbb{R} \to \mathbb{R}^2$ that is onto.
    b. Speculate about whether it is possible for any function $f : \mathbb{R} \to \mathbb{R}^2$ to be onto.

4. Suppose $T : V \to W$ is linear and one-to-one. Suppose $\{\mathbf{v}_1, \ldots, \mathbf{v}_n\}$ is a linearly independent subset of V. Show that $\{T(\mathbf{v}_1), \ldots, T(\mathbf{v}_n)\}$ is a linearly independent subset of W.

5. Suppose $T : V \to W$ is linear and onto. Suppose $\{\mathbf{v}_1, \ldots, \mathbf{v}_n\}$ spans V. Show that $\{T(\mathbf{v}_1), \ldots, T(\mathbf{v}_n)\}$ spans W.

6. Suppose $T : V \to W$ is linear. Suppose $\mathbf{v}_1, \ldots, \mathbf{v}_n$ are vectors in V such that $\{T(\mathbf{v}_1), \ldots, T(\mathbf{v}_n)\}$ is a linearly independent subset of W. Show that $\{\mathbf{v}_1, \ldots, \mathbf{v}_n\}$ is a linearly independent subset of V.

7. Suppose $B = \{\mathbf{u}_1, \ldots, \mathbf{u}_n\}$ is a basis for a vector space V. Suppose P is a nonsingular $n \times n$ matrix. For $j = 1, \ldots, n$, let $\mathbf{u}'_j = L_B(P\mathbf{e}_j)$, where $\mathbf{e}_j$ is the jth element of the standard basis for $\mathbb{R}^n$ and $L_B : \mathbb{R}^n \to V$ is the linear combination function defined in Section 6.1.

    a. Use Theorem 5.8 to show that P^{-1} is nonsingular.

    Use Theorem 6.14 to show that multiplication by P^{-1} defines an invertible function.

    Use Exercise 6 of Section 6.4 to show that L_B is an invertible function.

    Use Exercise 14 of Section 6.2 along with Exercises 4 and 5 above to conclude that $B' = \{\mathbf{u}'_1, \ldots, \mathbf{u}'_n\}$ is a basis for V.

b. Justify the steps in the derivation that for each $j = 1, \ldots, n$ we have

$$[\mathbf{u}_j']_B = [L_B(P\mathbf{e}_j)]_B$$
$$= P\mathbf{e}_j$$
$$= P[\mathbf{u}_j']_{B'}.$$

c. Use Theorem 6.3 to conclude that $P[\mathbf{v}]_{B'} = [\mathbf{v}]_B$ for all $\mathbf{v} \in V$ and hence that P is the change-of-basis matrix for changing from B' to B.

8. Show that an $n \times n$ matrix is invertible if and only if its columns form a basis for $\mathbb{R}^n$.

9. Show that an $n \times n$ matrix is invertible if and only if its rows form a basis for $\mathbb{R}^n$.

6.8 Isomorphism

We have run across several places where it was convenient to switch from writing elements in $\mathbb{R}^n$ as rows to writing them as columns. When we defined linear maps in terms of matrix multiplication, it was again convenient to identify $\mathbb{R}^n$ with $\mathbb{M}(n, 1)$. You have undoubtedly noticed strong similarities between other pairs of vector spaces.

For instance, we can set up a correspondence between $\mathbb{M}(2, 2)$ and $\mathbb{P}_3$ by pairing the generic 2×2 matrix $\begin{bmatrix} a & b \\ c & d \end{bmatrix}$ with the polynomial $ax^3 + bx^2 + cx + d$. Here are six important observations about the correspondence. Each is accompanied by an interpretation in more precise mathematical terminology.

1. No matrix corresponds to more than one polynomial (the correspondence is given by a well-defined function T).

2. Every matrix corresponds to a polynomial (the domain of T is $\mathbb{M}(2, 2)$).

3. No polynomial corresponds to more than one matrix (T is one-to-one).

4. Every polynomial corresponds to a matrix (T is onto).

5. The sum of two matrices corresponds to the sum of the polynomials that correspond to the two matrices (T is additive).

6. A scalar multiple of a matrix corresponds to the scalar times the polynomial that corresponds to the matrix (T is homogeneous).

More concisely, we have a function $T : \mathbb{M}(2, 2) \to \mathbb{P}_3$ that is one-to-one, onto, and linear. The fact that T is a function that is one-to-one and onto means that T preserves the underlying set-theoretical structure. The fact that T is linear means that it also preserves the vector space operations of addition and scalar multiplication.

Of course, there are other interesting mathematical operations on $\mathbb{M}(2, 2)$ (such as matrix multiplication and inverses) and on $\mathbb{P}_3$ (such as inner products, evaluation, differentiation, and integration). In the current context, however, we are interested in $\mathbb{M}(2, 2)$ and $\mathbb{P}_3$ only as vector spaces. So we do not require T to be in any way compatible with these other operations.

Here is the definition of the concept we have been using informally ever since we first switched from row to column notation for elements of $\mathbb{R}^2$.

Definition 6.29

An **isomorphism** is a linear map between vector spaces that is one-to-one and onto. A vector space V is **isomorphic** to a vector space W if and only if there is an isomorphism from V to W.

Quick Example *Show that $\mathbb{R}^2$ is isomorphic to the subspace $P = \{v \in \mathbb{R}^3 \mid (2, 3, 5) \cdot v = 0\}$ of $\mathbb{R}^3$.*

As you may have determined in Exercise 6 of Section 4.2, the set P can be written in the standard form of a plane through the origin:

$$P = \{r(-3, 2, 0) + s(-5, 0, 2) \mid r, s \in \mathbb{R}\}.$$

Let $L : \mathbb{R}^2 \to P$ be defined by $L\left(\begin{bmatrix} r \\ s \end{bmatrix}\right) = r(-3, 2, 0) + s(-5, 0, 2)$. By Theorem 6.9, L is linear. It is clearly onto. By the Dimension Theorem, nullity $L = \dim \mathbb{R}^2 - \operatorname{rank} L = 2 - 2 = 0$. By Theorem 6.28, L is one-to-one. Thus, L is an isomorphism from $\mathbb{R}^2$ to P. Therefore, $\mathbb{R}^2$ is isomorphic to P. ∎

Crossroads One of the fundamental problems in any area of mathematics is to classify the objects under study in terms of some simple, easily computed properties. There are very few cases in all of mathematics where a complete classification has been discovered. You may be interested in reading about some of these success stories: The classification of the five regular polyhedra dates back to ancient Greek civilization. The topological surfaces that can be formed from a finite number of triangles consist of the disk, the sphere, the torus (surface of a doughnut), the projective plane, and surfaces formed by joining multiple copies of these basic types. See Chapter 1 of *Algebraic Topology: An Introduction* by William S. Massey, (New York: Springer-Verlag, 1990). When you take a course in abstract algebra, be sure to watch for the classification theorem for Abelian groups that can be generated by a finite number of elements. The classification of finite simple groups is a monumental success story of twentieth-century mathematics. See "The Enormous Theorem" by Daniel Gorenstein in the December 1985 issue of *Scientific American*.

In a classification theorem we do not want to distinguish objects that are essentially the same as far as their mathematical properties are concerned. The concept of isomorphism gives a precise way of establishing whether or not two vector spaces are essentially the same. Our theorem states that finite-dimensional vector spaces can be classified by a single nonnegative integer, the dimension.

> ### Theorem 6.30 Classification Theorem for Finite-Dimensional Vector Spaces
>
> Suppose V and V' are finite-dimensional vector spaces. Then V is isomorphic to V' if and only if $\dim V = \dim V'$.

Proof Suppose first that V and V' are isomorphic. Let $T : V \to V'$ be an isomorphism. Since T is one-to-one, nullity $T = 0$. Since T is onto, rank $T = \dim V'$. By combining these two facts with the Dimension Theorem, we have $\dim V = \operatorname{rank} T + \operatorname{nullity} T = \dim V' + 0 = \dim V'$.

Conversely, suppose $\dim V = \dim V'$. Let $B = \{\mathbf{u}_1, \ldots, \mathbf{u}_n\}$ be a basis for V, and let $B' = \{\mathbf{u}'_1, \ldots, \mathbf{u}'_n\}$ be a basis for V'. By Theorem 6.9, there is a unique linear map $T : V \to V'$ that satisfies $T(\mathbf{u}_i) = \mathbf{u}'_i$ for $i = 1, \ldots, n$. Now any element of V' can be written $r_1 \mathbf{u}'_1 + \cdots + r_n \mathbf{u}'_n$ for some $r_1, \ldots, r_n \in \mathbb{R}$. It is easy to see that T maps the corresponding linear combination $r_1 \mathbf{u}_1 + \cdots + r_n \mathbf{u}_n$ of the vectors in B to this element:

$$T(r_1 \mathbf{u}_1 + \cdots + r_n \mathbf{u}_n) = r_1 T(\mathbf{u}_1) + \cdots + r_n T(\mathbf{u}_n)$$
$$= r_1 \mathbf{u}'_1 + \cdots + r_n \mathbf{u}'_n.$$

Hence, T is onto. By the Dimension Theorem,

$$\operatorname{nullity} T = \dim V - \operatorname{rank} T$$
$$= \dim V - \dim V'$$
$$= 0.$$

It follows from Theorem 6.28 that T is one-to-one. We have now verified all the conditions to conclude that T is an isomorphism. ∎

Quick Example *Use the Classification Theorem for Finite-Dimensional Vector Spaces to show that* $\mathbb{R}^2$ *is isomorphic to the subspace* $P = \{\mathbf{v} \in \mathbb{R}^3 \mid (2, 3, 5) \cdot \mathbf{v} = 0\}$ *of* $\mathbb{R}^3$.

We know that $\dim \mathbb{R}^2 = 2 = \dim P$. Hence, the vector space $\mathbb{R}^2$ is isomorphic to P. ∎

Exercises 6.8

1. Carry out the details to verify that the correspondence defined in this section between matrices in $\mathbb{M}(2, 2)$ and polynomials in $\mathbb{P}_3$ is an isomorphism.

2. What Euclidean space is isomorphic to the subspace S of all vectors in $\mathbb{R}^4$ orthogonal to $(1, -3, 0, 4)$? (Suggestion: We found a basis for S back in Section 4.2.)

3. Which pairs of the following vector spaces are isomorphic?

$$\mathbb{R}^7 \quad \mathbb{R}^{12} \quad \mathbb{M}(3, 3) \quad \mathbb{M}(3, 4) \quad \mathbb{M}(4, 3) \quad \mathbb{P}_6 \quad \mathbb{P}_8 \quad \mathbb{P}_{11} \quad \mathbb{P}$$

4. **a.** For any vector space V, show that $\mathrm{id}_V : V \to V$ is an isomorphism.

    b. Suppose $T : V \to V'$ is an isomorphism from the vector space V to the vector space V'. Prove that T is invertible and that T^{-1} is an isomorphism from V' to V.

    c. Suppose $T : V \to V'$ and $T' : V' \to V''$ are isomorphisms. Prove that $T' \circ T : V \to V''$ is an isomorphism.

5. **a.** Show that any vector space V is isomorphic to itself.

    b. Show that if a vector space V is isomorphic to a vector space V', then V' is isomorphic to V.

    c. Show that if the vector space V is isomorphic to the vector space V' and V' is isomorphic to the vector space V'', then V is isomorphic to V''.

6. Restate Exercise 8 of Section 1.7 in terms of the concept of isomorphism.

7. Show directly that the linear map T in the second part of the proof of the Classification Theorem has nullity $T = 0$. Use the Dimension Theorem to conclude that T is onto and hence is an isomorphism.

8. The rules of the fifteen game are very simple: Two players alternate in selecting integers from 1 to 9. No number can be selected more than once. Each player tries to select numbers so that three of them add up to 15. The first player to succeed wins the game. Try playing the fifteen game a few times. Show that the fifteen game is isomorphic to another familiar game.

9. Let B be a basis for an n-dimensional vector space V. Prove that the coordinate vector function $[\ ]_B : V \to \mathbb{R}^n$ is an isomorphism.

Project: Dual Spaces

This project deals with linear maps for which the range is the vector space $\mathbb{R}$, the real numbers. You will investigate a duality relation between a vector space and the space of real-valued linear maps defined on the vector space. When you apply this construction to the resulting dual space, you will be amazed to find the original space reemerge.

1. Suppose V is a vector space. Show that the **dual space** defined by

$$V^* = \{\alpha \in \mathbb{F}(V) \mid \alpha \text{ is linear}\}$$

is a subspace of $\mathbb{F}(V)$.

2. Suppose $T : V \to V'$ is a linear map between vector spaces V and V'. Show that the function $T^* : (V')^* \to V^*$ defined by $T^*(\alpha) = \alpha \circ T$ is linear. Show that $\mathrm{id}_V^* = \mathrm{id}_{V^*}$ and that if $T : V \to V'$ and $T' : V' \to V''$ are linear, then $(T \circ T')^* = (T')^* \circ T^*$. In the language of category theory, these properties show that the duality construction is a **contravariant functor**.

3. Let $\{\mathbf{u}_1, \ldots, \mathbf{u}_n\}$ be a basis for a vector space V. Define $\mathbf{u}_i^* : V \to \mathbb{R}$ by $\mathbf{u}_i^*(r_1\mathbf{u}_1 + \cdots + r_n\mathbf{u}_n) = r_i$. Show that $\mathbf{u}_i^* \in V^*$. Show that $\{\mathbf{u}_1^*, \ldots, \mathbf{u}_n^*\}$ is a basis for V^*. Conclude that V is isomorphic to its dual space V^*.

4. Let $\mathbb{R}^\infty = \{(x_1, x_2, \ldots\,) \mid x_1 \in \mathbb{R}, x_2 \in \mathbb{R}, \ldots\}$ be the vector space of infinite sequences of real numbers with the operations defined in Exercise 10 of Section 1.7. Let $\mathbf{e}_1, \mathbf{e}_2, \ldots$ denote the analogs of the standard basis elements in Euclidean spaces. Let

V be the subspace of $\mathbb{R}^\infty$ consisting of all linear combinations that can be formed from finite subsets of $\{\mathbf{e}_1, \mathbf{e}_2, \dots \}$. Define $\mathbf{e}_1^*, \mathbf{e}_2^*, \dots$ to be elements of V^* as was done in the finite-dimensional case. Show that not every element of V^* is a linear combination of $\mathbf{e}_1^*, \mathbf{e}_2^*, \dots$.

5. Let V denote an arbitrary vector space. For any $\mathbf{v} \in V$, define $f_\mathbf{v} : V^* \to \mathbb{R}$ by $f_\mathbf{v}(\alpha) = \alpha(\mathbf{v})$. Show that $f_\mathbf{v}$ is linear, and hence that $f_\mathbf{v}$ is an element of V^{**}, the dual space of V^*. Show that $\theta : V \to V^{**}$ defined by $\theta(\mathbf{v}) = f_\mathbf{v}$ is linear. Prove that if V is a finite-dimensional vector space, then θ is an isomorphism. What can you say about θ if V is not finite-dimensional?

6. Notice that the definition of θ makes no reference to a basis for its domain. This function is a **natural transformation** from any vector space to the double dual of the vector space in the sense made precise by the following observation. Suppose $T : V \to V'$ is a linear map between vector spaces V and V'. You previously constructed a linear map $T^* : (V')^* \to V^*$. Apply this construction again to obtain a linear map $T^{**} : V^{**} \to (V')^{**}$. With $\theta : V \to V^{**}$ and $\theta' : V' \to (V')^{**}$ as defined previously, show that $\theta' \circ T = T^{**} \circ \theta$. Be sure to draw a diagram to help keep track of the four vector spaces and four functions involved in this equation.

Project: Iterated Function Systems

Draw a triangle ABC. Mark a point P in the interior of the triangle. Roll an ordinary six-sided die. If a 1 or a 2 comes up, mark the point halfway between A and P. If a 3 or a 4 comes up, mark the point halfway between B and P. If a 5 or a 6 comes up, mark the point halfway between C and P. Relabel the new point P. Repeat the process of rolling the die, marking the midpoint, and relabeling. After a few dozen iterations (or, better yet, after a few thousand iterations plotted by a computer) you will see emerging, as from out of a fog, an intricate pattern of triangles within triangles.

This particular pattern is known as Sierpinski's triangle. Sets such as this and the snowflake curve described in Exercise 14 of Section 3.1 have fascinated mathematicians since their discovery around the beginning of the twentieth century. The complex structure we see when we look at the entire set is repeated as we zoom in to examine the details in any part of the set. In particular, Sierpinski's triangle consists of three sections, each similar to the entire set.

1. What changes occur to Sierpinski's triangle if you play the game with triangles of different sizes and shapes? How does the starting point influence the resulting figure? What happens when you change some of the rules for defining Sierpinski's triangle? For example, move toward the different vertices with unequal probabilities, move a distance other than halfway (perhaps a randomly chosen fraction), start with more than three vertices, arrange the vertices so they are not coplanar.

2. Analyze Sierpinski's triangle in terms of its self-similar property. Find three functions each of which shrinks Sierpinski's triangle to half its original size and maps it onto the portion near one of the vertices of the triangle. If you place one of the vertices at the origin, you can use a linear map for one of the functions. The other two will consist of a linear map followed by a translation.

3. In general, we can define an **affine map** to be the composition of a linear function followed by a translation. Show that the image of a line under an affine map is another

line. State and prove some theorems about the composition and inverses of affine maps, and try to find a way to represent affine maps in Euclidean spaces in terms of matrices and vector space operations.

4. Michael Barnsley, in his book *Fractals Everywhere* (San Diego: Academic Press, 1988), defines an **iterated function system**. This consists of several affine maps, and rather than choosing to apply these maps at random, we look at the union of the images of an initial set under all the maps. Apply all the affine maps to the resulting set and again look at the union of the images. Under certain circumstances iterations of this process will converge to a limiting set of points.

Write a computer program to plot the sets obtained from an iterated function system. Try creating Sierpinski's triangle from the three affine maps used above. Plot other sets determined by iterated function systems. Notice the extreme amount of data compression that results from describing these complex sets in terms of the the functions represented by matrices and vectors. The Collage Theorem in *Fractals Everywhere* gives a way of designing an iterated function system that will yield a given set. Barnsley's company, Iterated Systems Incorporated, has used these ideas to achieve data compression ratios of 500:1 for photographs of natural images. They anticipate applications of their methods in the transmission and storage of x-rays, photographic archives, and images from space.

Try your hand at using iterated function systems to create images of natural objects such as ferns, trees, clouds, or a mountain range.

Fractals Everywhere lists numerous references for additional information on the mathematics of fractals and chaos. Robert Devaney's book *Chaos, Fractals, and Dynamics: Computer Experiments in Mathematics* (Reading, MA: Addison-Wesley, 1990) gives specific suggestions for writing computer programs to display these fractal sets and other images arising from chaos.

Chapter 6 Summary

Chapter 6 relates pairs of vector spaces by functions that are compatible with the two vector space operations. Such a linear function can be represented by a matrix when the domain and range are finite-dimensional vector spaces. Composition of functions corresponds to matrix multiplication, and the inverse of a linear map corresponds to the inverse of its matrix. The kernel and image of a linear map give basic information about the function. These subspaces are complementary in a sense quantified by the Dimension Theorem. Their dimensions tell us whether the linear map is one-to-one or onto. If it is both, the domain and range are essentially the same vector space.

Computations

Matrix of a linear map
> Between Euclidean spaces
> With respect to bases for the domain and range
> Matrices of compositions and inverses
> Change-of-basis matrix
> Three principles for determining the matrix of a linear map
> Matrix of a linear map relative to a new basis

Defining a unique linear map by values on a basis

Subspaces associated with a linear map
Find a basis for the kernel
Find a basis for the image
Find a basis for the column space of a matrix
Find a basis for the row space of a matrix

Theory

Theory of functions
Domain, range, image, rule
One-to-one
Onto
Composition
Identity function
Inverse function

Linear functions
Composition and inverses are linear
One-to-one if and only if nullity is zero
Dimension Theorem: rank + nullity = dimension of domain
rank A = rank μ_A = dimension of column space of A

Classification of finite-dimensional vector spaces

Applications

Matrix representation of famous linear maps
Rotation
Differentiation
Shifting and scaling in the domain and range
Multiplying a polynomial by a power of x

Isomorphisms legitimize alternate notation for vectors

Review Exercises

1. Show that the function $T : \mathbb{C}([-\pi, \pi]) \to \mathbb{R}$ defined by

$$T(f) = \int_{-\pi}^{\pi} f(x) \cos x \, dx$$

is linear.

2. Show that the function $G : \mathbb{M}(2, 2) \to \mathbb{M}(2, 2)$ that transforms a matrix to its reduced row-echelon form is not linear.

3. Let $P \in \mathbb{M}(k, m)$ and $Q \in \mathbb{M}(n, p)$. Prove that $L : \mathbb{M}(m, n) \to \mathbb{M}(k, p)$ defined by $L(A) = PAQ$ is linear.

4. Consider the function $T : \mathbb{P}_2 \to \mathbb{P}_3$ defined by $T(p(x)) = 5x^2 p'(x)$.

   a. Prove that T is linear.

   b. Determine $T(x^2)$.

   c. Determine $T(x)$.

   d. Determine $T(1)$.

   e. Write down the matrix of T relative to the bases $\{x^2, x, 1\}$ for $\mathbb{P}_2$ and $\{x^3, x^2, x, 1\}$ for $\mathbb{P}_3$.

5. Suppose $A = \begin{bmatrix} 1 & 1 & 3 & 1 \\ 0 & 1 & 2 & 1 \\ -1 & 0 & 4 & 0 \end{bmatrix}$ is the matrix of the linear map $T : \mathbb{P}_3 \to \mathbb{P}_2$ relative to the bases $B = \{x, x+1, x^3, x^3 + x^2\}$ for $\mathbb{P}_3$ and $B' = \{x+1, x-2, x^2 + x\}$ for $\mathbb{P}_2$.

   a. Give the diagram of vector spaces and linear maps for this situation.

   b. Give the formula for A to be the matrix of T relative to B and B'.

   c. Write $T(x^3 + 3x - 2)$ as an element of $\mathbb{P}_2$.

6. Consider the function $T : \mathbb{P}_2 \to \mathbb{P}_3$ that transforms a polynomial p to a polynomial defined by $(x+1)p'(x^2)$.

   a. Show that T is linear.

   b. Determine the matrix A that represents T relative to the bases

   $$B = \{1, x+1, x^2 + x\} \quad \text{and} \quad B' = \{x^3, x^3 + x, x^2 + x, x + 1\}.$$

7. Consider the ordered bases for $\mathbb{R}^3$:

   $$B = \left\{ \begin{bmatrix} 1 \\ 2 \\ 3 \end{bmatrix}, \begin{bmatrix} 1 \\ 2 \\ 0 \end{bmatrix}, \begin{bmatrix} 1 \\ 0 \\ 0 \end{bmatrix} \right\} \quad \text{and} \quad B' = \left\{ \begin{bmatrix} 1 \\ 0 \\ 0 \end{bmatrix}, \begin{bmatrix} 1 \\ 1 \\ 0 \end{bmatrix}, \begin{bmatrix} 1 \\ 1 \\ 1 \end{bmatrix} \right\}.$$

   Find the change-of-basis matrix P for changing from the basis B to the basis B'.

8. Suppose $L : \mathbb{M}(4, 3) \to \mathbb{R}^7$ is linear and onto.

   a. Determine rank L.

   b. Determine nullity L.

9. Consider the function $T : \mathbb{P}_2 \to \mathbb{R}^2$ defined by $T(ax^2 + bx + c) = (a + 2c, b)$.

   a. Show that T is linear.

   b. Show that T is onto.

   c. Use the results of parts a and b to compute rank T and nullity T.

   d. Determine the matrix A of T relative to the bases $B = \{1, x, x^2\}$ for $\mathbb{P}_2$ and $E = \{(1, 0), (0, 1)\}$ for $\mathbb{R}^2$.

   e. Confirm that rank $T = \text{rank} A$.

10. Consider a linear map $T : \mathbb{R}^6 \to \mathbb{M}(2, 3)$.

    a. Show that if T is onto, then it is also one-to-one.

    b. Show that if T is one-to-one, then it is also onto.

11. Consider a linear map $T : \mathbb{M}(4, 3) \to \mathbb{P}_{10}$.

    a. If T is onto, what is the dimension of the kernel of T? Explain.

    b. Is it possible for T to be one-to-one? Explain.

12. Consider the linear map $T : \mathbb{R}^n \to \mathbb{R}^m$ defined by $T(\mathbf{v}) = A\mathbf{v}$, where

$$A = \begin{bmatrix} 1 & 2 & 1 & 1 \\ 2 & 4 & 3 & 6 \\ 1 & 2 & -1 & -7 \end{bmatrix}.$$

The reduced row-echelon form of this matrix is $\begin{bmatrix} 1 & 2 & 0 & -3 \\ 0 & 0 & 1 & 4 \\ 0 & 0 & 0 & 0 \end{bmatrix}.$

    a. Determine the value of m.

    b. Determine the value of n.

    c. Determine a basis for the row space of A.

    d. Determine a basis for the column space of A.

    e. Determine a basis for im T.

    f. Determine a basis for ker T.

    g. Confirm that the Dimension Theorem holds for T.

13. Consider the linear map $T : \mathbb{R}^n \to \mathbb{R}^m$ defined by $T(\mathbf{v}) = A\mathbf{v}$, where

$$A = \begin{bmatrix} 1 & -2 & 0 & 1 & 0 \\ 1 & -2 & -1 & 0 & -1 \\ -2 & 4 & 1 & -1 & 1 \\ 3 & -6 & 3 & 6 & 0 \end{bmatrix}.$$

The reduced row-echelon form of this matrix is $\begin{bmatrix} 1 & -2 & 0 & 1 & 0 \\ 0 & 0 & 1 & 1 & 0 \\ 0 & 0 & 0 & 0 & 1 \\ 0 & 0 & 0 & 0 & 0 \end{bmatrix}.$

    a. Determine the value of m.

    b. Determine the value of n.

    c. Determine a basis for the row space of A.

    d. Determine a basis for the column space of A.

    e. Determine a basis for im T.

    f. Determine a basis for ker T.

    g. Confirm that the Dimension Theorem holds for T.

14. Let $P : \mathbb{R}^3 \to \mathbb{R}^3$ denote the function that maps a vector to its projection onto the vector $\mathbf{u} = (1, 2, 3)$.

    a. Prove that P is linear.

    b. Give a basis for im P.

    c. Give a basis for ker P.

    d. Confirm that the Dimension Theorem holds for P.

15. Let $P : \mathbb{R}^3 \to \mathbb{R}^3$ denote the function that maps a vector to its projection onto the subspace $S = \text{span}\{(1, 1, 0), (1, -1, 1)\}$.

   a. Prove that P is linear.

   b. Give a basis for $\text{im} \, P$.

   c. Give a basis for $\ker P$.

   d. Confirm that the Dimension Theorem holds for P.

16. Let $T : \mathbb{R}^2 \to \mathbb{R}^2$ denote the function that maps $\mathbf{0}$ to $\mathbf{0}$ and maps a nonzero vector to the projection of $(1, 2)$ onto the vector. Show that T is not linear.

17. Suppose V, V', and V'' are vector spaces and V' is finite-dimensional. Suppose $T : V \to V'$ is linear and one-to-one. Suppose $T' : V' \to V''$ is linear and onto. Suppose $\text{im} \, T = \ker T'$.

   a. Prove that V is finite-dimensional.

   b. Prove that V'' is finite-dimensional.

   c. Prove that $\dim V' = \dim V + \dim V''$.

18. Consider the interval $(0, \infty)$ with vector addition defined to be ordinary multiplication and scalar multiplication defined by raising the vector (a positive number) to the power given by the scalar.

   a. Show that $(0, \infty)$ with these operations is a vector space.

   b. Show that the exponential function $\exp : \mathbb{R} \to (0, \infty)$ is an isomorphism from $\mathbb{R}$, with the operations of ordinary addition and multiplication, to $(0, \infty)$.

19. Suppose $T : V \to V'$ and $T' : V \to V'$ are isomorphisms.

   a. Give an example to show that $T + T' : V \to V'$ defined by $(T + T')(\mathbf{v}) = T(\mathbf{v}) + T'(\mathbf{v})$ need not be an isomorphism.

   b. Let c be a nonzero real number. Show that $cT : V \to V'$ defined by $(cT)(\mathbf{v}) = c(T(\mathbf{v}))$ is an isomorphism.

Chapter 7

Determinants

Imagine the possibility of distilling all the entries of a matrix down into one number that would determine the essential features of the matrix. For square matrices, the determinant fulfills this dream. It tells whether the matrix is invertible. Then, if the matrix is invertible, it gives formulas for the inverse and the solution of the associated system of linear equations. The determinant is also the key ingredient we will use in the next chapter to examine the detailed structure of a matrix that represents a linear transformation.

7.1 Mathematical Induction

This preliminary section of Chapter 7 introduces a mathematical technique for proving statements about the positive integers. As you will discover from the examples and exercises in this section, this technique is useful in virtually every branch of mathematics. It will be an indispensable tool in our work with matrices later in this chapter. We begin with a definition.

Definition 7.1

The **natural numbers** are the positive integers. The set $\{1, 2, 3, \ldots\}$ of all natural numbers is denoted $\mathbb{N}$.

Quite frequently, mathematical conjectures and theorems are statements that certain properties or formulas hold for all natural numbers. For example, if we consider sums of the first few odd natural numbers, we quickly observe a pattern:

$$
\begin{aligned}
1 &= 1 \\
1 + 3 &= 4 \\
1 + 3 + 5 &= 9 \\
1 + 3 + 5 + 7 &= 16 \\
1 + 3 + 5 + 7 + 9 &= 25
\end{aligned}
$$

This leads to the conjecture that $\sum_{k=1}^{n}(2k - 1) = n^2$. We have here a statement S_n that two algebraic expressions involving the natural number n are equal. We could continue testing this conjecture for a few more values of n or perhaps write a computer program to handle multiple-precision arithmetic to check the validity of S_n for enormously large values of n. This still falls far short of the endless task of checking S_n for the infinite number of possible values of n.

Crossroads   Beware of jumping to mathematical conclusions based on empirical evidence. Herbert Wilf has an interesting discussion of some strengthened versions of a conjecture that can be stated in terms of natural numbers known as the Riemann hypothesis. (See "A Greeting; and a View of Riemann's Hypothesis" in the January 1987 issue of the *American Mathematical Monthly*.) He points out that in recent years, counterexamples have been found to one of these strengthened conjectures, but the smallest value for which the formula fails is $n = 7,725,038,629$.

A more promising approach to proving a statement about the set of all natural numbers is based on a fundamental rule of inference known as **modus ponens**:

> If an implication $P \Longrightarrow Q$ is true and its hypothesis P is also true, then it follows that the conclusion Q is true.

In our situation, we want to apply modus ponens repeatedly in a fashion reminiscent of an infinite line of dominoes falling over.

> Suppose S_1 is true.
> If $S_1 \Longrightarrow S_2$ is also true, then S_2 is true.
> If $S_2 \Longrightarrow S_3$ is also true, then S_3 is true.
> If $S_3 \Longrightarrow S_4$ is also true, then S_4 is true.
> $\vdots$ $\qquad\qquad$ $\vdots$

Thus, we see that we need to verify the single case of the statement S_1 and provide a general proof of the implication $S_n \Longrightarrow S_{n+1}$ that the truth of the statement for one natural number n implies the truth of the statement for the succeeding natural number $n + 1$. This can be formulated as follows:

> **Principle 7.2   Principle of Mathematical Induction**
>
> Suppose S_n is a statement about the natural number n. Suppose S_1 is true and, for any natural number n, the implication $S_n \implies S_{n+1}$ is true. Then S_n is true for all natural numbers n.

Be sure to notice that in the implication $S_n \implies S_{n+1}$, the variable n refers to a specific, but arbitrary, natural number. This same value of n must appear in both the hypothesis S_n and the conclusion S_{n+1} of the implication. Do not confuse the proof of this implication, which will involve assuming the hypothesis S_n for a specific value of n, with the conclusion of the Principle of Mathematical Induction that S_n holds for all natural numbers n.

When we work with statements about natural numbers, variables such as k and n will often be understood from context to stand for natural numbers. We will frequently make use of this convention without further mention.

Crossroads   The Principle of Mathematical Induction can be adopted at a very fundamental level as an axiom for the system of natural numbers from which the integers, rational numbers, and real numbers can be developed in succession. In this text we have taken the real number system and its properties as a logical basis for the theory of vector spaces. From this point of view, we can consider the Principle of Mathematical Induction as one of the properties of the subset $\mathbb{N}$ of the real numbers. Later in this section we will mention a compromise position in which the Principle of Mathematical Induction is used to define the natural numbers as a subset of $\mathbb{R}$.

Let us return to the example involving the sums of odd natural numbers to illustrate a proof by mathematical induction. The statement S_n is the formula

$$\sum_{k=1}^{n}(2k - 1) = n^2.$$

We first verify S_1:

$$\sum_{k=1}^{1}(2k - 1) = 2 \cdot 1 - 1 = 1 = 1^2.$$

This establishes what is known as the **basis for the induction**.

Next we fix our attention on an arbitrary natural number n and assume S_n; that is, we assume the equality

$$\sum_{k=1}^{n}(2k - 1) = n^2$$

holds for this value of n. This assumption is known as the **induction hypothesis**. For this fixed value of n we want to use the induction hypothesis to derive S_{n+1}:

$$\sum_{k=1}^{n+1}(2k-1) = (n+1)^2.$$

The trick is to write the left-hand side in terms of $\sum_{k=1}^{n}(2k-1)$, a sum we know how to deal with. We do this by splitting off the last term:

$$\sum_{k=1}^{n+1}(2k-1) = \sum_{k=1}^{n}(2k-1) + 2(n+1) - 1$$
$$= n^2 + 2n + 1$$
$$= (n+1)^2.$$

We now conclude by the Principle of Mathematical Induction that

$$\sum_{k=1}^{n}(2k-1) = n^2$$

is true for all $n \geq 1$.

Sometimes the induction step is notationally simpler if we prove $S_{n-1} \Longrightarrow S_n$ for all $n > 1$. Starting with the smallest value, $n = 2$, this gives

$$S_1 \Longrightarrow S_2, \quad S_2 \Longrightarrow S_3, \quad \dots,$$

which is precisely the sequence of implications required for a proof by induction. Of course, we still must establish S_1 as a basis for the induction.

Quick Example　*Prove that the formula $\sum_{k=1}^{n} 2^{k-1} = 2^n - 1$ holds for all natural numbers n.*

For $n = 1$, we have $\sum_{k=1}^{1} 2^{k-1} = 2^0 = 1 = 2^1 - 1$.
For some fixed $n > 1$ assume that $\sum_{k=1}^{n-1} 2^{k-1} = 2^{n-1} - 1$. Then

$$\sum_{k=1}^{n} 2^{k-1} = \sum_{k=1}^{n-1} 2^{k-1} + 2^{n-1}$$
$$= (2^{n-1} - 1) + 2^{n-1}$$
$$= 2 \cdot 2^{n-1} - 1$$
$$= 2^n - 1.$$

Therefore, by the Principle of Mathematical Induction, the formula holds for all $n \geq 1$. ∎

The Principle of Mathematical Induction has applications far beyond proving that two algebraic expressions are equal. Consider, for example, the Tower of Hanoi puzzle. According to one legend, the Brahman priests have been working in a temple in Banares to transfer a tower of 64 golden rings, originally stacked in order of decreasing size, from one of three diamond pegs to another. The transfer must be made subject to the conditions that only one ring can be removed from the pegs at any time and that the rings on each of the pegs must always be in order from largest on the bottom to smallest on top. When the transfer of all 64 rings is completed, the world will end.

If you experiment with this puzzle (perhaps with a stack of coins if golden rings and diamond pegs are not readily available), you will be led to conjecture that n rings can be transferred in $2^n - 1$ moves. This conjecture, being a statement about natural numbers, is an ideal candidate for proof by the Principle of Mathematical Induction.

Quick Example   *Prove that the Tower of Hanoi puzzle with n rings can be solved in $2^n - 1$ moves.*

For $n = 1$, simply move the ring from one peg to another to solve the puzzle in $1 = 2^1 - 1$ moves.

   For some fixed $n \geq 1$, assume that the n-ring puzzle can be solved in $2^n - 1$ moves. For this value of n we want to derive from the induction hypothesis the conclusion that $n + 1$ rings can be transferred in $2^{n+1} - 1$ moves. Since the rules permit placing any of the n smaller rings on the largest ring, we can begin by ignoring the largest ring and using the n-ring solution to move the n smaller rings onto an empty peg. This can be done in $2^n - 1$ moves. Now use one move to transfer the large ring onto the third peg. Finally, ignore the large ring again and use $2^n - 1$ additional moves to transfer the n smaller rings onto the peg containing the large ring. Since the three pegs are indistinguishable, we can modify the original n-move solution to transfer these rings to the specified peg. This solves the $(n + 1)$-ring puzzle in

$$(2^n - 1) + 1 + (2^n - 1) = 2 \cdot 2^n - 1 = 2^{n+1} - 1$$

moves.

   Therefore, by the Principle of Mathematical Induction, for any natural number n the Tower of Hanoi puzzle with n rings can be solved in $2^n - 1$ moves.   ■

Definitions are often based on the Principle of Mathematical Induction. In such situations, the statement S_n is that the concept has been defined for the value n. We require an explicit definition of the concept for $n = 1$ and a definition of the concept for the value $n + 1$ in terms of the concept for the value n. The conclusion of the Principle of Mathematical Induction is that we have a definition of the concept for all values of the natural number n.

   Here is a simple example to illustrate the idea of inductive definitions. The formulas

$$x_1 = \sqrt{2}$$

and

$$x_{n+1} = \sqrt{2 + x_n} \quad \text{for } n \geq 1$$

define the sequence of real numbers

$$x_1 = \sqrt{2}$$

$$x_2 = \sqrt{2 + \sqrt{2}}$$

$$x_3 = \sqrt{2 + \sqrt{2 + \sqrt{2}}}$$

$$\vdots$$

Implicit in this definition is the understanding that the formula never involves taking the square root of a negative number. It is quite easy to verify that $2 + x_n$ is indeed positive for all $n \geq 1$. We use induction, of course.

For $n = 1$ we have that $2 + x_1 = 2 + \sqrt{2} > 0$.
For some fixed $n \geq 1$, assume that $2 + x_n > 0$.
Then $x_{n+1} = \sqrt{2 + x_n} > 0$, so $2 + x_{n+1} > 0$.

Crossroads   As was mentioned earlier, it is possible to use the Principle of Mathematical Induction to give a rigorous definition of the natural numbers as a subset of the real number system. Begin by calling a subset S of $\mathbb{R}$ **inductive** if and only if $1 \in S$ and for any real number x if $x \in S$, then $x + 1 \in S$. The set $\mathbb{N}$ of natural numbers can then be defined to be the intersection of all inductive subsets of $\mathbb{R}$.

Let us conclude this section with one variation on the theme of proof by mathematical induction. Suppose we want to prove that $2^n > n^2 + 1$ for all natural numbers n. For $n = 1$,

$$2^1 > 1^2 + 1$$

is false. This makes the proof of the result somewhat difficult. Rather than giving up entirely, we can try different values of n until we find one that works.

Here is how to verify the formula by mathematical induction using a different basis for the induction. If you understand how the Principle of Mathematical Induction works, you will agree that we can apply it here in a slightly modified form to handle our proof, which starts with the statement S_5. In Exercise 28 you are asked to show how the modified form of the Principle of Mathematical Induction can be derived from the original form.

Quick Example   *Prove that $2^n > n^2 + 1$ for all natural numbers $n \geq 5$.*

For $n = 5$,

$$2^5 = 32 > 26 = 5^2 + 1.$$

Now for some fixed $n \geq 5$, assume that $2^n > n^2 + 1$. For this value of n we want to prove that $2^{n+1} > (n + 1)^2 + 1$. The following string of equalities and inequalities results from some working backward and some planning ahead. The induction hypothesis is used in a very typical way in the second step. Notice also the use of the condition $n > 2$ (certainly true since $n \geq 5$) in the second-to-last step.

$$\begin{aligned}
2^{n+1} &= 2 \cdot 2^n \\
&> 2(n^2 + 1) \\
&= n^2 + n^2 + 2 \\
&= n^2 + n \cdot n + 2 \\
&> n^2 + 2n + 2 \\
&= (n + 1)^2 + 1.
\end{aligned}$$

We conclude from the modified version of the Principle of Mathematical Induction that $2^n > n^2 + 1$ for all natural numbers $n \geq 5$.   ∎

Exercises 7.1

Use mathematical induction to try to prove the statements in Exercises 1 to 20. Warning: Some of the statements are false, and some are conjectures that have defied all previous attempts at proof or disproof.

1. $\displaystyle\sum_{k=1}^{n} k = \frac{n(n+1)}{2}$

2. $\displaystyle\sum_{k=1}^{n} \frac{1}{k(k+1)} = 1 - \frac{1}{n+1}$

3. $\displaystyle\sum_{k=1}^{n} k^2 = \frac{n(n+1)(2n+1)}{6}$

4. $\displaystyle\sum_{k=1}^{n} k^3 = \frac{n^2(n+1)^2}{4}$

5. $2^n \geq n$

6. $n! \geq n^2$

7. $4^n - 1$ is divisible by 3.

8. $n^3 - n$ is divisible by 3.

9. $12^n - 5^n$ is divisible by 7.

10. $\displaystyle\sum_{k=1}^{n} k \cdot k! = (n+1)! - 1$

11. The sum of cubes of three consecutive natural numbers is divisible by 9.

12. For any real number x,
$$(1 + x + x^2 + \cdots + x^n)(1 - x) = 1 - x^{n+1}.$$

13. For any $r \times r$ matrix Q,
$$(I + Q + Q^2 + \cdots + Q^n)(I - Q) = I - Q^{n+1}.$$

14. For any natural number n, the expression $n^2 + n + 41$ is prime.

15. If $T : V \to W$ is a linear map, then for $\mathbf{v}_1, \ldots, \mathbf{v}_n \in V$ and $r_1, \ldots, r_n \in \mathbb{R}$,
$$T(r_1\mathbf{v}_1 + \cdots + r_n\mathbf{v}_n) = r_1 T(\mathbf{v}_1) + \cdots + r_n T(\mathbf{v}_n).$$

16. Suppose the $r \times r$ matrices $A_1, \ldots, A_n$ are nonsingular. Then
$$(A_1 \cdots A_n)^{-1} = A_n^{-1} \cdots A_1^{-1}.$$

17. Suppose A and P are $r \times r$ matrices and P is nonsingular. Then
$$(P^{-1}AP)^n = P^{-1}A^n P.$$

18. (Goldbach conjecture) Every even number greater than 2 is the sum of two prime numbers.

19. For any real number p

$$\begin{bmatrix} 1-p & p \\ p & 1-p \end{bmatrix}^n = \begin{bmatrix} \dfrac{1+(1-2p)^n}{2} & \dfrac{1-(1-2p)^n}{2} \\ \dfrac{1-(1-2p)^n}{2} & \dfrac{1+(1-2p)^n}{2} \end{bmatrix}.$$

20. The n-ring Tower of Hanoi puzzle cannot be solved with fewer than $2^n - 1$ moves.

21. If the priests move one ring every second, how many years will it take until the end of the world?

22. For any real number $x \geq -1$, the inequality $(1+x)^n \geq 1 + nx$ holds for all $n \in \mathbb{N}$.

23. Let $s_1 = \sqrt{1}$, $s_2 = \sqrt{1 + \sqrt{1}}$, $s_3 = \sqrt{1 + \sqrt{1 + \sqrt{1}}}$, ... Prove that $s_n \leq 2$ and $s_n \leq s_{n+1}$ for any n. (Hence, by the completeness property of the real number system, the sequence (s_n) converges. Can you determine the value of $\lim_{n \to \infty} s_n$?)

24. Suppose the real numbers r and s satisfy $0 < r < s$. Suppose the sequence $(a_0, a_1, a_2, \dots)$ of real numbers satisfies $ra_{n-1} < a_n < sa_{n-1}$ for $n \geq 1$. Show for any natural number n, that $r^n a_0 < a_n < s^n a_0$.

25. The Fibonacci sequence is defined inductively by $F_0 = 1$, $F_1 = 1$, and, for $n \geq 1$,

$$F_{n+1} = F_n + F_{n-1}.$$

a. Prove that if x satisfies the equation $x^2 = x + 1$, then $x^n = xF_n + F_{n-1}$ for $n \geq 1$.

b. Let r and s denote the two roots of $x^2 = x + 1$. Use the two equations $r^n = rF_n + F_{n-1}$ and $s^n = sF_n + F_{n-1}$ to derive the explicit formula

$$F_n = \frac{1}{\sqrt{5}}\left[\left(\frac{1+\sqrt{5}}{2}\right)^n - \left(\frac{1-\sqrt{5}}{2}\right)^n\right].$$

26. Let $x_0 = 1$, and for $n \geq 0$ let $x_{n+1} = \dfrac{x_n^2 + 2}{2x_n}$.

a. For any $n \geq 0$, show that x_n is a rational number with $1 \leq x_n \leq 2$.

b. For any $n \geq 0$, show that $|x_n^2 - 2| \leq \dfrac{1}{4^{2^n - 1}}$.

c. Conclude that $\lim_{n \to \infty} x_n = \sqrt{2}$.

27. Let $x_1 = 1$, and for $n \geq 1$ let $x_{n+1} = \dfrac{1}{1 + x_n}$.

a. For any $n \geq 2$, show that $\frac{1}{2} \leq x_n \leq \frac{2}{3}$.

b. For any $n \geq 2$, show that $|x_{n+1} - x_n| \leq \left(\frac{4}{9}\right)^n$.

c. Look up (or prove for yourself) a theorem on sequences to justify the conclusion that $\lim_{n \to \infty} x_n$ exists.

d. Show that $\lim_{n \to \infty} x_n$ equals the golden mean $\dfrac{\sqrt{5} - 1}{2}$.

28. Prove the following modification of the Principle of Mathematical Induction that allows us to use a basis for induction other than $n = 1$.

Let n_0 be an integer. Suppose for any integer $n \geq n_0$ that S_n is a statement about n. Suppose S_{n_0} is true and for any integer $n \geq n_0$ the implication $S_n \implies S_{n+1}$ is true. Then S_n is true for all integers $n \geq n_0$.

(Suggestion: Apply the Principle of Mathematical Induction in its original form to the statement $S'_n = S_{n+n_0-1}$.)

29. Use the Principle of Mathematical Induction to prove the following principle, known as the **Principle of Complete Mathematical Induction**.

Suppose S_n is a statement about the natural number n. Suppose S_1 is true and for any natural number n the implication

$$S_k \text{ for all } k \leq n \implies S_{n+1}$$

is true. Then S_n is true for all natural numbers.

30. Prove that the intersection of any collection of inductive subsets of $\mathbb{R}$ is an inductive set.

7.2 Definition

According to Exercise 9 of Section 2.2, the system

$$ax + by = r$$
$$cx + dy = s$$

has a unique solution if and only if the entries of the coefficient matrix $\begin{bmatrix} a & b \\ c & d \end{bmatrix}$ satisfy $ad - bc \neq 0$.

According to Exercise 12 of Section 5.2, the matrix $\begin{bmatrix} a & b \\ c & d \end{bmatrix}$ has an inverse if and only if $ad - bc \neq 0$.

These two exercises lead us to suspect that we can determine some decisive information about a 2×2 matrix $\begin{bmatrix} a & b \\ c & d \end{bmatrix}$ from the number $ad - bc$. Think of each term of the expression $ad - bc$ as the product of an entry in the first row, a or b, times the quantity from the part of the matrix not in the row or column of that entry, d or c, respectively. The following definition extends this idea to square matrices of arbitrary size.

Definition 7.3

The **determinant** of an $n \times n$ matrix A, denoted $\det A$, is defined inductively as follows:

For a 1×1 matrix $A = [a_{11}]$, $\det A = a_{11}$.

For an $n \times n$ matrix A with $n > 1$, let A_{ij} denote the $(n - 1) \times (n - 1)$ submatrix obtained from A by deleting the ith row and the jth column. Then

$$\det A \;=\; \sum_{j=1}^{n}(-1)^{1+j}a_{1j}\det A_{1j}.$$

In other words, we compute the determinant of an $n \times n$ matrix A by expansion along the first row of the matrix, multiplying the various entries a_{1j} in turn by the determinants of the $(n-1) \times (n-1)$ submatrices A_{1j} and then adding these products together with alternating signs. Of course, if $n > 2$, we will need to compute the determinants $\det A_{1j}$ by expansion along the first rows of the submatrices A_{1j} for $j = 1, \ldots, n$. After $n-1$ such levels of repeated expansion, we will be dealing with 1×1 submatrices where the determinant is equal to the single entry in the matrix.

This definition gives us what we wanted for a 2×2 matrix $A = \begin{bmatrix} a & b \\ c & d \end{bmatrix}$, namely,

$$\det \begin{bmatrix} a & b \\ c & d \end{bmatrix} = a \det A_{11} - b \det A_{12}$$
$$= a \det[d] - b \det[c]$$
$$= ad - bc.$$

The expansion of a 3×3 matrix $A = \begin{bmatrix} a & b & c \\ d & e & f \\ g & h & i \end{bmatrix}$ is a little more complicated. Since the formula for the determinant of a 2×2 matrix is relatively simple, we might as well use it as soon as we reach that stage in the computation. The explicit formula derived below for the determinant of a 3×3 matrix is not one that many people commit to memory.

$$\det \begin{bmatrix} a & b & c \\ d & e & f \\ g & h & i \end{bmatrix} = a \det A_{11} - b \det A_{12} + c \det A_{13}$$

$$= a \det \begin{bmatrix} e & f \\ h & i \end{bmatrix} - b \det \begin{bmatrix} d & f \\ g & i \end{bmatrix} + c \det \begin{bmatrix} d & e \\ g & h \end{bmatrix}$$

$$= a(ei - fh) - b(di - fg) + c(dh - eg)$$

$$= aei - afh - bdi + bfh + cdh - ceg.$$

Quick Example *Use the definition to compute* $\det \begin{bmatrix} 2 & 1 & 3 & 4 \\ 0 & 1 & 1 & 0 \\ 1 & 0 & 1 & 1 \\ 4 & 2 & 1 & -1 \end{bmatrix}.$

$$\det \begin{bmatrix} 2 & 1 & 3 & 4 \\ 0 & 1 & 1 & 0 \\ 1 & 0 & 1 & 1 \\ 4 & 2 & 1 & -1 \end{bmatrix} = 2 \det \begin{bmatrix} 1 & 1 & 0 \\ 0 & 1 & 1 \\ 2 & 1 & -1 \end{bmatrix} - 1 \det \begin{bmatrix} 0 & 1 & 0 \\ 1 & 1 & 1 \\ 4 & 1 & -1 \end{bmatrix}$$

$$+ 3 \det \begin{bmatrix} 0 & 1 & 0 \\ 1 & 0 & 1 \\ 4 & 2 & -1 \end{bmatrix} - 4 \det \begin{bmatrix} 0 & 1 & 1 \\ 1 & 0 & 1 \\ 4 & 2 & 1 \end{bmatrix}$$

$$= 2\big(1(-1 - 1) - 1(0 - 2) + 0\big) - 1\big(0 - 1(-1 - 4) + 0\big)$$

$$+ 3\big(0 - 1(-1 - 4) + 0\big) - 4\big(0 - 1(1 - 4) + 1(2 - 0)\big)$$
$$= 2 \cdot 0 - 1 \cdot 5 + 3 \cdot 5 - 4 \cdot 5$$
$$= -10. \quad \blacksquare$$

As you can see, even with an occasional zero to help simplify things, computing determinants of large matrices from the definition can quickly get out of hand. Even a computer will hesitate noticeably when using the definition to compute the determinant of an $n \times n$ matrix when n is appreciably larger than 10.

Fortunately, determinants obey properties stated in the following theorem that make their computation more tractable. Notice that properties b, c, and e tell what happens to the determinant of a matrix as row operations are applied to the matrix.

Theorem 7.4

Suppose A is an $n \times n$ matrix.

a. If A has a row of zeros, then $\det A = 0$.

b. If two rows of A are interchanged, then the determinant of the resulting matrix is $-\det A$.

c. If one row of A is multiplied by a constant c, then the determinant of the resulting matrix is $c \det A$.

d. If two rows of A are identical, then $\det A = 0$.

e. If a multiple of one row of A is added to another row of A, the determinant of the resulting matrix is $\det A$.

f. $\det I_n = 1$.

Properties a through e follow easily from a basic theorem about determinants that will be stated and proved in the next section. Exercise 3 at the end of this section asks you to prove property f. This is a nice application of the Principle of Mathematical Induction to the inductive definition of determinants.

For now, consider the following example as an illustration of the use of these properties in computing a determinant as we follow the steps in the Gauss-Jordan reduction algorithm. Watch how the sign of the determinant changes when we interchange two rows, how a constant multiple $c \neq 0$ appears to be factored out of a row when we multiply a row by $\frac{1}{c}$, and how the determinant is unaffected when a multiple of one row is added to another row.

Quick Example *Use Theorem 7.4 to compute* $\det \begin{bmatrix} 2 & 1 & 3 & 4 \\ 0 & 1 & 1 & 0 \\ 1 & 0 & 1 & 1 \\ 4 & 2 & 1 & -1 \end{bmatrix}$.

$$\det \begin{bmatrix} 2 & 1 & 3 & 4 \\ 0 & 1 & 1 & 0 \\ 1 & 0 & 1 & 1 \\ 4 & 2 & 1 & -1 \end{bmatrix} = -\det \begin{bmatrix} 1 & 0 & 1 & 1 \\ 0 & 1 & 1 & 0 \\ 2 & 1 & 3 & 4 \\ 4 & 2 & 1 & -1 \end{bmatrix} = -\det \begin{bmatrix} 1 & 0 & 1 & 1 \\ 0 & 1 & 1 & 0 \\ 0 & 1 & 1 & 2 \\ 0 & 2 & -3 & -5 \end{bmatrix}$$

$$= -\det \begin{bmatrix} 1 & 0 & 1 & 1 \\ 0 & 1 & 1 & 0 \\ 0 & 0 & 0 & 2 \\ 0 & 0 & -5 & -5 \end{bmatrix} = \det \begin{bmatrix} 1 & 0 & 1 & 1 \\ 0 & 1 & 1 & 0 \\ 0 & 0 & -5 & -5 \\ 0 & 0 & 0 & 2 \end{bmatrix}$$

$$= -5\det \begin{bmatrix} 1 & 0 & 1 & 1 \\ 0 & 1 & 1 & 0 \\ 0 & 0 & 1 & 1 \\ 0 & 0 & 0 & 2 \end{bmatrix} = -10\det \begin{bmatrix} 1 & 0 & 1 & 1 \\ 0 & 1 & 1 & 0 \\ 0 & 0 & 1 & 1 \\ 0 & 0 & 0 & 1 \end{bmatrix}$$

$$= -10\det \begin{bmatrix} 1 & 0 & 0 & 0 \\ 0 & 1 & 0 & 0 \\ 0 & 0 & 1 & 0 \\ 0 & 0 & 0 & 1 \end{bmatrix} = -10\det I = -10. \quad \blacksquare$$

Several of the steps in the preceding example could easily have been combined to reduce the amount of paperwork. For matrices of size 5×5 or larger, this method is extremely valuable in reducing the amount of arithmetic in computing determinants.

Exercises 7.2

1. Compute the determinants of the matrices

   a. $\begin{bmatrix} -3 & 2 & 4 \\ 1 & -2 & 5 \\ 3 & 1 & 7 \end{bmatrix}$ **b.** $\begin{bmatrix} 4 & 0 & 0 \\ 1 & -3 & 0 \\ 2 & 1 & 5 \end{bmatrix}$

   c. $\begin{bmatrix} \lambda - 1 & 4 \\ -3 & -\lambda - 2 \end{bmatrix}$ **d.** $\begin{bmatrix} 2 & 0 & 1 & 1 \\ 1 & -1 & 3 & 0 \\ 1 & 2 & 1 & 3 \\ 0 & 1 & -1 & 2 \end{bmatrix}$

2. **a.** Use Definition 7.3 to compute $\det \begin{bmatrix} -2 & 2 & 7 & 0 \\ 0 & 9 & 0 & 3 \\ 0 & 5 & 0 & 0 \\ 4 & 1 & 8 & 2 \end{bmatrix}$.

   b. Use Theorem 7.4 to compute this determinant by keeping track of the changes that occur as you apply row operations to put the matrix in reduced row-echelon form.

3. Use the Principle of Mathematical Induction to prove that the determinant of the $n \times n$ identity matrix I_n is equal to 1.

4. Show that $\det \begin{bmatrix} a & b & c \\ d & e & f \\ g & h & i \end{bmatrix} = \det \begin{bmatrix} a & d & g \\ b & e & h \\ c & f & i \end{bmatrix}$.

5. Prove that the determinant of the $n \times n$ zero matrix is equal to 0.

6. On the planet Matrix every month is seven weeks long. Compute the determinant of the 7×7 array of integers that appears on their calendar every month. (Suggestion: Use Theorem 7.4 to simplify the computations.)

7. Compute the determinants of the matrices

a. $\begin{bmatrix} -5 & 9 \\ 0 & 3 \end{bmatrix}$

b. $\begin{bmatrix} 7 & 4 & -3 \\ 0 & -3 & 9 \\ 0 & 0 & 2 \end{bmatrix}$

c. $\begin{bmatrix} -2 & 8 & 7 & 4 \\ 0 & 3 & 5 & 17 \\ 0 & 0 & 1 & 9 \\ 0 & 0 & 0 & 5 \end{bmatrix}$

d. $\begin{bmatrix} a & b & c \\ 0 & d & e \\ 0 & 0 & f \end{bmatrix}$

e. What feature of these matrices makes it relatively easy to compute their determinants?

f. Formulate a general property of determinants suggested by this observation.

g. Use the Principle of Mathematical Induction to prove your conjecture.

8. Use the Principle of Mathematical Induction to show that the determinant of an $n \times n$ matrix can be computed by adding and subtracting $n!$ terms consisting of products of entries of the matrix.

9. Refine the result of the previous exercise to show that each of the terms consists of exactly one factor from each row and one factor from each column of the matrix.

10. Show that the number of arithmetic operations (additions, subtractions, multiplications, and divisions) needed to put an $n \times n$ matrix in reduced row-echelon form is less than a constant times n^3.

11. Find examples of 2×2 matrices A and B for which

$$\det(A + B) \neq \det A + \det B.$$

12. Consider a parallelogram in $\mathbb{R}^2$ with vertices at $(0, 0)$, (a, b), (c, d), and $(a + c, b + d)$.

a. Show that the area of this parallelogram is $\left| \det \begin{bmatrix} a & b \\ c & d \end{bmatrix} \right|$.

b. What happens if $\det \begin{bmatrix} a & b \\ c & d \end{bmatrix} = 0$?

c. What is the significance of the sign of $\det \begin{bmatrix} a & b \\ c & d \end{bmatrix}$?

d. Interpret the determinant of a 3×3 matrix as the volume of a three-dimensional object.

7.3 Properties of Determinants

In this section we want to get down to the business of proving some properties of the determinant function. Since our definition of determinant was an inductive definition, it is natural to expect that the proofs of the basic theorems will also be proofs by mathematical induction. Some of these proofs consist of computations of formidable appearance. The truth of the matter is that these computations are straightforward expansions of various determinants. Once you catch on to the pattern, the masses of subscripts and summations will not be so intimidating.

We will use the notation introduced in the previous section for the $(n-1) \times (n-1)$ submatrix A_{ij} obtained by deleting the ith row and the jth column from the $n \times n$ matrix A. For $r \neq i$ and $k \neq j$ we will also let $A_{ij,rk}$ denote the $(n-2) \times (n-2)$ submatrix obtained by deleting from A rows i and r and columns j and k. Notice that row r of A corresponds to row $r-1$ of A_{1j}. Likewise, column k of A corresponds to column k of A_{1j} if $k < j$, and it corresponds to column $k-1$ of A_{1j} if $k > j$. The following diagram compares the indexing system in A_{1j} with that of the original matrix A.

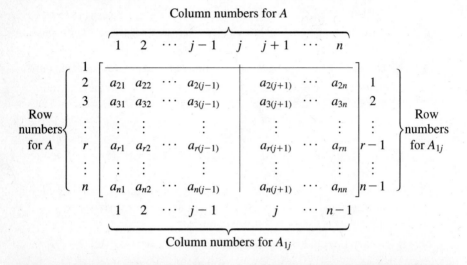

Theorem 7.5

The determinant of a square matrix can be computed by expansion along any row. That is, if A is an $n \times n$ matrix and $r = 1, \ldots, n$, then

$$\det A = \sum_{k=1}^{n} (-1)^{r+k} a_{rk} \det A_{rk}.$$

Proof The proof of this formula is by induction on the size of the matrix A. The smallest value of n we need to consider is $n = 2$. In this case there are two possible values of r. Computing $\det A$ by expansion along the second row gives $-a_{21}a_{12} + a_{22}a_{11}$, which agrees with the expansion of $\det A$ along the first row. Now for some fixed $n > 2$, suppose that A is an $n \times n$ matrix and that the result has been verified for square matrices of size $(n-1) \times (n-1)$. We can assume that $r \neq 1$. We will expand $\det A$ along the first row and then apply the induction hypothesis to expand the determinant of each A_{1j} along row $r-1$ (corresponding to row r of A).

$$\det A = \sum_{j=1}^{n} (-1)^{1+j} a_{1j} \det A_{1j}$$

$$= \sum_{j=1}^{n} (-1)^{1+j} a_{1j} \left(\sum_{k=1}^{j-1} (-1)^{(r-1)+k} a_{rk} \det A_{1j,rk} \right.$$

$$\left. + \sum_{k=j+1}^{n} (-1)^{(r-1)+(k-1)} a_{rk} \det A_{1j,rk} \right)$$

$$= \sum_{k<j} (-1)^{j+r+k} a_{1j} a_{rk} \det A_{1j,rk} + \sum_{k>j} (-1)^{j+r+k-1} a_{1j} a_{rk} \det A_{1j,rk}.$$

Let us work a little on the expression on the other side of the equation. Begin by expanding the determinant of each A_{rk} along the first row.

$$\sum_{k=1}^{n} (-1)^{r+k} a_{rk} \det A_{rk}$$

$$= \sum_{k=1}^{n} (-1)^{r+k} a_{rk} \left(\sum_{j=1}^{k-1} (-1)^{1+j} a_{1j} \det A_{rk,1j} + \sum_{j=k+1}^{n} (-1)^{1+(j-1)} a_{1j} \det A_{rk,1j} \right)$$

$$= \sum_{j<k} (-1)^{r+k+j+1} a_{rk} a_{1j} \det A_{rk,1j} + \sum_{j>k} (-1)^{r+k+j} a_{rk} a_{1j} \det A_{rk,1j}.$$

The final expressions in these two derivations can be matched up since $A_{1j,rk} = A_{rk,1j}$. ∎

The proof of this theorem is quite a workout. In addition to the practice in using subscripts and summation notation in a proof by induction, the usefulness of this result will justify the effort necessary to follow through the derivation. In particular, this theorem reduces the proof of Theorem 7.4 to a few simple computations. Exercises 1 through 5 provide some suggestions to guide you through the verifications of the remaining five parts of Theorem 7.4.

You should be aware of a similar result that allows you to compute a determinant by expansion along any column. Exercise 6 at the end of this section challenges you to try your hand at proving the following result.

Theorem 7.6

The determinant of a square matrix can be computed by expansion along any column. That is, if A is an $n \times n$ matrix and $c = 1, \ldots, n$, then

$$\det A = \sum_{i=1}^{n} (-1)^{i+c} a_{ic} \det A_{ic}.$$

The following theorem says that the determinant function is compatible with matrix multiplication. This will be an important result in the next chapter. Theorem 5.5 and Theorem 7.4 form the foundation of the proof. With these theorems in place, the rest of the proof is easy.

Theorem 7.7

Suppose A and B are $n \times n$ matrices. Then

$$\det(AB) = (\det A)(\det B).$$

Proof Consider a sequence of elementary row operations that change A into reduced row-echelon form A'. As we know from Theorem 7.4, the row operations used to reduce A to A' will determine a constant K with $\det A = K \det A'$. By Theorem 5.5, the result of applying this sequence of row operations to AB will be $A'B$. Thus, for this same constant K, we have $\det(AB) = K \det(A'B)$.

Since A is a square matrix, A' will either be the identity matrix or it will have at least one row of zeros along the bottom. In the first case we have $\det A = K \det A' = K \det I = K$. Hence,

$$\det(AB) = K \det(A'B) = K \det(IB) = K \det B = (\det A)(\det B).$$

In the second case we have $\det A = K \det A' = K \cdot 0 = 0$. Also, $A'B$ will have at least one row of zeros along the bottom. Hence,

$$\det(AB) = K \det(A'B) = K \cdot 0 = 0 = 0 \det B = (\det A)(\det B).$$

In either case we have the desired result. ∎

As an antidote to the intricate proofs of these basic properties of determinants, we conclude this section with a simple operation that interchanges the rows and columns of a matrix.

Definition 7.8

The **transpose** of an $m \times n$ matrix $A = [a_{ij}]$ is the $n \times m$ matrix, denoted A^t, whose ij-entry is a_{ji}. That is,

$$\begin{bmatrix} a_{11} & a_{12} & \cdots & a_{1n} \\ a_{21} & a_{22} & \cdots & a_{2n} \\ \vdots & \vdots & & \vdots \\ a_{m1} & a_{m2} & \cdots & a_{mn} \end{bmatrix}^t = \begin{bmatrix} a_{11} & a_{21} & \cdots & a_{m1} \\ a_{12} & a_{22} & \cdots & a_{m2} \\ \vdots & \vdots & & \vdots \\ a_{1n} & a_{2n} & \cdots & a_{mn} \end{bmatrix}.$$

When working with this concept, we will not be able to rely on our convention that the first subscript for the entry of a matrix is the row index and the second subscript is the column index. For example, if a_{12} is the 12-entry of a matrix A, it is the 21-entry of A^t. In such a situation, let us adopt the explicit notation that

$$A_{[ij]}$$

denotes the entry in row i and column j of the matrix A. Then $A^t_{[ij]} = A_{[ji]}$.

Here is a theorem that lists some of the basic facts about the transpose operation. The product rule for transposes is undoubtedly the trickiest of the five to prove. This proof is given below. You are asked to prove the other four properties in Exercise 7 at the end of this section.

Theorem 7.9

Suppose $A, A' \in \mathbb{M}(m, n)$, $B \in \mathbb{M}(n, p)$, and $r \in \mathbb{R}$. Then

  a. $(A^t)^t = A$
  b. $(rA)^t = r(A^t)$
  c. $(A + A')^t = A^t + (A')^t$
  d. $(AB)^t = B^t A^t$
  e. If A is a square invertible matrix, then $(A^{-1})^t = (A^t)^{-1}$.

Proof of d   Since AB is an $m \times p$ matrix, $(AB)^t$ is a $p \times m$ matrix. Since B^t is a $p \times n$ matrix and A^t is a $n \times m$ matrix, $B^t A^t$ is a $p \times m$ matrix. We need to prove that each entry of $(AB)^t$ matches up with the corresponding entry of $B^t A^t$. To do this, let $i = 1, \ldots, p$ be a row index and let $j = 1, \ldots, m$ be a column index. Then

$$(AB)^t_{[ij]} = (AB)_{[ji]}$$

$$= \sum_{k=1}^{n} A_{[jk]} B_{[ki]}$$

$$= \sum_{k=1}^{n} A^t_{[kj]} B^t_{[ik]}$$

$$= \sum_{k=1}^{n} B^t_{[ik]} A^t_{[kj]}$$

$$= \left(B^t A^t \right)_{[ij]}. \quad \blacksquare$$

The concluding theorem of this section states that interchanging the rows and columns of a matrix by taking its transpose yields a matrix with the same determinant. Exercise 8 at the end of this section asks you to prove this result.

Theorem 7.10

Suppose A is an $n \times n$ matrix. Then $\det A = \det A^t$.

Exercises 7.3

1. Prove Theorem 7.4, part a. (Suggestion: Expand $\det A$ along the row that contains only zeros.)

2. Prove Theorem 7.4, part b. (Suggestion: Use induction on n. Check the result for 2×2 matrices as the basis of the induction. For larger values of n, expand $\det A$ along one of the rows not involved in the interchange.)

3. Prove Theorem 7.4, part c. (Suggestion: Expand $\det A$ along the row that is being multiplied by the constant.)

4. Prove Theorem 7.4, part d. (Suggestion: Interchange the two rows that are identical and ask yourself what number is equal to its negative.)

5. Prove Theorem 7.4, part e. (Suggestion: Expand $\det A$ along the row that is being modified and use part d of the theorem.)

6. Prove Theorem 7.6. (You may want to get warmed up for this proof by first proving the result for the case of expansion along the first column. This will provide you with the essence of the proof of the general result.)

7. Prove statements a, b, c, and e of Theorem 7.9.

8. Prove Theorem 7.10.

9. Suppose A is an $n \times n$ matrix and r is a real number. Find a simple formula for $\det(rA)$ in terms of r and $\det A$. Prove your conjecture.

10. Suppose A is a square matrix. Use induction to prove for any integer $n \geq 0$ that

$$\det A^n = (\det A)^n.$$

11. Prove that if the square matrix A is nonsingular, then

$$\det A^{-1} = (\det A)^{-1}.$$

12. Prove that if A and P are $n \times n$ matrices and P is nonsingular, then

$$\det(P^{-1}AP) = \det A.$$

13. Suppose A, B, and C are $n \times n$ matrices. Suppose the rth row of A is the sum of the rth row of B and the rth row of C, and all other rows of A, B, and C are identical. Show that

$$\det A = \det B + \det C.$$

7.4 Cramer's Rule

In Chapter 5 we accumulated several conditions that were equivalent to a matrix having an inverse. We are now in a position to add to the list a condition involving the determinant of the matrix. The following theorem summarizes our earlier results and incorporates this new condition.

Theorem 7.11

For an $n \times n$ matrix A, the following conditions are equivalent:

a. A is nonsingular.

b. A has a right inverse.

c. A has a left inverse.

d. $\operatorname{rank} A = n$.

e. A can be row-reduced to I.

f. For any $\mathbf{b} \in \mathbb{R}^n$, the equation $A\mathbf{v} = \mathbf{b}$ has a unique solution $\mathbf{v} \in \mathbb{R}^n$.

g. $\det A \neq 0$.

Proof Our earlier work contains all the essential ingredients for showing that conditions a through f are equivalent. To see that conditions e and g are equivalent, notice that the reduced form of the square matrix A will either be I (with determinant 1) or have a row of zeros (with determinant 0). By Theorem 7.4, the matrices obtained in the reduction process are nonzero multiples of $\det A$. Hence, $\det A \neq 0$ if and only if A can be reduced to I. ∎

Now that we can tell whether or not a matrix is invertible by checking its determinant, we might harbor hope of using the determinant to obtain an explicit formula for the inverse of a square matrix. Such a formula does indeed exist. Furthermore, with the machinery we have developed in this chapter, it is quite easy to prove that the formula does the job. The following definition contains the two essential ingredients.

Definition 7.12

Suppose A is an $n \times n$ matrix. Then for integers i and j between 1 and n, the *ij*-**cofactor** of A is the real number

$$(-1)^{i+j} \det A_{ij}.$$

The **adjoint** of A is the $n \times n$ matrix, denoted $\operatorname{adj} A$, whose *ij*-entry is the *ji*-cofactor of A.

Theorem 7.13

Suppose the square matrix A is nonsingular. Then

$$A^{-1} = \frac{1}{\det A} \operatorname{adj} A.$$

Proof Let c_{ij} denote the *ij*-cofactor of A. Expansion of $\det A$ along row i gives

$$\det A = \sum_{k=1}^{n} (-1)^{i+k} a_{ik} \det A_{ik} = \sum_{k=1}^{n} a_{ik} c_{ik}.$$

For $i \neq j$, let B be the matrix obtained from A by replacing row j by row i. Since rows i and j of B are identical, we know by part d of Theorem 7.4 that $\det B = 0$. On the other hand, the expansion of $\det B$ along row j gives

$$0 = \det B = \sum_{k=1}^{n} (-1)^{j+k} a_{ik} \det A_{jk} = \sum_{k=1}^{n} a_{ik} c_{jk}.$$

As we consider the various possibilities for i and j, we find that

$$\begin{bmatrix} a_{11} & \cdots & a_{1n} \\ \vdots & & \vdots \\ a_{i1} & \cdots & a_{in} \\ \vdots & & \vdots \\ a_{n1} & \cdots & a_{nn} \end{bmatrix} \begin{bmatrix} c_{11} & \cdots & c_{j1} & \cdots & c_{n1} \\ \vdots & & \vdots & & \vdots \\ c_{1n} & \cdots & c_{jn} & \cdots & c_{nn} \end{bmatrix}$$

$$
= \begin{bmatrix} \det A & 0 & \cdots & 0 \\ 0 & \det A & \cdots & 0 \\ \vdots & \vdots & & \vdots \\ 0 & 0 & \cdots & \det A \end{bmatrix} = (\det A)I.
$$

Theorem 7.11 ensures that $\det A \neq 0$. Once we recognize that $c_{ji} = (\operatorname{adj} A)_{[ij]}$, we have

$$
A\left(\frac{1}{\det A}\operatorname{adj} A\right) = \frac{1}{\det A}A(\operatorname{adj} A)
$$

$$
= \frac{1}{\det A}(\det A)I
$$

$$
= I.
$$

The result follows by Theorem 5.13. ∎

Your admiration of this elegant formula for the inverse of a matrix should be tempered by the computational complexity of evaluating it. Even for a 3×3 matrix such as in the following example, this formula involves determinants of one 3×3 matrix and nine 2×2 matrices.

Quick Example   *Use Theorem 7.13 to compute the inverse of the 3×3 matrix*

$$
A = \begin{bmatrix} 1 & 0 & 2 \\ -1 & 3 & 0 \\ 0 & 3 & 1 \end{bmatrix}.
$$

We first compute the determinants of the nine submatrices:

$$
\begin{array}{lll}
\det A_{11} = 3 & \det A_{12} = -1 & \det A_{13} = -3 \\
\det A_{21} = -6 & \det A_{22} = 1 & \det A_{23} = 3 \\
\det A_{31} = -6 & \det A_{32} = 2 & \det A_{33} = 3
\end{array}
$$

Then $\det A = 1\det A_{11} - 0\det A_{12} + 2\det A_{13} = 1 \cdot 3 - 0 \cdot (-1) + 2 \cdot (-3) = -3$. The ij-entry of $\operatorname{adj} A$ is the ji-cofactor $(-1)^{j+i}\det A_{ji}$. So

$$
A^{-1} = \frac{1}{\det A}\operatorname{adj} A = \frac{1}{-3}\begin{bmatrix} 3 & 6 & -6 \\ 1 & 1 & -2 \\ -3 & -3 & 3 \end{bmatrix} = \begin{bmatrix} -1 & -2 & 2 \\ -\frac{1}{3} & -\frac{1}{3} & \frac{2}{3} \\ 1 & 1 & -1 \end{bmatrix}. \quad ∎
$$

We can use Theorem 7.13 to derive an explicit formula for solving a system of n equations in n unknowns when there is a unique solution.

Theorem 7.14   Cramer's Rule

Suppose A is a nonsingular $n \times n$ matrix and $\mathbf{b} = \begin{bmatrix} b_1 \\ \vdots \\ b_n \end{bmatrix}$ is a vector in $\mathbb{R}^n$. Then the solution $\mathbf{x} = \begin{bmatrix} x_1 \\ \vdots \\ x_n \end{bmatrix} \in \mathbb{R}^n$ of the system $A\mathbf{x} = \mathbf{b}$ is given by

$$x_j = \frac{1}{\det A} \det A_j$$

where A_j is the matrix obtained by replacing the jth column of A with $\mathbf{b}$.

Proof Let c_{ij} be the ij-cofactor of A. We know that

$$\mathbf{x} = A^{-1}\mathbf{b} = \frac{1}{\det A}(\mathrm{adj}\,A)\mathbf{b} = \frac{1}{\det A} \begin{bmatrix} c_{11} & \cdots & c_{n1} \\ \vdots & & \vdots \\ c_{1j} & \cdots & c_{nj} \\ \vdots & & \vdots \\ c_{1n} & \cdots & c_{nn} \end{bmatrix} \begin{bmatrix} b_1 \\ \vdots \\ b_n \end{bmatrix}.$$

Thus,

$$x_j = \frac{1}{\det A} \sum_{i=1}^{n} c_{ij}b_i = \frac{1}{\det A} \sum_{i=1}^{n} (-1)^{i+j} b_i \det A_{ij} = \frac{1}{\det A} \det A_j$$

since by Theorem 7.6 the sum is the expansion of $\det A_j$ along its jth column. ∎

Quick Example *Use Cramer's rule to find the solution of the system*

$$3x + y = 4$$
$$2x - 2y = 1$$

The coefficient matrix corresponding to the system is $A = \begin{bmatrix} 3 & 1 \\ 2 & -2 \end{bmatrix}$, for which $\det A = -8$. Thus,

$$x = \tfrac{1}{-8} \det \begin{bmatrix} 4 & 1 \\ 1 & -2 \end{bmatrix} = \tfrac{1}{-8}(-9) = \tfrac{9}{8},$$

$$y = \tfrac{1}{-8} \det \begin{bmatrix} 3 & 4 \\ 2 & 1 \end{bmatrix} = \tfrac{1}{-8}(-5) = \tfrac{5}{8}. \quad ∎$$

Again you can see that the computational complexity limits the practical usefulness of Cramer's rule even under the restricted conditions when it applies. Nevertheless, the need may arise for an explicit formula for the value of one of the unknowns of a linear system in terms of a parameter in the coefficients. The Gauss-Jordan reduction algorithm would probably lead to several cases depending on the value of the parameter and whether row interchanges are necessary. Cramer's rule may then be a reasonable alternative.

Quick Example *Estimate the error in the solution x of the following system if the coefficient a is known to be within 0.01 of 6.8. What if a is known to be within 0.01 of −6.8?*

$$4x + y - 2z = 3$$
$$-x \quad\quad + z = 2$$
$$ax - 3y + z = 0$$

Cramer's rule gives an explicit formula for x in terms of a:

$$x = \frac{\det \begin{bmatrix} 3 & 1 & -2 \\ 2 & 0 & 1 \\ 0 & -3 & 1 \end{bmatrix}}{\det \begin{bmatrix} 4 & 1 & -2 \\ -1 & 0 & 1 \\ a & -3 & 1 \end{bmatrix}} = \frac{3 \cdot 3 - 1 \cdot 2 - 2 \cdot (-6)}{4 \cdot 3 - 1(-1 - a) - 2 \cdot 3} = \frac{19}{a + 7}.$$

The derivative $\dfrac{dx}{da} = \dfrac{-19}{(a + 7)^2}$ gives the rate of change in x relative to a change in a. That is,

$$\Delta x \approx \frac{dx}{da} \Delta a = \frac{-19}{(a + 7)^2} \Delta a.$$

If $a \approx 6.8$ with a variation of 0.01, then

$$\Delta x \approx \frac{-19}{(6.8 + 7)^2}(0.01) \approx -0.000998.$$

This is a relatively modest error in the quantity $x \approx \dfrac{19}{6.8 + 7} \approx 1.38$. The magnitude of the error is approximately 0.07% of the magnitude of x.

If a is within 0.01 of -6.8, then

$$\Delta x \approx \frac{-19}{(-6.8 + 7)^2}(0.01) \approx -4.75.$$

This is a deviation 475 times as large as the deviation in a. Although $x \approx \dfrac{19}{-6.8 + 7} = 95.0$ is also larger than in the previous case, the magnitude of the error here is 5% of the magnitude of x.   ∎

Exercises 7.4

1.   Given a square matrix A with randomly determined entries, what is the probability that A is invertible?

2.   Use Theorem 7.13 to find inverses of

   a. $\begin{bmatrix} 5 & 2 \\ -1 & 3 \end{bmatrix}$

   b. $\begin{bmatrix} 2 & -5 & 0 \\ 3 & 1 & 2 \\ 2 & 2 & -3 \end{bmatrix}$

3.   Let f be the function that assigns to any real number $x \neq 6$ the 11-entry of the inverse of the matrix $\begin{bmatrix} x & 3 \\ 4 & 2 \end{bmatrix}$. Show that f is continuous. (Suggestion: Find an explicit formula for $f(x)$.)

4.   Use Cramer's rule to solve the system
$$7x + 3y = 2$$
$$2x - 4y = -3$$

5.   Use Cramer's rule to solve the system
$$3x - 2y + 2z = 1$$
$$x + 4z = 2$$
$$2x + y - z = 0$$

6.   Let f be the function that assigns to any real number $c \neq -4$ the value of y that satisfies
$$cx - 2y + 2z = 1$$
$$x + 4z = 2$$
$$2x + y - z = 0$$

a. Show that f is differentiable.

b. Use the derivative $\dfrac{dy}{dc}$ to estimate the error Δy in y if c is known to be within 0.002 of 4.1. Estimate the relative error $\dfrac{\Delta y}{y}$.

c. Estimate the error Δy in y if c is known to be within 0.002 of -4.1. Estimate the relative error $\dfrac{\Delta y}{y}$.

7.   **a.** Show that the determinant of the 3×3 Vandermonde matrix $\begin{bmatrix} 1 & a & a^2 \\ 1 & b & b^2 \\ 1 & c & c^2 \end{bmatrix}$

is $(c - a)(c - b)(b - a)$.

b. Conclude that the 3×3 Vandermonde matrix is nonsingular if and only if a, b, and c are distinct real numbers.

8.   Fill in the details of this alternative derivation of Cramer's rule for solving $Ax = \mathbf{b}$. It is based on Stephen Robinson's article "A Short Proof of Cramer's Rule" in *Mathematics Magazine*, March–April 1979, pages 94–95.

a. Let X_j denote the matrix obtained by replacing the jth column of I with $\mathbf{x}$. Let A_j be the matrix obtained by replacing the jth column of A with $\mathbf{b}$. Verify that $AX_j = A_j$.

b. Show that $\det X_j = x_j$, the jth component of $\mathbf{x}$.

c. Apply Theorem 7.7 to the matrix equation in part a and solve for x_j.

9.   Fill in the details of yet another proof of Cramer's rule for solving $Ax = \mathbf{b}$.

a. Use the equation $Ax = \mathbf{b}$ to write $\mathbf{b}$ as a linear combination of the columns of A.

b. Multiply column j of A by x_j, the jth component of $\mathbf{x}$. Prove that the determinant of the resulting matrix is $x_j \det A$.

c. For each $k = 1, \ldots, j - 1, j + 1, \ldots, n$, add x_k times the kth column of A to the matrix obtained in part a. Prove that these operations do not change the determinant of the matrix.

d. Notice that the resulting matrix is A_j, the matrix obtained by replacing the jth column of A with $\mathbf{b}$. Conclude that $\det A_j = x_j \det A$.

7.5 Cross Product

In the previous section we used determinants in an explicit formula for the inverse of a matrix and in Cramer's rule for solving linear systems. The next chapter demonstrates further applications of determinants in finding numerical quantities that characterize certain essential properties for square matrices. In this section we will exploit the formal properties of determinants to develop a new kind of multiplication of vectors in three-dimensional Euclidean space. Although specialized exclusively to $\mathbb{R}^3$, this operation is very useful for geometry and mechanics in the space we are most familiar with from everyday experience.

When working with vectors in $\mathbb{R}^3$, the traditional notation for the standard basis vectors is $\mathbf{i} = (1, 0, 0)$, $\mathbf{j} = (0, 1, 0)$, and $\mathbf{k} = (0, 0, 1)$. We will adopt this notation for our discussion of the cross product. This not only minimizes the appearance of subscripts, but it also reminds us that the cross product is restricted to the vector space $\mathbb{R}^3$.

Definition 7.15

Suppose $\mathbf{v} = (v_1, v_2, v_3)$ and $\mathbf{w} = (w_1, w_2, w_3)$ are vectors in $\mathbb{R}^3$. The **cross product** of $\mathbf{v}$ with $\mathbf{w}$ is the vector in $\mathbb{R}^3$ denoted $\mathbf{v} \times \mathbf{w}$ and defined by

$$\mathbf{v} \times \mathbf{w} = (v_2 w_3 - v_3 w_2)\mathbf{i} - (v_1 w_3 - v_3 w_1)\mathbf{j} + (v_1 w_2 - v_2 w_1)\mathbf{k}.$$

The formula for the cross product $\mathbf{v} \times \mathbf{w}$ is easy to remember if you think of it as the formal expansion of the determinant

$$\det \begin{bmatrix} \mathbf{i} & \mathbf{j} & \mathbf{k} \\ v_1 & v_2 & v_3 \\ w_1 & w_2 & w_3 \end{bmatrix}.$$

Of course, the entries in the first row are vectors rather than real numbers, so scalar multiplication must be judiciously used to multiply real numbers by vectors.

The cross product exhibits some strange behavior. For example, compute

$$(1, -1, 2) \times (-3, 1, 1) = \det \begin{bmatrix} \mathbf{i} & \mathbf{j} & \mathbf{k} \\ 1 & -1 & 2 \\ -3 & 1 & 1 \end{bmatrix}$$
$$= (-1 - 2)\mathbf{i} - (1 + 6)\mathbf{j} + (1 - 3)\mathbf{k}$$
$$= (-3, -7, -2),$$

and then take the cross product of this result with some other vector, say $(0, 0, 1)$:

$$(-3, -7, -2) \times (0, 0, 1) = \det \begin{bmatrix} \mathbf{i} & \mathbf{j} & \mathbf{k} \\ -3 & -7 & -2 \\ 0 & 0 & 1 \end{bmatrix}$$

$$= -7\mathbf{i} - (-3)\mathbf{j} + 0\mathbf{k}$$
$$= (-7, 3, 0).$$

This is not the same as

$$(1, -1, 2) \times \left((-3, 1, 1) \times (0, 0, 1)\right) = (1, -1, 2) \times \det \begin{bmatrix} \mathbf{i} & \mathbf{j} & \mathbf{k} \\ -3 & 1 & 1 \\ 0 & 0 & 1 \end{bmatrix}$$

$$= (1, -1, 2) \times (1, 3, 0)$$

$$= \det \begin{bmatrix} \mathbf{i} & \mathbf{j} & \mathbf{k} \\ 1 & -1 & 2 \\ 1 & 3 & 0 \end{bmatrix}$$

$$= (-6, 2, 4).$$

Thus, we see that the associative law fails for the cross product.

Rather than dwelling on the failures of the cross product, let us consider a list of algebraic properties that are true of the cross product.

Theorem 7.16 Algebraic Properties of the Cross Product

Suppose $\mathbf{v} = (v_1, v_2, v_3)$, $\mathbf{w} = (w_1, w_2, w_3)$, and $\mathbf{x} = (x_1, x_2, x_3)$ are elements of $\mathbb{R}^3$, and r is a real number. Then

    a. $\mathbf{v} \times \mathbf{w} = -(\mathbf{w} \times \mathbf{v})$

    b. $\mathbf{v} \times (\mathbf{w} + \mathbf{x}) = (\mathbf{v} \times \mathbf{w}) + (\mathbf{v} \times \mathbf{x})$

    c. $(\mathbf{v} + \mathbf{w}) \times \mathbf{x} = (\mathbf{v} \times \mathbf{x}) + (\mathbf{w} \times \mathbf{x})$

    d. $r(\mathbf{v} \times \mathbf{w}) = (r\mathbf{v}) \times \mathbf{w} = \mathbf{v} \times (r\mathbf{w})$

    e. $\mathbf{v} \times \mathbf{0} = \mathbf{0} \times \mathbf{v} = \mathbf{0}$

    f. $\mathbf{v} \times \mathbf{v} = \mathbf{0}$

    g. $\mathbf{v} \cdot (\mathbf{w} \times \mathbf{x}) = (\mathbf{v} \times \mathbf{w}) \cdot \mathbf{x} = \det \begin{bmatrix} v_1 & v_2 & v_3 \\ w_1 & w_2 & w_3 \\ x_1 & x_2 & x_3 \end{bmatrix}$

Proof   Instead of spending a lot of time verifying the formal properties of the determinant when vectors appear in the first row, we can use the properties of determinants of the 2×2 matrices that arise as the coordinates of a cross product. For example, to verify the **anticommutative law,** listed as formula a, we write

$$\mathbf{v} \times \mathbf{w} = \det \begin{bmatrix} v_2 & v_3 \\ w_2 & w_3 \end{bmatrix} \mathbf{i} - \det \begin{bmatrix} v_1 & v_3 \\ w_1 & w_3 \end{bmatrix} \mathbf{j} + \det \begin{bmatrix} v_1 & v_2 \\ w_1 & w_2 \end{bmatrix} \mathbf{k}$$

$$= -\det \begin{bmatrix} w_2 & w_3 \\ v_2 & v_3 \end{bmatrix} \mathbf{i} + \det \begin{bmatrix} w_1 & w_3 \\ v_1 & v_3 \end{bmatrix} \mathbf{j} - \det \begin{bmatrix} w_1 & w_2 \\ v_1 & v_2 \end{bmatrix} \mathbf{k}$$

$$= -(\mathbf{w} \times \mathbf{v}).$$

You are asked to verify the other properties as exercises at the end of this section. ∎

Although the algebraic properties of the cross product leave something to be desired, the geometric properties listed below are quite useful in working with geometric objects in $\mathbb{R}^3$.

Theorem 7.17 Geometric Properties of the Cross Product

Suppose $\mathbf{v}$ and $\mathbf{w}$ are vectors in $\mathbb{R}^3$. Let θ be the angle between $\mathbf{v}$ and $\mathbf{w}$. Then

  a. $\mathbf{v} \cdot (\mathbf{v} \times \mathbf{w}) = 0$

  b. $\mathbf{w} \cdot (\mathbf{v} \times \mathbf{w}) = 0$

  c. $\|\mathbf{v} \times \mathbf{w}\|^2 = \|\mathbf{v}\|^2\|\mathbf{w}\|^2 - (\mathbf{v} \cdot \mathbf{w})^2$

  d. $\|\mathbf{v} \times \mathbf{w}\| = \|\mathbf{v}\|\,\|\mathbf{w}\| \sin \theta$

You are asked to carry out the straightforward computations to prove these four results as exercises at the end of this section.

You should recognize the first two properties as saying that $\mathbf{v} \times \mathbf{w}$ is orthogonal to $\mathbf{v}$ and to $\mathbf{w}$. The last property says that the length of $\mathbf{v} \times \mathbf{w}$ is equal to the area of the parallelogram determined by the arrows that represent $\mathbf{v}$ and $\mathbf{w}$. (See Figure 7.1.) To see this, simply notice that $\|\mathbf{v}\|$ is the length of the base and $\|\mathbf{w}\| \sin \theta$ is the altitude of the parallelogram.

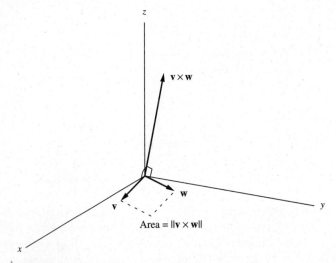

Figure 7.1 Geometric properties of the cross product.

Given two vectors $\mathbf{v}$ and $\mathbf{w}$ in $\mathbb{R}^3$, the geometric properties of the cross product prescribe the direction of $\mathbf{v} \times \mathbf{w}$ (along the line orthogonal to $\mathbf{v}$ and $\mathbf{w}$) and the length of $\mathbf{v} \times \mathbf{w}$ (equal in magnitude to the area of a certain parallelogram). This leaves at most two possibilities for $\mathbf{v} \times \mathbf{w}$. To distinguish geometrically between $\mathbf{v} \times \mathbf{w}$ and $-(\mathbf{v} \times \mathbf{w})$ requires a discussion of orientation. Whereas most people can distinguish their right hand from their left, mathematicians find this concept somewhat elusive. In the next section we will resolve this problem to a certain extent. In particular, we will show that if $\mathbf{v} \times \mathbf{w} \neq \mathbf{0}$, then when you hold your right hand so that your fingers curve from **v** to **w**, your thumb will point in the general direction of **v** × **w**. With this final bit of information, we have a complete geometric characterization of the cross product.

In Section 1.9 we defined a plane in terms of two direction vectors. In Section 4.2 the dot product enabled us to describe a plane in $\mathbb{R}^3$ in terms of a vector orthogonal to the plane. Recall the example of the plane defined by $3x + y - 2z = 4$. We read off a normal vector $(3, 1, -2)$ and then chose two vectors $(1, -3, 0)$ and $(0, 2, 1)$ orthogonal to $(3, 1, -2)$ as direction vectors for this plane.

The cross product provides a direct way to reverse this process. Given the two direction vectors in this example, their cross product

$$(1, -3, 0) \times (0, 2, 1) = \det \begin{bmatrix} \mathbf{i} & \mathbf{j} & \mathbf{k} \\ 1 & -3 & 0 \\ 0 & 2 & 1 \end{bmatrix} = (-3, -1, 2)$$

is a normal vector to the plane. This is, of course, a nonzero multiple of the normal vector we began with.

Exercises 7.5

1. Use the cross product to find a vector in $\mathbb{R}^3$ that is orthogonal to $(2, 5, 1)$ and $(-1, 2, -1)$.

2. Find a normal vector for the plane

$$\{r(5, 1, 7) + s(-3, 2, 2) + (1, 0, -4) \mid r, s \in \mathbb{R}\}.$$

3. Find a normal vector for the plane passing through the three points $(1, 2, 1)$, $(2, -1, 3)$, and $(0, 1, 5)$. Write an equation that defines this plane.

4. When the normal vector for a plane is computed by taking the cross product of a pair of direction vectors for the plane, what assures us that the normal vector will be nonzero?

5. Prove Theorem 7.16, part b: $\mathbf{v} \times (\mathbf{w} + \mathbf{x}) = (\mathbf{v} \times \mathbf{w}) + (\mathbf{v} \times \mathbf{x})$.

6. Prove Theorem 7.16, part c: $(\mathbf{v} + \mathbf{w}) \times \mathbf{x} = (\mathbf{v} \times \mathbf{x}) + (\mathbf{w} \times \mathbf{x})$.

7. Prove Theorem 7.16, part d: $r(\mathbf{v} \times \mathbf{w}) = (r\mathbf{v}) \times \mathbf{w} = \mathbf{v} \times (r\mathbf{w})$.

8. Prove Theorem 7.16, part e: $\mathbf{v} \times \mathbf{0} = \mathbf{0} \times \mathbf{v} = \mathbf{0}$.

9. Prove Theorem 7.16, part f: $\mathbf{v} \times \mathbf{v} = \mathbf{0}$.

10. Prove Theorem 7.16, part g: $\mathbf{v} \cdot (\mathbf{w} \times \mathbf{x}) = (\mathbf{v} \times \mathbf{w}) \cdot \mathbf{x} = \det \begin{bmatrix} v_1 & v_2 & v_3 \\ w_1 & w_2 & w_3 \\ x_1 & x_2 & x_3 \end{bmatrix}$.

11. For three vectors $\mathbf{v} = \begin{bmatrix} v_1 \\ v_2 \\ v_3 \end{bmatrix}$, $\mathbf{w} = \begin{bmatrix} w_1 \\ w_2 \\ w_3 \end{bmatrix}$, and $\mathbf{x} = \begin{bmatrix} x_1 \\ x_2 \\ x_3 \end{bmatrix}$ in $\mathbb{R}^3$, show by direct computation that

$$\det \begin{bmatrix} v_1 & w_1 & x_1 \\ v_2 & w_2 & x_2 \\ v_3 & w_3 & x_3 \end{bmatrix} = (\mathbf{v} \times \mathbf{w}) \cdot \mathbf{x}.$$

12. Under what conditions on the vectors $\mathbf{v}$ and $\mathbf{w}$ in $\mathbb{R}^3$ will $\mathbf{v} \times \mathbf{w} = \mathbf{0}$?

13. Prove Theorem 7.17, part a: $\mathbf{v} \cdot (\mathbf{v} \times \mathbf{w}) = 0$.

14. Prove Theorem 7.17, part b: $\mathbf{w} \cdot (\mathbf{v} \times \mathbf{w}) = 0$.

15. Prove Theorem 7.17, part c: $\|\mathbf{v} \times \mathbf{w}\|^2 = \|\mathbf{v}\|^2\|\mathbf{w}\|^2 - (\mathbf{v} \cdot \mathbf{w})^2$.

16. Prove Theorem 7.17, part d: $\|\mathbf{v} \times \mathbf{w}\| = \|\mathbf{v}\|\|\mathbf{w}\|\sin\theta$.

17. Explain why no ambiguity would arise if someone were so sloppy as to omit the parentheses in the expression $(\mathbf{v} \times \mathbf{w}) \cdot \mathbf{x}$.

7.6 Orientation

This section covers two related topics. First, we want to use the determinant to distinguish between the two possible orientations for an ordered basis for $\mathbb{R}^3$. Second, we want to establish the right-hand rule for the cross product.

For vectors $\mathbf{v}$, $\mathbf{w}$, and $\mathbf{x}$ in $\mathbb{R}^3$, let $[\,\mathbf{v} \quad \mathbf{w} \quad \mathbf{x}\,]$ denote the 3×3 matrix whose columns consist of the coordinates of the three vectors in the indicated order. Here is the key definition.

Definition 7.18

An ordered basis $\{\mathbf{u}_1, \mathbf{u}_2, \mathbf{u}_3\}$ for $\mathbb{R}^3$ is **right-handed** if and only if $\det[\,\mathbf{u}_1 \quad \mathbf{u}_2 \quad \mathbf{u}_3\,] > 0$. It is **left-handed** if and only if $\det[\,\mathbf{u}_1 \quad \mathbf{u}_2 \quad \mathbf{u}_3\,] < 0$. An **orientation** of $\mathbb{R}^3$ is the selection of a right-handed or a left-handed ordered basis.

Notice that any ordered basis $\{\mathbf{u}_1 \;\; \mathbf{u}_2 \;\; \mathbf{u}_3\}$ for $\mathbb{R}^3$ is either right-handed or left-handed. (See Figure 7.2.) Indeed, the matrix $[\,\mathbf{u}_1, \quad \mathbf{u}_2, \quad \mathbf{u}_3\,]$ has rank 3, so by Theorem 7.11, the determinant of this matrix is nonzero.

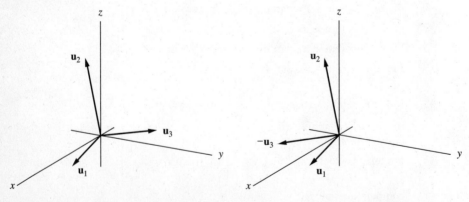

Figure 7.2 $\{\mathbf{u}_1, \mathbf{u}_2, \mathbf{u}_3\}$ is a left-handed basis. $\{\mathbf{u}_1, \mathbf{u}_2, -\mathbf{u}_3\}$ is a right-handed basis.

Quick Example *Show that the standard basis $\{\mathbf{i}, \mathbf{j}, \mathbf{k}\}$ for $\mathbb{R}^3$ is right-handed. Let $\mathbf{v}$ and $\mathbf{w}$ be any vectors in $\mathbb{R}^3$ with $\mathbf{v} \times \mathbf{w} \neq \mathbf{0}$. Show that $\{\mathbf{v}, \mathbf{w}, \mathbf{v} \times \mathbf{w}\}$ is right-handed.*

Since $\det[\,\mathbf{i} \quad \mathbf{j} \quad \mathbf{k}\,] = \det I_3 = 1 > 0$, the standard basis $\{\mathbf{i}, \mathbf{j}, \mathbf{k}\}$ for $\mathbb{R}^3$ is right-handed. If $\mathbf{v}$ and $\mathbf{w}$ are any vectors in $\mathbb{R}^3$ with $\mathbf{v} \times \mathbf{w} \neq \mathbf{0}$, then by Exercise 11 of Section 7.5 we have $\det[\,\mathbf{v} \quad \mathbf{w} \quad \mathbf{v} \times \mathbf{w}\,] = (\mathbf{v} \times \mathbf{w}) \cdot (\mathbf{v} \times \mathbf{w}) > 0$. Hence, the ordered basis $\{\mathbf{v}, \mathbf{w}, \mathbf{v} \times \mathbf{w}\}$ is right-handed. ∎

Our next goal is to establish the geometric significance of the orientation of an ordered basis.

First we want to consider the effect of an invertible linear map $T : \mathbb{R}^3 \rightarrow \mathbb{R}^3$ on an orientation $\{\mathbf{u}_1, \mathbf{u}_2, \mathbf{u}_3\}$. Let A be the matrix of T relative to the standard basis for $\mathbb{R}^3$. Since T is invertible, $\{T(\mathbf{u}_1), T(\mathbf{u}_2), T(\mathbf{u}_3)\}$ is also a basis for $\mathbb{R}^3$ (see Exercises 4 and 5 of Section 6.7). Now

$$
\begin{aligned}
\det\left[\, T(\mathbf{u}_1) \quad T(\mathbf{u}_2) \quad T(\mathbf{u}_3)\,\right] &= \det[\, A\mathbf{u}_1 \quad A\mathbf{u}_2 \quad A\mathbf{u}_3\,] \\
&= \det\!\left(A[\,\mathbf{u}_1 \;\mathbf{u}_2 \;\mathbf{u}_3\,]\right) \\
&= (\det A)\!\left(\det[\,\mathbf{u}_1 \;\mathbf{u}_2 \;\mathbf{u}_3\,]\right).
\end{aligned}
$$

Thus, we see that T preserves the orientation if $\det A > 0$ and T reverses the orientation if $\det A < 0$. Of course, since T is invertible, the possibility that $\det A = 0$ is excluded.

By Exercise 7 of Section 6.3, the matrices that represent rotations about the x-axis, the y-axis, and the z-axis are

$$
\begin{bmatrix} 1 & 0 & 0 \\ 0 & \cos\alpha & -\sin\alpha \\ 0 & \sin\alpha & \cos\alpha \end{bmatrix}, \quad
\begin{bmatrix} \cos\beta & 0 & -\sin\beta \\ 0 & 1 & 0 \\ \sin\beta & 0 & \cos\beta \end{bmatrix}, \quad
\begin{bmatrix} \cos\gamma & -\sin\gamma & 0 \\ \sin\gamma & \cos\gamma & 0 \\ 0 & 0 & 1 \end{bmatrix},
$$

respectively. Notice that the determinants of these three matrices are positive. This reinforces our suspicion that rotations are orientation-preserving.

We now have done all the preliminary work to make sense of the right-hand rule for distinguishing between $\mathbf{v} \times \mathbf{w}$ and $-(\mathbf{v} \times \mathbf{w})$. If either of the vectors $\mathbf{v}$ and $\mathbf{w}$ is a multiple of the other, then by Theorem 7.16, parts d and f, we have $\mathbf{v} \times \mathbf{w} = \mathbf{0}$. Otherwise, neither $\mathbf{v}$ nor $\mathbf{w}$ is equal to $\mathbf{0}$, and the angle θ between $\mathbf{v}$ and $\mathbf{w}$ satisfies $0 < \theta < \pi$. Hence, Theorem 7.17, part d, yields $\mathbf{v} \times \mathbf{w} \neq \mathbf{0}$. For the remainder of this discussion, let us restrict our attention to the case in which neither $\mathbf{v}$ nor $\mathbf{w}$ is a multiple of the other.

Apply a rotation R_z about the z-axis so that the image of $\mathbf{v}$ lies in the xz-plane. Apply a rotation R_y about the y-axis to map $R_z(\mathbf{v})$ into the positive x-axis. We have $R_y \circ R_z(\mathbf{v}) = (a, 0, 0)$ for some real number a with $a > 0$. Apply a third rotation R_x, this time about the x-axis, to map $R_y \circ R_z(\mathbf{w})$ into the xy-plane with a positive y-coordinate. We have $R_x \circ R_y \circ R_z(\mathbf{w}) = (b, c, 0)$ for some real numbers b and c with $c > 0$. Since R_x does not move points on the x-axis, we also have $R_x \circ R_y \circ R_z(\mathbf{v}) = R_y \circ R_z(\mathbf{v}) = (a, 0, 0)$. See Figure 7.3. By Exercise 7 at the end of this section, the cross product is compatible with rotations about the axes. Thus,

$$
\begin{aligned}
R_x \circ R_y \circ R_z\!\left(\mathbf{v} \times \mathbf{w}\right) &= R_x \circ R_y\!\left(R_z(\mathbf{v}) \times R_z(\mathbf{w})\right) \\
&= R_x\!\left(R_y \circ R_z(\mathbf{v}) \times R_y \circ R_z(\mathbf{w})\right) \\
&= R_x \circ R_y \circ R_z(\mathbf{v}) \times R_x \circ R_y \circ R_z(\mathbf{w}) \\
&= (a, 0, 0) \times (b, c, 0) \\
&= (0, 0, ac).
\end{aligned}
$$

Since $ac > 0$, this vector lies on the positive z-axis.

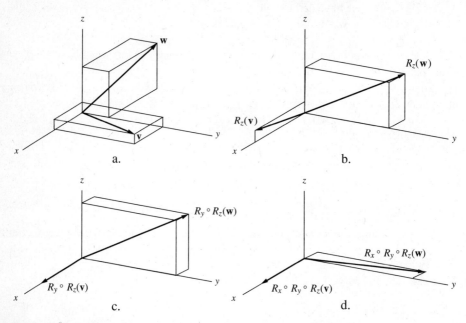

Figure 7.3 (a) The vectors **v** and **w**. (b) Rotate about the z-axis. (c) Rotate about the y-axis. (d) Rotate about the x-axis.

The coordinate system in $\mathbb{R}^3$ is traditionally set up so that when you orient your right hand with your fingers curving from the positive x-axis to the positive y-axis, then your thumb points in the general direction of the positive z-axis. With this same orientation of your right hand, your fingers curve from $R_x \circ R_y \circ R_z(\mathbf{v})$ (on the x-axis) through the smallest angle (in the xy-plane) to $R_x \circ R_y \circ R_z(\mathbf{w})$, and your thumb points in the general direction of $R_x \circ R_y \circ R_z(\mathbf{v} \times \mathbf{w})$. Apply inverses of the rotations R_x, R_y, and R_z to your hand. You will find your fingers curving from **v** through the smallest possible angle to **w** and your thumb pointing in the general direction of $\mathbf{v} \times \mathbf{w}$. This, then, is the **right-hand rule**. It is not a law of nature, but rather a consequence of our conventional choice of orientation of the axis system in $\mathbb{R}^3$.

Finally, consider any ordered basis $\{\mathbf{u}_1, \mathbf{u}_2, \mathbf{u}_3\}$ of $\mathbb{R}^3$. By Exercise 11 of Section 7.5, $\det[\,\mathbf{u}_1 \quad \mathbf{u}_2 \quad \mathbf{u}_3\,] = (\mathbf{u}_1 \times \mathbf{u}_2) \cdot \mathbf{u}_3$. Therefore, if $\{\mathbf{u}_1, \mathbf{u}_2, \mathbf{u}_3\}$ is right-handed, the cosine of the angle between $\mathbf{u}_1 \times \mathbf{u}_2$ and $\mathbf{u}_3$ is positive. Hence, the angle is less than $\pi/2$. Since $\mathbf{u}_1 \times \mathbf{u}_2$ is orthogonal to $\mathbf{u}_1$ and $\mathbf{u}_2$, both $\mathbf{u}_1 \times \mathbf{u}_2$ and $\mathbf{u}_3$ are on the same side of the plane spanned by $\mathbf{u}_1$ and $\mathbf{u}_2$. Even though $\mathbf{u}_3$ is not necessarily orthogonal to $\mathbf{u}_1$ and $\mathbf{u}_2$, when the fingers of your right hand curve from $\mathbf{u}_1$ to $\mathbf{u}_2$, your thumb will point in the general direction of $\mathbf{u}_3$. On the other hand, if $\{\mathbf{u}_1, \mathbf{u}_2, \mathbf{u}_3\}$ is left-handed, then the cosine of the angle between $\mathbf{u}_1 \times \mathbf{u}_2$ and $\mathbf{u}_3$ is negative. Hence, the angle is greater than $\pi/2$, so $\mathbf{u}_1 \times \mathbf{u}_2$ and $\mathbf{u}_3$ are on opposite sides of the plane spanned by $\mathbf{u}_1$ and $\mathbf{u}_2$. In this case, when the fingers of your left hand curve from $\mathbf{u}_1$ to $\mathbf{u}_2$, your thumb will point in the general direction of $\mathbf{u}_3$.

Exercises 7.6

1.   For each of the following ordered sets, determine whether it is a right-handed basis, a left-handed basis, or not a basis. Compare the results obtained from the definition and from the right-hand rule.

   a. $\{(0, 0, 1), (0, 1, 1), (1, 1, 1)\}$

   b. $\{(1, 0, 1), (1, 1, 0), (0, 1, 1)\}$

   c. $\{(1, 1, 1), (1, 0, 1), (0, 1, 0)\}$

2.   **a.** Show that if any two elements of an ordered basis for $\mathbb{R}^3$ are interchanged, the resulting basis has the orientation opposite that of the original.

   b. What happens if the three elements are shifted one place to the right, with the last one taking the place of the first?

3.   Show that the composition of orientation-preserving linear maps is orientation-preserving. What about the composition of orientation-reversing linear maps? What if an orientation-preserving linear map is composed with an orientation-reversing linear map? Be sure to consider the two possible orders of composing the two maps.

4.   Show that the inverse of an orientation-preserving linear map is orientation-preserving. What about the inverse of an orientation-reversing linear map?

5.   Define a concept of orientation for $\mathbb{R}^2$. Investigate a criterion that determines whether a linear operator on $\mathbb{R}^2$ preserves or reverses the orientation.

6.   Generalize the concept of orientation to Euclidean spaces of any dimension.

7.   Suppose $R : \mathbb{R}^3 \to \mathbb{R}^3$ is a rotation about one of the three axes. Use the definition of the cross product to show that for $\mathbf{v}, \mathbf{w} \in \mathbb{R}^3$, we have

$$R(\mathbf{v}) \times R(\mathbf{w}) = R(\mathbf{v} \times \mathbf{w}).$$

   That is, the cross product is compatible with rotations about the coordinate axes.

8.   Check to make sure that when the fingers of your two hands curve in the same direction, your thumbs point in opposite directions.

Project: Alternative Definitions

There are several ways to introduce the concept of the determinant of a square matrix. The inductive definition in Section 7.2 in terms of expansion by minors leads fairly directly to the results we needed. However, the computational manipulations in many of the proofs tend to obscure the underlying concepts. In this project you are invited to consider alternative definitions of the determinant of a square matrix.

   One popular definition of determinants follows from some preliminary work on permutation functions. You will discover that the main results about determinants are closely related to some simple properties of permutations. Since permutations are useful in other contexts, you will be doubly rewarded when you run across them as old friends in subsequent work in mathematics.

1. A **permutation** of a set S is a function $\sigma : S \to S$ that is one-to-one and onto. In this project we need only consider permutations defined on finite sets of the form $\{1, 2, \ldots, n\}$ where n is a positive integer. The ordered set $\{\sigma(1), \sigma(2), \ldots, \sigma(n)\}$ is a convenient way to specify a permutation σ defined on $\{1, 2, \ldots, n\}$. Use this notation to write down all permutations of $\{1, 2, 3\}$. How many permutations are there on a set of n elements?

Given a permutation σ defined on a set $\{1, 2, \ldots, n\}$, define an **inversion** to be a pair of integers i and j in $\{1, 2, \ldots, n\}$ with $i < j$ and $\sigma(i) > \sigma(j)$. Find a systematic method of counting the number of inversions of a permutation σ. Determine the number of inversions of each of the permutations of $\{1, 2, 3\}$. What are the minimum and maximum numbers of inversions possible for a permutation of $\{1, 2, \ldots, n\}$? Define a permutation to be **even** if and only if the number of inversions is even. Similarly define an **odd** permutation.

Define a **transposition** to be a permutation that interchanges two elements and leaves the other elements fixed. Suppose σ is any permutation of $\{1, 2, \ldots, n\}$ and τ is a transposition defined on $\{1, 2, \ldots, n\}$. Show that the number of inversions of σ and the number of inversions of $\tau \circ \sigma$ differ by 1. You might want to consider first the case where τ is a transposition of adjacent elements. Show that any permutation on $\{1, 2, \ldots, n\}$ can be written as a composition of transposition. Investigate the relation between the number of transpositions in such a composition and the number of inversions that occur in the permutation.

Given a permutation σ on $\{1, 2, \ldots, n\}$, define the **sign** of σ by

$$\text{sign } \sigma = \begin{cases} 1 & \text{if } \sigma \text{ is even,} \\ -1 & \text{if } \sigma \text{ is odd.} \end{cases}$$

Show that if ρ and σ are two permutations on $\{1, 2, \ldots, n\}$, then

$$\text{sign}(\rho \circ \sigma) = (\text{sign } \rho)(\text{sign } \sigma).$$

Restate this result in terms of even and odd permutations. Show that $\text{sign}(\sigma^{-1}) = \text{sign } \sigma$.

2. You may be interested in investigating the properties of permutations more thoroughly. This is a standard topic in beginning texts on abstract algebra. The article "Defining the Sign of a Permutation" by Stuart Nelson in the June–July 1987 issue of the *American Mathematical Monthly* presents an alternative to the preceding definition of sign σ.

3. Let S_n denote the set of all permutations of $\{1, 2, \ldots, n\}$. Here is the alternative definition of the determinant of an $n \times n$ matrix $A = [a_{ij}]$:

$$\det A = \sum_{\sigma \in S_n} (\text{sign } \sigma) a_{1\sigma(1)} a_{2\sigma(2)} \cdots a_{n\sigma(n)}.$$

Write out the terms of this summation and verify that it agrees with our original definition for matrices of sizes 3×3 and smaller. Prove that it agrees with our original definition for square matrices of all sizes.

4. Use the new definition to prove Theorem 7.4 and Exercise 13 of Section 7.3.

5. Several other approaches to determinants are in common use. Here are two interesting alternatives. You should have little trouble finding references in your library if you prefer reading someone else's exposition to embarking on your

own discovery mission. Your goal is to use the alternative definition to derive the basic results of determinants. Be sure to weigh the advantages of not having to prove the properties used in the alternative definition against the work needed to derive the formula for the expansion of the determinant by minors.

The results of Exercise 11 of Section 7.2 can be formulated in a more general setting. This leads to a definition of the determinant of an $n \times n$ matrix in terms of the n-dimensional volume of the region in $\mathbb{R}^n$ determined by the rows of the matrix. Some convention is needed regarding the sign of the determinant.

Perhaps the most sophisticated approach is to define the determinant in terms of certain characteristic properties. As background information, you may first want to show that the determinant is the only function from $\mathbb{M}(n, n)$ to $\mathbb{R}$ that is linear as a function of each of the n rows, changes sign when two rows are interchanged, and maps the identity matrix to 1. Hence, it is reasonable to take these three properties as a definition of the determinant. With this approach, the first order of business is to show that such a function does in fact exist and that it is unique. A clever way is to show that the set of all functions from $\mathbb{M}(n, n)$ to $\mathbb{R}$ that satisfy the first two conditions is a one-dimensional subspace of $\mathbb{F}(\mathbb{M}(n, n))$. Then all such functions are scalar multiples of the one that maps the identity matrix to 1.

Project: Curve Fitting with Determinants

In Section 2.4 we saw how to use systems of linear equations to determine the coefficients of a curve satisfying certain conditions. In Section 7.4 we encountered a formula involving determinants to solve a system of linear equations. This project explores the possibility of combining these two ideas and using determinants directly to derive the formula of the desired curve.

1. There are many standard forms of an equation for a line in the xy-coordinate plane. The point-slope form and the slope-intercept form are useful in many situations, but they exclude vertical lines. The general equation $ax + by + c = 0$, on the other hand, will cover all the possibilities. The only restriction is that a and b are not both equal to 0.

Suppose we want to find the equation of the line through $(2, -3)$ and $(7, 4)$. Putting these pairs of values of x and y in the general equation of a line gives the two equations

$$2a - 3b + c = 0$$
$$7a + 4b + c = 0$$

Adjoin the general equation $ax + by + c = 0$ to this system. The resulting system of three equations in the unknowns a, b, and c is known to have a nontrivial solution (remember the condition on a and b). By Theorem 7.11 the determinant of the coefficient matrix must be zero. Thus,

$$0 = \det \begin{bmatrix} x & y & 1 \\ 2 & -3 & 1 \\ 7 & 4 & 1 \end{bmatrix}$$

$$= x \det \begin{bmatrix} -3 & 1 \\ 4 & 1 \end{bmatrix} - y \det \begin{bmatrix} 2 & 1 \\ 7 & 1 \end{bmatrix} + 1 \det \begin{bmatrix} 2 & -3 \\ 7 & 4 \end{bmatrix}$$

$$= -7x + 5y - 13$$

is an equation of the line containing the two given points.

Compare this method with other methods of finding an equation for this line. How many operations of arithmetic does each require? How much work would be required to implement the methods as a computer program that will handle any case? How gracefully do the methods fail if the given points are not distinct?

2. We can use this determinant method even for curves defined by more complicated equations. For example, Exercise 4 of Section 2.9 asked for an equation of a circle passing through $(1, 1)$, $(-2, 0)$, and $(0, 1)$. Instead of working with the specific equation of the form $x^2 + y^2 + ax + by + c = 0$, we need to have coefficients in all terms so as to create a system of equations with more than one solution. Try an equation of the form

$$r_1(x^2 + y^2) + r_2x + r_3y + r_4 = 0.$$

Combining this general equation with the specific equations that must hold in order for the given points to lie on the curve leads to the system

$$r_1(x^2 + y^2) + \quad r_2x + r_3y + r_4 = 0$$
$$r_1(1^2 + 1^2) + \quad r_21 + r_31 + r_4 = 0$$
$$r_1((-2)^2 + 0^2) + r_2(-2) + r_30 + r_4 = 0$$
$$r_1(0^2 + 1^2) + \quad r_20 + r_31 + r_4 = 0$$

Since this system will have more than one solution, the determinant of the coefficient matrix will equal zero. Write out this equation to obtain the coefficients of the equation for the circle.

Adapt this idea to obtain equations for other standard curves in $\mathbb{R}^2$. Try getting a polynomial of degree n to pass through $n + 1$ specific points. How many points are needed to determine a general conic section with equation involving xy-terms as well as a constant term and first- and second-degree terms in x and y? What about linear combinations of trigonometric functions? Try to extend this method to surfaces in $\mathbb{R}^3$.

In each case, compare the determinant method with other methods. Check the number of arithmetic operations, the complexity arising from handling special cases, and the problem of degenerate curves or surfaces.

3. The determinant formulas for curves and surfaces show that the coefficients are differentiable functions of the coordinates of the points that define the sets. We can thus use the derivative to measure how sensitive the coefficients are to small changes in the defining points.

For a differentiable function $f : \mathbb{R} \to \mathbb{R}$, the **sensitivity** of the dependent variable y to changes in the independent variable x is defined to be

$$\lim_{\Delta x \to 0} \frac{\frac{\Delta y}{y}}{\frac{\Delta x}{x}} = \frac{x}{y} \lim_{\Delta x \to 0} \frac{\Delta y}{\Delta x} = \frac{x}{y} \frac{dy}{dx}.$$

This is the limiting value of the ratio of the relative change in y to the relative change in x.

Investigate the sensitivity of the coefficients to changes in the coordinates of the given points. In what situations is the sensitivity high? In what cases is it low?

Chapter 7 Summary

Chapter 7 uses the proof technique of mathematical induction to define determinants and derive some basic properties. Determinants allow us to tell if a matrix is invertible, they give a formula for the inverse, and they are used in Cramer's rule for solving certain linear systems. The formal definition of a determinant gives the cross product on $\mathbb{R}^3$. The sign of a determinant distinguishes right from left.

Computations

Determinant of a matrix
Definition in terms of expansion along the first row
Expansion along any row or column
Effect of row operations

Transpose of a matrix

Formulas involving the determinant
Adjoint of a matrix
Formula for A^{-1}
Cramer's rule for solving $A\mathbf{x} = \mathbf{b}$

The determinant in $\mathbb{R}^3$
Cross product
Orientation of an ordered basis
Orientation-preserving and orientation-reversing linear operators

Theory

Principle of Mathematical Induction
Basis for induction, inductive hypothesis
Modification for basis other than $n = 1$
Inductive definitions

Properties of determinants
Expansion along any row or column
$\det(AB) = (\det A)(\det B)$

Properties of transpose

A is nonsingular if and only if $\det A \neq 0$

Properties of the cross product
Algebraic properties
Geometric properties

Applications

Sensitivity of solutions to errors in coefficients

Vector in $\mathbb{R}^3$ orthogonal to two given vectors

Right-hand rule for cross products

Review Exercises

1. Prove that $13^n - 8^n$ is divisible by 5 for any natural number n.

2. Prove that $\displaystyle\sum_{k=0}^{n} \frac{1}{2^k} = \frac{2^{n+1} - 1}{2^n}$ for nonnegative integer n.

3. **a.** Illustrate the definition of determinant to compute $\det \begin{bmatrix} 0 & 3 & 2 \\ 1 & 5 & 1 \\ -4 & 2 & -1 \end{bmatrix}$.

    b. With no further computation, what is $\det \begin{bmatrix} 0 & 3 & 2 \\ 10 & 50 & 10 \\ -4 & 2 & -1 \end{bmatrix}$?

    c. With no further computation, what is $\det \begin{bmatrix} 0 & 3 & 2 \\ 10 & 56 & 14 \\ -4 & 2 & -1 \end{bmatrix}$?

4. **a.** Illustrate the definition of determinant to compute $\det \begin{bmatrix} 5 & 0 & 0 & 0 \\ 8 & 1 & 2 & 3 \\ 0 & 7 & 4 & 3 \\ 6 & 5 & 3 & 1 \end{bmatrix}$.

    b. With no further computation, what is $\det \begin{bmatrix} 5 & 0 & 0 & 0 \\ 888 & 111 & 222 & 333 \\ 0 & 7 & 4 & 3 \\ 6 & 5 & 3 & 1 \end{bmatrix}$?

    c. With no further computation, what is $\det \begin{bmatrix} 805 & 100 & 200 & 300 \\ 8 & 1 & 2 & 3 \\ 0 & 7 & 4 & 3 \\ 6 & 5 & 3 & 1 \end{bmatrix}$?

5. Consider the general 2×2 matrix $A = \begin{bmatrix} a & b \\ c & d \end{bmatrix}$.

    a. Use Theorem 7.13 to give an explicit formula for A^{-1}.

    b. Determine scalars r and s so that A^{-1} is a linear combination $rA + sI$.

6. Suppose A is an $n \times n$ matrix. Consider the function $T : \mathbb{R}^n \to \mathbb{R}$ that assigns to any vector $\mathbf{v} \in \mathbb{R}^n$ the determinant of the matrix obtained by replacing the first column A with $\mathbf{v}$. Show that T is linear.

7. Let $A = \begin{bmatrix} a & b \\ c & d \end{bmatrix}$ and $A' = \begin{bmatrix} a' & b' \\ c' & d' \end{bmatrix}$. Express $\det(A + A')$ as a sum of the determinants of the four matrices formed by taking one of the columns of A and one of the columns of A'.

8. Consider the matrix $A = \begin{bmatrix} 2 & 0 & 1 \\ 0 & 3 & 1 \\ -1 & 1 & 1 \end{bmatrix}$.

   a. Compute the determinant of A.

   b. Compute the adjoint of A.

   c. Use the results of parts a and b to write down the inverse of A.

9. Use Cramer's rule to solve the system

$$5x - y + 2z = 1$$
$$x + 3y + z = 2$$
$$2x - y + 3z = -1$$

10. Consider the function f that assigns to a real number x the 11-entry of the inverse of the matrix $A = \begin{bmatrix} 1 & 1 & 2 \\ x & 3 & 1 \\ 1 & x & 2 \end{bmatrix}$.

   a. What is the domain of f?

   b. Show that f is differentiable.

   c. Use the value of $f'(3)$ to estimate the error in the 11-entry of the inverse of A if x is known to be within 0.05 of 3.

11. Consider the function f that assigns to a real number a the value of z that satisfies the following system of linear equations

$$x + ay + z = 3$$
$$ax - y - z = -1$$
$$x + y + z = 1$$

   a. For what values of a is f defined?

   b. Show that f is differentiable.

   c. Use the value of $f'(2.15)$ to estimate the error in z if a is known to be within 0.004 of 2.15.

12. Suppose $\mathbf{u}$, $\mathbf{v}$, and $\mathbf{w}$ are vectors in $\mathbb{R}^3$.

   a. Give a counterexample using the standard basis vectors for $\mathbb{R}^3$ to show that the associative law $(\mathbf{u} \times \mathbf{v}) \times \mathbf{w} = \mathbf{u} \times (\mathbf{v} \times \mathbf{w})$ does not hold in general.

   b. Prove that $(\mathbf{u} \times \mathbf{v}) \times \mathbf{w} = (\mathbf{u} \cdot \mathbf{w})\mathbf{v} - (\mathbf{v} \cdot \mathbf{w})\mathbf{u}$.

   c. Prove that $\mathbf{u} \times (\mathbf{v} \times \mathbf{w}) = (\mathbf{u} \cdot \mathbf{w})\mathbf{v} - (\mathbf{u} \cdot \mathbf{v})\mathbf{w}$.

13.   **a.** Use the cross product to give a formula in terms of the variable a for a vector orthogonal to (a, a^2, a^3) and $(1, 0, 0)$.

      b. Show that the three coordinates of your vector are continuous functions of a.

14.   Let $\mathbf{v}$ and $\mathbf{w}$ be two vectors in $\mathbb{R}^3$. Describe a sequence of rotations about the coordinate axes so that the image of $\mathbf{v}$ is on the positive z-axis and the image of $\mathbf{w}$ is in the yz-plane with a positive y-coordinate.

Eigenvalues and Eigenvectors

This *final chapter brings together all the topics we have covered in the course. Our goal is to analyze the characteristic features of any given linear transformation. Of course, we will be dealing with vectors and vector spaces. We will be using the concepts of dimension in our quest for a basis that is especially compatible with the linear transformation. In certain cases we will even be able to choose an orthogonal basis. Matrices will provide concrete representations of the linear transformations. The Gauss-Jordan reduction algorithm and the determinant function will be our basic tools in carrying out numerous computational details.*

8.1 Definitions

Consider the moderately unpleasant-looking linear transformation $T : \mathbb{R}^2 \to \mathbb{R}^2$ defined by

$$T\left(\begin{bmatrix} x \\ y \end{bmatrix}\right) = \begin{bmatrix} 23 & -12 \\ 40 & -21 \end{bmatrix}\begin{bmatrix} x \\ y \end{bmatrix} = \begin{bmatrix} 23x - 12y \\ 40x - 21y \end{bmatrix}$$

This formula provides little in the way of insight as to how T transforms vectors in the plane. Notice, however, that

$$T\left(\begin{bmatrix} 3 \\ 5 \end{bmatrix}\right) = \begin{bmatrix} 9 \\ 15 \end{bmatrix} = 3\begin{bmatrix} 3 \\ 5 \end{bmatrix} \quad \text{and}$$

$$T\left(\begin{bmatrix} 1 \\ 2 \end{bmatrix}\right) = \begin{bmatrix} -1 \\ -2 \end{bmatrix} = -1\begin{bmatrix} 1 \\ 2 \end{bmatrix}.$$

Hence, relative to the basis $B = \left\{ \begin{bmatrix} 3 \\ 5 \end{bmatrix}, \begin{bmatrix} 1 \\ 2 \end{bmatrix} \right\}$ for $\mathbb{R}^2$, the matrix of T is particularly simple. According to Theorem 6.11, the columns of this matrix are

$$\left[T\left(\begin{bmatrix}3\\5\end{bmatrix}\right)\right]_B = \begin{bmatrix}9\\15\end{bmatrix}_B = \begin{bmatrix}3\\0\end{bmatrix} \quad \text{and} \quad \left[T\left(\begin{bmatrix}1\\2\end{bmatrix}\right)\right]_B = \begin{bmatrix}-1\\-2\end{bmatrix}_B = \begin{bmatrix}0\\-1\end{bmatrix}.$$

Transforming a vector in $\mathbb{R}^2$ by multiplication by this matrix $D = \begin{bmatrix}3 & 0\\0 & -1\end{bmatrix}$ is comparatively easy to describe. Looking at this transformation geometrically, we see that the vector gets stretched by a factor of 3 in the x-direction and reflected through the x-axis.

This insight helps us picture what the original transformation T does. It can be described as a three-step process. First warp the plane linearly so that $\begin{bmatrix}3\\5\end{bmatrix}$ and $\begin{bmatrix}1\\2\end{bmatrix}$ coincide with the original locations of $\begin{bmatrix}1\\0\end{bmatrix}$ and $\begin{bmatrix}0\\1\end{bmatrix}$, respectively. Then perform the stretching and reflecting as described above. Finally, let the plane snap back to its unwarped configuration. A person with extremely bad astigmatism who sees B as an orthonormal basis might even be able to visualize the stretching and reflecting without bothering with the change of basis.

The goal of this chapter is to show how to simplify the matrix of a linear map by choosing a basis that is especially compatible with the map. As illustrated in the introductory example, the key is to try to find basis vectors that are mapped to scalar multiples of themselves. This leads to the following definition. Recall that a linear operator is a linear map whose domain is equal to its range.

Definition 8.1

Suppose $T: V \rightarrow V$ is a linear operator. Suppose λ is a real number and $\mathbf{v} \in V$ is a nonzero vector such that

$$T(\mathbf{v}) = \lambda\mathbf{v}.$$

Then λ is an **eigenvalue** of T, and $\mathbf{v}$ is an **eigenvector** of T associated with λ. The set $E_T(\lambda) = \{\mathbf{v} \in V \mid T(\mathbf{v}) = \lambda\mathbf{v}\}$ is the **eigenspace** of T associated with λ.

Thus, for the linear operator T discussed at the beginning of this section, $\begin{bmatrix}3\\5\end{bmatrix}$ is an eigenvector associated with the eigenvalue 3, and $\begin{bmatrix}1\\2\end{bmatrix}$ is an eigenvector associated with the eigenvalue -1.

Crossroads   We ran across eigenvalues and eigenvectors in another setting earlier in this text. Recall from Theorem 5.17 that a regular Markov chain with transition matrix P has an equilibrium vector $\mathbf{s}$ satisfying $P\mathbf{s} = \mathbf{s}$. This can be rephrased as saying that the linear map defined in terms of multiplication by the matrix P has an eigenvector $\mathbf{s}$ associated with the eigenvalue 1.

The theory of differential equations makes extensive use of eigenvalues and eigenvectors. As a simple example that hints at this interplay, let $\mathbb{D}^{(\infty)}$

be the subspace of $\mathbb{F}(\mathbb{R})$ of all infinitely differentiable functions, and consider the differentiation operator $D: \mathbb{D}^{(\infty)} \to \mathbb{D}^{(\infty)}$. The formula $De^x = e^x$ tells us that the exponential function is an eigenvector of D associated with the eigenvalue 1.

The study of linear operators on infinite-dimensional spaces is an important part of mathematics known as functional analysis. Nevertheless, in this text we will be primarily concerned with finding eigenvalues and eigenvectors of linear operators that are defined in terms of matrix multiplication. Thus, we will concentrate on linear operators defined on finite-dimensional vector spaces. The following theorem serves as a first step in this direction.

Theorem 8.2

Suppose A is an $n \times n$ matrix and $T: \mathbb{R}^n \to \mathbb{R}^n$ is defined by $T(\mathbf{v}) = A\mathbf{v}$. Then the real number λ is an eigenvalue of T if and only if $\det(\lambda I - A) = 0$.

Proof   Rather than provide separate proofs for the two directions of the implication in the statement of this theorem, we can use the machinery we have developed to put together a string of logically equivalent statements:

$$\lambda \text{ is an eigenvalue of } T \Longleftrightarrow T(\mathbf{v}) = \lambda\mathbf{v} \text{ for some nonzero } \mathbf{v} \in \mathbb{R}^n$$
$$\Longleftrightarrow A\mathbf{v} = \lambda\mathbf{v} \text{ for some nonzero } \mathbf{v} \in \mathbb{R}^n$$
$$\Longleftrightarrow \lambda\mathbf{v} - A\mathbf{v} = \mathbf{0} \text{ for some nonzero } \mathbf{v} \in \mathbb{R}^n$$
$$\Longleftrightarrow (\lambda I - A)\mathbf{v} = \mathbf{0} \text{ for some nonzero } \mathbf{v} \in \mathbb{R}^n$$
$$\Longleftrightarrow \det(\lambda I - A) = 0$$

where the last step follows from the equivalence of conditions f and g of Theorem 7.11. ∎

For a linear map $T: \mathbb{R}^n \to \mathbb{R}^n$ defined in terms of multiplication by an $n \times n$ matrix A, this theorem says that the eigenvalues of T are precisely the real solutions of the equation $\det(\lambda I - A) = 0$. As you will soon discover, the formula $\det(\lambda I - A)$ expands to a sum of various constant multiples of $1, \lambda, \lambda^2, \lambda^3, \dots, \lambda^n$. Therefore, the formula $\det(\lambda I - A)$ defines a polynomial of degree n in the variable λ. See Exercise 6 at the end of this section. Here is some relevant terminology.

Definition 8.3

Suppose A is an $n \times n$ matrix. The nth degree polynomial in the variable λ defined by

$$\det(\lambda I - A)$$

is the **characteristic polynomial** of A. The real zeros of the characteristic polynomial (real numbers that are solutions of the **characteristic equation**

$\det(\lambda I - A) = 0$) are called the **eigenvalues** of the matrix A. A nonzero vector $\mathbf{v} \in \mathbb{R}^n$ such that $(\lambda I - A)\mathbf{v} = \mathbf{0}$ is an **eigenvector** of A associated with λ. The solution space

$$E_A(\lambda) = \{\mathbf{v} \in \mathbb{R}^n \mid (\lambda I - A)\mathbf{v} = \mathbf{0}\}$$

of this homogeneous system is the **eigenspace** of A associated with λ.

Be sure to notice that eigenvalues of a linear operator are required to be real numbers even though the characteristic equation may have solutions that are complex numbers with nonzero imaginary parts. Consider, for example, the linear operator $R_{\pi/2}: \mathbb{R}^2 \to \mathbb{R}^2$ that rotates vectors counterclockwise through an angle of $\pi/2$. Clearly, no vector other than $\mathbf{0}$ will be rotated onto a multiple of itself. So we are not surprised that $R_{\pi/2}$ has no eigenvalues. On the other hand, recall from Section 6.3 that the matrix of $R_{\pi/2}$ is

$$A = \begin{bmatrix} 0 & -1 \\ 1 & 0 \end{bmatrix}.$$

The characteristic polynomial is $\det(\lambda I - A) = \lambda^2 + 1$, which has only the complex roots i and $-i$. Since neither of these solutions is real, the matrix A has no eigenvalues.

Crossroads It is possible to develop the theory of vector spaces and linear transformations using the complex numbers as scalars in place of the real numbers. With that approach, all solutions to the characteristic equation will be eigenvalues. In fact, the Fundamental Theorem of Algebra states that any polynomial (with real or complex coefficients) of degree n has n complex roots counted according to multiplicity. Thus, every $n \times n$ matrix will have n eigenvalues when we allow complex numbers. In this text, however, we will stick with scalars and eigenvalues as real numbers as much as possible. Eventually, in Section 8.4, we will need to consider briefly the possibility of complex solutions to the characteristic equation.

Quick Example *Find the eigenvalues of the matrix* $A = \begin{bmatrix} 1 & 2 & -1 \\ 1 & 0 & 1 \\ 4 & -4 & 5 \end{bmatrix}$. *For each eigenvalue, find a basis for the associated eigenspace.*

The first step is to find the eigenvalues of A by solving the characteristic equation.

$$0 = \det(\lambda I - A) = \det \begin{bmatrix} \lambda - 1 & -2 & 1 \\ -1 & \lambda & -1 \\ -4 & 4 & \lambda - 5 \end{bmatrix}$$

$$= (\lambda - 1)(\lambda(\lambda - 5) + 4) + 2(-(\lambda - 5) - 4) + 1(-4 + 4\lambda)$$

$$= (\lambda - 1)(\lambda^2 - 5\lambda + 4) - 2(\lambda - 1) + 4(\lambda - 1)$$

$$= (\lambda - 1)(\lambda^2 - 5\lambda + 6)$$

$$= (\lambda - 1)(\lambda - 2)(\lambda - 3)$$

For each of the three eigenvalues, we proceed to find the associated eigenspace. In each case this amounts to solving a homogeneous system of equations where the coefficient matrix is $\lambda I - A$.

For $\lambda = 1$,

$$\begin{bmatrix} 0 & -2 & 1 \\ -1 & 1 & -1 \\ -4 & 4 & -4 \end{bmatrix} \longrightarrow \begin{bmatrix} 1 & -1 & 1 \\ 0 & 1 & -\frac{1}{2} \\ 0 & 0 & 0 \end{bmatrix} \longrightarrow \begin{bmatrix} 1 & 0 & \frac{1}{2} \\ 0 & 1 & -\frac{1}{2} \\ 0 & 0 & 0 \end{bmatrix}.$$

So the general solution is $\begin{bmatrix} x \\ y \\ z \end{bmatrix} = \begin{bmatrix} -\frac{1}{2}r \\ \frac{1}{2}r \\ r \end{bmatrix} = r\begin{bmatrix} -\frac{1}{2} \\ \frac{1}{2} \\ 1 \end{bmatrix}$. Hence, $\left\{ \begin{bmatrix} -1 \\ 1 \\ 2 \end{bmatrix} \right\}$ is a basis

for the eigenspace $E_A(1)$.

For $\lambda = 2$,

$$\begin{bmatrix} 1 & -2 & 1 \\ -1 & 2 & -1 \\ -4 & 4 & -3 \end{bmatrix} \longrightarrow \begin{bmatrix} 1 & -2 & 1 \\ 0 & 0 & 0 \\ 0 & -4 & 1 \end{bmatrix} \longrightarrow \begin{bmatrix} 1 & 0 & \frac{1}{2} \\ 0 & 1 & -\frac{1}{4} \\ 0 & 0 & 0 \end{bmatrix}.$$

So the general solution is $\begin{bmatrix} x \\ y \\ z \end{bmatrix} = \begin{bmatrix} -\frac{1}{2}r \\ \frac{1}{4}r \\ r \end{bmatrix} = r\begin{bmatrix} -\frac{1}{2} \\ \frac{1}{4} \\ 1 \end{bmatrix}$. Hence, $\left\{ \begin{bmatrix} -2 \\ 1 \\ 4 \end{bmatrix} \right\}$ is basis for

the eigenspace $E_A(2)$.

For $\lambda = 3$,

$$\begin{bmatrix} 2 & -2 & 1 \\ -1 & 3 & -1 \\ -4 & 4 & -2 \end{bmatrix} \longrightarrow \begin{bmatrix} 1 & -1 & \frac{1}{2} \\ 0 & 2 & -\frac{1}{2} \\ 0 & 0 & 0 \end{bmatrix} \longrightarrow \begin{bmatrix} 1 & 0 & \frac{1}{4} \\ 0 & 1 & -\frac{1}{4} \\ 0 & 0 & 0 \end{bmatrix}.$$

So the general solution is $\begin{bmatrix} x \\ y \\ z \end{bmatrix} = \begin{bmatrix} -\frac{1}{4}r \\ \frac{1}{4}r \\ r \end{bmatrix} = r\begin{bmatrix} -\frac{1}{4} \\ \frac{1}{4} \\ 1 \end{bmatrix}$. Hence, $\left\{ \begin{bmatrix} -1 \\ 1 \\ 4 \end{bmatrix} \right\}$ is a basis

for the eigenspace $E_A(3)$. ∎

Exercises 8.1

1. Find out the meaning of the German word *eigen*. How does this apply in the context of eigenvalues and eigenvectors of linear algebra?

2. **a.** Find a nonzero function $f \in \mathbb{D}^{(\infty)}$ to show that 2 is an eigenvalue of the differentiation operator $D: \mathbb{D}^{(\infty)} \to \mathbb{D}^{(\infty)}$.

   b. Show that all real numbers are eigenvalues of D.

3. Suppose λ is an eigenvalue of a linear operator $T : V \to V$. Show that the eigenspace $E_T(\lambda)$ is a subspace of V.

4. Suppose λ is an eigenvalue of the matrix A. What element of the eigenspace $E_A(\lambda)$ is not an eigenvector of A associated with λ?

5. Suppose λ is an eigenvalue of a linear operator $T: \mathbb{R}^n \to \mathbb{R}^n$. Let A be the matrix of T relative to the standard basis for $\mathbb{R}^n$. Show that the eigenspace $E_T(\lambda)$ of the operator T is equal to the eigenspace $E_A(\lambda)$ of the matrix A.

6. Prove that if A is an $n \times n$ matrix, then $\det(\lambda I - A)$ defines an nth-degree polynomial in the variable λ.

7. Find the eigenvalues of the following matrices. For each eigenvalue, find a basis for the corresponding eigenspace.

a. $\begin{bmatrix} 6 & -24 & -4 \\ 2 & -10 & -2 \\ 1 & 4 & 1 \end{bmatrix}$

b. $\begin{bmatrix} 7 & -24 & -6 \\ 2 & -7 & -2 \\ 0 & 0 & 1 \end{bmatrix}$

c. $\begin{bmatrix} 3 & 1 & 0 \\ 0 & 3 & 1 \\ 0 & 0 & 3 \end{bmatrix}$

d. $\begin{bmatrix} 3 & -7 & -4 \\ -1 & 9 & 4 \\ 2 & -14 & -6 \end{bmatrix}$

e. $\begin{bmatrix} -1 & -1 & 10 \\ -1 & -1 & 6 \\ -1 & -1 & 6 \end{bmatrix}$

f. $\begin{bmatrix} \frac{1}{2} & -5 & 5 \\ \frac{3}{2} & 0 & -4 \\ \frac{1}{2} & -1 & 0 \end{bmatrix}$

8. Compute the eigenvalues of the matrices

a. $\begin{bmatrix} -5 & 9 \\ 0 & 3 \end{bmatrix}$

b. $\begin{bmatrix} 7 & 4 & -3 \\ 0 & -3 & 9 \\ 0 & 0 & 2 \end{bmatrix}$

c. $\begin{bmatrix} -2 & 8 & 7 & 4 \\ 0 & 3 & 5 & 17 \\ 0 & 0 & 1 & 9 \\ 0 & 0 & 0 & 5 \end{bmatrix}$

d. $\begin{bmatrix} a & b & c \\ 0 & d & e \\ 0 & 0 & f \end{bmatrix}$

e. What feature of these matrices makes it relatively easy to compute their eigenvalues?

f. Formulate a general result suggested by this observation.

g. Prove your conjecture.

9. a. Give a 2×2 matrix with eigenvalues 2 and 5. (Suggestion: Keep things simple. Use the standard basis for $\mathbb{R}^2$ as eigenvectors.)

b. Describe how to construct an $n \times n$ matrix with any prescribed eigenvalues $\lambda_1, \ldots, \lambda_r$, for $1 \le r \le n$.

10. Prove that a square matrix is singular if and only if 0 is one of its eigenvalues.

11. a. Show that any 2×2 matrix of the form $\begin{bmatrix} a & b \\ b & a \end{bmatrix}$ has eigenvalues.

b. Under what conditions on a and b will such a matrix have two distinct eigenvalues?

12. Show that every nonzero vector in a vector space V is an eigenvector of the identity operator $\mathrm{id}_V : V \to V$. With what eigenvalues are they associated?

13. Suppose $\mathbf{v}$ is an eigenvector of an $n \times n$ matrix A associated with the eigenvalue λ. Suppose P is a nonsingular $n \times n$ matrix. Show that $P^{-1}\mathbf{v}$ is an eigenvector of $P^{-1}AP$ associated with the eigenvalue λ.

14. Suppose that A and A' are $n \times n$ matrices. Suppose $\mathbf{v}$ is an eigenvector of A associated with the eigenvalue λ. Suppose $\mathbf{v}$ is also an eigenvector of A' associated with the eigenvalue λ'. Show that $\mathbf{v}$ is an eigenvector of $A + A'$ associated with $\lambda + \lambda'$.

15. Suppose $\mathbf{v}$ is an eigenvector of an $n \times n$ matrix A associated with the eigenvalue λ.

   a. Show that $\mathbf{v}$ is an eigenvector of A^2. With what eigenvalue is it associated?

   b. State and prove a generalization of your result in part a to higher powers of A.

   c. What can you say about eigenvalues and eigenvectors of A^{-1} and other negative powers of A?

16. Suppose $T : V \to V$ and $T' : V \to V$ are linear operators on a vector space V. Suppose $\mathbf{v}$ is an eigenvector of T associated with the eigenvalue λ. Suppose $\mathbf{v}$ is also an eigenvector of T' associated with the eigenvalue λ'. Show that $\mathbf{v}$ is an eigenvector of the composition $T' \circ T$. What is the eigenvalue?

17. **a.** The position of the hour hand on a clock determines the position of the minute hand. How many times in each twelve-hour period do the hands point in exactly the same direction? How many times do they line up in exactly opposite directions?

   b. Implement this functional relation between the position of the hands of a clock graphically on a computer or graphing calculator. Allow the user to have some convenient way of specifying the position of the hour hand (perhaps as an angle). Then graph the position of the two hands and give a signal if the hands line up (at least to the resolution of the screen).

   c. Generalize your program so the relation between the two hands can be any function whose domain and range are $\mathbb{R}^2$. You may wish to normalize the lengths of the domain variable (the hour hand) and the range variable (the minute hand) in order to make them easier to display on the screen.

   d. Set up your program for a linear relation between the two variables. Try to get the hands to point in the same direction. Verify that the corresponding vectors are eigenvectors of the linear map.

8.2 Similarity

Given a linear operator $T : V \to V$ defined on a finite-dimensional vector space V, we know that the matrix that represents T depends on the choice of an ordered basis for V. By Theorem 6.18, if A is the matrix of T relative to the ordered basis B and A' is the matrix of T relative to the ordered basis B', then $A' = P^{-1}AP$, where P is the matrix for changing from the basis B' to the basis B.

On the other hand, suppose $B = \{\mathbf{u}_1, \dots, \mathbf{u}_n\}$ is a basis for a vector space V and A is the matrix of a linear operator $T : V \to V$ relative to B. For any nonsingular $n \times n$ matrix P, Exercise 7 of Section 6.7 guarantees that there is a basis $B' = \{\mathbf{u}'_1, \dots, \mathbf{u}'_n\}$ such that P is the matrix for changing from the basis B' to the basis B. Thus, $A' = P^{-1}AP$ is the matrix of T relative to this new basis.

This discussion leads us to focus our attention on the relation among square matrices defined as follows.

Definition 8.4

An $n \times n$ matrix A is **similar** to an $n \times n$ matrix B if and only if $A = P^{-1}BP$ for some nonsingular $n \times n$ matrix P. This relation is denoted

$$A \sim B.$$

The following theorem lists some simple properties of this relation. From an abstract point of view, this theorem shows that the similarity relation among square matrices works very much like the geometric relation of similarity among triangles. You are asked to verify these three properties in Exercise 1 at the end of this section.

Theorem 8.5

For any $n \times n$ matrices A, B, and C,
   a. $A \sim A$ (reflexivity)
   b.   if $A \sim B$, then $B \sim A$ (symmetry)
   c.   if $A \sim B$ and $B \sim C$, then $A \sim C$ (transitivity)

For a given $n \times n$ matrix A, we can consider the set of all $n \times n$ matrices that are similar to A. Theorem 8.5 makes it easy to verify that A is in this set and that any two matrices in this set are similar to each other. Let us refer to such a set of matrices as a **similarity class**.

Although it is easy to write down a nonsingular matrix P and compute another matrix $P^{-1}AP$ similar to A, it is not so easy to look at two $n \times n$ matrices and determine whether or not they are similar. We do have several tools to help us decide.

To begin with, if A and $B = P^{-1}AP$ are similar, then by Exercise 12 of Section 7.3,

$$\det B = \det(P^{-1}AP) = \det A.$$

Thus, the determinant function assigns the same value to all matrices in a similarity class. It follows that if two $n \times n$ matrices have different determinants, they cannot be similar. Of course, just because two $n \times n$ matrices have the same determinant, we cannot jump to the conclusion that they are similar matrices. We may need to call into service other invariants such as the function defined next.

Definition 8.6

The **main diagonal** of an $n \times n$ matrix $A = [a_{ij}]$ consists of those positions where the row index is equal to the column index. The **trace** of A, denoted tr A, is the sum of the entries on the main diagonal of A. That is,

$$\mathrm{tr}\, A = \sum_{k=1}^{n} a_{kk}.$$

Here are some of the basic properties of the trace function.

Theorem 8.7

Suppose A and B are $n \times n$ matrices, P is an invertible $n \times n$ matrix, and r is a real number. Then

   a. $\text{tr}(A + B) = \text{tr}\,A + \text{tr}\,B$,
   b. $\text{tr}(rA) = r\,\text{tr}\,A$,
   c. $\text{tr}(AB) = \text{tr}(BA)$,
   d. $\text{tr}\,A = \text{tr}(P^{-1}AP)$.

Properties a and b are the conditions for $\text{tr} : \mathbb{M}(n, n) \to \mathbb{R}$ to be a linear function. The primary value of property c is to facilitate the proof of property d. Property d can be rephrased to say that similar matrices have the same trace. Hence, the trace, like the determinant, is a function that is constant on each similarity class. You are asked to prove these properties in Exercise 10 at the end of this section.

What if two $n \times n$ matrices have the same determinant and the same trace? We still cannot conclude that the matrices are similar. In such a situation the characteristic polynomial is a more refined invariant that is often useful in deciding whether or not the matrices are similar. Exercises 13 and 14 at the end of this section contain some suggestions to help you show that the determinant and the trace can be read off by looking at the coefficients of the characteristic polynomial. The other coefficients of the characteristic polynomial provide additional invariants that may be of help in determining that two matrices are not similar.

Theorem 8.8

If two matrices are similar, then they have the same characteristic polynomial.

Proof   Suppose the matrices A and $B = P^{-1}AP$ are similar. For any real number λ,

$$
\begin{aligned}
\det(\lambda I - B) &= \det(\lambda I - P^{-1}AP) \\
&= \det(\lambda P^{-1}IP - P^{-1}AP) \\
&= \det\!\left(P^{-1}(\lambda I)P - P^{-1}AP\right) \\
&= \det\!\left(P^{-1}(\lambda I - A)P\right) \\
&= \det(\lambda I - A). \quad \blacksquare
\end{aligned}
$$

This theorem says that all the matrices in any similarity class have the same characteristic polynomial. In particular, all the matrices in any similarity class have the same eigenvalues. Furthermore, the multiplicity with which a given eigenvalue occurs in the characteristic polynomial will be the same for all these matrices. The characteristic polynomial goes a long way toward indicating whether two matrices are similar. Yet there are still matrices that are not similar even though they have the same characteristic polynomial. You may want to take an advanced course in linear algebra to learn more of the story of the classification of similarity classes of matrices.

Theorem 8.8 also allows us to define the characteristic polynomial of a linear operator $T : V \to V$ on a finite-dimensional vector space V. Since the matrices of T with respect to any basis are similar, they all have the same characteristic polynomial. As you might expect (see Exercise 15 at the end of this section), the real roots of the characteristic polynomial are the eigenvalues of T, and the eigenvectors of the matrix are the coordinate vectors of the eigenvectors of T.

Definition 8.9

Suppose $T : V \to V$ is a linear operator on a finite-dimensional vector space V. The **characteristic polynomial** of T is the characteristic polynomial of a matrix that represents T relative to some basis for V.

We now want to turn our attention to a related problem. Among all the matrices in a given similarity class, we would like to find a particularly simple representative. For example, we might hope to find a matrix of the form described in the following definition.

Definition 8.10

A **diagonal** matrix is a square matrix whose only nonzero entries occur on the main diagonal of the matrix.

The next section will show how the problem of finding eigenvalues and eigenvectors is related to the problem of finding a diagonal matrix similar to a given matrix. We will soon discover, though, that not every square matrix is similar to a diagonal matrix. Thus, the following definition makes a useful distinction among matrices in $\mathbb{M}(n, n)$.

Definition 8.11

An $n \times n$ matrix A is **diagonalizable** if and only if there is a diagonal matrix D with $D \sim A$.

Of course, if an $n \times n$ matrix A is diagonalizable, then so is every matrix that is similar to A.

Exercises 8.2

1. Prove Theorem 8.5.

2. Show that the zero matrix $\mathbf{0} \in \mathbb{M}(n, n)$ is the only matrix in its similarity class.

3. Find all matrices similar to the multiplicative identity matrix $I \in \mathbb{M}(n, n)$.

4. **a.** Show that $\begin{bmatrix} 2 & 0 \\ 0 & 3 \end{bmatrix}$ is similar to $\begin{bmatrix} 3 & 0 \\ 0 & 2 \end{bmatrix}$.

b. Show that $\begin{bmatrix} 2 & 0 & 0 \\ 0 & 3 & 0 \\ 0 & 0 & 7 \end{bmatrix}$ is similar to $\begin{bmatrix} 7 & 0 & 0 \\ 0 & 2 & 0 \\ 0 & 0 & 3 \end{bmatrix}$.

5. Suppose the $n \times n$ matrices A and B are similar. What similarity relations hold among the powers of A and the powers of B? What about inverses and other negative powers of A and B?

6. Find two matrices that are of the same size and have the same determinant but are not similar. (Suggestion: Keep things simple. Look at 2×2 diagonal matrices.)

7. Find two matrices that have the same trace but different determinants.

8. Find two matrices that have the same determinant and same trace but different characteristic polynomials.

9. Find two matrices that have the same characteristic polynomial but are not similar. (Suggestion: Exercise 3 will help.)

10. Prove Theorem 8.7.

11. Find examples of square matrices A and B for which $\mathrm{tr}(AB) \neq (\mathrm{tr}\,A)(\mathrm{tr}\,B)$.

12. **a.** Prove that the characteristic polynomial of a 2×2 matrix A is $\lambda^2 - (\mathrm{tr}A)\lambda + \det A$.

b. The Cayley-Hamilton Theorem states that any square matrix satisfies its characteristic polynomial. Prove the 2×2 version of this theorem. (Suggestion: Set up notation for the four entries of a 2×2 matrix A, compute $A^2 - (\mathrm{tr}\,A)A + (\det A)I$, and show that this simplifies to the 2×2 zero matrix.)

13. Show that the value of the determinant of a square matrix can be obtained from the characteristic polynomial of the matrix. (Suggestion: Evaluate the characteristic polynomial at $\lambda = 0$. Notice that this gives the constant term of the polynomial. Be careful to make an adjustment so the sign comes out right.)

14. Show that the value of the trace of an $n \times n$ matrix A can be obtained from the characteristic polynomial of the matrix. (Suggestion: Try a few 2×2 and 3×3 examples to see what is going on. Concentrate on the coefficient of λ^{n-1}. In the expansion of the $\det(\lambda I - A)$ as in Exercise 9 of Section 7.2, only one of the $n!$ terms will contribute to this coefficient.)

15. Suppose A is the matrix of a linear map $T : V \to V$ relative to a basis B for the finite–dimensional vector space V.

a. Prove that if $\mathbf{v} \in V$ is an eigenvector of T associated with the eigenvalue λ, then $[\mathbf{v}]_B$ is an eigenvector of A associated with the eigenvalue λ.

b. Prove that if $\mathbf{v} \in \mathbb{R}^n$ is an eigenvector of A associated with the eigenvalue λ, then $L_B(\mathbf{v})$ is an eigenvector of T associated with the eigenvalue λ. (Recall that L_B is the linear combination function as defined in Section 6.1.)

16. **a.** Compute the characteristic polynomial of the diagonal matrix $\begin{bmatrix} 2 & 0 & 0 \\ 0 & -5 & 0 \\ 0 & 0 & 1 \end{bmatrix}$.

b. Compute the characteristic polynomial of the diagonal matrix $\begin{bmatrix} 3 & 0 & 0 \\ 0 & 3 & 0 \\ 0 & 0 & -2 \end{bmatrix}$.

c. Determine the characteristic polynomial of an arbitrary diagonal matrix

$$\begin{bmatrix} r_1 & 0 & \cdots & 0 \\ 0 & r_2 & \cdots & 0 \\ \vdots & \vdots & & \vdots \\ 0 & 0 & \cdots & r_n \end{bmatrix}.$$

d. State the relation between the diagonal entries of a diagonal matrix and the eigenvalues of the matrix. Be sure to mention what happens if the same number occurs more than once.

17. **a.** Determine the eigenvalues and corresponding eigenvectors of the diagonal matrix $\begin{bmatrix} 2 & 0 & 0 \\ 0 & -5 & 0 \\ 0 & 0 & 1 \end{bmatrix}$. Determine the dimensions of the eigenspaces.

b. Determine the eigenvalues and corresponding eigenvectors of the diagonal matrix $\begin{bmatrix} 3 & 0 & 0 \\ 0 & 3 & 0 \\ 0 & 0 & -2 \end{bmatrix}$. Determine the dimensions of the eigenspaces.

c. Generalize the results of parts a and b to a statement about the eigenvalues and eigenvectors of an arbitrary diagonal matrix. Be sure to mention the connection between the dimension of an eigenspace and the multiplicity of the associated eigenvalue.

d. Prove your conjecture.

8.3 Diagonalization

In this section we continue our quest for a basis such that a given linear operator has a particularly simple matrix representation. Keep in mind that for a linear operator represented by a matrix A, finding such a basis is equivalent to finding an invertible matrix P for which $P^{-1}AP$ is especially simple.

Here is a theorem that relates the occurrence of an eigenvector in a basis to the form of one of the columns of the matrix of the linear operator.

Theorem 8.12

Suppose $B = \{\mathbf{u}_1, \ldots, \mathbf{u}_n\}$ is an ordered basis for a vector space V and $T : V \to V$ is a linear operator. Let A be the matrix of T relative to B. Then $\mathbf{u}_j$ is an eigenvector of T if and only if the jth column of A is equal to $\lambda \mathbf{e}_j$ for some real number λ. In this case, λ is the eigenvalue that $\mathbf{u}_j$ is associated with.

Proof Suppose first that $\mathbf{u}_j$ is an eigenvector of T. Let λ denote the eigenvalue that $\mathbf{u}_j$ is associated with. Then the jth column of A is

$$\left[T(\mathbf{u}_j)\right]_B = [\lambda \mathbf{u}_j]_B = \lambda \mathbf{e}_j.$$

Conversely, suppose the jth column of A is $\lambda \mathbf{e}_j$ for some real number λ. Then

$$\left[T(\mathbf{u}_j)\right]_B = A[\mathbf{u}_j]_B = A\mathbf{e}_j = \lambda \mathbf{e}_j = [\lambda \mathbf{u}_j]_B.$$

Since the coordinate vector map is one-to-one, it follows that

$$T(\mathbf{u}_j) = \lambda \mathbf{u}_j.$$

That is, $\mathbf{u}_j$ is an eigenvector of T associated with the eigenvalue λ. ∎

As an immediate consequence of Theorem 8.12, we have the following result.

Theorem 8.13

A linear operator $T : V \to V$ defined on a finite-dimensional vector space V can be represented by a diagonal matrix if and only if V has an ordered basis $\{\mathbf{u}_1, \ldots, \mathbf{u}_n\}$ with all elements being eigenvectors of T. In this situation, the entries on the main diagonal of the matrix are the eigenvalues of T and $\mathbf{u}_j$ is an eigenvector associated with the jj-entry of the matrix.

Mathematical Strategy Session Ideally we would like to represent a given linear operator on an n-dimensional vector space by a diagonal matrix. This would give us an enormous amount of information about the linear operator. In such a situation,

1. The eigenvalues are the entries on the diagonal of the matrix.
2. The characteristic polynomial has n real roots, again the entries on the diagonal of the matrix.
3. The eigenvectors of the diagonal matrix are the standard basis vectors $\mathbf{e}_1, \ldots, \mathbf{e}_n$ of $\mathbb{R}^n$. These are the coordinate vectors of eigenvectors of T.
4. The dimension of each eigenspace is equal to the number of times the associated eigenvalue appears on the diagonal of the matrix. This is also the multiplicity of the eigenvalue as a root of the characteristic polynomial.

For a linear operator $T : V \to V$ defined on a finite-dimensional vector space V, we might naively hope to find a basis for which each vector is an eigenvector. However, we already know of examples such as the rotation $R_{\pi/2} : \mathbb{R}^2 \to \mathbb{R}^2$ that have no eigenvectors. Hence, no basis for $\mathbb{R}^2$ will yield a diagonal matrix for $R_{\pi/2}$.

Thus, we see that the problem of finding a diagonal matrix to represent a linear operator on an n-dimensional vector space has two distinct aspects. First, the characteristic polynomial must have n real roots. Second, the dimension of each eigenspace must equal the multiplicity of the associated eigenvalue as a root of the characteristic polynomial.

In working out explicit examples, we often discover whether a basis of eigenvectors exists as the result of our success or failure in being able to find enough solutions to certain linear systems. Here is a theorem that provides some insight into what we can expect.

Theorem 8.14

Suppose $\lambda_1, \ldots, \lambda_n$ are distinct eigenvalues of a linear operator $T : V \to V$. For $i = 1, \ldots, n$, suppose $\{\mathbf{u}_{i1}, \ldots, \mathbf{u}_{ir_i}\}$ is a linearly independent subset of $E_T(\lambda_i)$. Then

$$\{\mathbf{u}_{11}, \ldots, \mathbf{u}_{1r_1}, \ \mathbf{u}_{21}, \ldots, \mathbf{u}_{2r_2}, \ \ldots, \ \mathbf{u}_{n1}, \ldots, \mathbf{u}_{nr_n}\}$$

is a linearly independent set.

Proof We first prove the theorem for the special case where we have one eigenvector $\mathbf{u}_i$ associated with each eigenvalue λ_i. The proof is by mathematical induction applied to n, the number of distinct eigenvalues.

If $n = 1$, then $\{\mathbf{u}_1\}$ is a linearly independent set since it consists of a single nonzero vector.

For some fixed $n \geq 1$ suppose that $\{\mathbf{u}_1, \ldots, \mathbf{u}_n\}$ is linearly independent. For this fixed value of n we want to show that $\{\mathbf{u}_1, \ldots, \mathbf{u}_{n+1}\}$ is linearly independent. Suppose we have a linear combination of these vectors that equals $\mathbf{0}$:

$$a_1\mathbf{u}_1 + \cdots + a_{n+1}\mathbf{u}_{n+1} = \mathbf{0}.$$

Apply T to both sides of this equation. This gives

$$\begin{aligned}
\mathbf{0} &= T(\mathbf{0}) \\
&= T(a_1\mathbf{u}_1 + \cdots + a_{n+1}\mathbf{u}_{n+1}) \\
&= a_1 T(\mathbf{u}_1) + \cdots + a_{n+1}T(\mathbf{u}_{n+1}) \\
&= a_1\lambda_1\mathbf{u}_1 + \cdots + a_{n+1}\lambda_{n+1}\mathbf{u}_{n+1}.
\end{aligned}$$

Now multiply both sides of the original equation by λ_{n+1}. This gives

$$\begin{aligned}
\mathbf{0} &= \lambda_{n+1}\mathbf{0} \\
&= \lambda_{n+1}(a_1\mathbf{u}_1 + \cdots + a_{n+1}\mathbf{u}_{n+1}) \\
&= a_1\lambda_{n+1}\mathbf{u}_1 + \cdots + a_{n+1}\lambda_{n+1}\mathbf{u}_{n+1}.
\end{aligned}$$

Look at the difference between the two expressions we have derived.

$$\begin{aligned}
\mathbf{0} &= \mathbf{0} - \mathbf{0} \\
&= a_1(\lambda_1 - \lambda_{n+1})\mathbf{u}_1 + \cdots + a_n(\lambda_n - \lambda_{n+1})\mathbf{u}_n + a_{n+1}(\lambda_{n+1} - \lambda_{n+1})\mathbf{u}_{n+1} \\
&= a_1(\lambda_1 - \lambda_{n+1})\mathbf{u}_1 + \cdots + a_n(\lambda_n - \lambda_{n+1})\mathbf{u}_n.
\end{aligned}$$

This linear combination involves only the vectors in the linearly independent set mentioned in the induction hypothesis. Hence,

$$a_1(\lambda_1 - \lambda_{n+1}) = \cdots = a_n(\lambda_n - \lambda_{n+1}) = \mathbf{0}.$$

Since the eigenvalues are distinct, we must have

$$a_1 = \cdots = a_n = \mathbf{0}.$$

Our original equation now becomes simply

$$a_{n+1}\mathbf{u}_{n+1} = \mathbf{0}.$$

Since $\mathbf{u}_{n+1}$ is nonzero, we also have $a_{n+1} = 0$. Thus, all the coefficients in our original linear combination must be zero. It follows that the vectors form a linearly independent set.

By the Principle of Mathematical Induction, we conclude that the special case holds for any number n of eigenvalues.

We are now ready to prove the general result where for each $i = 1, \ldots, n$ we are given a linearly independent subset $\{\mathbf{u}_{i1}, \ldots, \mathbf{u}_{ir_i}\}$ of the eigenspace $E_T(\lambda_i)$.

Suppose we have a linear combination of all these vectors that equals $\mathbf{0}$:

$$(a_{11}\mathbf{u}_{11} + \cdots + a_{1r_1}\mathbf{u}_{1r_1}) + \cdots + (a_{n1}\mathbf{u}_{n1} + \cdots + a_{nr_n}\mathbf{u}_{nr_n}) = \mathbf{0}.$$

Notice that the vectors

$$\mathbf{u}_1 = a_{11}\mathbf{u}_{11} + \cdots + a_{1r_1}\mathbf{u}_{1r_1}$$
$$\vdots \qquad \vdots$$
$$\mathbf{u}_n = a_{n1}\mathbf{u}_{n1} + \cdots + a_{nr_n}\mathbf{u}_{nr_n}$$

are elements of the respective eigenspaces. Our equation can be written

$$1\mathbf{u}_1 + \cdots + 1\mathbf{u}_n = \mathbf{0}.$$

Let us drop any of the terms with $\mathbf{u}_i = \mathbf{0}$. The remaining equation involves a linear combination of eigenvectors associated with distinct eigenvalues with all of the coefficients equal to 1. As we know from the special case of this theorem, these eigenvectors are linearly independent. Hence, we must have dropped all of the terms of the left-hand side of the preceding equation. That is,

$$\mathbf{0} = \mathbf{u}_1 = a_{11}\mathbf{u}_{11} + \cdots + a_{1r_1}\mathbf{u}_{1r_1}$$
$$\vdots \quad \vdots \qquad \qquad \vdots$$
$$\mathbf{0} = \mathbf{u}_n = a_{n1}\mathbf{u}_{n1} + \cdots + a_{nr_n}\mathbf{u}_{nr_n}$$

From these equations, the linear independence of the sets of vectors listed in the hypothesis of the theorem implies that all the coefficients must equal zero. Hence, the set

$$\{\mathbf{u}_{11}, \ldots, \mathbf{u}_{1r_1}, \ldots, \mathbf{u}_{n1}, \ldots, \mathbf{u}_{nr_n}\}$$

is linearly independent. ∎

Quick Example   *Which of the following matrices can be diagonalized?*

$$A = \begin{bmatrix} 4 & 1 & 1 \\ 0 & 7 & 1 \\ 0 & 0 & 2 \end{bmatrix}, \quad B = \begin{bmatrix} 4 & 1 & 1 \\ 0 & 7 & 1 \\ 0 & 0 & 7 \end{bmatrix}, \quad C = \begin{bmatrix} 4 & 1 & 1 \\ 0 & 7 & 0 \\ 0 & 0 & 7 \end{bmatrix}$$

For the matrix A, the characteristic polynomial

$$\det(\lambda I - A) = (\lambda - 4)(\lambda - 7)(\lambda - 2)$$

has three distinct roots, $\lambda = 4$, $\lambda = 7$, and $\lambda = 2$. Choose three eigenvectors associated with each of these eigenvalues. They will be linearly independent and hence form a basis for $\mathbb{R}^3$. Therefore, A can be diagonalized.

For the matrix B, the characteristic polynomial

$$\det(\lambda I - B) = (\lambda - 4)(\lambda - 7)(\lambda - 7)$$

has two distinct roots, $\lambda = 4$ and $\lambda = 7$. The repeated root has a one-dimensional eigenspace. Indeed,

$$7I - B = \begin{bmatrix} -3 & -1 & -1 \\ 0 & 0 & -1 \\ 0 & 0 & 0 \end{bmatrix} \longrightarrow \begin{bmatrix} 1 & \frac{1}{3} & 0 \\ 0 & 0 & 1 \\ 0 & 0 & 0 \end{bmatrix},$$

so the eigenspace will have a basis with one element. Since a double eigenvalue of a diagonal matrix would be associated with a two-dimensional eigenspace, it follows that B cannot be diagonalized. ∎

For the matrix C, the characteristic polynomial

$$\det(\lambda I - C) = (\lambda - 4)(\lambda - 7)(\lambda - 7)$$

has a repeated root. This time, however, the eigenvalue is associated with a two-dimensional eigenspace. Indeed,

$$7I - C = \begin{bmatrix} -3 & -1 & -1 \\ 0 & 0 & 0 \\ 0 & 0 & 0 \end{bmatrix} \longrightarrow \begin{bmatrix} 1 & \frac{1}{3} & \frac{1}{3} \\ 0 & 0 & 0 \\ 0 & 0 & 0 \end{bmatrix},$$

so the eigenspace has a basis with two elements. Together with an eigenvector associated with the eigenvalue 4, this will form a basis for $\mathbb{R}^3$. Hence, C can be diagonalized. ∎

The previous two theorems lead to the following basic condition for a linear operator to be represented by a diagonal matrix.

Theorem 8.15

A linear operator $T : V \rightarrow V$ defined on a finite-dimensional vector space V can be represented by a diagonal matrix if and only if the sum of the dimensions of all the eigenspaces is equal to dim V.

Proof Suppose T can be represented by a diagonal matrix. Let $\lambda_1, \ldots, \lambda_r$ denote the distinct entries on the main diagonal of such a matrix. By Theorem 8.13, V has a basis of eigenvectors corresponding to the eigenvalues $\lambda_1, \ldots, \lambda_r$. Of the eigenvectors in such a basis, those that correspond to λ_i for any $i = 1, \ldots, r$, must span the eigenspace $E_T(\lambda_i)$. Otherwise, by the Expansion Theorem, we could adjoin vectors to obtain a basis for $E_T(\lambda_i)$. By Theorem 8.14, these vectors could be adjoined to the original basis to yield a linearly independent set with more than dim V elements. Hence, we must have

$$\sum_{i=1}^{r} \dim E_T(\lambda_i) = \dim V.$$

A similar argument shows that there cannot be any other eigenspaces. Hence, the sum of the dimensions of all the eigenspaces is equal to dim V.

Conversely, suppose the sum of the dimensions of all the eigenspaces is equal to dim V. Choose a basis for each of the eigenspaces. By Theorem 8.14, the union of all these bases will be linearly independent. Since the set contains dim V elements, we know by Theorem 3.15 that it will be a basis for V. Hence, by Theorem 8.13, T can be represented by a diagonal matrix. ■

Exercise 8 at the end of this section asks you to prove the following corollary. It is a simple but important special case of the previous theorem.

Corollary 8.16

Suppose $T : V \to V$ is a linear operator defined on an n-dimensional vector space V. Prove that if T has n distinct eigenvalues, then T can be represented by a diagonal matrix.

Mathematical Strategy Session To summarize the results so far in this chapter, here is an outline of the procedure for trying to diagonalize an $n \times n$ matrix A.

1. Solve the characteristic equation $\det(\lambda I - A) = 0$ to find the eigenvalues of A.

2. For each real eigenvalue λ, find a basis for the corresponding eigenspace $E_A(\lambda)$.

3. If fewer than n basis vectors are obtained, A is not diagonalizable. If n basis vectors are found, use them as the columns of an $n \times n$ matrix P. The matrix P will be the matrix for changing from the basis of eigenvectors to the standard basis for $\mathbb{R}^n$.

4. $P^{-1}AP$ will be a diagonal matrix whose jj-entry is the eigenvalue with which the eigenvector in the jth column of P is associated.

In the next section we will examine an important but easily recognized type of matrix that can always be diagonalized.

Exercises 8.3

1. None of the following matrices can be diagonalized. In each case, determine whether the characteristic polynomial does not have enough real roots or whether the dimensions of the eigenspaces are not large enough.

    a. $\begin{bmatrix} 5 & 2 & 4 \\ 0 & 5 & 1 \\ 0 & 0 & 3 \end{bmatrix}$ **b.** $\begin{bmatrix} 1 & -1 & 0 \\ 1 & 1 & 0 \\ 0 & 0 & 1 \end{bmatrix}$

    c. $\begin{bmatrix} 2 & 0 & 2 \\ 0 & 2 & 0 \\ 0 & 0 & 2 \end{bmatrix}$ **d.** $\begin{bmatrix} 2 & 2 & 0 \\ 0 & 2 & 2 \\ 0 & 0 & 2 \end{bmatrix}$

2. For each of the following matrices, use the eigenvalues and eigenvectors you found in Exercise 7 of Section 8.1 to diagonalize the matrix or to conclude that the matrix is not diagonalizable.

    a. $\begin{bmatrix} 6 & -24 & -4 \\ 2 & -10 & -2 \\ 1 & 4 & 1 \end{bmatrix}$ **b.** $\begin{bmatrix} 7 & -24 & -6 \\ 2 & -7 & -2 \\ 0 & 0 & 1 \end{bmatrix}$

    c. $\begin{bmatrix} 3 & 1 & 0 \\ 0 & 3 & 1 \\ 0 & 0 & 3 \end{bmatrix}$ **d.** $\begin{bmatrix} 3 & -7 & -4 \\ -1 & 9 & 4 \\ 2 & -14 & -6 \end{bmatrix}$

    e. $\begin{bmatrix} -1 & -1 & 10 \\ -1 & -1 & 6 \\ -1 & -1 & 6 \end{bmatrix}$ **f.** $\begin{bmatrix} \frac{1}{2} & -5 & 5 \\ \frac{3}{2} & 0 & -4 \\ \frac{1}{2} & -1 & 0 \end{bmatrix}$

3. Prove Theorem 8.14 by applying the Principle of Mathematical Induction directly to the number n of distinct eigenvalues in the general case where there is a set of r_i linearly independent eigenvectors associated with each eigenvalue λ_i.

4. Consider the matrix $A = \begin{bmatrix} 16 & -6 \\ 45 & -17 \end{bmatrix}$.

    a. Find the eigenvalues of A.

    b. Find an eigenvector associated with each eigenvalue.

    c. Find a diagonal matrix similar to A.

    d. Use Exercise 17 of Section 7.1 as a simple way to compute A^9.

5. Consider the matrix $A = \begin{bmatrix} 1-p & p \\ p & 1-p \end{bmatrix}$, where $p \in \mathbb{R}$,

    a. Find the eigenvalues of A.

    b. Find an eigenvector associated with each eigenvalue.

    c. Find a diagonal matrix similar to A.

    d. Use Exercise 17 of Section 7.1 to give a derivation of the formula

$$A^n = \begin{bmatrix} \dfrac{1 + (1 - 2p)^n}{2} & \dfrac{1 - (1 - 2p)^n}{2} \\ \dfrac{1 - (1 - 2p)^n}{2} & \dfrac{1 + (1 - 2p)^n}{2} \end{bmatrix}$$

    established in Exercise 19 of Section 7.1.

6.   Suppose the $n \times n$ matrix A satisfies $A^2 = I$.

    a. Show that if λ is an eigenvalue of A, then $\lambda = 1$ or $\lambda = -1$.

    b. Show that any vector $\mathbf{v} \in \mathbb{R}^n$ can be written as a sum $\mathbf{v} = \mathbf{w} + \mathbf{x}$ for some $\mathbf{w} \in E_A(1)$ and $\mathbf{x} \in E_A(-1)$. (Suggestion: Let $\mathbf{w} = \frac{1}{2}(\mathbf{v} + A\mathbf{v})$ and $\mathbf{x} = \frac{1}{2}(\mathbf{v} - A\mathbf{v})$.)

    c. Conclude that A is diagonalizable.

7.   Suppose the $n \times n$ matrix A satisfies $A^2 = A$.

    a. Show that if λ is an eigenvalue of A, then $\lambda = 1$ or $\lambda = 0$.

    b. Show that any vector $\mathbf{v} \in \mathbb{R}^n$ can be written as a sum $\mathbf{v} = \mathbf{w} + \mathbf{x}$ for some $\mathbf{w} \in E_A(1)$ and $\mathbf{x} \in E_A(0)$. (Suggestion: Let $\mathbf{w} = A\mathbf{v}$ and $\mathbf{x} = \mathbf{v} - A\mathbf{v}$.)

    c. Conclude that A is diagonalizable.

8.   Prove Corollary 8.16.

9.   **a.** Generalize Exercise 4 of Section 8.2 to show that if two diagonal matrices have the same values occurring on their main diagonals, with each value occurring the same number of times in both matrices but with the values possibly occurring in a different order, then the two matrices are similar.

    b. Prove that if two diagonal matrices are similar, then they have the same values occurring on their main diagonals, with each value occurring the same number of times in both matrices but with the values possibly occurring in a different order.

8.4 Symmetric Matrices

We have encountered matrices that are not diagonalizable. Some matrices do not have enough real eigenvalues; others have real eigenvalues but with eigenspaces that do not provide enough linearly independent eigenvectors. In this section we will consider a class of matrices that have neither of these defects and hence can always be counted on to be diagonalizable.

Definition 8.17

A matrix is **symmetric** if and only if it equals its transpose.

Notice that symmetric matrices are necessarily square. The symmetry arises from the appearance of each entry a_{ij} in the position a_{ji} reflected across the main diagonal of the matrix.

Crossroads   Symmetric matrices arise in many areas of mathematics. For example, the conic sections you studied in analytic geometry can all be described in terms of a quadratic equation of the form

$$ax^2 + 2bxy + cy^2 + dx + ey + f = 0.$$

The terms of degree 2 can be written

$$ax^2 + 2bxy + cy^2 = \begin{bmatrix} x & y \end{bmatrix} \begin{bmatrix} a & b \\ b & c \end{bmatrix} \begin{bmatrix} x \\ y \end{bmatrix}.$$

The change of basis that diagonalizes the symmetric matrix $\begin{bmatrix} a & b \\ b & c \end{bmatrix}$ has the effect of eliminating the xy-term in the equation. In this situation, the change of basis can be accomplished by a rotation of the axes. With respect to the new coordinates, we can read off from the equation the foci, vertices, directrices, asymptotes, and other important features of the curve.

In multivariate calculus, the Hessian matrix of a function $f : \mathbb{R}^n \to \mathbb{R}$ is defined to be the $n \times n$ matrix $\left[\dfrac{\partial^2 f}{\partial x_i \partial x_j} \right]$. You are probably familiar with the theorem that if the second partial derivatives of f are continuous, then $\dfrac{\partial^2 f}{\partial x_i \partial x_j} = \dfrac{\partial^2 f}{\partial x_j \partial x_i}$. That is, the Hessian matrix is symmetric. Choosing a basis that diagonalizes the Hessian matrix essentially replaces partial derivatives with directional derivatives. In the new coordinate system, the function can be approximated by an especially simple quadratic function. This determines the behavior of f to the extent that, for example, we can often determine whether a critical point of f is a local maximum or a local minimum.

Symmetric matrices are important in statistics in the form of the covariance matrix. Here the change of basis to diagonalize the matrix leads to a decomposition of a multivariate distribution as a product of independent distributions in the directions of the new basis vectors.

You will also encounter symmetric matrices and their eigenvalues in the project on graph theory at the end of this chapter.

The fact that any symmetric matrix can be diagonalized is quite amazing. But even more is true. We can always choose the new basis to be orthonormal. This, of course, requires an inner product; we will use the standard dot product on $\mathbb{R}^n$. In this section we will use the column notation for vectors in $\mathbb{R}^n$. Then, with $\mathbf{v} \in \mathbb{R}^n$ identified with the corresponding element of $\mathbb{M}(n, 1)$, the transpose $\mathbf{v}^t$ denotes the matrix in row notation. If we further agree to identify a 1×1 matrix with a real number, we can write the dot product of two vectors $\mathbf{v}$ and $\mathbf{w}$ in $\mathbb{R}^n$ in terms of matrix multiplication: $\mathbf{v} \cdot \mathbf{w} = \mathbf{v}^t \mathbf{w}$. Having established these conventions, we can state our first result.

Theorem 8.18

Suppose A is an $n \times n$ matrix. For any vectors $\mathbf{v}$ and $\mathbf{w}$ in $\mathbb{R}^n$,

$$(A\mathbf{v}) \cdot \mathbf{w} = \mathbf{v} \cdot (A^t\mathbf{w}).$$

Proof   The proof is a simple application of the formula of part d of Theorem 7.9 concerning the transpose of a product of matrices.

$$\begin{aligned}
(A\mathbf{v}) \cdot \mathbf{w} &= (A\mathbf{v})^t \mathbf{w} \\
&= (\mathbf{v}^t A^t)\mathbf{w} \\
&= \mathbf{v}^t(A^t\mathbf{w}) \\
&= \mathbf{v} \cdot (A^t\mathbf{w}). \quad \blacksquare
\end{aligned}$$

This result immediately suggests another theorem.

Theorem 8.19

An $n \times n$ matrix A is symmetric if and only if

$$(A\mathbf{v}) \cdot \mathbf{w} = \mathbf{v} \cdot (A\mathbf{w})$$

for all vectors $\mathbf{v}, \mathbf{w} \in \mathbb{R}^n$.

Proof   If A is symmetric, then by definition $A^t = A$. Hence,

$$(A\mathbf{v}) \cdot \mathbf{w} = \mathbf{v} \cdot (A^t\mathbf{w}) = \mathbf{v} \cdot (A\mathbf{w}).$$

Conversely, suppose the formula holds for all vectors in $\mathbb{R}^n$. Since $A\mathbf{e}_j$ is the jth column of $A = [a_{ij}]$, and $a_{ij} = (A\mathbf{e}_j) \cdot \mathbf{e}_i$ is the ith entry of this column, we have

$$\begin{aligned}
a_{ij} &= (A\mathbf{e}_j) \cdot \mathbf{e}_i \\
&= \mathbf{e}_j \cdot (A\mathbf{e}_i) \\
&= (A\mathbf{e}_i) \cdot \mathbf{e}_j \\
&= a_{ji}.
\end{aligned}$$

Thus, A is symmetric. $\blacksquare$

Theorem 8.19 uses the dot product on $\mathbb{R}^n$ to give a condition for a matrix to be symmetric. We can use this condition to extend the idea of a symmetric matrix to any linear operator on an inner product space. Here is the definition followed by a theorem relating the extended idea back to matrices.

Definition 8.20

A linear operator $T : V \rightarrow V$ defined on an inner product space V is **symmetric** if and only if

$$\langle T(\mathbf{v}), \mathbf{w} \rangle = \langle \mathbf{v}, T(\mathbf{w}) \rangle$$

for all vectors $\mathbf{v}, \mathbf{w} \in V$.

Theorem 8.21

Suppose $T : V \rightarrow V$ is a symmetric linear operator defined on a finite-dimensional inner product space V. Suppose B is an orthonormal basis for V. Then the matrix of T relative to B is symmetric.

Proof   Let A be the matrix of T relative to B. That is, $[T(\mathbf{v})]_B = A[\mathbf{v}]_B$ for all $\mathbf{v} \in V$. Now an arbitrary pair of vectors in $\mathbb{R}^n$ can be written as coordinate vectors $[\mathbf{v}]_B$ and $[\mathbf{w}]_B$ for some $\mathbf{v}, \mathbf{w} \in V$. To verify the condition of Theorem 8.19, we use the result of Exercise 16a of Section 4.4 to switch between the dot product on $\mathbb{R}^n$ and the inner product on V:

$$\begin{aligned}
\left(A[\mathbf{v}]_B \right) \cdot [\mathbf{w}]_B &= [T(\mathbf{v})]_B \cdot [\mathbf{v}]_B \\
&= \langle T(\mathbf{v}), \mathbf{w} \rangle \\
&= \langle \mathbf{v}, T(\mathbf{w}) \rangle \\
&= [\mathbf{v}]_B \cdot [T(\mathbf{w})]_B \\
&= [\mathbf{v}]_B \cdot \left(A[\mathbf{w}]_B \right).
\end{aligned}$$

Thus, the matrix A is symmetric.   ■

Recall from Exercise 3 of Section 6.5 that the matrix P for changing from a basis $\{\mathbf{u}_1, \ldots, \mathbf{u}_n\}$ for $\mathbb{R}^n$ to the standard basis for $\mathbb{R}^n$ consists of the columns $\mathbf{u}_1, \ldots, \mathbf{u}_n$. Thus, if $\{\mathbf{u}_1, \ldots, \mathbf{u}_n\}$ is an orthonormal basis, then the ij-entry of the product $P^t P$ is

$$\mathbf{u}_i \cdot \mathbf{u}_j = \begin{cases} 1 & \text{if } i = j, \\ 0 & \text{if } i \neq j. \end{cases}$$

It follows that $P^t P = I$. Conversely, if an $n \times n$ matrix P satisfies the equation $P^t P = I$, then the computation of the ij-entry is 0 or 1 as required to verify that the columns of P form an orthonormal set. These simple observations lead to the introduction of a new term.

Definition 8.22

A matrix P is **orthogonal** if and only if $P^t P = I$.

It is especially easy to find the inverse of an orthogonal matrix. Simply take the transpose. As a consequence, when an orthogonal matrix P is used to change bases, a symmetric matrix A is conjugated into another symmetric matrix $P^{-1}AP$. Indeed,

$$(P^{-1}AP)^t = P^t A^t (P^{-1})^t$$
$$= P^{-1} A (P^t)^t$$
$$= P^{-1} AP.$$

We now come to the first major theorem of this section. It is a generalization of Exercise 11 of Section 8.1, in which you were asked to verify that the eigenvalues of a 2×2 symmetric matrix are real numbers. The proof of the theorem requires us to extend our results about matrices and determinants to the setting of matrices whose entries are complex numbers. In addition to some basic laws of matrix algebra, we need the fact that the determinant of the product of two matrices equals the product of their determinants. The proof of this result (Theorem 7.7) depends in turn on the effects row operations have on the determinant of a matrix (Theorem 7.4, parts b, c, and e). You are encouraged to look back over the derivations of these results. Rest assured, however, that the proofs translate word for word into proofs involving scalars that are complex numbers and matrices with complex entries. Some of the exercises at the end of this section ask you to derive a few basic results about complex numbers. You may want to use them to refresh your memory before getting involved in the details of the proof of the following theorem.

Theorem 8.23

The characteristic polynomial of an $n \times n$ symmetric matrix has n real roots.

Proof   The Fundamental Theorem of Algebra states that any polynomial of degree n has n complex roots counted according to multiplicity. Thus, the characteristic polynomial of any $n \times n$ matrix has n complex roots. We need to show for an $n \times n$ symmetric matrix that each of these roots is a real number.

Let the complex number $\alpha + \beta i$ with $\alpha, \beta \in \mathbb{R}$ be a root of the characteristic polynomial $\det(\lambda I - A)$. Since

$$\det\big((\alpha + \beta i)I - A\big) = 0,$$

it follows that

$$\det\big((\alpha + \beta i)I - A\big)\det\big((\alpha - \beta i)I - A\big) = 0.$$

A straightforward computation shows that

$$((\alpha + \beta i)I - A)((\alpha - \beta i)I - A) = (\alpha I - A)^2 + \beta^2 I.$$

Thus,

$$0 = \det\big((\alpha + \beta i)I - A\big)\det\big((\alpha - \beta i)I - A\big)$$
$$= \det\big(((\alpha + \beta i)I - A)((\alpha - \beta i)I - A)\big)$$
$$= \det\big((\alpha I - A)^2 + \beta^2 I\big).$$

By Theorem 7.11 there is a nonzero vector $\mathbf{v} \in \mathbb{R}^n$ such that $\big((\alpha I - A)^2 + \beta^2 I\big)\mathbf{v} = \mathbf{0}$. Since $\alpha I - A$ is a symmetric matrix, we have

$$0 = \big(((\alpha I - A)^2 + \beta^2 I)\mathbf{v}\big) \cdot \mathbf{v}$$
$$= \big((\alpha I - A)^2 \mathbf{v}\big) \cdot \mathbf{v} + (\beta^2 \mathbf{v}) \cdot \mathbf{v}$$
$$= \big((\alpha I - A)\mathbf{v}\big) \cdot \big((\alpha I - A)\mathbf{v}\big) + \beta^2 (\mathbf{v} \cdot \mathbf{v})$$
$$= \|(\alpha I - A)\mathbf{v}\|^2 + \beta^2 \|\mathbf{v}\|^2.$$

Since $\|\mathbf{v}\| \neq 0$, this is possible only if $\beta = 0$. Hence, the root $\alpha + \beta i = \alpha$ is a real number. ■

The second major result of this section derives its name from the set of eigenvalues of a linear operator. This set, called the **spectrum** of the linear operator, characterizes the operator much as a pattern of light-wave frequencies characterizes a chemical compound.

Theorem 8.24   Spectral Theorem

Suppose $T : V \to V$ is a symmetric linear operator defined on a finite-dimensional inner product space V. Then V has an orthonormal basis of eigenvectors of T.

Proof   We will apply the Principle of Mathematical Induction to the dimension n of the vector space V.

The empty set is an orthonormal basis for any zero-dimensional vector space. So the result holds for $n = 0$.

Suppose for some fixed integer $n > 0$ that for a symmetric linear operator on an inner product space of dimension $n - 1$, there is an orthonormal basis of eigenvectors. Let $T : V \to V$ be a symmetric linear operator on the n-dimensional inner product space V. We can apply the Gram-Schmidt orthonormalization process to get an orthonormal basis for V. By Theorem 8.21 the matrix of T relative to this basis is symmetric, and by Theorem 8.23 there is a real root λ of the characteristic polynomial of this matrix. By Exercise 15 of Section 8.2, the eigenvector associated with λ will be the coordinate vector of an eigenvector $\mathbf{v}_1$ for T associated with the eigenvalue λ. Let

$$S = \{\mathbf{v} \in V \mid \langle \mathbf{v}, \mathbf{v}_1 \rangle = 0\}.$$

By Exercises 11 and 12 of Section 4.4, S is an $(n - 1)$-dimensional subspace of V. For any $\mathbf{v} \in S$, we have

$$\langle T(\mathbf{v}), \mathbf{v}_1 \rangle = \langle \mathbf{v}, T(\mathbf{v}_1) \rangle = \langle \mathbf{v}, \lambda \mathbf{v}_1 \rangle = \lambda \langle \mathbf{v}, \mathbf{v}_1 \rangle = \lambda \cdot 0 = 0.$$

Thus, $T(\mathbf{v}) \in S$. Since T satisfies the symmetry condition of Definition 8.20 for all vectors in V, it certainly satisfies the condition for all vectors in S. Thus, the restriction of T to the domain S is a symmetric linear operator on the inner product space S of dimension $n - 1$. By the induction hypothesis, S has an orthonormal basis $\{\mathbf{v}_2, \ldots, \mathbf{v}_n\}$ of eigenvectors of T. Since the vectors in this set are orthogonal to $\mathbf{v}_1$, the set $\{\mathbf{v}_1, \mathbf{v}_2, \ldots, \mathbf{v}_n\}$ is an orthogonal set of vectors. By Theorem 4.14, this set is a linearly independent subset of V. By Theorem 3.15, $\{\mathbf{v}_1, \mathbf{v}_2, \ldots, \mathbf{v}_n\}$ is a basis for V. Once we normalize $\mathbf{v}_1$, we will have an orthonormal basis for V consisting of eigenvectors of T.

By the Principle of Mathematical Induction, the result holds for inner product spaces of any finite dimension. ■

Here is what the Spectral Theorem says for symmetric matrices.

Corollary 8.25

Suppose A is a symmetric matrix. Then there is an orthogonal matrix P such that $P^{-1}AP$ is diagonal.

Proof   Multiplication by an $n \times n$ symmetric matrix A defines a symmetric linear map $\mu_A : \mathbb{R}^n \to \mathbb{R}^n$. By the Spectral Theorem, $\mathbb{R}^n$ has an orthonormal basis of eigenvectors of μ_A. We can use these vectors as the columns of an orthogonal matrix P. Then $P^{-1}AP$ will be diagonal. ■

Although the proof of the Spectral Theorem indicates an inductive procedure for building up an orthogonal basis of eigenvectors for a symmetric linear operator, the next theorem often simplifies the process a great deal.

Theorem 8.26

Suppose a symmetric linear operator has eigenvectors $\mathbf{v}_1$ and $\mathbf{v}_2$ that are associated with distinct eigenvalues. Then $\mathbf{v}_1$ and $\mathbf{v}_2$ are orthogonal.

Proof   Let T be a symmetric linear operator. Suppose the eigenvector $\mathbf{v}_1$ is associated with the eigenvalue λ_1 and the eigenvector $\mathbf{v}_2$ is associated with the eigenvalue λ_2. Then

$$\begin{aligned} 0 &= \langle T(\mathbf{v}_1), \mathbf{v}_2 \rangle - \langle T(\mathbf{v}_1), \mathbf{v}_2 \rangle \\ &= \langle T(\mathbf{v}_1), \mathbf{v}_2 \rangle - \langle \mathbf{v}_1, T(\mathbf{v}_2) \rangle \\ &= \langle \lambda_1 \mathbf{v}_1, \mathbf{v}_2 \rangle - \langle \mathbf{v}_1, \lambda_2 \mathbf{v}_2 \rangle \\ &= \lambda_1 \langle \mathbf{v}_1, \mathbf{v}_2 \rangle - \lambda_2 \langle \mathbf{v}_1, \mathbf{v}_2 \rangle \\ &= (\lambda_1 - \lambda_2) \langle \mathbf{v}_1, \mathbf{v}_2 \rangle \end{aligned}$$

Since $\lambda_1 \neq \lambda_2$, we must have $\mathbf{v}_1 \cdot \mathbf{v}_2 = 0$. ■

Mathematical Strategy Session   The process for diagonalizing a symmetric matrix with an orthogonal change-of-basis matrix is somewhat lengthy and computationally intensive. However, it is entirely mechanical. You are encouraged to take advantage of a computer algebra system or specialized software to perform the computations. Here is a summary of the process.

With assurance from the Spectral Theorem that a symmetric matrix A is diagonalizable, we can apply the standard procedure as outlined at the end of Section 8.3 to find a basis for $\mathbb{R}^n$ consisting of eigenvectors of A. For each eigenspace apply the Gram-Schmidt process to produce an orthonormal basis of eigenvectors for that subspace. By Theorem 8.26 the set of all these basis vectors will form an orthonormal set of n vectors. Hence, we have an orthonormal basis for $\mathbb{R}^n$ consisting of eigenvectors for A. The matrix P whose columns are the coordinates of these eigenvectors is therefore an orthogonal matrix such that $P^{-1}AP$ is diagonal.

Exercises 8.4

1. Suppose P and Q are $n \times n$ symmetric matrices and $r \in \mathbb{R}$. Prove or give a counterexample for each of the following claims.

   a. $P + Q$ is symmetric.

   b. rP is symmetric.

   c. P^{-1} is symmetric.

   d. PQ is symmetric.

   e. P^t is symmetric.

2. Suppose A is an $n \times n$ matrix.

   a. Show that $A + A^t$ is symmetric.

   b. Show that AA^t is symmetric.

3. Suppose P and Q are $n \times n$ orthogonal matrices and $r \in \mathbb{R}$. Prove or give a counterexample for each of the following claims.

   a. $P + Q$ is orthogonal.

   b. rP is orthogonal.

   c. P^{-1} is orthogonal.

   d. PQ is orthogonal.

   e. P^t is orthogonal.

4. Verify that the following matrices are orthogonal.

   a. $\begin{bmatrix} \frac{1}{\sqrt{2}} & \frac{1}{\sqrt{2}} \\ -\frac{1}{\sqrt{2}} & \frac{1}{\sqrt{2}} \end{bmatrix}$

   b. $\begin{bmatrix} \frac{1}{2} & -\frac{\sqrt{3}}{2} \\ \frac{\sqrt{3}}{2} & \frac{1}{2} \end{bmatrix}$

   c. $\begin{bmatrix} 0 & 0 & 1 \\ 1 & 0 & 0 \\ 0 & 1 & 0 \end{bmatrix}$

   d. $\begin{bmatrix} \frac{3}{7} & \frac{6}{7} & -\frac{2}{7} \\ -\frac{2}{7} & \frac{3}{7} & \frac{6}{7} \\ \frac{6}{7} & -\frac{2}{7} & \frac{3}{7} \end{bmatrix}$

5. What are the possible values for the determinant of an orthogonal matrix?

6. Diagonalize the following symmetric matrices.

   a. $\begin{bmatrix} 2 & 3 \\ 3 & 2 \end{bmatrix}$

   b. $\begin{bmatrix} -1 & 4 \\ 4 & -1 \end{bmatrix}$

   c. $\begin{bmatrix} -2 & -2 & 2 \\ -2 & 1 & -4 \\ 2 & -4 & 1 \end{bmatrix}$

   d. $\begin{bmatrix} 1 & 1 & 1 \\ 1 & 1 & 1 \\ 1 & 1 & 1 \end{bmatrix}$

e. $\begin{bmatrix} 2 & 3 & 0 & 0 \\ 3 & 2 & 0 & 0 \\ 0 & 0 & -1 & 4 \\ 0 & 0 & 4 & -1 \end{bmatrix}$ **f.** $\begin{bmatrix} 9 & 0 & 0 & 0 \\ 0 & -7 & 0 & 0 \\ 0 & 0 & 5 & 0 \\ 0 & 0 & 0 & -3 \end{bmatrix}$

7. Diagonalize the matrices in the previous exercise with orthogonal matrices.

8. Let $\bar{z}$ denote the complex conjugate of the complex number z. That is, if $z = \alpha + \beta i$, where $\alpha, \beta \in \mathbb{R}$, then $\bar{z} = \alpha - \beta i$. Suppose z_1 and z_2 are complex numbers.

   a. Show that z_1 is real if and only if $z_1 = \bar{z}_1$.

   b. Show that $\overline{z_1 + z_2} = \bar{z}_1 + \bar{z}_2$.

   c. Show that $\overline{z_1 z_2} = \bar{z}_1 \bar{z}_2$.

   d. Suppose p is a polynomial with real coefficients. Show that if z is a root of p, then $\bar{z}$ is also a root of p.

9. Examine the proof of Theorem 7.7 in light of extending it to matrices with complex entries. Trace all the results used in this proof back to basic properties of real numbers. Convince yourself that these results hold for complex numbers.

10. The proof of Theorem 8.23 uses some basic laws of matrix algebra for scalars that are complex numbers and matrices whose entries are complex numbers. Examine the proof to see which of the laws are needed. Verify these laws in this setting.

8.5 Systems of Differential Equations

We have seen how the notation and algebraic operations of matrices simplify our work with systems of linear equations. In this section we want to apply the theory of matrices and determinants to systems of linear equations where the coefficients and unknowns represent functions rather than real numbers.

In your calculus course you undoubtedly encountered simple differential equations such as $y' = ay$. A solution is a function in $\mathbb{D}(\mathbb{R})$ whose derivative is equal to the constant a times the function itself. You can easily check that the function defined by e^{ax} is a solution to this equation. Furthermore, if f is any solution, then the ratio $r(x) = \dfrac{f(x)}{e^{ax}} = f(x)e^{-ax}$ satisfies

$$r'(x) = f'(x)e^{-ax} - af(x)e^{-ax} = (f'(x) - af(x))e^{-ax} = 0e^{-ax} = 0.$$

Hence, $r(x) = k$ for some constant k. It follows that $f(x) = ke^{ax}$ is a multiple of the basic solution e^{ax}.

We want to generalize this to a system of equations in which the derivatives of the unknown functions are expressed as linear combinations of the functions themselves. To begin with, we will even allow the coefficients of the linear combinations to be functions of the independent variable. The technique we develop for solving such systems, however, will be limited to the case of constant coefficients.

Here is the notation we will use in this section. For $i = 1, \ldots, n$ and $j = 1, \ldots, n$ let a_{ij} denote a real-valued function of a real variable. For a given array of functions a_{ij}, we want to find differentiable, real-valued functions $\varphi_1, \ldots, \varphi_n$ of a real variable that satisfy the system

$$\varphi_1' = a_{11}\varphi_1 + \cdots + a_{1n}\varphi_n$$

$$\vdots \qquad\qquad \vdots$$

$$\varphi_n' = a_{n1}\varphi_1 + \cdots + a_{nn}\varphi_n$$

Quick Example   *For the system of differential equations*

$$\varphi_1'(t) = t\varphi_2(t) + \varphi_3(t)$$
$$\varphi_2'(t) = \qquad\quad - \varphi_3(t)$$
$$\varphi_3'(t) = \;\; \varphi_2(t)$$

identify the coefficient functions and show that one solution consists of the functions defined by $\phi_1(t) = t\sin t$, $\phi_2(t) = \cos t$, and $\phi_3(t) = \sin t$.

The coefficient functions are

$$a_{11}(t) = 0 \qquad a_{12}(t) = t \qquad a_{13}(t) = \;\;\; 1$$
$$a_{21}(t) = 0 \qquad a_{22}(t) = 0 \qquad a_{23}(t) = -1$$
$$a_{31}(t) = 0 \qquad a_{32}(t) = 1 \qquad a_{33}(t) = \;\;\; 0$$

With the given definitions for the functions φ_1, φ_2, and φ_3, we have

$$\varphi'(t) = t\cos t + \sin t = t\varphi_2(t) + \varphi_3(t)$$
$$\varphi_2'(t) = \qquad\quad - \sin t = \qquad\quad - \varphi_3(t)$$
$$\varphi_3'(t) = \;\; \cos t \qquad\quad = \;\; \varphi_2(t) \qquad\qquad\blacksquare$$

We can simplify the notation by extending the concepts of matrix algebra to matrices whose entries depend on the value of some real number. That is, the entries will be real-valued functions of a real variable. In particular, we let $\varphi = \begin{bmatrix} \varphi_1 \\ \vdots \\ \varphi_n \end{bmatrix}$ denote the $n \times 1$ matrix function whose entries are the unknown functions in the system of differential equations, and we let $A = [a_{ij}]$ denote the $n \times n$ matrix function whose entries are the coefficient functions for the system. With our usual identification of $\mathbb{M}(n, 1)$ and $\mathbb{R}^n$, we have $\varphi : \mathbb{R} \rightarrow \mathbb{R}^n$ and $A : \mathbb{R} \rightarrow \mathbb{M}(n, n)$. Thus, for any real number t, we have an ordinary vector $\varphi(t) \in \mathbb{R}^n$ and an ordinary matrix $A(t) \in \mathbb{M}(n, n)$.

Since the laws of matrix algebra hold when the matrices are evaluated at any real number, we can rest assured that the corresponding laws hold for matrix functions. In particular, the set of all $m \times n$ matrix functions forms a vector space. Furthermore, the laws corresponding to matrix multiplication (Theorem 5.4) hold for matrix functions.

A **continuous** matrix-valued function is one for which the entries are continuous real-valued functions. A **differentiable** matrix-valued function is one for which the entries are differentiable real-valued functions.

We define the **derivative** of a matrix-valued function to be the matrix function formed by differentiating the entries of the original matrix function. This operation is denoted by the prime notation or, alternatively, by the differential notation. For example, if $\varphi = \begin{bmatrix} \varphi_1 \\ \vdots \\ \varphi_n \end{bmatrix}$ is a differentiable vector function, then

$$\varphi'(t) = \frac{d}{dt}\varphi(t) = \begin{bmatrix} \varphi_1'(t) \\ \vdots \\ \varphi_n'(t) \end{bmatrix}.$$

With this notation, we can write our earlier example as

$$\frac{d}{dt}\begin{bmatrix} \varphi_1(t) \\ \varphi_2(t) \\ \varphi_3(t) \end{bmatrix} = \begin{bmatrix} 0 & t & 1 \\ 0 & 0 & -1 \\ 0 & 1 & 0 \end{bmatrix}\begin{bmatrix} \varphi_1(t) \\ \varphi_2(t) \\ \varphi_3(t) \end{bmatrix},$$

or more simply as

$$\varphi' = A\varphi.$$

We will refer to an equation of this form as a **linear differential equation**. In general, such a matrix equation represents a system of n linear differential equations involving n real-valued functions.

The following theorem presents some basic rules of differentiation in the context of matrix functions. These formulas follow directly from the definition of the matrix operations in terms of the entries and the corresponding results for derivatives of real-valued functions. You are asked to verify these results in Exercise 4 at the end of this section.

Theorem 8.27

Suppose $A : \mathbb{R} \to \mathbb{M}(m, n)$, $B : \mathbb{R} \to \mathbb{M}(m, n)$, and $C : \mathbb{R} \to \mathbb{M}(n, p)$ are differentiable matrix functions, and r is a real number. Then
  a. $(A + B)' = A' + B'$
  b. $(rA)' = rA'$
  c. $(AC)' = A'C + AC'$

We also need a formula for the derivative of the determinant of a differentiable matrix function. Exercise 10 at the end of this section gives suggestions for a proof that is less formal but perhaps more enlightening than the proof given here.

Theorem 8.28

Suppose $A : \mathbb{R} \to \mathbb{M}(n, n)$ is a differentiable matrix function. Let $A = [a_{ij}]$. Then

$$\frac{d}{dt}\det A(t) = \det\begin{bmatrix} a_{11}' & \cdots & a_{1n}' \\ a_{21} & \cdots & a_{2n} \\ \vdots & & \vdots \\ a_{n1} & \cdots & a_{nn} \end{bmatrix} + \det\begin{bmatrix} a_{11} & \cdots & a_{1n} \\ a_{21}' & \cdots & a_{2n}' \\ a_{31} & \cdots & a_{3n} \\ \vdots & & \vdots \\ a_{n1} & \cdots & a_{nn} \end{bmatrix}$$

$$+ \cdots + \det \begin{bmatrix} a_{11} & \cdots & a_{1n} \\ a_{21} & \cdots & a_{2n} \\ \vdots & & \vdots \\ a_{(n-1)1} & \cdots & a_{(n-1)n} \\ a'_{n1} & \cdots & a'_{nn} \end{bmatrix}.$$

Proof   We use induction on the size of the matrix. The result is trivial for a 1×1 matrix. Suppose the result holds for matrix functions of size $(n-1) \times (n-1)$.

$$\frac{d}{dt} \det A(t) = \frac{d}{dt} \sum_{j=1}^{n} (-1)^{1+j} a_{1j} \det A_{1j}$$

$$= \sum_{j=1}^{n} (-1)^{1+j} a'_{1j} \det A_{1j} + \sum_{j=1}^{n} (-1)^{1+j} a_{1j} \frac{d}{dt} (\det A_{1j})$$

$$= \det \begin{bmatrix} a'_{11} & \cdots & a'_{1n} \\ a_{21} & \cdots & a_{2n} \\ \vdots & & \vdots \\ a_{n1} & \cdots & a_{nn} \end{bmatrix} + \sum_{j=1}^{n} (-1)^{1+j} a_{1j} \det \begin{bmatrix} a'_{21} & \cdots & a'_{2(j-1)} & a'_{2(j+1)} & \cdots & a'_{2n} \\ a_{31} & \cdots & a_{3(j-1)} & a_{3(j+1)} & \cdots & a_{3n} \\ \vdots & & \vdots & \vdots & & \vdots \\ a_{n1} & \cdots & a_{n(j-1)} & a_{n(j+1)} & \cdots & a_{nn} \end{bmatrix}$$

$$+ \cdots + \sum_{j=1}^{n} (-1)^{1+j} a_{1j} \det \begin{bmatrix} a_{21} & \cdots & a_{2(j-1)} & a_{2(j+1)} & \cdots & a_{2n} \\ \vdots & & \vdots & \vdots & & \vdots \\ a_{(n-1)1} & \cdots & a_{(n-1)(j-1)} & a_{(n-1)(j+1)} & \cdots & a_{(n-1)n} \\ a'_{n1} & \cdots & a'_{n(j-1)} & a'_{n(j+1)} & \cdots & a'_{nn} \end{bmatrix}$$

$$= \det \begin{bmatrix} a'_{11} & \cdots & a'_{1n} \\ a_{21} & \cdots & a_{2n} \\ \vdots & & \vdots \\ a_{n1} & \cdots & a_{nn} \end{bmatrix} + \det \begin{bmatrix} a_{11} & \cdots & a_{1n} \\ a'_{21} & \cdots & a'_{2n} \\ a_{31} & \cdots & a_{3n} \\ \vdots & & \vdots \\ a_{n1} & \cdots & a_{nn} \end{bmatrix} + \cdots + \det \begin{bmatrix} a_{11} & \cdots & a_{1n} \\ \vdots & & \vdots \\ a_{(n-1)1} & \cdots & a_{(n-1)n} \\ a'_{n1} & \cdots & a'_{nn} \end{bmatrix}. \quad \blacksquare$$

Having established some notation and basic facts, we now turn our attention to the problem of solving linear systems of differential equations. The following theorem guarantees that a system with a continuous coordinate matrix always has a solution and furthermore that the solution is unique. This is a standard theorem in a course on differential equations.

Theorem 8.29   Existence and Uniqueness of Solutions for a System of Linear Differential Equations

Suppose $A : [c, d] \to \mathbb{M}(n, n)$ is a continuous matrix function defined on some interval $[c, d]$ with $c < d$. Let $t_0 \in [c, d]$. For any $\mathbf{v}_0 \in \mathbb{R}^n$ there is exactly one differentiable vector function $\boldsymbol{\varphi} : [c, d] \to \mathbb{R}^n$ that satisfies

$$\boldsymbol{\varphi}' = A\boldsymbol{\varphi} \qquad \text{and} \qquad \boldsymbol{\varphi}(t_0) = \mathbf{v}_0.$$

The point t_0 in the interval $[c, d]$ is commonly chosen to be the left-hand endpoint c. In this case the specified value of $\varphi(c)$ is called the **initial condition**. The problem of finding a vector function φ that satisfies the initial condition as well as the differential equation is called an **initial-value problem**.

Rather than delve into the proof of the Existence and Uniqueness Theorem, we proceed with investigating its application in the proof of the next theorem. This result is an excellent illustration of the algebraic aspects of differential equations.

Theorem 8.30

Suppose $A : [c, d] \to \mathbb{M}(n, n)$ is a continuous matrix function defined on some interval $[c, d]$ with $c < d$. The set of all differentiable functions $\varphi : [c, d] \to \mathbb{R}^n$ that satisfy $\varphi' = A\varphi$ is an n-dimensional subspace of all vector functions from $[c, d]$ into $\mathbb{R}^n$.

Proof   In Exercise 1 at the end of this section, you are asked to carry out the straightforward application of the Subspace Theorem to verify that the solution set is a subspace.

The heart of the proof of this theorem lies in the construction of a basis for the solution space. For $i = 1, \ldots, n$ let φ_i denote the solution to the differential equation that satisfies the initial condition $\varphi_i(c) = \mathbf{e}_i$ where $\mathbf{e}_i$ is the ith standard basis element for $\mathbb{R}^n$.

It is easy to show that the set $\{\varphi_1, \ldots, \varphi_n\}$ is linearly independent. Indeed, if $r_1\varphi_1 + \cdots + r_n\varphi_n = \mathbf{0}$, then when we evaluate both sides at the point c, we have

$$\mathbf{0} = \mathbf{0}(c)$$
$$= r_1\varphi_1(c) + \cdots + r_n\varphi_n(c)$$
$$= r_1\mathbf{e}_1 + \cdots + r_n\mathbf{e}_n.$$

This yields the desired result that $r_1 = \cdots = r_n = 0$.

To show that $\{\varphi_1, \ldots, \varphi_n\}$ spans the subspace of solutions, let φ denote an arbitrary solution. Since $\varphi(c) \in \mathbb{R}^n$, we can find real numbers $r_1, \ldots, r_n$ so that $\varphi(c) = r_1\mathbf{e}_1 + \cdots + r_n\mathbf{e}_n$. Now $r_1\varphi_1 + \cdots + r_n\varphi_n$ is a solution to the differential equation whose value at c is

$$r_1\varphi_1(c) + \cdots + r_n\varphi_n(c) = r_1\mathbf{e}_1 + \cdots + r_n\mathbf{e}_n$$
$$= \varphi(c).$$

By the uniqueness of solutions to the initial-value problem, we conclude that $\varphi = r_1\varphi_1 + \cdots + r_n\varphi_n$.

This establishes that $\{\varphi_1, \ldots, \varphi_n\}$ is a basis for the solution space, and hence that this space is of dimension n.   ∎

Theorem 8.30 shows that we need to find n linearly independent solutions to a linear system of n differential equations in order to have a basis for the complete solution space. We can use determinants to detect whether a set of vector functions is linearly independent.

Definition 8.31

Suppose X is any set and n is a natural number. The **Wronskian** of the vector functions $\varphi_i : X \to \mathbb{R}^n$ for $i = 1, \ldots, n$ is the function $W : \mathbb{R} \to \mathbb{R}$ defined as the determinant of the matrix whose columns consist of $\varphi_1, \ldots, \varphi_n$. We write

$$W(t) = \det[\varphi_1(t) \; \cdots \; \varphi_n(t)].$$

From our work with determinants and linear independence of sets of functions, it is easy to see that if the Wronskian is nonzero at some point, then the vector functions are linearly independent. For vector functions that are solutions of a system of linear differential equations, this leads to the following result. Exercise 13 at the end of this section asks you to write out the details of the proof.

Theorem 8.32

Suppose $A : [c, d] \to \mathbb{M}(n, n)$ is a continuous matrix function. For $i = 1, \ldots, n$, suppose $\varphi_i : [c, d] \to \mathbb{R}^n$ are solutions to the differential equation $\varphi' = A\varphi$. If the Wronskian is nonzero at some point of $[c, d]$, then $\{\varphi_1, \ldots, \varphi_n\}$ is a basis for the solution space of this differential equation.

Exercise 11 at the end of this section gives an example of a linearly independent set of vector functions whose Wronskian is always equal to zero. For linearly independent vector functions that are solutions of a linear differential equation, however, the Wronskian must be nonzero not just at some point of the interval $[c, d]$, but at every point. Here is a complete statement of this surprisingly strong result.

Theorem 8.33

Suppose $A : [c, d] \to \mathbb{M}(n, n)$ is a continuous matrix function. For $i = 1, \ldots, n$, suppose $\varphi_i : [c, d] \to \mathbb{R}^n$ are solutions to the differential equation $\varphi' = A\varphi$. If the Wronskian is equal to zero at some point of $[c, d]$, then $\{\varphi_1, \ldots, \varphi_n\}$ is linearly dependent.

Proof   Suppose the Wronskian is equal to zero at some point $t_0 \in [c, d]$. Then the columns of the matrix $[\varphi_1(t_0) \; \cdots \; \varphi_n(t_0)]$ are linearly dependent. That is, there are constants $r_1, \ldots, r_n$, not all of which are zero, such that

$$r_1\varphi_1(t_0) + \cdots + r_n\varphi_n(t_0) = \mathbf{0}.$$

Since $r_1\varphi_1 + \cdots + r_n\varphi_n$ and the constant function $\mathbf{0}$ are both solutions of the differential equation $\varphi' = A\varphi$ with $\varphi(t_0) = \mathbf{0}$, the uniqueness statement of Theorem 8.29 implies that $r_1\varphi_1(t) + \cdots + r_n\varphi_n(t) = \mathbf{0}$ for all $t \in [c, d]$. Thus, we have a nontrivial linear combination of $\varphi_1, \ldots, \varphi_n$ equal to the zero function. Hence, $\{\varphi_1, \ldots, \varphi_n\}$ is a linearly dependent set.   ■

Of course, if the set of functions $\{\varphi_1, \ldots, \varphi_n\}$ is linearly dependent, then the Wronskian is equal to zero at all points of $[c, d]$. Thus, if the Wronskian is zero at any one point of $[c, d]$, then it is zero at all points of $[c, d]$. This fact is also a consequence of the following formula, which relates the value of the Wronskian at any two points of $[c, d]$. Simply notice that the exponential function is never zero.

Theorem 8.34   Abel's Formula

Suppose $A : [c, d] \to \mathbb{M}(n, n)$ is a continuous matrix function. For $j = 1, \ldots, n$, suppose $\varphi_j : [c, d] \to \mathbb{R}^n$ are solutions to the differential equation $\varphi' = A\varphi$. Then for $t \in [c, d]$, the Wronskian of these solutions satisfies

$$W(t) = W(c) \exp\left(\int_c^t \operatorname{tr} A(s)\, ds\right).$$

Proof For $j = 1, \ldots, n$, we introduce notation for the coordinates of φ_j by letting

$$\varphi_j(t) = \begin{bmatrix} \varphi_{1j}(t) \\ \vdots \\ \varphi_{nj}(t) \end{bmatrix}.$$ The main work consists of computing the derivative of the Wron-

skian. The key result is the fact established as part e of Theorem 7.4 that the value of a determinant does not change when a multiple of one row is added to another row. This comes into play immediately after we apply Theorem 8.28 to expand $W'(t)$ and use the fact that the ith entry of $\varphi'_j = A\varphi_j$ is $\varphi'_{ij}(t) = a_{i1}(t)\varphi_{1j}(t) + \cdots + a_{in}(t)\varphi_{nj}(t)$. Now

$$W'(t) = \frac{d}{dt}(\det[\varphi_1(t) \cdots \varphi_n(t)])$$

$$= \sum_{i=1}^n \det \begin{bmatrix} \varphi_{11} & \cdots & \varphi_{1n}(t) \\ \vdots & & \vdots \\ \varphi'_{i1} & \cdots & \varphi'_{in}(t) \\ \vdots & & \vdots \\ \varphi_{n1} & \cdots & \varphi_{nn}(t) \end{bmatrix}$$

$$= \sum_{i=1}^n \det \begin{bmatrix} \varphi_{11} & \cdots & \varphi_{1n}(t) \\ \vdots & & \vdots \\ a_{i1}(t)\varphi_{11}(t) + \cdots + a_{in}(t)\varphi_{n1}(t) & \cdots & a_{i1}(t)\varphi_{1n}(t) + \cdots + a_{in}(t)\varphi_{nn}(t) \\ \vdots & & \vdots \\ \varphi_{n1} & \cdots & \varphi_{nn}(t) \end{bmatrix}$$

$$= \sum_{i=1}^n \det \begin{bmatrix} \varphi_{11} & \cdots & \varphi_{1n}(t) \\ \vdots & & \vdots \\ a_{ii}(t)\varphi_{i1}(t) & \cdots & a_{ii}(t)\varphi_{in}(t) \\ \vdots & & \vdots \\ \varphi_{n1} & \cdots & \varphi_{nn}(t) \end{bmatrix}$$

$$= \sum_{i=1}^{n} a_{ii}(t) \det[\boldsymbol{\varphi}_1(t) \cdots \boldsymbol{\varphi}_n(t)]$$

$$= (\operatorname{tr} A(t))W(t).$$

By Exercise 2 at the end of this section, the solution to this differential equation is given by

$$W(t) = W(c)\exp\left(\int_c^t \operatorname{tr} A(s)\,ds\right). \quad \blacksquare$$

At long last we are ready to look at a technique for solving a linear system $\boldsymbol{\varphi}' = A\boldsymbol{\varphi}$ of differential equations. For the remainder of this section, we will restrict our attention to the case where the entries of the coefficient matrix are constant functions. That is, we will assume $A \in \mathbb{M}(n, n)$ is an ordinary matrix of real numbers.

If the coefficient matrix is a diagonal matrix, the differential equation

$$\begin{bmatrix} \varphi_1' \\ \vdots \\ \varphi_n' \end{bmatrix} = \begin{bmatrix} \lambda_1 & \cdots & 0 \\ \vdots & \ddots & \vdots \\ 0 & \cdots & \lambda_n \end{bmatrix} \begin{bmatrix} \varphi_1 \\ \vdots \\ \varphi_n \end{bmatrix}$$

reduces to solving the equations $\varphi_i' = \lambda_i\varphi_i$ for $i = 1, \ldots, n$. Thus, in this case, $e^{\lambda_i t}\mathbf{e}_i$ for $i = 1, \ldots, n$ are n solutions. Since the Wronskian of these solutions is the product $e^{\lambda_1 t} \cdots e^{\lambda_n t}$, which is not equal to 0, these solutions form a basis for the solution space.

The case where the coefficient matrix A is diagonalizable is nearly as easy. Suppose P is an invertible matrix such that $P^{-1}AP$ is diagonal. For any vector function $\boldsymbol{\psi}$ that satisfies $\boldsymbol{\psi}' = P^{-1}AP\boldsymbol{\psi}$, we have (by Exercise 5 at the end of this section)

$$(P\boldsymbol{\psi})' = P\boldsymbol{\psi}' = AP\boldsymbol{\psi}.$$

Thus, $\boldsymbol{\varphi} = P\boldsymbol{\psi}$ is a solution to the original equation $\boldsymbol{\varphi}' = A\boldsymbol{\varphi}$. Therefore, we can solve the differential equation $\boldsymbol{\varphi}' = A\boldsymbol{\varphi}$ once we find eigenvalues $\lambda_1, \ldots, \lambda_n$ of A and associated eigenvectors $\mathbf{v}_1, \ldots, \mathbf{v}_n$ that form a basis for $\mathbb{R}^n$. Then the vector functions $\boldsymbol{\varphi}_i = e^{\lambda_i t}\mathbf{v}_i$ for $i = 1, \ldots, n$ form a basis for the solution space. Indeed, with the eigenvectors as columns of the change-of-basis matrix $P = [\mathbf{v}_1 \cdots \mathbf{v}_n]$, we have

$$P(e^{\lambda_i t}\mathbf{e}_i) = e^{\lambda_i t}(P\mathbf{e}_i) = e^{\lambda_i t}\mathbf{v}_i = \boldsymbol{\varphi}_i.$$

These solutions form a basis for the solution space since the Wronskian is nonzero:

$$W(t) = \det[\boldsymbol{\varphi}_1 \cdots \boldsymbol{\varphi}_n]$$

$$= \det\left(P\begin{bmatrix} e^{\lambda_1 t} & \cdots & 0 \\ \vdots & \ddots & \vdots \\ 0 & \cdots & e^{\lambda_n t} \end{bmatrix}\right)$$

$$= (\det P)\det\begin{bmatrix} e^{\lambda_1 t} & \cdots & 0 \\ \vdots & \ddots & \vdots \\ 0 & \cdots & e^{\lambda_n t} \end{bmatrix}$$

$$\neq 0.$$

Quick Example   *Find a basis for the solution space of the system*

$$\varphi_1' = -2\varphi_1 + \varphi_2$$
$$\varphi_2' = 4\varphi_1 + \varphi_2$$

Find the solution that satisfies the initial condition $\varphi(0) = \begin{bmatrix} 7 \\ 3 \end{bmatrix}$.

The coefficient matrix of the system is $A = \begin{bmatrix} -2 & 1 \\ 4 & 1 \end{bmatrix}$. The characteristic polynomial is

$$\det(\lambda I - A) = \det \begin{bmatrix} \lambda + 2 & -1 \\ -4 & \lambda - 1 \end{bmatrix}$$

$$= (\lambda + 2)(\lambda - 1) - 4$$
$$= \lambda^2 + \lambda - 6$$
$$= (\lambda - 2)(\lambda + 3).$$

For the eigenvalue $\lambda = 2$, the matrix $2I - A = \begin{bmatrix} 4 & -1 \\ -4 & 1 \end{bmatrix}$ reduces to $\begin{bmatrix} 1 & -\frac{1}{4} \\ 0 & 0 \end{bmatrix}$, which yields the eigenvector $\begin{bmatrix} 1 \\ 4 \end{bmatrix}$.

For the eigenvalue $\lambda = -3$, the matrix $-3I - A = \begin{bmatrix} -1 & -1 \\ -4 & -4 \end{bmatrix}$ reduces to $\begin{bmatrix} 1 & 1 \\ 0 & 0 \end{bmatrix}$, which yields the eigenvector $\begin{bmatrix} 1 \\ -1 \end{bmatrix}$.

Thus, the vector functions defined by

$$\varphi_1(t) = e^{2t} \begin{bmatrix} 1 \\ 4 \end{bmatrix} = \begin{bmatrix} e^{2t} \\ 4e^{2t} \end{bmatrix} \quad \text{and} \quad \varphi_2(t) = e^{-3t} \begin{bmatrix} 1 \\ -1 \end{bmatrix} = \begin{bmatrix} e^{-3t} \\ -e^{-3t} \end{bmatrix}$$

form a basis for the solution space.

We can now determine the coefficients r_1 and r_2 of the linear combination $\varphi(t) = r_1\varphi_1(t) + r_2\varphi_2(t)$ that satisfies the initial condition. The requirement is that

$$\begin{bmatrix} 7 \\ 3 \end{bmatrix} = r_1\varphi_1(0) + r_2\varphi_2(0) = r_1 \begin{bmatrix} 1 \\ 4 \end{bmatrix} + r_2 \begin{bmatrix} 1 \\ -1 \end{bmatrix} = \begin{bmatrix} 1 & 1 \\ 4 & -1 \end{bmatrix} \begin{bmatrix} r_1 \\ r_2 \end{bmatrix}.$$

Notice the appearance of the change-of-basis matrix that diagonalizes A. Since this is invertible, we are assured that there exist unique values for r_1 and r_2. In this case, the solution $r_1 = 2$ and $r_2 = 5$ leads to the solution

$$\varphi(t) = 2 \begin{bmatrix} e^{2t} \\ 4e^{2t} \end{bmatrix} + 5 \begin{bmatrix} e^{-3t} \\ -e^{-3t} \end{bmatrix} = \begin{bmatrix} 2e^{2t} + 5e^{-3t} \\ 8e^{2t} - 5e^{-3t} \end{bmatrix}$$

of the initial-value problem. ■

More complicated situations arise when the coefficient matrix is not diagonalizable or when the system is not homogeneous. Nevertheless, the concepts of linear algebra we have developed in this section form a foundation for much of the further work with systems of differential equations. Watch for these techniques when you take a course in differential equations.

Exercises 8.5

1. By Exercise 12 of Section 1.7, the set of all functions from an interval $[c, d]$ with $c < d$ to $\mathbb{R}^n$ forms a vector space. Show that the solution set described in Theorem 8.30 is a subspace of this vector space.

2. Consider the differential equation $y' = a(x)y$, where $a : [c, d] \to \mathbb{R}$ is a continuous function.

    a. Show that the function defined by $y(x) = \exp\left(\int_c^x a(t)\, dt\right)$ is a solution of the differential equation.

    b. Show that any solution of the differential equation is a constant multiple of this basic solution.

    c. Show for any $y_0 \in \mathbb{R}$ that the initial-value problem

$$y' = a(x)y, \quad y(c) = y_0$$

    has a unique solution $y(x) = y_0 \exp\left(\int_c^x a(t)\, dt\right)$.

3. A **second-order homogeneous linear differential equation** has the form

$$y'' + a(x)y' + b(x)y = 0$$

where the real-valued functions a and b are defined on some interval $[c, d]$ with $c < d$. Show that such an equation can be reformulated as a system of two linear differential equations involving the unknown functions $\varphi_1 = y$, $\varphi_2 = y'$, and their first derivatives. Generalize this result to show that an nth-order linear homogeneous differential equation is equivalent to a first-order system of n linear differential equations.

4. Verify the three derivative formulas of Theorem 8.27.

5. State and prove simplified versions of the product rule for differentiating matrix functions in the cases where the left or right factor is a constant matrix.

6. Suppose $A : \mathbb{R} \to \mathbb{M}(n, n)$ is a continuous matrix function. Suppose for some $t_0 \in \mathbb{R}$ that $A(t_0)$ is invertible. Show that there is an open interval containing t_0 such that for any t in the interval, $A(t)$ is invertible.

7. Suppose $A : \mathbb{R} \to \mathbb{M}(n, n)$ is a differentiable matrix function. Suppose for some $t_0 \in \mathbb{R}$ that $A(t_0)$ is invertible. Prove that on some open interval containing t_0 the inverse-matrix function A^{-1} is differentiable and that

$$(A^{-1})' = -A^{-1}A'A^{-1}.$$

8. Give an example of a differentiable matrix function $A : \mathbb{R} \to \mathbb{M}(n, n)$ such that $AA' \neq A'A$.

9. Investigate the possibility of a power rule for the derivative of A^n.

10. By Exercise 9 of Section 7.2 the determinant of an $n \times n$ matrix function A is the sum of terms consisting of products of the form $a_1 a_2 \cdots a_n$, where a_i

is one of the entries in row i of A. Show that $(a_1a_2\cdots a_n)' = a_1'a_2\cdots a_n + a_1a_2'\cdots a_n + \cdots + a_1a_2\cdots a_n'$. Thus, the derivative of each term of $\det A$ is a sum of n terms each involving the derivative of an entry from a different row of A. Regroup the terms to obtain the result of Theorem 8.28.

11. Let $\varphi_1 : [-1, 1] \to \mathbb{R}^2$ be defined by

$$\varphi_1(t) = \begin{cases} \mathbf{e}_1 & \text{if } t < 0 \\ \mathbf{0} & \text{if } t \geq 0 \end{cases}$$

and let $\varphi_2 : [-1, 1] \to \mathbb{R}^2$ be defined by

$$\varphi_2(t) = \begin{cases} \mathbf{0} & \text{if } t < 0 \\ \mathbf{e}_2 & \text{if } t \geq 0 \end{cases}$$

Show that the Wronskian $W(t) = \det[\, \varphi_1(t) \quad \varphi_2(t)\,]$ is equal to zero for all $t \in [-1, 1]$. Show that $\{\varphi_1, \varphi_2\}$ is nevertheless a linearly independent set.

12. Give an example of a pair of functions $\varphi_1 : [0, 1] \to \mathbb{R}^2$ and $\varphi_2 : [0, 1] \to \mathbb{R}^2$ such that the Wronskian is zero at some points of $[0, 1]$ and nonzero at other points of $[0, 1]$.

13. Write out the details of the proof of Theorem 8.32.

14. Use the technique developed in this section to find a basis for the solution space of the following systems of differential equations.

a. $\begin{aligned} y_1' &= 5y_1 \\ y_2' &= -4y_2 \end{aligned}$

b. $\begin{aligned} y_1' &= -2y_1 - 2y_2 \\ y_2' &= 3y_1 + 5y_2 \end{aligned}$

c. $\varphi' = \begin{bmatrix} -1 & 1 & 0 \\ 0 & -2 & 0 \\ 0 & -6 & 1 \end{bmatrix} \varphi$

d. $\begin{aligned} y_1' &= 3y_1 - 3y_2 - 2y_3 \\ y_2' &= -y_1 + 5y_2 + 2y_3 \\ y_3' &= y_1 - 3y_2 \end{aligned}$

15. For each of the systems of differential equations in the previous exercise, find the solution that satisfies the respective initial condition.

a. $\begin{bmatrix} y_1(3) \\ y_2(3) \end{bmatrix} = \begin{bmatrix} 1 \\ 2 \end{bmatrix}$

b. $\begin{bmatrix} y_1(0) \\ y_2(0) \end{bmatrix} = \begin{bmatrix} 2 \\ -3 \end{bmatrix}$

c. $\varphi(0) = \begin{bmatrix} 1 \\ 1 \\ 1 \end{bmatrix}$

d. $\begin{bmatrix} y_1(-1) \\ y_2(-1) \\ y_3(-1) \end{bmatrix} = \begin{bmatrix} 4 \\ 0 \\ -1 \end{bmatrix}$

16. The technique for solving linear systems of differential equations extends to the setting where we consider scalars to be complex numbers and admit eigenvectors with complex entries. Illustrate this extension in finding a basis for the solution space of the system

$$\begin{aligned} y_1' &= -y_2 \\ y_2' &= y_1 \end{aligned}$$

Use Euler's formula $e^{\alpha+i\beta} = e^{\alpha}(\cos \beta + i \sin \beta)$, where $\alpha, \beta \in \mathbb{R}$, to find linear combinations (with complex coefficients) of the solutions that yield a basis consisting of real-valued functions.

Project: Graph Theory

Although mathematicians have studied graphs for over 250 years, only recently have the applications of graphs to computer science and mathematical models of discrete phenomena resulted in intense research in graph theory. This project introduces you to this fascinating branch of mathematics. The use of eigenvalues in graph theory will give you an alternative way of looking at this concept. The article "Applications of Graph Theory in Linear Algebra" by Michael Doob in the March 1984 issue of *Mathematics Magazine* examines additional relations between graph theory and linear algebra.

Graph theory is the study of the combinatorial properties of collections of points for which certain pairs are designated as being adjacent. More formally, we say that a **graph** consists of a finite set V of points and a set E of unordered pairs of distinct elements of V. We call the elements of V the **vertices**, and we often represent a vertex by a point in the plane. We call the elements of E the **edges** of the graph and represent an edge by a line segment between the points that represent the two vertices involved. In such a representation, points of intersection among the line segments other than those that define the graph are artifacts and are not to be considered part of the graph. We say that an edge $\{x, y\}$ **joins** two vertices x and y. Also, two vertices are **adjacent** if and only if there is an edge that joins them.

1. What are the minimum and maximum number of edges possible for a graph with n vertices? A graph with the maximum number of edges is called a **complete** graph on n vertices. Give a definition of the concept of isomorphism for graphs. Based on your definition, classify all graphs with four or fewer vertices.

2. Propose a reasonable definition of a **path** in a graph from one vertex to another. Propose a definition of a **connected** graph. For a set of n vertices, suppose that edges are inserted at random. In terms of the likelihood of an edge existing between two vertices, investigate the probability that the graph is connected. Write a computer simulation of the process of inserting edges to obtain an estimate of this probability.

3. Define the **degree** of a vertex to be the number of other vertices adjacent to it. Show that in any graph, the number of vertices of odd degree is even. Show that a graph contains a path that passes through each vertex exactly once if and only if the number of vertices of odd degree is 0 or 2. Leonhard Euler founded the subject of graph theory when, in 1735, he used this result to solve the popular problem of the Seven Bridges of Königsberg. You may enjoy reading his argument in volume 1 of *The World of Mathematics*, edited by James R. Newman (New York: Simon & Schuster, 1956, pp. 570–580).

4. Suppose a graph with v vertices and e edges is represented in the plane by points and line segments that intersect only at their endpoints. Let r denote the number of bounded regions formed in the plane by these line segments. Show that the alternating sum $r - e + v$ (known as the **Euler characteristic**) does not depend on the graph or the particular representation of the graph in the plane.

5. For a graph with an ordered set of n vertices, the **adjacency matrix** is the $n \times n$ matrix $A = [a_{ij}]$, where

$$a_{ij} = \begin{cases} 1 & \text{if vertex } i \text{ is adjacent to vertex } j, \\ 0 & \text{otherwise.} \end{cases}$$

Compute the adjacency matrices of some simple graphs. The **Peterson graph** has the adjacency matrix

$$\begin{bmatrix} 0 & 1 & 0 & 0 & 1 & 1 & 0 & 0 & 0 & 0 \\ 1 & 0 & 1 & 0 & 0 & 0 & 1 & 0 & 0 & 0 \\ 0 & 1 & 0 & 1 & 0 & 0 & 0 & 1 & 0 & 0 \\ 0 & 0 & 1 & 0 & 1 & 0 & 0 & 0 & 1 & 0 \\ 1 & 0 & 0 & 1 & 0 & 0 & 0 & 0 & 0 & 1 \\ 1 & 0 & 0 & 0 & 0 & 0 & 0 & 1 & 1 & 0 \\ 0 & 1 & 0 & 0 & 0 & 0 & 0 & 0 & 1 & 1 \\ 0 & 0 & 1 & 0 & 0 & 1 & 0 & 0 & 0 & 1 \\ 0 & 0 & 0 & 1 & 0 & 1 & 1 & 0 & 0 & 0 \\ 0 & 0 & 0 & 0 & 1 & 0 & 1 & 1 & 0 & 0 \end{bmatrix}.$$

Show that this matrix contains enough information to reconstruct the graph given an ordered set of 10 vertices. In particular, draw a simple representation of the Peterson graph as a five-pointed star in a pentagon, with vertices of the star joined to corresponding vertices of the pentagon.

What properties must a matrix have in order for it to be the adjacency matrix of a graph? Show that any matrix with these properties determines a unique graph on a given ordered set of vertices.

Let A be the adjacency matrix of a graph with n vertices. Choose a vector $\mathbf{v} \in \mathbb{R}^n$ and label the ith vertex with the ith coordinate of $\mathbf{v}$. Now relabel the vertices of the graph with the corresponding coordinates of $A\mathbf{v}$. Explain how to determine this relabeling from the representation of the graph.

Show that the eigenvalues of an adjacency matrix do not depend on the ordering of the vertices. What happens to the eigenvectors when the vertices are reordered? Based on these observations, we can define the **eigenvalues** of a graph to be the eigenvalues of the adjacency matrix of the graph.

Show that the complete graph on four vertices has eigenvalues 3 (with multiplicity 1) and -1 (with multiplicity 3). Find bases for the corresponding eigenspaces. Examine the patterns obtained when the vertices are labeled with an eigenvector and then relabeled with the product of the adjacency matrix times the eigenvector. Based on these combinatorial ideas, determine the eigenvalues of complete graphs on two, three, and five vertices. Generalize your results to arbitrary complete graphs.

6. Show that the eigenvalues of the Peterson graph are 3, 1, and -2 with multiplicities 1, 5, and 4, respectively. Define other simple families of graphs and explore their eigenvalues.

7. Show that the absolute value of any eigenvalue of a graph is less than or equal to the maximum degree of any vertex of the graph. (Suggestion: Consider the entry in an eigenvector of largest absolute value.) Show that the maximum degree of a connected graph is an eigenvalue if and only if every vertex has the same degree. In this case show that this eigenvalue is of multiplicity 1.

Project: Numerical Methods for Eigenvalues and Eigenvectors

The definition of eigenvalue and the use of determinants to detect singular matrices provide the theoretical method of determining eigenvalues as described in Section 8.1.

This direct method is fairly reasonable for paper-and-pencil computation of eigenvalues and eigenvectors of a 2×2 matrix. It is less reasonable for a 3×3 matrix.

Several factors contribute to the burdensome computations involved in this process. First you have to compute the characteristic polynomial. This involves evaluation of a determinant with a variable entry. Thus, you cannot expect any help from row operations. Next you have to find the roots of this polynomial. Even for cubic polynomials, you had better hope the matrix is rigged so that the roots are integers or at least rational numbers. Otherwise, you might be better off using Newton's method to obtain numerical approximations to the roots. Once you have the eigenvalues, you can then proceed to find the corresponding eigenvectors.

For matrices of size much larger than 3×3, this process gets completely out of hand even with a computer to assist in the computations. Mathematicians have come up with several simple but ingenious alternatives to the direct method for computing eigenvalues and eigenvectors. These numeric methods give approximations rather than exact values. However, for practical applications this is usually adequate. In fact, an exact formula involving complicated fractions and radicals would most likely be approximated by a decimal value before the result is put to use.

1. Suppose λ is an eigenvalue of a square matrix A. If $\mathbf{v}$ is close to an eigenvector associated with λ, then we expect that $A\mathbf{v}$ will be close to λ times the eigenvector, which in turn will be close to $\lambda\mathbf{v}$. Experiment with a 2×2 matrix A with known eigenvalues. Write a computer program that allows the user to choose a vector $\mathbf{v}$ and then plots $\mathbf{v}$ in comparison with the image $A\mathbf{v}$. Notice how much better this idea works if the eigenvalue λ is substantially larger in magnitude than the other eigenvalue.

This idea leads to the **power method** for approximating an eigenvalue of a diagonalizable matrix A. Suppose the eigenvalue λ is larger in magnitude than any other eigenvalue of A. Such an eigenvalue is called the **dominant** eigenvalue. If $A\mathbf{v}$ is closer to an eigenvector than $\mathbf{v}$, why not repeat the process and consider $A(A\mathbf{v}) = A^2\mathbf{v}$ and higher powers of A applied to $\mathbf{v}$? Modify your program to show these vectors. You may need to change the scale of your coordinate systems as these vectors move outside your viewing rectangle. Better yet, scale the result after each computation so that the largest component is 1.

Also, look at the numerical values of the successive iterates of the power method. Compare what happens when the vector is scaled each time you multiply by A. Compute the **Raleigh quotients** $\dfrac{\langle \mathbf{v}, A\mathbf{v} \rangle}{\langle \mathbf{v}, \mathbf{v} \rangle}$. Prove that these values converge to the dominant eigenvalue as the vector $\mathbf{v}$ gets closer to an eigenvector associated with λ.

2. Write $\mathbf{v}$ as a linear combination of the eigenvectors of A. Use the linearity of matrix multiplication to obtain a theoretical explanation of the convergence you observed for the power method.

What happens if the original vector is an eigenvector associated with an eigenvalue other than the dominant eigenvalue? What happens if roundoff error in the computation (usually something we would rather avoid) causes the result to be slightly different from an eigenvector associated with this eigenvalue?

What hope is there for the power method if A is not diagonalizable?

3. Consider repeatedly squaring A to obtain first A^{2^k} and then $A^{2^k}\mathbf{v}$. Compare the number of multiplications involved here with the number needed for iterative multiplication by A to compute $A^{2^k}\mathbf{v}$.

4. Obtain a standard text on numerical analysis and look up Aitken's method for accelerating the convergence of a sequence. Try this on the power method for approximating the dominant eigenvalue of a matrix.

5. You can easily check that the reciprocal of an eigenvalue of a nonsingular matrix A is an eigenvalue of the inverse matrix A^{-1} (see Exercise 15 of Section 8.1). The **inverse power method** can be used to find the eigenvalue of smallest absolute value as the reciprocal of the dominant eigenvalue of the inverse of the matrix. Try this method with a matrix with known eigenvalues.

Start with a vector $\mathbf{v}_0$ and define $\mathbf{v}_k = A^{-k}\mathbf{v}_0$. Notice that $A\mathbf{v}_{k+1} = \mathbf{v}_k$. Thus, we can use the Gauss-Jordan reduction process to determine $\mathbf{v}_{k+1}$ in terms of $\mathbf{v}_k$. Find a way to keep track of the row operation used in reducing A so that you can easily compute $\mathbf{v}_{k+1}$ from $\mathbf{v}_k$.

6. For a real number r and an eigenvalue λ of a square matrix A, show that $\lambda - r$ is an eigenvalue of $A - rI$. What happens if you apply the power method or the inverse power method to $A - rI$? Experiment with the resulting **shifted power method** and **shifted inverse power method** to find all the eigenvalues of a square matrix.

Chapter 8 Summary

Chapter 8 ties together all of the major themes of linear algebra. We look for eigenvectors and their eigenvalues as characteristic features of linear operators. These help simplify the matrix representation of a linear operator, in some cases to a diagonal matrix, the ultimate in simplicity.

Computations

Eigenvalues and eigenvectors
Definition
Eigenvalues as real roots of the characteristic polynomial
Eigenvectors as solutions of a homogeneous system
Eigenspaces

Invariants of similarity classes
Determinant
Trace
Characteristic polynomial

Diagonalization
Characteristic polynomial must have all real roots
Dimensions of eigenspaces must agree with multiplicities of eigenvalues
Diagonalize a symmetric matrix with an orthonormal basis

Theory

Similarity
Reflexive, symmetric, and transitive properties
Similarity classes
Similar matrices represent the same linear operator

Bases of eigenvectors
Finding a basis of eigenvectors is equivalent to diagonalizability
Eigenvectors with distinct eigenvalues are linearly independent

Symmetric operator
Represented by a symmetric matrix relative to an orthonormal basis
Characteristic polynomial has all real roots
Spectral Theorem
Eigenvectors with distinct eigenvalues are orthogonal

Applications

Diagonalization simplifies computing powers of a matrix

Symmetric matrices
Conic sections
Hessian matrix of second-order partial derivatives
Covariance matrix in statistics

Systems of differential equations
Solution spaces are finite-dimensional function spaces
Wronskian for determining linear independence
Diagonalization leads to easy solutions

Review Exercises

1. Consider the matrix $A = \begin{bmatrix} -10 & -14 & 13 \\ 5 & 7 & -5 \\ -2 & -4 & 5 \end{bmatrix}$.

a. Find the eigenvalues of A.

b. Give an eigenvector associated with each eigenvalue.

c. Find an invertible matrix P and a diagonal matrix D so that $P^{-1}AP = D$.

2. Consider the matrix $A = \begin{bmatrix} 10 & -3 & -4 \\ -12 & 3 & 6 \\ 20 & -6 & -8 \end{bmatrix}$.

a. Find the eigenvalues of A.

b. For each eigenvalue find a corresponding eigenvector.

c. Find an invertible matrix P and a diagonal matrix D so that $P^{-1}AP = D$.

3. Albrect Dürer's engraving *Melancholia* of 1514 contains the 4×4 magic square

$$A = \begin{bmatrix} 16 & 3 & 2 & 13 \\ 5 & 10 & 11 & 8 \\ 9 & 6 & 7 & 12 \\ 4 & 15 & 14 & 1 \end{bmatrix}.$$

    a. Confirm that the sum of the entries in every row is 34.

    b. Reformulate this result in terms of a certain vector being an eigenvector of A associated with the eigenvalue 34.

    c. Find the other eigenvalues of A.

    d. Find an eigenvector associated with each eigenvalue.

    e. Find an invertible matrix P and a diagonal matrix D so that $P^{-1}AP = D$.

4. Suppose $\mathbf{v}$ is an eigenvector of an $n \times n$ matrix A associated with the eigenvalue λ. For any real number r, show that $\mathbf{v}$ is an eigenvector of $A + rI$. What is the associated eigenvalue?

5.   **a.** Find the characteristic polynomial of $\begin{bmatrix} 0 & 1 & 0 \\ 0 & 0 & 1 \\ -2 & 4 & 5 \end{bmatrix}$.

    b. Find the characteristic polynomial of $\begin{bmatrix} 0 & 1 & 0 \\ 0 & 0 & 1 \\ -a & -b & -c \end{bmatrix}$.

    c. Give a 4×4 matrix whose characteristic polynomial is $\lambda^4 + 2\lambda^3 - 7\lambda + 6$.

6. Consider the matrix $A = \begin{bmatrix} 19 & -24 \\ 12 & -15 \end{bmatrix}$.

    a. Find the eigenvalues of A.

    b. Find an eigenvector associated with each eigenvalue.

    c. Find a diagonal matrix similar to A.

    d. Use Exercise 17 of Section 7.1 as a simple way to compute A^{12}.

7. Suppose the square matrix A satisfies $A^k = \mathbf{0}$ for some nonnegative integer k. Prove that $\lambda = 0$ is the only possible eigenvalue of A.

8. Consider the transpose operator $T : \mathbb{M}(2, 2) \to \mathbb{M}(2, 2)$ defined by

$$T\left(\begin{bmatrix} a & b \\ c & d \end{bmatrix}\right) = \begin{bmatrix} a & c \\ b & d \end{bmatrix}.$$

    a. Show that 1 and -1 are eigenvalues of T.

    b. Show that 1 and -1 are the only eigenvalues of T.

    c. Find the dimension of the eigenspace associated with the eigenvalue 1.

    d. Find the dimension of the eigenspace associated with the eigenvalue -1.

9. Consider the linear operator $T : \mathbb{P}_2 \to \mathbb{P}_2$ defined by $T(ax^2 + bx + c) = a(x - 1)^2 + b(x - 1) + c$. Show that 1 is the only eigenvalue of T. Find a basis for $E_T(1)$.

10. Consider the linear operator $T : \mathbb{F}(\mathbb{R}) \to \mathbb{F}(\mathbb{R})$ defined by $T(f(x)) = f(x - 1)$.

    a. Show that the function $y = e^x$ is an eigenvector of T associated with the eigenvalue e^{-1}.

    b. Show that the function $y = e^{2x}$ is an eigenvector of T. What is the associated eigenvalue?

    c. Show that any positive number is an eigenvalue of T.

    d. Show that the function $y = \sin(\pi x)$ is an eigenvector of T. What is the associated eigenvalue?

    e. Show that any negative number is an eigenvalue of T.

    f. Show that 0 is not an eigenvalue of T.

11. Determine which of the following matrices can be diagonalized. Give reasons for your decisions.

    a. $\begin{bmatrix} 4 & 0 & 0 \\ 0 & -3 & 1 \\ 0 & 0 & -3 \end{bmatrix}$
    b. $\begin{bmatrix} 4 & 0 & 0 \\ 0 & -3 & 0 \\ 0 & 0 & -3 \end{bmatrix}$.

    c. $\begin{bmatrix} 6 & -3 & 8 \\ 0 & 2 & 9 \\ 0 & 0 & -5 \end{bmatrix}$
    d. $\begin{bmatrix} \sqrt{2} & 52 & -9 & 1.7 \\ 52 & 2 & 3 & \pi \\ -9 & 3 & 0 & 0.9 \\ 1.7 & \pi & 0.9 & 12 \end{bmatrix}$

12. Verify that the following 2×2 matrices are orthogonal.

    a. $\begin{bmatrix} \frac{5}{13} & \frac{12}{13} \\ -\frac{12}{13} & \frac{5}{13} \end{bmatrix}$
    b. $\begin{bmatrix} 0 & 1 \\ 1 & 0 \end{bmatrix}$
    c. $\begin{bmatrix} \frac{1}{\sqrt{5}} & \frac{2}{\sqrt{5}} \\ \frac{2}{\sqrt{5}} & -\frac{1}{\sqrt{5}} \end{bmatrix}$

    d. Show that every 2×2 orthogonal matrix is of the form $\begin{bmatrix} a & -b \\ b & a \end{bmatrix}$ or $\begin{bmatrix} a & b \\ b & -a \end{bmatrix}$, where $a^2 + b^2 = 1$.

    e. Show that every 2×2 orthogonal matrix is of the form $\begin{bmatrix} \cos\theta & -\sin\theta \\ \sin\theta & \cos\theta \end{bmatrix}$ or $\begin{bmatrix} \cos\theta & \sin\theta \\ \sin\theta & -\cos\theta \end{bmatrix}$ for some $\theta \in [0, 2\pi)$.

13. Verify that the following matrices are orthogonal.

    a. $\begin{bmatrix} \frac{2}{3} & -\frac{2}{3} & \frac{1}{3} \\ \frac{2}{3} & \frac{1}{3} & -\frac{2}{3} \\ \frac{1}{3} & \frac{2}{3} & \frac{2}{3} \end{bmatrix}$
    b. $\begin{bmatrix} \frac{1}{\sqrt{2}} & -\frac{1}{\sqrt{2}} & 0 & 0 \\ \frac{1}{\sqrt{2}} & \frac{1}{\sqrt{2}} & 0 & 0 \\ 0 & 0 & \frac{1}{2} & \frac{\sqrt{3}}{2} \\ 0 & 0 & -\frac{\sqrt{3}}{2} & \frac{1}{2} \end{bmatrix}$
    c. $\begin{bmatrix} 0 & 1 & 0 \\ 0 & 0 & -1 \\ -1 & 0 & 0 \end{bmatrix}$

14. Consider the symmetric matrix $A = \begin{bmatrix} 9 & -6 & 18 \\ -6 & 4 & -12 \\ 18 & -12 & 36 \end{bmatrix}$

    a. Find the eigenvalues of A.

    b. Find a basis for the eigenspaces associated with each eigenvalue.

    c. Find an orthonormal basis of $\mathbb{R}^3$ consisting of eigenvectors of A.

    d. Find an orthogonal matrix P and a diagonal matrix D so that $P^{-1}AP = D$.

15. Consider the symmetric matrix $A = \begin{bmatrix} 1 & 1 & 1 & 1 \\ 1 & 1 & 1 & 1 \\ 1 & 1 & 1 & 1 \\ 1 & 1 & 1 & 1 \end{bmatrix}$.

   a. Find the eigenvalues of A.

   b. Find a basis for the eigenspaces associated with each eigenvalue.

   c. Find an orthonormal basis of $\mathbb{R}^4$ consisting of eigenvectors of A.

   d. Find an orthogonal matrix P and a diagonal matrix D so that $P^{-1}AP = D$.

16.   **a.** Find the eigenvalues of the matrix $\begin{bmatrix} 3 & 6 \\ 2 & 2 \end{bmatrix}$ and find an eigenvector associated with each eigenvalue.

   b. Use the result of part a to find a basis for the solution space of the differential equation

$$y_1' = 3y_1 + 6y_2$$
$$y_2' = 2y_1 + 2y_2$$

   c. Find the solution of this system of differential equations that satisfies the
   initial condition $\begin{bmatrix} y_1(0) \\ y_2(0) \end{bmatrix} = \begin{bmatrix} -2 \\ 1 \end{bmatrix}$.

17.   Let k be a nonnegative integer. Let $\varphi_1 : [-1, 1] \to \mathbb{R}^2$ be defined by

$$\varphi_1(t) = \begin{cases} t^k e_1 & \text{if } t < 0 \\ 0 & \text{if } t \geq 0 \end{cases}$$

   and let $\varphi_2 : [-1, 1] \to \mathbb{R}^2$ be defined by

$$\varphi_2(t) = \begin{cases} 0 & \text{if } t < 0 \\ t^k e_2 & \text{if } t \geq 0 \end{cases}$$

   a. Show that φ_1 and φ_2 are $k - 1$ times differentiable functions.

   b. Show that the Wronskian $W(t) = \det[\,\varphi_1(t) \quad \varphi_2(t)\,]$ is equal to zero for all
   $t \in [-1, 1]$.

   c. Show that $\{\varphi_1, \varphi_2\}$ is a linearly independent set.

   d. Why do the results of parts b and c not contradict Theorem 8.33?

Answers to Selected Exercises

Section 1.1

1. **a.** $x \in S \cap T \implies x \in S$ and $x \in T \implies x \in S$
    b. $x \in S \cap T \implies x \in S$ and $x \in T \implies x \in T$
    c. $x \in S \implies x \in S$ or $x \in T \implies x \in S \cup T$
    d. $x \in T \implies x \in S$ or $x \in T \implies x \in S \cup T$

4. **a.** The empty set is an identity element for unions in the same way that 0 is an identity element for addition of numbers. The law $S \cap \varnothing = \varnothing$ for any set S is analogous to the corresponding law $x \cdot 0 = 0$ for any real number x.

    b. $R \cap (S \cup T) = (R \cap S) \cup (R \cap T)$ and $(S \cup T) \cap (S \cup T) = (S \cap S) \cup (S \cap T) \cup (S \cap T) \cup (T \cap T)$, for example.

    c. Both of these are true. For real numbers, if $xy = 0$, then $x = 0$ or $y = 0$. For sets, however, it is possible for $S \cap T = \varnothing$ with $S \neq \varnothing$ and $T \neq \varnothing$.

7. **a.** $\varnothing$ and $\{0\}$ are the two subsets of $\{0\}$.

    b. 4

    c. 8

    d. The s subsets are still subsets of the enlarged set. The new element can be adjoined to these subsets to create new subsets. Any subset of the enlarged set that does not contain the new element is in the first group of s subsets; any subset that contains the new element is in the second group of s subsets. Hence, there are exactly $2s$ subsets of the enlarged set.

    e. The number of subsets is $s = 2^k$. This works even for the empty set, which has $1 = 2^0$ subset, namely, $\varnothing$ itself.

Section 1.2

2. For example, $\mathbf{0} + \mathbf{v} = \mathbf{v}$. Also, if $r\mathbf{v} = \mathbf{0}$, then $r = 0$ or $\mathbf{v} = \mathbf{0}$. These will be established in the next section. Subtraction is not built in as part of the axioms. Soon we will be able to define subtraction in terms of addition and additive inverses. Order relations and square roots are not normally available in vector spaces.

5.   **a.** 1. $\mathbf{v} \oplus \mathbf{w} = \mathbf{w} \oplus \mathbf{v}$

          2. $(\mathbf{v} \oplus \mathbf{w}) \oplus \mathbf{x} = \mathbf{v} \oplus (\mathbf{w} \oplus \mathbf{x})$

          3. There is a vector in V, denoted $\mathbf{0}$, such that $\mathbf{v} \oplus \mathbf{0} = \mathbf{v}$.

          4. For each vector $\mathbf{v}$ in V there is a vector in V, denoted $-\mathbf{v}$, such that $\mathbf{v} \oplus (-\mathbf{v}) = \mathbf{0}$.

          5. $r \odot (\mathbf{v} \oplus \mathbf{w}) = r \odot \mathbf{v} \oplus r \odot \mathbf{w}$

          6. $(r + s) \odot \mathbf{v} = r \odot \mathbf{v} \oplus s \odot \mathbf{v}$

          7. $r \odot (s \odot \mathbf{v}) = (rs) \odot \mathbf{v}$

          8. $1 \odot \mathbf{v} = \mathbf{v}$

  b. The type of elements that are being operated on will determine whether to use a vector space operation or an operation of real numbers.

6.   **a.** The minus sign in $(-r)\mathbf{v}$ denotes the inverse of the real number r with respect to addition of real numbers. The minus sign in $r(-\mathbf{v})$ denotes the inverse of the vector $\mathbf{v}$ relative to the addition of vectors in the vector space.

  b. It is not clear whether $-r\mathbf{v}$ is the additive inverse of the real number r times the vector $\mathbf{v}$ or the additive inverse of the vector $r\mathbf{v}$.

9.   **a.** $\left\{ \frac{1}{2}, 1, \frac{3}{2}, 2, \frac{5}{2}, \ldots \right\}$ **b.** $\left\{ \frac{1}{2}, \frac{1}{4}, \frac{1}{8}, \frac{1}{16}, \ldots \right\}$

Section 1.3

3. $\frac{1}{2}\mathbf{v} + \frac{1}{2}\mathbf{v} = \left(\frac{1}{2} + \frac{1}{2} \right)\mathbf{v} = 1\mathbf{v} = \mathbf{v}$

7. Suppose $r \neq 0$ and $r\mathbf{v} = \mathbf{0}$. Then $\mathbf{v} = 1\mathbf{v} = \left(\frac{1}{r}r \right)\mathbf{v} = \frac{1}{r}(r\mathbf{v}) = \frac{1}{r}\mathbf{0} = \mathbf{0}$.

8.
$$\mathbf{v} + (-\mathbf{v}) = \mathbf{0}$$
$$(-1)\mathbf{v} + (\mathbf{v} + (-\mathbf{v})) = (-1)\mathbf{v} + \mathbf{0}$$
$$((-1)\mathbf{v} + \mathbf{v}) + (-\mathbf{v}) = (-1)\mathbf{v}$$
$$((-1)\mathbf{v} + 1\mathbf{v}) + (-\mathbf{v}) = (-1)\mathbf{v}$$
$$(-1 + 1)\mathbf{v} + (-\mathbf{v}) = (-1)\mathbf{v}$$
$$0\mathbf{v} + (-\mathbf{v}) = (-1)\mathbf{v}$$
$$\mathbf{0} + (-\mathbf{v}) = (-1)\mathbf{v}$$
$$-\mathbf{v} = (-1)\mathbf{v}$$

15. **a.** There are 14 groupings: $\mathbf{u} + (\mathbf{v} + (\mathbf{w} + (\mathbf{x} + \mathbf{y})))$, $\mathbf{u} + (\mathbf{v} + ((\mathbf{w} + \mathbf{x}) + \mathbf{y}))$, $\mathbf{u} + ((\mathbf{v} + \mathbf{w}) + (\mathbf{x} + \mathbf{y}))$, $\mathbf{u} + ((\mathbf{v} + (\mathbf{w} + \mathbf{x})) + \mathbf{y}$, $\mathbf{u} + (((\mathbf{v} + \mathbf{w}) + \mathbf{x}) + \mathbf{y}$, $(\mathbf{u} + \mathbf{v}) + (\mathbf{w} + (\mathbf{x} + \mathbf{y}))$, $(\mathbf{u} + \mathbf{v}) + ((\mathbf{w} + \mathbf{x}) + \mathbf{y})$, $(\mathbf{u} + (\mathbf{v} + \mathbf{w})) + (\mathbf{x} + \mathbf{y})$, $((\mathbf{u} + \mathbf{v}) + \mathbf{w}) + (\mathbf{x} + \mathbf{y})$, $(\mathbf{u} + (\mathbf{v} + (\mathbf{w} + \mathbf{x}))) + \mathbf{y}$, $(\mathbf{u} + ((\mathbf{v} + \mathbf{w}) + \mathbf{x})) + \mathbf{y}$, $((\mathbf{u} + \mathbf{v}) + (\mathbf{w} + \mathbf{x})) + \mathbf{y}$, $((\mathbf{u} + (\mathbf{v} + \mathbf{w})) + \mathbf{x}) + \mathbf{y}$, $(((\mathbf{u} + \mathbf{v}) + \mathbf{w}) + \mathbf{x}) + \mathbf{y}$.

b. There are 14 groupings of 1 and 5; there are 5 groupings of 2 and 4; there are $2 \cdot 2 = 4$ groupings of 3 and 3; there are 5 groupings of 4 and 2; there are 14 groupings of 5 and 1. This gives a total of 42 groupings.

c. The number of groupings of n terms is known as the nth Catalan number $C(n) \cdot C(n) = \sum_{k=1}^{n-1} C(k)C(n-k)$.

Section 1.4

2. $\mathbf{v} - \mathbf{w} = \mathbf{v} + (-\mathbf{w}) = -\mathbf{w} + \mathbf{v}$

7. $\mathbf{v} - (\mathbf{w} + \mathbf{x}) = \mathbf{v} + (-(\mathbf{w} + \mathbf{x})) = \mathbf{v} + (-1)(\mathbf{w} + \mathbf{x}) = \mathbf{v} + ((-1)\mathbf{w} + (-1)\mathbf{x}) = \mathbf{v} + (-\mathbf{w} + (-\mathbf{x})) = (\mathbf{v} + (-\mathbf{w})) + (-\mathbf{x}) = (\mathbf{v} - \mathbf{w}) - \mathbf{x}$

9. $\mathbf{v} - (-\mathbf{w} + \mathbf{x}) = \mathbf{v} + (-(-\mathbf{w} + \mathbf{x})) = \mathbf{v} + (-1)((-1)\mathbf{w} + \mathbf{x}) = \mathbf{v} + ((-1)((-1)\mathbf{w}) + (-1)\mathbf{x}) = (\mathbf{v} + ((-1)(-1))\mathbf{w}) + (-\mathbf{x}) = (\mathbf{v} + 1\mathbf{w}) - \mathbf{x} = (\mathbf{v} + \mathbf{w}) - \mathbf{x}$

12. $(r - s)\mathbf{v} = (r + (-s))\mathbf{v} = r\mathbf{v} + (-s)\mathbf{v} = r\mathbf{v} + (-(s\mathbf{v})) = r\mathbf{v} - (s\mathbf{v})$

Section 1.5

3. All eight axioms are satisfied; this is a vector space.

5. The only vector (w_1, w_2) that satisfies $(1, 1) + (w_1, w_2) = (1 + w_1, 1 + w_1 + 1 + w_2) = (1, 1)$ is $(w_1, w_2) = (0, -1)$. However, $(0, 2) + (0, -1) = (0 + 0, 0 + 2 + (-1)) = (0, 1) \neq (0, 2)$. Thus, there is no vector that acts an additive identity, as required by Axiom 3. Axiom 8 also fails.

8. This is not a vector space since S is not closed under scalar multiplication. For example, scalar multiplication of $\frac{1}{2} \in \mathbb{R}$ times $(1, 1) \in S$ is $\frac{1}{2}(1, 1) = (\frac{1}{2}, \frac{1}{2})$, which is not an element of S.

10. This is a vector space. First, it is closed under addition: if $(v_1, v_2, v_3) \in P$ and $(w_1, w_2, w_3) \in P$, then $v_1 = v_2 + v_3$ and $w_1 = w_2 + w_3$, so $v_1 + w_1 = (v_2 + v_3) + (w_2 + w_3) = (v_2 + w_2) + (v_3 + w_3)$. Therefore, $(v_1, v_2, v_3) + (w_1, w_2, w_3) = (v_1 + w_1, v_2 + w_2, v_3 + w_3) \in P$. Second, it is closed under scalar multiplication: if $r \in \mathbb{R}$ and $(v_1, v_2, v_3) \in P$, then $v_1 = v_2 + v_3$, so $rv_1 = rv_2 + rv_3$. Therefore, $r(v_1, v_2, v_3) = (rv_1, rv_2, rv_3) \in P$. Third, $(0, 0, 0) \in P$ since $0 = 0 + 0$, and if $(v_1, v_2, v_3) \in P$, then $v_1 = v_2 + v_3$; so $-v_1 = -v_2 + (-v_3)$, and hence $(-v_1, -v_2, -v_3) \in P$. Also, the identities of the eight axioms hold for any vectors in $\mathbb{R}^3$, so they certainly hold for the elements of P with $(0, 0, 0) \in P$ as the additive identity and $(-v_1, -v_2, -v_3) \in P$ as the additive inverse of any element $(v_1, v_2, v_3) \in P$.

13. $a = 1, b = 2$, and $c = -1$.

15. For any values of a, b, c, and d, the last coordinate on the left side of the equation will be 0. Since the last coordinate of the vector on the right side of the equation is not 0, it is impossible to find scalars that will satisfy this equation.

16. Some typical examples other than those mentioned in the text: an inventory listing the quantities of 1000 different items in a warehouse, the prices of 20 stocks in an investment portfolio, or the level of stimulation of the four sensations of taste. Addition and scalar multiplication might be useful to get a total for all the warehouses, to compute an average price vector of the stocks over a period of time, or the result of a 10% nerve blockage.

Section 1.6

3.
$$\begin{bmatrix} x \\ y \\ z \end{bmatrix} = a \begin{bmatrix} 1 \\ 3 \\ 0 \end{bmatrix} + b \begin{bmatrix} 4 \\ 0 \\ 1 \end{bmatrix} + c \begin{bmatrix} 1 \\ -2 \\ 5 \end{bmatrix}$$

5. **a.** $a = 0, b = 0, c = 0$, and $d = 0$.

b. $a = 4, b = 1, c = 2$, and $d = 2$.

c. No, in each case the 11-entry forces the value of a, the 12-entry forces the value of b, the 21-entry forces the value of c, and the 22-entry forces the value of d.

9. **a.** If $A = \begin{bmatrix} a_1 \cdots a_n \end{bmatrix}$ corresponds to $\mathbf{a} = (a_1, \ldots, a_n)$ and $B = \begin{bmatrix} b_1 \cdots b_n \end{bmatrix}$ corresponds to $\mathbf{b} = (b_1, \ldots, b_n)$, then $A + B = \begin{bmatrix} a_1 \cdots a_n \end{bmatrix} + \begin{bmatrix} b_1 \cdots b_n \end{bmatrix} = \begin{bmatrix} a_1 + b_1 \cdots a_n + b_n \end{bmatrix}$ corresponds to $(a_1 + b_1, \ldots, a_n + b_n) = (a_1, \ldots, a_n) + (b_1, \ldots, b_n) = \mathbf{a} + \mathbf{b}$.

b. If $A = \begin{bmatrix} a_1 \cdots a_n \end{bmatrix}$ corresponds to $\mathbf{a} = (a_1, \ldots, a_n)$ and $r \in \mathbb{R}$, then $rA = r \begin{bmatrix} a_1 \cdots a_n \end{bmatrix} = \begin{bmatrix} ra_1 \cdots ra_n \end{bmatrix}$ corresponds to $(ra_1, \ldots, ra_n) = r(a_1, \ldots, a_n) = r\mathbf{a}$.

10. Suppose A and B are any two matrices. Augment A or B with columns of zeros so that they both have the same number of columns. Similarly, pad the resulting matrices with rows of zeros so that they have the same number of rows. Let A' and B' be the resulting matrices. Define the sum of A and B to be $A' + B'$. Among the eight axioms, Axiom 4 raises the only problem. The additive identity will need to be an all-zero matrix of some specific size. The sum of a matrix and its additive inverse will not necessarily be the zero matrix of the specified size. One way to overcome this problem is to identify all matrices that can be obtained from one another by augmenting or deleting rows and columns of zeros. The set of classes of equivalent matrices will then form a vector space.

Section 1.7

1. **a.** $0.01000 = 0.00999\cdots$, but f assigns 0 to the first form of this number and 9 to the second.

   b. $\frac{1}{2} = \frac{2}{4}$, but g assigns $1 + 2 = 3$ to the first form of this number and $2 + 4 = 6$ to the second.

   c. h is a well-defined function. Even though a rational number can be written in many forms as the quotient of integers, the difference between the number of factors of 2 in the numerator and denominator will not change.

   d. For example, s will assign $y = 1$ and $y = -1$ to $x = 0$.

4. $\left[r[f+g]\right](x) = r\left[[f+g](x)\right] = r[f(x)+g(x)] = r[f(x)]+r[g(x)] = [rf](x)+[rg](x) = [rf + rg](x)$

5. **a.** $(x+1)^3 = a(x^3+1) + b(x^2+x)$

   b. $a = 1, b = 3$

8. **a.** If f corresponds to $(f(1),\ldots,f(n))$ and g corresponds to $(g(1),\ldots,g(n))$, then $f + g$ corresponds to $((f+g)(1),\ldots,(f+g)(n)) = (f(1)+g(1),\ldots, f(n)+g(n)) = (f(1),\ldots,f(n)) + (g(1),\ldots,g(n))$.

   b. If f corresponds to $(f(1),\ldots,f(n))$ and $r \in \mathbb{R}$, then rf corresponds to $((rf)(1),\ldots,(rf)(n)) = (r(f(1)),\ldots,r(f(n))) = r(f(1),\ldots,f(n))$.

11. **a.** Multiply the number of choices for a value to assign to 1 by the number of choices for a value to assign to 2 to obtain the number of possible functions: $3 \cdot 3 = 9$.

    b. n^m

    c. There is only $1 = n^0$ function (the empty function) from $\varnothing$ to any other set Y. There are $0 = 0^m$ functions from a nonempty set X to $\varnothing$. The formula is indeterminate if $m = 0$ and $n = 0$. However, there is still one function from $\varnothing$ to $\varnothing$.

Section 1.8

3. $(1, 1) \in S$, but $\frac{1}{2}(1, 1) = \left(\frac{1}{2}, \frac{1}{2}\right) \notin S$. Thus, S is not closed under scalar multiplication. Hence, S is not a subspace.

5. $\{(x, y) \in \mathbb{R}^2 \mid x = 0 \text{ or } y = 0\}$ is the union of the x-axis and the y-axis. This set is closed under scalar multiplication but is not closed under addition. For example, $(1, 0)$ and $(0, 1)$ are in the set, but their sum $(1, 0) + (0, 1) = (1, 1)$ is not in the set.

7. $\mathbf{0} \in S$ since $\mathbf{0}(0) = 0$. Thus, $S \neq \varnothing$. Let f and g be elements of S. That is, $f(0) = 0$ and $g(0) = 0$. Then $f + g \in S$ because $(f + g)(0) = f(0) + g(0) = 0+0 = 0$. Also, let $r \in \mathbb{R}$. Then $rf \in S$ because $(rf)(0) = r(f(0)) = r\,0 = 0$. By the Subspace Theorem, S is a subspace.

11. Consider the constant function $\mathbf{1}$ defined by $\mathbf{1}(x) = 1$ for all $x \in \mathbb{R}$. Now $\mathbf{1} \in S$ since $\mathbf{1}'(x) + \mathbf{1}(x) = 0 + 1 = 0$. However, $2\mathbf{1} \notin S$ because

$(21)'(x) + (21)(x) = 2 \neq 1$. Thus, S is not closed under scalar multiplication. Hence, S is not a subspace.

16. **a.** The additive identity $\mathbf{0}$ of V satisfies $\mathbf{v} + \mathbf{0} = \mathbf{v}$ for all $\mathbf{v} \in V$. Hence, it satisfies this condition for all $\mathbf{v} \in S$. Thus, $\mathbf{0}$ is an additive identity of S. By Exercise 5 of Section 1.3, this is the unique vector in S satisfying this condition. Thus, the additive identity of S is the additive identity of V.

b. Let $\mathbf{v}$ be an element of S. The additive inverse $-\mathbf{v}$ of $\mathbf{v}$ in the vector space V satisfies $\mathbf{v} + (-\mathbf{v}) = \mathbf{0}$. Thus, $-\mathbf{v}$ is an additive identity of $\mathbf{v}$ in the vector space S. By Theorem 1.3, this is the unique vector in S satisfying this condition. Thus, the additive inverse of $\mathbf{v}$ in the vector space S is the additive inverse of $\mathbf{v}$ in the vector space V.

19. Since S and T are subspaces of V, they both contain the additive identity (see Exercise 16). That is, $\mathbf{0} \in S \cap T$. Hence, $S \cap T \neq \varnothing$. Let $\mathbf{v} \in S \cap T$ and $\mathbf{w} \in S \cap T$. That is, $\mathbf{v} \in S$ and $\mathbf{v} \in T$, and $\mathbf{w} \in S$ and $\mathbf{w} \in T$. Since S and T are vector spaces, they are closed under addition. Thus, $\mathbf{v} + \mathbf{w} \in S$ and $\mathbf{v} + \mathbf{w} \in T$. That is, $\mathbf{v} + \mathbf{w} \in S \cap T$. Also, let $r \in \mathbb{R}$. Again S and T are vector spaces, so they are closed under scalar multiplication. Thus, $r\mathbf{v} \in S$ and $r\mathbf{v} \in T$. That is, $r\mathbf{v} \in S \cap T$. By the Subspace Theorem, $S \cap T$ is a subspace.

Section 1.9

1. **a.** If $v_1 \neq 0$, then both lines have slope equal to $\frac{v_2}{v_1}$. If $v_1 = 0$, then both lines are vertical. In either case they are parallel.

b. The distance between both pairs of points is $\sqrt{v_1^2 + v_2^2}$.

c. Interchange the roles of (v_1, v_2) and (w_1, w_2) to see that the other two sides of the quadrilateral are parallel and of the same length.

3. $\mathbf{v} = 1\mathbf{v} + 0\mathbf{w}$, $\mathbf{w} = 0\mathbf{v} + 1\mathbf{w}$, and $\mathbf{0} = 0\mathbf{v} + 0\mathbf{w}$.

6. Let $(0.8, -0.6)$ be the direction vector and let $(0.6, 0.8)$ be the point selected on the line. Then the line is $\{r(0.8, -0.6) + (0.6, 0.8) \mid r \in \mathbb{R}\}$.

9. Suppose $a \neq 0$. A point (x, y) is in the set if and only if $x = -\frac{b}{a}y + \frac{c}{a}$. If we introduce the variable r to denote an arbitrary value of y, this can be rewritten as $(x, y) = (-\frac{b}{a}r + \frac{c}{a}, r) = r(-\frac{b}{a}, 1) + (\frac{c}{a}, 0)$. Thus, the set can be written in the standard form of line $\{r(-\frac{b}{a}, 1) + (\frac{c}{a}, 0) \mid r \in \mathbb{R}\}$. A similar argument holds if $b \neq 0$.

11. Let $(1, 1, 1)$ be the direction vector and let $(0, 0, 0)$ be the point selected on the line. Then the line is $\{r(1, 1, 1) \mid r \in \mathbb{R}\}$.

14. Let $(1, -1, 0)$ and $(1, 0, -1)$ be the direction vectors and let $\left(\frac{1}{\sqrt{3}}, \frac{1}{\sqrt{3}}, \frac{1}{\sqrt{3}}\right)$ be the point selected on the line. Then the plane is $\{r(1, -1, 0) + s(1, 0, -1) + \left(\frac{1}{\sqrt{3}}, \frac{1}{\sqrt{3}}, \frac{1}{\sqrt{3}}\right) \mid r, s \in \mathbb{R}\}$.

18. $\left\{ a \begin{bmatrix} 2 & 0 \\ 1 & 1 \end{bmatrix} + b \begin{bmatrix} -1 & 1 \\ 0 & 1 \end{bmatrix} + \begin{bmatrix} 0 & 5 \\ 2 & 0 \end{bmatrix} \,\middle|\, a, b \in \mathbb{R} \right\}$

19. Every solution of the differential equation is a multiple of the exponential function. Thus, the line is $\{r \exp \mid r \in \mathbb{R}\}$.

Review Exercises: Chapter 1

2. If a set contains more than one element, it will contain a nonzero number x. If $x > 0$, then x itself is the desired positive number in the set. If $x < 0$, then x^2 is positive, and since the set is closed under multiplication, x^2 is in the set.

7. **a.** The additive identity is $(3, -5)$. Indeed, for any $\mathbf{v} = (v_1, v_2) \in \mathbb{R}^2$, we have $(v_1, v_2) + (3, -5) = (v_1 + 3 - 3, v_2 + (-5) + 5) = (v_1, v_2)$.

b. Let $\mathbf{v} = (v_1, v_2)$ and $\mathbf{w} = (w_1, w_2)$ be arbitrary elements in $\mathbb{R}^2$. Then $r(\mathbf{v} + \mathbf{w}) = r((v_1, v_2) + (w_1, w_2)) = r(v_1 + w_1 - 3, v_2 + w_2 + 5) = (r(v_1 + w_1 - 3) - 3r + 3, r(v_2 + w_2 + 5) + 5r - 5) = ((rv_1 - 3r + 3) + (rw_1 - 3r + 3) - 3, (rv_2 + 5r - 5) + (rw_2 + 5r - 5) + 5) = (rv_1 - 3r + 3, rv_2 + 5r - 5) + (rw_1 - 3r + 3, rw_2 + 5r - 5) = r(v_1, v_2) + r(w_1, w_2) = r\mathbf{v} + r\mathbf{w}$.

9. $a = 1, b = 1, c = 1$, and $d = -5$.

13. $(1, 1, 1)$ and $(1, 1, -1)$ are elements of the set. However, the sum $(1, 1, 1) + (1, 1, -1) = (2, 2, 0)$ is not an element of the set. Thus, the set is not closed under addition. Hence, it is not a subspace.

17. **a.** If $\mathbf{0} \in P$, then $\mathbf{0} = r_0\mathbf{v} + s_0\mathbf{w} + \mathbf{x}$ for some $r_0, s_0 \in \mathbb{R}$. Thus, an arbitrary element $r\mathbf{v} + s\mathbf{w} + \mathbf{x}$ of P can be written $r\mathbf{v} + s\mathbf{w} + (-r_0\mathbf{v} - s_0\mathbf{w}) = (r - r_0)\mathbf{v} + (s - s_0)\mathbf{w}$. Also, an arbitrary element of the form $r\mathbf{v} + s\mathbf{w}$ can be written as an element of P. Indeed, $r\mathbf{v} + s\mathbf{w} = (r + r_0)\mathbf{v} + (s + s_0)\mathbf{w} + (-r_0\mathbf{v} - s_0\mathbf{w}) = (r + r_0)\mathbf{v} + (s + s_0)\mathbf{w} + \mathbf{x}$. Thus, P is equal to the set $\{r\mathbf{v} + s\mathbf{w} \mid r, s \in \mathbb{R}\}$.

b. If a plane is a subspace of V, then it must contain the additive identity $\mathbf{0}$. Conversely, if the plane contains $\mathbf{0}$, then it is of the form $\{r\mathbf{v} + s\mathbf{w} \mid r, s \in \mathbb{R}\}$, which is a subspace of V.

Section 2.1

3. **a.** $\{r(-2, 1, 0, 0) + s(-3, 0, 1, 1) + (1, 0, 2, 0) \mid r, s \in \mathbb{R}\}$

b. $\{r(-3, 1, 0, 0) + s(1, 0, 1, 0) + (-6, 0, 0, 5) \mid r, s \in \mathbb{R}\}$

c. $\{r(1, 2, 1, 0) + s(2, 1, 0, 1) + (4, 3, 0, 0) \mid r, s \in \mathbb{R}\}$

d. $\{r(-3, 1, 0, 0, 0) + s(1, 0, -2, -1, 1) + (0, 0, -1, 2, 0) \mid r, s \in \mathbb{R}\}$

4. **a.** Divide through by a nonzero coefficient and solve for the corresponding unknown. Assign arbitrary variables to the other two unknowns to write the solution set in the standard form of a plane. For example, if $a \neq 0$, then the solution set is $\{r(-\frac{b}{a}, 1, 0) + s(-\frac{c}{a}, 0, 1) + (\frac{d}{a}, 0, 0) \mid r, s \in \mathbb{R}\}$.

b. The plane passes through the origin if and only if $(0, 0, 0)$ is a solution. This is true if and only if $d = 0$.

c. If $d = 0$ also, then the solution set is $\mathbb{R}^3$. If $d \neq 0$, then the solution set is $\varnothing$.

5. The solution set of each equation is $\varnothing$, a plane, or all of $\mathbb{R}^3$. The intersection of two such sets can be $\varnothing$ (typically from two parallel planes), a line (as the intersection of two planes), a plane (typically from two identical planes), or all of $\mathbb{R}^3$ (all coefficients equal to zero).

Section 2.2

1. This matrix has three leading 1s. They occur in the first, second, and fourth columns. Hence, three of the columns contain leading 1s. Three rows contain leading 1s.

5. **a.** $\begin{bmatrix} 1 & 0 & 0 \\ 0 & 1 & 0 \\ 0 & 0 & 1 \end{bmatrix}$ **b.** $\begin{bmatrix} 1 & 0 & 0 & 2 \\ 0 & 1 & 0 & -1 \\ 0 & 0 & 1 & 4 \end{bmatrix}$

   c. $\begin{bmatrix} 1 & 0 & -2 & 1 \\ 0 & 1 & 1 & 3 \\ 0 & 0 & 0 & 0 \end{bmatrix}$ **d.** $\begin{bmatrix} 1 & -2 & 0 & 7 \\ 0 & 0 & 1 & 2 \\ 0 & 0 & 0 & 0 \end{bmatrix}$

7. **a.** $\begin{bmatrix} 1 & 0 & * \\ 0 & 1 & * \end{bmatrix}, \begin{bmatrix} 1 & * & 0 \\ 0 & 0 & 1 \end{bmatrix}, \begin{bmatrix} 0 & 1 & 0 \\ 0 & 0 & 1 \end{bmatrix}, \begin{bmatrix} 1 & * & * \\ 0 & 0 & 0 \end{bmatrix}, \begin{bmatrix} 0 & 1 & * \\ 0 & 0 & 0 \end{bmatrix},$
   $\begin{bmatrix} 0 & 0 & 1 \\ 0 & 0 & 0 \end{bmatrix}, \begin{bmatrix} 0 & 0 & 0 \\ 0 & 0 & 0 \end{bmatrix}$

   b. There are 10 patterns with three leading 1s, 10 with two leading 1s, 5 with one leading 1, and 1 with zero leading 1s. There is a total of 26 patterns.

9. **a.** If $a = 0$, then $c \neq 0$ and $b \neq 0$. Thus, the augmented matrix $\begin{bmatrix} 0 & b & r \\ c & d & s \end{bmatrix}$ of the system reduces to a matrix with leading 1s in the first two columns. If $a \neq 0$, then multiply the first row of the augmented matrix by $\frac{1}{a}$ and then add $-c$ times the first row to the second row to obtain the matrix $\begin{bmatrix} 1 & \frac{b}{a} & \frac{r}{a} \\ 0 & \frac{ad-bc}{a} & \frac{as-cr}{a} \end{bmatrix}$. Since $ad - bc \neq 0$, this reduces to a matrix with leading 1s in the first two columns. In both cases, we see that the system has a unique solution.

   b. First suppose $a = 0$. If $b = 0$ also, then the solution set of the first equation will be $\varnothing$ (if $r \neq 0$) or $\mathbb{R}^2$ (if $r = 0$). In the first case, the solution set of the system is $\varnothing$. In the second case, the solution set of the system will be $\varnothing$, a line, or $\mathbb{R}^2$, depending on the second equation. If $b \neq 0$, then $c = 0$. Thus, x can be assigned an arbitrary value. Now suppose $a \neq 0$. Apply row operations as in part a. The second row of the matrix begins with two 0s. Hence, the solution set of the system will be $\varnothing$ (if $\frac{as-cr}{a} \neq 0$) or a line (if $\frac{as-cr}{a} = 0$). In all cases, there is not a unique solution.

Section 2.3

1. **a.** $\{(-2, 2, -1)\}$
   b. $\{(-6, 6, -3)\}$
   c. $\{(0, 0, 0)\}$
   d. $\{(-1, 1, -\frac{1}{2})\}$
   e. If the constants on the right side of the equation are multiplied by a constant, then the solution is multiplied by the same constant.

5. **a.** $\{(\frac{16}{15}, \frac{1}{5}, -2, -\frac{1}{3})\}$
   b. $\{r(3, 1, -4, 2) + (\frac{7}{2}, \frac{7}{2}, -4, 0) \mid r \in \mathbb{R}\}$
   c. $\{r(-1, 4, -1, 1) + (1, -4, 1, 0) \mid r \in \mathbb{R}\}$
   d. $\{r(0, 1, -1, -1, 1, 0) + s(0, 0, 0, -1, 0, 1) + (2, 2, -2, 4, 0, 0) \mid r, s \in \mathbb{R}\}$

7. **a.** $\{r(-0.2, 0.6, 1, 0, 0) + s(-0.6, -0.2, 0, 1, 0)$
   $\qquad\qquad\qquad + t(0.4, 0.8, 0, 0, 1) + (2.6, 2.2, 0, 0, 0) \mid r, s, t \in \mathbb{R}\}$
   b. $\varnothing$
   c. $\{r(-0.2, 0.6, 1, 0, 0) + s(0.4, 0.8, 0, 0, 1) + (2.6, 2.2, 0, 0, 0) \mid r, s \in \mathbb{R}\}$
   d. $\{r(-0.2, 0.6, 1, 0, 0) + s(-0.6, -0.2, 0, 1, 0)$
   $\qquad\qquad\qquad + t(0.4, 0.8, 0, 0, 1) \mid r, s, t \in \mathbb{R}\}$

Section 2.4

3. **a.** $p(x) = \frac{1}{2}(e^2 - 2e + 1)x^2 - \frac{1}{2}(e^2 - 4e3)x + 1$
   $\approx 1.47625x^2 + 0.242036x + 1$
   b. $q(x) = -\frac{1}{4}(e^2 - 4e + 3)x^4 + \frac{1}{2}(2e^2 - 8e + 7)x^3$
   $\qquad\quad - \frac{1}{4}(3e^2 - 16e + 19)x^2 + x + 1$
   $\approx 0.121017x^4 + 0.015929x^3 + 0.581335x^2 + x + 1$

5. $a = \frac{2}{4-\pi}$ and $b = c = \frac{2-\pi}{4-\pi}$.

7. Let a, b, and c denote the fraction of the total population of types A, B, and C, respectively. Then the equilibrium conditions together with the fact that $a + b + c = 1$ give the system

$$
\begin{aligned}
.5a + .05b + .1c &= a \\
.25a + .95b + .2c &= b \\
.25a \qquad\quad + .7c &= c \\
a + \quad b + \quad c &= 1
\end{aligned}
$$

The solution is $a = \frac{6}{61} \approx .098$, $b = \frac{50}{61} \approx .820$, and $c = \frac{5}{61} \approx .082$.

9. $a = \frac{1}{2}$, $b = 0$, and $c = \frac{1}{2}$.

Review Exercises: Chapter 2

1. **a.** In order to reverse this row operation, we need to be able to multiply the row by the reciprocal of the constant.

b. If the multiplier is zero, the row operation produces no change in the matrix. This worthless operation still has an inverse.

4. **a.** The solution set is $\varnothing$.

b. We can assign arbitrary values to the three free variables and solve for the four leading variables in terms of these arbitrary parameters. The solution set will be a three-dimensional hyperplane in $\mathbb{R}^7$.

9. **a.** $\{(5, 1, 1, -2)\}$

b. $\{r(6, 2, 5, -3) + (1, -\frac{1}{3}, -\frac{7}{3}, 0) \mid r \in \mathbb{R}\}$

c. $\varnothing$

10. $\{r(-1, 1, 0, 0) + s(-5, 0, -1, 1) \mid r, s \in \mathbb{R}\}$

12. **a.** Yes **b.** No **c.** No **d.** No

  e. Yes **f.** No **g.** Yes **h.** No

16. $\frac{1}{3}$ are Leftists, $\frac{2}{5}$ are Centrists, and $\frac{4}{15}$ are Rightists.

Section 3.1

3. **a.** $(1, 5, 3, 0) = 3(1, 1, 1, 0) - 2(1, 0, 0, 1) + 2(0, 1, 0, 1)$

b. $(0, 0, 1, -2) = 1(1, 1, 1, 0) + (-1)(1, 0, 0, 1) + (-1)(0, 1, 0, 1)$

c. $(1, 1, 1, -3)$ is not a linear combination of these vectors.

d. $(-3, -5, -5, 2) = -5(1, 1, 1, 0) + 2(1, 0, 0, 1) + 0(0, 1, 0, 1)$

e. $(1, 0, 2, -1)$ is not a linear combination of these vectors.

f. $(1, 1, \frac{1}{2}, 1) = \frac{1}{2}(1, 1, 1, 0) + \frac{1}{2}(1, 0, 0, 1) + \frac{1}{2}(0, 1, 0, 1)$

7. If such a linear combination exists, we could find scalars a and b so that the identity $x^2 e^x = ae^x + bxe^x$ holds for all $x \in \mathbb{R}$. We can take $x = -1, x = 0$, and $x = 1$, for example, to obtain an inconsistent system

$$ae^{-1} - be^{-1} = e^{-1}$$
$$a = 0$$
$$ae + be = e$$

Hence, no such linear combination is possible.

11. $\mathbf{v} = 5\mathbf{w}_1 - 2\mathbf{w}_2 = 5(8\mathbf{x}_1 + \mathbf{x}_2) - 2(-3\mathbf{x}_1 + 2\mathbf{x}_2) = 46\mathbf{x}_1 + 1\mathbf{x}_2$

14. Let L_n and A_n denote the length and enclosed area of the star at the nth stage of the construction. For the original triangle, we have $L_0 = 3$ and $A_0 = \frac{\sqrt{3}}{4}$.

a. $L_1 = 3(\frac{4}{3})$, $A_1 = \frac{\sqrt{3}}{4}(1 + \frac{3}{9})$

b. $L_2 = 3\left(\frac{4}{3}\right)^2$, $A_2 = \frac{\sqrt{3}}{4}\left(1 + \frac{3}{9} + \frac{4\cdot3}{9^2}\right)$

c. At each stage in the construction, we replace each side with four segments $\frac{1}{3}$ as long. Thus, the number of sides increases by a factor of 4, and the length increases by a factor of $\frac{4}{3}$. One triangle is adjoined to each side, so the number of new triangles increase by a factor of 4, but the area of each triangle is $\frac{1}{9}$ the area of one of the triangles adjoined at the previous stage. Thus, there are $3 \cdot 4^n$ sides, $L_n = 3\left(\frac{4}{3}\right)^n$, and

$$
\begin{aligned}
A_n &= \frac{\sqrt{3}}{4}\left(1 + \frac{3}{9} + \frac{4\cdot3}{9^2} + \cdots + \frac{4^{n-1}\cdot3}{9^n}\right) \\
&= \frac{\sqrt{3}}{4}\left(1 + \frac{\frac{1}{3} - \frac{4^n\cdot3}{9^{n+1}}}{1 - \frac{4}{9}}\right) = \frac{\sqrt{3}}{20}\left(8 - 3\left(\frac{4}{9}\right)^n\right).
\end{aligned}
$$

d. The length increases without bound, and the limiting area is $\frac{2\sqrt{3}}{5}$.

Section 3.2

3. A linear combination of the five vectors set equal to an arbitrary vector in $\mathbb{R}^3$ yields a system of linear equations with coefficient matrix $\begin{bmatrix} 3 & 1 & -2 & 5 & 1 \\ 1 & 5 & 2 & 4 & 1 \\ 1 & 3 & 1 & 3 & 1 \end{bmatrix}$.
Since this reduces to a matrix with three leading 1s, it will always be possible to find the coefficients for the linear combination.

6. A matrix $\begin{bmatrix} a & b \\ c & d \end{bmatrix}$ in the span of this set must satisfy $a + d - b - c = 0$. Indeed, each of the four matrices satisfies this condition, and the condition is preserved by the operations of addition and scalar multiplication used to form linear combinations. A matrix such as $\begin{bmatrix} 1 & 0 \\ 0 & 0 \end{bmatrix}$ that does not satisfy this condition is therefore not in the span of the set.

9. Given any finite set of polynomials, let n denote the highest degree of the polynomials in the set. No linear combination of these polynomials will contain powers of x^{n+1}. Thus, the set cannot span $\mathbb{P}$.

13. Since span$\{v_1, \ldots, v_m\}$ is a subspace of the vector space, the previous exercise yields that span$\{w_1, \ldots, w_n\} \subseteq$ span$\{v_1, \ldots, v_m\}$. Similarly, since span$\{w_1, \ldots, w_n\}$ is a subspace of the vector space, span$\{v_1, \ldots, v_m\} \subseteq$ span$\{w_1, \ldots, w_n\}$. Hence, span$\{v_1, \ldots, v_m\}$ = span$\{w_1, \ldots, w_n\}$.

15. In $\mathbb{R}^2$, the vectors $(1, 0)$ and $(2, 0)$ are both linear combinations of $(1, 0)$ and $(0, 1)$. Even though $\{(1, 0), (0, 1)\}$ spans $\mathbb{R}^2$, the set $\{(1, 0), (2, 0)\}$ does not span $\mathbb{R}^2$. Thus, the claim does not hold in general. What about the converse?

Section 3.3

1. a. The hypothesis is $2x + 1 = 5$; the conclusion is $x = 2$. The statement is true. The converse is the statement "if $x = 2$, then $2x + 1 = 5$." The converse is true.

x	Hypothesis	Conclusion	Implication	Converse
1	False	False	True	True
2	True	True	True	True

b. The hypothesis is $x = y$; the conclusion is $|x| = |y|$. The statement is true. The converse is the statement "if $|x| = |y|$, then $x = y$." The converse is false.

x	y	Hypothesis	Conclusion	Implication	Converse
1	2	False	False	True	True
1	1	True	True	True	True
1	-1	False	True	True	False

c. The hypothesis is $\sqrt{x + 3} = 1$; the conclusion is $x = 4$. The statement is true (because the hypothesis is always false). The converse is the statement "if $x = 4$, then $\sqrt{x + 3} = 1$." The converse is false.

x	Hypothesis	Conclusion	Implication	Converse
1	False	False	True	True
4	False	True	True	False

d. The hypothesis is $\tan x = 1$; the conclusion is $x = \frac{\pi}{4}$. The statement is false. The converse is the statement "if $x = \frac{\pi}{4}$, then $\tan x = 1$." The converse is true.

x	Hypothesis	Conclusion	Implication	Converse
0	False	False	True	True
$\frac{5\pi}{4}$	True	False	False	True
$\frac{\pi}{4}$	True	True	True	True

e. The hypothesis is "f is differentiable"; the conclusion is "f is continuous." The statement is true. The converse is the statement "if f is continuous, then f is differentiable." The converse is false.

$f(x)$	Hypothesis	Conclusion	Implication	Converse
$[x]$	False	False	True	True
$[x]$	False	True	True	False
x^2	True	True	True	True

f. The hypothesis is "today is February 28"; the conclusion is "tomorrow is March 1." The statement is false. The converse is the statement "if tomorrow is March 1, then today is February 28." The converse is false.

Today	Tomorrow	Hypothesis	Conclusion	Implication	Converse
February 1	February 2	False	False	True	True
February 28	February 29	True	False	False	True
February 29	March 1	False	True	True	False
February 28	March 1	True	True	True	True

g. The hypothesis is "today is February 29"; the conclusion is "tomorrow is March 1." The statement is true. The converse is the statement "if tomorrow is March 1, then today is February 29." The converse is false.

Today	Tomorrow	Hypothesis	Conclusion	Implication	Converse
February 1	February 2	False	False	True	True
February 28	March 1	False	True	True	False
February 29	March 1	True	True	True	True

5. The set is linearly independent.

7. The polynomials form a linearly independent set.

9. Suppose $ae^x + be^{-x} = 0$ for all $x \in \mathbb{R}$. Letting $x = 0$ and $x = 1$ yields $a + b = 0$ and $ae + be^{-1} = 0$. The only solution to these two equations is $a = b = 0$. Thus, the functions form a linearly independent set.

10. $e^2 e^x + 0e^{2x} + (-1)e^{x+2} = 0$ for all $x \in \mathbb{R}$. Hence, the functions form a linearly dependent set.

13. Suppose $a_1 \mathbf{v}_1 + \cdots + a_m \mathbf{v}_m = \mathbf{0}$. Then $a_1 \mathbf{v}_1 + \cdots + a_m \mathbf{v}_m + 0\mathbf{v}_{m+1} + \cdots + 0\mathbf{v}_{m+k} = \mathbf{0} + 0\mathbf{v}_{m+1} + \cdots + 0\mathbf{v}_{m+k} = \mathbf{0}$. Since the vectors in the terms of this sum are linearly independent, all the coefficients must be zero. In particular, $a_1 = \cdots = a_m = 0$. Thus, $\{\mathbf{v}_1, \ldots, \mathbf{v}_m\}$ is a linearly independent subset.

17. a. The set is linearly independent.

b. The set is linearly dependent.

c. The set is linearly independent.

d. The set is linearly dependent.

e. The set is linearly dependent.

Section 3.4

2. Yes, for example, $\{(0, 1), (1, 1), (2, 1), (3, 1), (4, 1)\}$. No, $\mathbb{R}^2$ does not have a unique basis.

4. The set is a basis for $\mathbb{R}^4$.

6. The set is a basis for $\mathbb{M}(2, 2)$.

9. a. Since $\mathbf{0}(2) = 0$, the zero polynomial is in S; so $S \neq \emptyset$. If $p, q \in S$ and $r \in \mathbb{R}$, then $p(2) = q(2) = 0$, so $(p + q)(2) = p(2) + q(2) = 0$ and $(rp)(2) = r(p(2)) = r0 = 0$. Thus, S is closed under addition and scalar multiplication. By the Subspace Theorem, S is a subspace of $\mathbb{P}_3$.

b. Suppose $r(x-2) + sx(x-2) + tx^2(x-2) = 0$ for all $x \in \mathbb{R}$. Let $x = 0$ to find that $r = 0$. Compare coefficients of the first and third power terms on both sides of the equation to find that $s = t = 0$ also. Thus, the set is linearly independent. Let p denote an arbitrary element of S. Since $p(2) = 0$, we know that $x - 2$ is a factor of p. That is, $p(x) = (x - 2)(a + bx + cx^2)$ for some real numbers a, b, and c. We can rewrite this as $p(x) = a(x - 2) + bx(x - 2) + cx^2(x - 2)$ to see that p is a linear combination of the three polynomials. Thus, the set spans S. Hence, it is a basis for S.

  c. Suppose $r(x - 2) + s(x - 2)^2 + t(x - 2)^3 = 0$ for all $x \in \mathbb{R}$. Take the first, second, and third derivatives of both sides and evaluate at $x = 2$ to find that $r = s = t = 0$. Thus, the set is linearly independent. Let p denote an arbitrary element of S. As in part b, we can write $p(x) = (x - 2)(a + bx + cx^2) = (x - 2)\big((a + 2b + 4c) + (b + 4c)(x - 2) + c(x - 2)^2\big) = (a + 2b + 4c)(x - 2) + (b + 4c)(x - 2)^2 + c(x - 2)^3$. Thus, the set spans S. Hence, it is a basis for S.

  d. $\dim S = 3$

12. If $\{\mathbf{v}, \mathbf{w}\}$ is a basis for a vector space, then it spans the space and is linearly independent. We can conclude from exercises in the previous two sections that $\{\mathbf{v} + \mathbf{w}, \mathbf{w}\}$ also spans the vector space and is linearly independent. Therefore, it is a basis.

Section 3.5

1.   **a.** Suppose $\sqrt{2} = \frac{p}{q}$, where p and q are positive integers with no common prime factors.

  b. $2q^2 = p^2$

  c. Since the square of an odd number is odd, the only way p^2 can be even is for p itself to be even.

  d. Thus, we can write $p = 2k$.

  e. Now $2q^2 = p^2 = 4k^2$, or $q^2 = 2k^2$. Since q^2 is even, we can conclude as above that q is even.

  f. Thus, p and q have a common factor of 2. This contradicts the assumption that p and q have no common prime factors. Therefore, it is impossible to write $\sqrt{2}$ as a rational number.

7. Assume $m > 4$. Write $\mathbf{v}_1, \ldots, \mathbf{v}_m$ as linear combinations of $\mathbf{w}_1, \mathbf{w}_2, \mathbf{w}_3$, and $\mathbf{w}_4$. Use the coefficients in a homogeneous system of four equations in m unknowns. Let $(c_1, \ldots, c_m)$ be a nontrivial solution. Then $c_1\mathbf{v}_1 + \cdots c_m\mathbf{v}_m$ will be a nontrivial linear combination that is equal to zero. This contradiction to the linear independence of $\{\mathbf{v}_1, \ldots, \mathbf{v}_m\}$ shows that our initial assumption was incorrect. That is, we must have $m \leq 4$.

10. Assume that $r_1\mathbf{v}_1 + \cdots + r_n\mathbf{v}_n + r_{n+1}\mathbf{v}_{n+1} = \mathbf{0}$. First we must have $r_{n+1} = 0$; if not, $\mathbf{v}_{n+1} = -\frac{r_1}{r_{n+1}}\mathbf{v}_1 - \cdots - \frac{r_n}{r_{n+1}}\mathbf{v}_n$, a linear combination of $\{\mathbf{v}_1, \ldots, \mathbf{v}_n\}$. This contradicts the hypothesis that $\mathbf{v}_{n+1} \notin \operatorname{span}\{\mathbf{v}_1, \ldots, \mathbf{v}_n\}$. Now the original equation reduces to $r_1\mathbf{v}_1 + \cdots + r_n\mathbf{v}_n = \mathbf{0}$. Since $\{\mathbf{v}_1, \ldots, \mathbf{v}_n\}$ is linearly independent, we have that $r_1 = 0, \ldots, r_n = 0$. Since all $n + 1$ coefficients have been shown to be equal to zero, we conclude that $\{v_1, \ldots, v_n, v_{n+1}\}$ is linearly independent.

12. We cannot apply the Expansion Theorem directly to S because we do not know that S is finite-dimensional. Instead we use the construction of a basis given in the proof of that theorem. Start with $\varnothing$ as a linearly independent subset of S. If this spans S, we have a basis for S. Otherwise, consider the set $\{v_1\}$ where $v_1 \in S$ and $v_1 \notin \text{span}\, \varnothing = \{0\}$. This set is linearly independent. If it spans S, we have a basis. Otherwise, there is $v_2 \in S$ with $v_2 \notin \text{span}\{v_1\}$. By the Expansion Lemma, $\{v_1, v_2\}$ is linearly independent. Repeat this process of adjoining vectors to this set as long as the set does not span S. At each stage, choose the vector not in the span of the previous vectors. Thus, the sets are linearly independent. This process must stop before we reach dim V elements. Otherwise, we would have an independent set with more elements than a set that spans V. This would contradict the Comparison Theorem. It follows that we will eventually obtain a basis for S with a number of elements less than or equal to dim V. That is, S is finite-dimensional, and dim $S \le$ dim V.

17. Set a linear combination of the polynomials equal to 0. Comparison of the coefficients of the powers of x gives a homogeneous system of equations whose

coefficient matrix is $\begin{bmatrix} 0 & 1 & 1 & 0 & 0 & 1 \\ 1 & 0 & 1 & 0 & 1 & 0 \\ 2 & 3 & 0 & 1 & 3 & 2 \\ 0 & 0 & 0 & 1 & 1 & 1 \end{bmatrix}$. This matrix reduces to a ma-

trix with leading 1s in the first four columns. Thus, we can eliminate the last two vectors. The resulting set $\{x^2 + 2x, x^3 + 3x, x^3 + x^2, x + 1\}$ will be linearly independent and still span the same subspace of $\mathbb{P}_3$. Since dim $\mathbb{P}_3 = 4$, Theorem 3.15 yields that this linearly independent set of four elements is a basis for $\mathbb{P}_3$.

19. The vector $(1, 0, 0)$ is not a linear combination of $(2, 3, 1)$ and $(4, -2, 1)$. Hence, we can adjoin it to form a linearly independent set with $3 = $ dim $\mathbb{R}^3$ elements. We cannot form a larger linearly independent subset of $\mathbb{R}^3$. Hence, $\{(2, 3, 1), (4, -2, 1), (1, 0, 0)\}$ must be a basis for $\mathbb{R}^3$.

21. Let $\{v_1, \ldots, v_n\}$ be a basis for S. If there were a vector $v \in V$ with $v \notin S = \text{span}\{v_1, \ldots, v_n\}$, we could adjoin v to the basis to get a linearly independent set with $n + 1$ elements. By the Comparison Theorem, this cannot happen in the vector space V with dim $V = $ dim $S = n$.

Section 3.6

1. a. $\begin{bmatrix} 1 \\ 0 \\ 0 \end{bmatrix}$ **b.** $\begin{bmatrix} -2 \\ 3 \\ 6 \end{bmatrix}$ **c.** $\begin{bmatrix} 0 \\ \frac{3}{2} \\ -\frac{5}{2} \end{bmatrix}$

5. **a.** $\begin{bmatrix} 0 \\ 1 \\ 0 \end{bmatrix}$ **b.** $\begin{bmatrix} \frac{7}{4} \\ -\frac{3}{4} \\ \frac{1}{4} \end{bmatrix}$ **c.** $\begin{bmatrix} 1 \\ 0 \\ 1 \end{bmatrix}$

d. $\begin{bmatrix} -2 \\ 3 \\ -2 \end{bmatrix}$ **e.** $\begin{bmatrix} \frac{1}{2} \\ \frac{1}{2} \\ \frac{1}{2} \end{bmatrix}$ **f.** $\begin{bmatrix} \frac{33}{2} \\ -\frac{15}{2} \\ \frac{1}{2} \end{bmatrix}$

7. $\sin(x + \frac{\pi}{3}) = \sin x \cos \frac{\pi}{3} + \cos x \sin \frac{\pi}{3} = \frac{1}{2} \sin x + \frac{\sqrt{3}}{2} \cos x$. So the coordinate vector is $\begin{bmatrix} \frac{1}{2} \\ \frac{\sqrt{3}}{2} \end{bmatrix}$.

9. Let $\begin{bmatrix} a \\ b \end{bmatrix}$ and $\begin{bmatrix} c \\ d \end{bmatrix}$ denote the two vectors in B. The first condition gives $\begin{bmatrix} a \\ b \end{bmatrix} + \begin{bmatrix} c \\ d \end{bmatrix} = \begin{bmatrix} 3 \\ 2 \end{bmatrix}$, and the second condition gives $2\begin{bmatrix} a \\ b \end{bmatrix} + \begin{bmatrix} c \\ d \end{bmatrix} = \begin{bmatrix} -1 \\ 4 \end{bmatrix}$. These equations lead to the system

$$\begin{array}{rcl} a \quad + c & = & 3 \\ b \quad + d & = & 2 \\ 2a \quad + c & = & -1 \\ 2b \quad + d & = & 4 \end{array}$$

with solution $a = -4, b = 2, c = 7, d = 0$. Hence, $B = \left\{ \begin{bmatrix} -4 \\ 2 \end{bmatrix}, \begin{bmatrix} 7 \\ 0 \end{bmatrix} \right\}$.

Review Exercises: Chapter 3

3. **a.** Suppose $a\begin{bmatrix} 0 & 0 \\ 0 & 1 \end{bmatrix} + b\begin{bmatrix} 0 & 0 \\ 1 & 2 \end{bmatrix} + c\begin{bmatrix} 0 & 1 \\ 2 & 3 \end{bmatrix} + d\begin{bmatrix} 1 & 2 \\ 3 & 4 \end{bmatrix} = \begin{bmatrix} 0 & 0 \\ 0 & 0 \end{bmatrix}$.
Then $d = 0, c + 2d = 0, b + 2c + 3d = 0$, and $a + 2b + 3c + 4d = 0$. It follows that $a = b = c = d = 0$. Thus, the set is linearly independent.

b. $\dim \mathbb{M}(2, 2) = 4$, and B has four elements. By Theorem 3.15, since B is linearly independent, it is a basis for $\mathbb{M}(2, 2)$.

c. $\begin{bmatrix} 2 & 3 \\ 0 & -1 \end{bmatrix} = 2\begin{bmatrix} 0 & 0 \\ 0 & 1 \end{bmatrix} + (-4)\begin{bmatrix} 0 & 0 \\ 1 & 2 \end{bmatrix} + (-1)\begin{bmatrix} 0 & 1 \\ 2 & 3 \end{bmatrix} + 2\begin{bmatrix} 1 & 2 \\ 3 & 4 \end{bmatrix}$

d. $\begin{bmatrix} 2 & 3 \\ 0 & -1 \end{bmatrix}_B = \begin{bmatrix} 2 \\ -4 \\ -1 \\ 2 \end{bmatrix}$

5. Since $\sin 2x = 2 \sin x \cos x$, we have a nontrivial linear combination of these three functions that equals zero: $0f + g - 2h = \mathbf{0}$. Thus, the functions form a linearly dependent set.

7. **a.** Since $\mathbf{v}+\mathbf{w}$ and $\mathbf{v}-\mathbf{w}$ are elements of V, we know that $\text{span}\{\mathbf{v}+\mathbf{w}, \mathbf{v}-\mathbf{w}\} \subseteq V$. Now, let $\mathbf{x}$ be an arbitrary element of V. We know that $\mathbf{x} = a\mathbf{v} + b\mathbf{w}$ for some $a, b \in \mathbb{R}$. Thus, $\mathbf{x} = \left(\frac{a+b}{2}\right)(\mathbf{v} + \mathbf{w}) + \left(\frac{a-b}{2}\right)(\mathbf{v} - \mathbf{w})$. Therefore, $\{\mathbf{v} + \mathbf{w}, \mathbf{v} - \mathbf{w}\}$ spans V.

b. Suppose $a(\mathbf{v} + \mathbf{w}) + b(\mathbf{v} - \mathbf{w}) = \mathbf{0}$. Then $(a + b)\mathbf{v} + (a - b)\mathbf{w} = \mathbf{0}$. Since $\{\mathbf{v}, \mathbf{w}\}$ is linearly independent, $a + b = 0$ and $a - b = 0$. From these equations, it follows that $a = b = 0$. Thus, $\{\mathbf{v} + \mathbf{w}, \mathbf{v} - \mathbf{w}\}$ is linearly independent.

9.

a.	No	**b.**	No	**c.**	Yes
d.	No	**e.**	Yes	**f.**	No
g.	No	**h.**	No	**i.**	Yes
j.	Yes	**k.**	Yes	**l.**	Yes

13. **a.** $\begin{bmatrix} 1 \\ 1 \\ \ln 2 \end{bmatrix}$ **b.** $\begin{bmatrix} 1 \\ -1 \\ 0 \end{bmatrix}$ **c.** $\begin{bmatrix} 0 \\ -1 \\ 0 \end{bmatrix}$ **d.** $\begin{bmatrix} 0 \\ -2 \\ 0 \end{bmatrix}$

15. **a.** Let $\mathbf{x} \in S \cup T$. Either $\mathbf{x} \in S$ or $\mathbf{x} \in T$. In the first case, $\mathbf{x} = a_1\mathbf{v}_1 + \cdots + a_m\mathbf{v}_m = a_1\mathbf{v}_1 + \cdots + a_m\mathbf{v}_m + 0\mathbf{w}_1 + \cdots + 0\mathbf{w}_n$. In the second case, $\mathbf{x} = b_1\mathbf{w}_1 + \cdots + b_n\mathbf{w}_n = 0\mathbf{v}_1 + \cdots + 0\mathbf{v}_m + b_1\mathbf{w}_1 + \cdots + b_n\mathbf{w}_n$. In either case, $\mathbf{x} \in \text{span}\{\mathbf{v}_1, \ldots, \mathbf{v}_m, \mathbf{w}_1, \ldots, \mathbf{w}_n\}$.

b. Suppose W is a subspace containing $S \cup T$. Since W contains the vectors $\mathbf{v}_1, \ldots, \mathbf{v}_m$ and $\mathbf{w}_1, \ldots, \mathbf{w}_n$ and since W is closed under addition and scalar multiplication, it follows that W contains all linear combinations of the vectors $\mathbf{v}_1, \ldots, \mathbf{v}_m, \mathbf{w}_1, \ldots, \mathbf{w}_n$. Thus, $\text{span}\{\mathbf{v}_1, \ldots, \mathbf{v}_m, \mathbf{w}_1, \ldots, \mathbf{w}_n\} \subseteq W$.

17. **a.** Suppose $a_1\mathbf{x}_1 + \cdots + a_k\mathbf{x}_k + a_{k+1}\mathbf{v}_{k+1} + \cdots + a_m\mathbf{v}_m + b_{k+1}\mathbf{w}_{k+1} + \cdots + b_n\mathbf{w}_n = \mathbf{0}$. Then $a_1\mathbf{x}_1 + \cdots + a_k\mathbf{x}_k + a_{k+1}\mathbf{v}_{k+1} + \cdots + a_m\mathbf{v}_m = -b_{k+1}\mathbf{w}_{k+1} - \cdots - b_n\mathbf{w}_n$ is in both S and T. Thus, it is a linear combination of $\mathbf{x}_1, \ldots, \mathbf{x}_k$. By Theorem 3.17, the coefficients for writing a vector as a linear combination of basis vectors are unique. Thus, $a_{k+1} = \cdots = a_m = b_{k+1} = \cdots = b_n = 0$. The original equation now becomes $a_1\mathbf{x}_1 + \cdots + a_k\mathbf{x}_k = \mathbf{0}$. Since $\{\mathbf{x}_1, \ldots, \mathbf{x}_k\}$ is linearly independent, we also conclude that $a_1 = \cdots = a_k = 0$. Hence, all the coefficients are equal to zero. It follows that $\{\mathbf{x}_1, \ldots, \mathbf{x}_k, \mathbf{v}_{k+1}, \ldots, \mathbf{v}_m, \mathbf{w}_{k+1}, \ldots, \mathbf{w}_n\}$ is linearly independent.

b. Any element of $S \cup T$ can be written as a linear combination of elements of $\{\mathbf{x}_1, \ldots, \mathbf{x}_k, \mathbf{v}_{k+1}, \ldots, \mathbf{v}_m, \mathbf{w}_{k+1}, \ldots, \mathbf{w}_n\}$. Hence, any linear combination of elements of $S \cup T$ can also be written as a linear combination of elements of $\{\mathbf{x}_1, \ldots, \mathbf{x}_k, \mathbf{v}_{k+1}, \ldots, \mathbf{v}_m, \mathbf{w}_{k+1}, \ldots, \mathbf{w}_n\}$. Thus, $\{\mathbf{x}_1, \ldots, \mathbf{x}_k, \mathbf{v}_{k+1}, \ldots, \mathbf{v}_m, \mathbf{w}_{k+1}, \ldots, \mathbf{w}_n\}$ spans $\text{span}(S \cup T)$. Therefore, this set with $m + n - k$ elements is a basis for $\text{span}(S \cup T)$. Hence, $\dim\big(\text{span}(S \cup T)\big) = m + n - k = \dim S + \dim T - \dim(S \cap T)$.

Section 4.1

4. a. $\frac{1}{4}$

   b. 0

   c. $-4\pi^2$

   d. 0

   e. 0

   f. 0

8. a. $\langle -\mathbf{v}, \mathbf{w} \rangle = \langle (-1)\mathbf{v}, \mathbf{w} \rangle = (-1)\langle \mathbf{v}, \mathbf{w} \rangle = \langle \mathbf{v}, (-1)\mathbf{w} \rangle = \langle \mathbf{v}, -\mathbf{w} \rangle$, and
$\langle \mathbf{v}, -\mathbf{w} \rangle = \langle \mathbf{v}, (-1)\mathbf{w} \rangle = (-1)\langle \mathbf{v}, \mathbf{w} \rangle = -\langle \mathbf{v}, \mathbf{w} \rangle$.

   b. $\langle -\mathbf{v}, -\mathbf{w} \rangle = \langle (-1)\mathbf{v}, (-1)\mathbf{w} \rangle = (-1)(-1)\langle \mathbf{v}, \mathbf{w} \rangle = \langle \mathbf{v}, \mathbf{w} \rangle$

10. $\langle \mathbf{v}+\mathbf{w}, \mathbf{v}+\mathbf{w} \rangle = \langle \mathbf{v}, \mathbf{v}+\mathbf{w} \rangle + \langle \mathbf{w}, \mathbf{v}+\mathbf{w} \rangle = \langle \mathbf{v}, \mathbf{v} \rangle + \langle \mathbf{v}, \mathbf{w} \rangle + \langle \mathbf{w}, \mathbf{v} \rangle + \langle \mathbf{w}, \mathbf{w} \rangle =$
$\langle \mathbf{v}, \mathbf{v} \rangle + \langle \mathbf{v}, \mathbf{w} \rangle + \langle \mathbf{v}, \mathbf{w} \rangle + \langle \mathbf{w}, \mathbf{w} \rangle = \|\mathbf{v}\|^2 + 2\langle \mathbf{v}, \mathbf{w} \rangle + \|\mathbf{w}\|^2$

15. 1. $\langle \mathbf{v}, \mathbf{v} \rangle = 2v_1^2 - v_1 v_2 - v_2 v_1 + 5v_2^2 = v_1^2 + (v_1 - v_2)^2 + 4v_2^2 \geq 0$. Also, the only way the sum of nonnegative real numbers can equal zero is if all the terms are zero. Hence, $v_1 = v_2 = 0$; that is, $\mathbf{v} = (0,0) = \mathbf{0}$.

   2. $\langle \mathbf{v}, \mathbf{w} \rangle = 2v_1 w_1 - v_1 w_2 - v_2 w_1 + 5v_2 w_2 = 2w_1 v_1 - w_1 v_2 - w_2 v_1 + 5w_2 v_2 = \langle \mathbf{w}, \mathbf{v} \rangle$

   3. $\langle r\mathbf{v}, \mathbf{w} \rangle = 2(rv_1)w_1 - (rv_1)w_2 - (rv_2)w_1 + 5(rv_2)w_2 = r(2v_1 w_1 - v_1 w_2 - v_2 w_1 + 5v_2 w_2) = r\langle \mathbf{v}, \mathbf{w} \rangle$

   4. $\langle \mathbf{v} + \mathbf{w}, \mathbf{x} \rangle = 2(v_1 + w_1)x_1 - (v_1 + w_1)x_2 - (v_2 + w_2)x_1 + 5(v_2 + w_2)x_2 = 2v_1 x_1 + 2w_1 x_1 - v_1 x_2 - w_1 x_2 - v_2 x_1 - w_2 x_1 + 5v_2 x_2 + 5w_2 x_2 = 2v_1 x_1 - v_1 x_2 - v_2 x_1 + 5v_2 x_2 + 2w_1 x_1 - w_1 x_2 - w_2 x_1 + 5w_2 x_2 = \langle \mathbf{v}, \mathbf{x} \rangle + \langle \mathbf{w}, \mathbf{x} \rangle$

17. a. 1. $\langle p, p \rangle = (p(-1))^2 + (p(0))^2 + (p(1))^2 \geq 0$. If $\langle p, p \rangle = (p(-1))^2 + (p(0))^2 + (p(1))^2 = 0$, then $p(-1) = p(0) = p(1) = 0$. But the only polynomial in $\mathbb{P}_2$ that has three zeros is the constant zero function. Thus, $p = \mathbf{0}$.

   2. $\langle p, q \rangle = p(-1)q(-1) + p(0)q(0) + p(1)q(1) = q(-1)p(-1) + q(0)p(0) + q(1)p(1) = \langle q, p \rangle$

   3. $\langle rp, q \rangle = (rp(-1))q(-1) + (rp(0))q(0) + (rp(1))q(1) = r(p(-1)q(-1) + p(0)q(0) + p(1)q(1)) = r\langle p, q \rangle$

   4. $\langle p_1 + p_2, q \rangle = (p_1(-1) + p_2(-1))q(-1) + (p_1(0) + p_2(0))q(0) + (p_1(1) + p_2(1))q(1) = p_1(-1)q(-1) + p_2(-1)q(-1) + p_1(0)q(0) + p_2(0)q(0) + p_1(1)q(1) + p_2(1)q(1) = p_1(-1)q(-1) + p_1(0)q(0) + p_1(1)q(1) + p_2(-1)q(-1) + p_2(0)q(0) + p_2(1)q(1) = \langle p_1, q \rangle + \langle p_2, q \rangle$

   b. For $p(x) = q(x) = 1$, this inner product gives 3, whereas the standard inner product gives $\int_{-1}^{1} 1 \, dx = 2$.

   c. Consider $p(x) = x(x - 1)(x + 1)$, for example. This is a nonzero element of $\mathbb{P}_3$ although $\langle p, p \rangle = (p(-1))^2 + (p(0))^2 + (p(1))^2 = 0$. Thus, the positive-definite law fails.

21. $\|v-w\|^2 = \langle v-w, v-w \rangle = \langle v+(-1)w, v+(-1)w \rangle = \langle v, v \rangle + (-1)\langle v, w \rangle + (-1)\langle w, v \rangle + (-1)^2\langle w, w \rangle = \langle v, v \rangle - \langle v, w \rangle - \langle v, w \rangle + \langle w, w \rangle = \|v\|^2 - 2\langle v, w \rangle + \|w\|^2$

Section 4.2

3. Consider the unit hypercube with one vertex at the origin and edges along the four axes. The angles between the pairs of vectors chosen from $(1, 0, 0, 0)$, $(1, 1, 0, 0)$, $(1, 1, 1, 0)$ are as in the previous exercise. The angles between these vectors and $(1, 1, 1, 1)$ are $60°$, $45°$, and $30°$, respectively.

7. $\{r(2, -4, 1) \mid r \in \mathbb{R}\}$

9. Let $S = \{w \in \mathbb{R}^n \mid w \cdot v = 0\}$. Since $0 \cdot v = 0$, we have $0 \in S$, and so $S \neq \emptyset$. Let $w_1, w_2 \in S$ and $r \in \mathbb{R}$. Then $w_1 \cdot v = 0$ and $w_2 \cdot v = 0$. Thus, $(w_1 + w_2) \cdot v = w_1 \cdot v + w_2 \cdot v = 0 + 0 = 0$ and $(rw_1) \cdot v = r(w_1 \cdot v) = r0 = 0$. It follows that $w_1 + w_2 \in S$ and $rw_1 \in S$. That is, S is closed under addition and scalar multiplication. By the Subspace Theorem, S is a subspace of $\mathbb{R}^n$.

11. $(2, 1, 3)$ is normal to the first plane, and $(1, -5, 1)$ is normal to the second plane. Since $(2, 1, 3) \cdot (1, -5, 1) = 0$, these vectors are orthogonal. It follows that the planes are orthogonal.

15. The orthogonal projection is $-\frac{28}{17}(-3, 2, 2) = \left(\frac{84}{17}, -\frac{56}{17}, -\frac{56}{17}\right)$. We can write $(4, -3, -5) = \left(\frac{84}{17}, -\frac{56}{17}, -\frac{56}{17}\right) + \left(-\frac{16}{17}, \frac{5}{17}, -\frac{29}{17}\right)$, where the first vector is parallel to $(-3, 2, 2)$ and the second is orthogonal to $(-3, 2, 2)$.

17. **a.** $c_1 u_1 \cdot u_1 + c_2 u_2 \cdot u_1 = v \cdot u_1$ and $c_1 u_1 \cdot u_2 + c_2 u_2 \cdot u_2 = v \cdot u_2$.

b. The equations $21c_1 + 21c_2 = -11$ and $21c_1 + 35c_2 = -10$ have solution $c_1 = -\frac{25}{42}$ and $c_2 = \frac{1}{14}$. The projection is $-\frac{25}{42}(2, -4, -1) + \frac{1}{14}(5, -3, 1) = \left(-\frac{5}{6}, \frac{13}{6}, \frac{2}{3}\right)$. We can write $(1, 4, -3) = \left(-\frac{5}{6}, \frac{13}{6}, \frac{2}{3}\right) + \left(\frac{11}{6}, \frac{11}{6}, -\frac{11}{3}\right)$, where the first vector is in the plane spanned by $(2, -4, -1)$ and $(5, -3, 1)$, and the second is orthogonal to this plane.

c. The equations $21c_1 = -11$ and $14c_2 = 1$ have solution $c_1 = -\frac{11}{21}$ and $c_2 = \frac{1}{14}$. The projection is $-\frac{11}{21}(2, -4, -1) + \frac{1}{14}(3, 1, 2) = \left(-\frac{5}{6}, \frac{13}{6}, \frac{2}{3}\right)$. We can write $(1, 4, -3) = \left(-\frac{5}{6}, \frac{13}{6}, \frac{2}{3}\right) + \left(\frac{11}{6}, \frac{11}{6}, -\frac{11}{3}\right)$, where the first vector is in the plane spanned by $(2, -4, -1)$ and $(5, -3, 1)$, and the second is orthogonal to this plane.

d. The two vectors that span the plane are orthogonal.

Section 4.3

3. $\|v\| = \|(v-w)+w\| \leq \|v-w\| + \|w\|$, and $\|w\| = \|(w-v)+v\| \leq \|w-v\| + \|v\|$. The conclusion follows from the inequalities $-\|v-w\| \leq \|v\| - \|w\| \leq \|v-w\|$.

5. Apply the Cauchy-Schwarz inequality in $\mathbb{R}^n$ to the vectors $v = (a_1, \ldots, a_n)$ and $w = (1, \ldots, 1)$. This gives $a_1 + \cdots + a_n \leq \sqrt{a_1^2 + \cdots + a_n^2}\sqrt{n}$. Divide both sides of this inequality by n to obtain the result.

7. **a.** Apply the Cauchy-Schwarz inequality to the functions 1 and f in $\mathbb{C}([a, b])$. This gives $\left| \int_a^b f(x) \, dx \right| \leq \sqrt{\int_a^b 1 \, dx} \sqrt{\int_a^b (f(x))^2 \, dx} = \sqrt{b - a} \sqrt{\int_a^b (f(x))^2 \, dx}$. Square the nonnegative quantities on both sides to obtain the result.

   b. Let $x \in [a, b]$. Apply the result of part a to f' on the interval $[a, x]$ to obtain $\left(\int_a^x f'(t) \, dt \right)^2 \leq (x - a) \int_a^x (f'(t))^2 \, dt \leq (x - a) \int_a^b (f'(x))^2 \, dx$. By the Fundamental Theorem of Calculus, the integral on the left side of this inequality is equal to $f(x) - f(a) = f(x) - 0 = f(x)$.

   c. Integrate both sides of the inequality in part b, and notice that $\int_a^b (f'(x))^2 \, dx$ is a constant in this integration. This gives
   $$\int_a^b (f(x))^2 \, dx \leq \left(\int_a^b (x - a) \, dx \right) \left(\int_a^b (f'(x))^2 \, dx \right) = \frac{(b - a)^2}{2} \int_a^b (f'(x))^2 \, dx.$$

9. **a.** $\cos \theta = \frac{\sqrt{15}}{4}$, so $\theta \approx 14° \, 29'$.

   b. $\frac{3}{4} x$

   c. $g(x) = \frac{3}{4} x + (x^2 - \frac{3}{4} x)$

11. $\cos \theta = \frac{a}{|a|} \sqrt{\frac{6}{e^2 - 1}} \approx \pm 0.969$. So if $a > 0$, then $\theta \approx 14°$, and if $a < 0$, then $\theta \approx 166°$.

Section 4.4

1. The second longest word in the text is *orthonormalization*. The longest word in the text is *antidifferentiation*.

5. $(\frac{3}{7}, \frac{6}{7}, -\frac{2}{7}) \cdot (-\frac{2}{7}, \frac{3}{7}, \frac{6}{7}) = 0$, $(\frac{3}{7}, \frac{6}{7}, -\frac{2}{7}) \cdot (\frac{6}{7}, -\frac{2}{7}, \frac{3}{7}) = 0$, and $(-\frac{2x}{7}, \frac{3}{7}, \frac{6}{7}) \cdot (\frac{6}{7}, -\frac{2}{7}, \frac{3}{7}) = 0$. Also, $\|(\frac{3}{7}, \frac{6}{7}, -\frac{2}{7})\| = 1$, $\|(-\frac{2}{7}, \frac{3}{7}, \frac{6}{7})\| = 1$, and $\|(\frac{6}{7}, -\frac{2}{7}, \frac{3}{7})\| = 1$. By Theorem 4.14, this set is linearly independent. Since dim $\mathbb{R}^3 = 3$, Theorem 3.15 yields the result that this set spans $\mathbb{R}^3$ and hence is a basis.

7. Start with the basis $\{1, x, x^2\}$ for $\mathbb{P}_2$. Then $e_1 = \frac{\sqrt{2}}{2}$. Next, $v_2 = x$ and $e_2 = \frac{\sqrt{6}}{2} x$. Finally, $v_3 = x^2 - \frac{1}{3}$ and $e_3 = \frac{3\sqrt{10}}{4} x^2 - \frac{\sqrt{10}}{4}$.

12. By the Expansion Theorem we can adjoin vectors to the linearly independent set $\{\mathbf{w}\}$ to obtain a basis for V of $n = \dim V$ elements. Use this basis to generate an orthonormal basis $\{e_1, \ldots, e_n\}$ by the Gram-Schmidt process. The first vector e_1 will be a scalar multiple of $\mathbf{w}$. Therefore, the remaining vectors will be orthogonal to $\mathbf{w}$. That is, $e_2, \ldots, e_n \in \mathbf{w}^\perp$. We can write any vector $\mathbf{v} \in \mathbf{w}^\perp$ as a linear combination of $\{e_1, \ldots, e_n\}$. Since the coefficient of e_1 is $\langle \mathbf{v}, e_1 \rangle = 0$, we see that $\{e_2, \ldots, e_n\}$ spans $\mathbf{w}^\perp$. Of course, this set is linearly independent. Hence, it is a basis for $\mathbf{w}^\perp$. It follows that $\dim \mathbf{w}^\perp = n - 1$.

15. $\|\mathbf{v} + \mathbf{w}\|^2 = \|\mathbf{v}\|^2 + 2\langle \mathbf{v}, \mathbf{w} \rangle + \|\mathbf{w}\|^2 = \|\mathbf{v}\|^2 + 2 \cdot 0 + \|\mathbf{w}\|^2 = \|\mathbf{v}\|^2 + \|\mathbf{w}\|^2$

18. $\|\mathbf{u}_1\| = 1$, so $e_1 = \frac{1}{\|\mathbf{u}_1\|} \mathbf{u}_1 = \mathbf{u}_1$. The terms subtracted from $\mathbf{u}_2, \ldots, \mathbf{u}_k$ to produce $\mathbf{v}_2, \ldots, \mathbf{v}_k$ all have coefficients equal to zero. Hence, $\mathbf{v}_i = \mathbf{u}_i$ for $i = 2, \ldots, k$. For $i = 2, \ldots, k$, we have $\|\mathbf{u}_i\| = 1$, so $e_i = \frac{1}{\|\mathbf{v}_i\|} \mathbf{v}_i = \mathbf{u}_i$.

Section 4.5

3. **a.** $\left(\frac{1}{\sqrt{7}}, \frac{2}{\sqrt{7}}, -\frac{1}{\sqrt{7}}, \frac{1}{\sqrt{7}}\right) \cdot \left(\frac{1}{\sqrt{3}}, -\frac{1}{\sqrt{3}}, -\frac{1}{\sqrt{3}}, 0\right) = 0,$

$\left(\frac{1}{\sqrt{7}}, \frac{2}{\sqrt{7}}, -\frac{1}{\sqrt{7}}, \frac{1}{\sqrt{7}}\right) \cdot \left(\frac{2}{\sqrt{23}}, -\frac{1}{\sqrt{23}}, \frac{3}{\sqrt{23}}, \frac{3}{\sqrt{23}}\right) = 0,$

$\left(\frac{1}{\sqrt{3}}, -\frac{1}{\sqrt{3}}, -\frac{1}{\sqrt{3}}, 0\right) \cdot \left(\frac{2}{\sqrt{23}}, -\frac{1}{\sqrt{23}}, \frac{3}{\sqrt{23}}, \frac{3}{\sqrt{23}}\right) = 0,$

$\left(\frac{1}{\sqrt{7}}, \frac{2}{\sqrt{7}}, -\frac{1}{\sqrt{7}}, \frac{1}{\sqrt{7}}\right) \cdot \left(\frac{1}{\sqrt{7}}, \frac{2}{\sqrt{7}}, -\frac{1}{\sqrt{7}}, \frac{1}{\sqrt{7}}\right) = 1,$

$\left(\frac{1}{\sqrt{3}}, -\frac{1}{\sqrt{3}}, -\frac{1}{\sqrt{3}}, 0\right) \cdot \left(\frac{1}{\sqrt{3}}, -\frac{1}{\sqrt{3}}, -\frac{1}{\sqrt{3}}, 0\right) = 1,$ and

$\left(\frac{2}{\sqrt{23}}, -\frac{1}{\sqrt{23}}, \frac{3}{\sqrt{23}}, \frac{3}{\sqrt{23}}\right) \cdot \left(\frac{2}{\sqrt{23}}, -\frac{1}{\sqrt{23}}, \frac{3}{\sqrt{23}}, \frac{3}{\sqrt{23}}\right) = 1.$

b. The coefficients of $\text{proj}_S(1, 0, 0, 1)$ with respect to the given basis are
$(1, 0, 0, 1) \cdot \left(\frac{1}{\sqrt{7}}, \frac{2}{\sqrt{7}}, -\frac{1}{\sqrt{7}}, \frac{1}{\sqrt{7}}\right) = \frac{2}{\sqrt{7}}, (1, 0, 0, 1) \cdot \left(\frac{1}{\sqrt{3}}, -\frac{1}{\sqrt{3}}, -\frac{1}{\sqrt{3}}, 0\right) = \frac{1}{\sqrt{3}},$ and $(1, 0, 0, 1) \cdot \left(\frac{2}{\sqrt{23}}, -\frac{1}{\sqrt{23}}, \frac{3}{\sqrt{23}}, \frac{3}{\sqrt{23}}\right) = \frac{5}{\sqrt{23}}.$ Thus,

$\text{proj}_S(1, 0, 0, 1) = \frac{2}{\sqrt{7}}\left(\frac{1}{\sqrt{7}}, \frac{2}{\sqrt{7}}, -\frac{1}{\sqrt{7}}, \frac{1}{\sqrt{7}}\right) + \frac{1}{\sqrt{3}}\left(\frac{1}{\sqrt{3}}, -\frac{1}{\sqrt{3}}, -\frac{1}{\sqrt{3}}, 0\right) +$
$\frac{5}{\sqrt{23}}\left(\frac{2}{\sqrt{23}}, -\frac{1}{\sqrt{23}}, \frac{3}{\sqrt{23}}, \frac{3}{\sqrt{23}}\right) = \left(\frac{509}{483}, \frac{10}{483}, \frac{16}{483}, \frac{151}{161}\right).$

c. $(1, 0, 0, 1) = \left(\frac{509}{483}, \frac{10}{483}, \frac{16}{483}, \frac{151}{161}\right) + \left(-\frac{26}{483}, -\frac{10}{483}, -\frac{16}{483}, \frac{10}{161}\right)$

5. Apply the Gram-Schmidt process to $\{f, g\}$ to generate an orthonormal basis for S. $e_1 = \frac{\sqrt{6}}{2}x$, $v_2 = x^2$, and $e_2 = \frac{\sqrt{10}}{2}x^2$.

a. $(\text{proj}_S(h))(x) = \frac{3}{5}x$

b. $h(x) = \frac{3}{5}x + (x^3 - \frac{3}{5}x)$

c. $\|x^3 - \frac{3}{5}x\| = \frac{2\sqrt{14}}{35}$

7. $a_0 = \langle x^2, \frac{1}{\sqrt{2\pi}} \rangle = \frac{\sqrt{2}}{3}\pi^{5/2}$, $a_1 = \langle x^2, \frac{1}{\sqrt{\pi}} \cos x \rangle = -4\sqrt{\pi}$, and $a_2 = \langle x^2, \frac{1}{\sqrt{\pi}} \cos 2x \rangle = \sqrt{\pi}$ are the first three nonzero coefficients. The Fourier approximation is $\frac{1}{3}\pi^2 - 4\cos x + \cos 2x$.

9. **a.** The graph of f_1 has a minimum of -2 and a maximum of 2. The graph of f_2 has a smaller minimum and a larger maximum. The graph of f_3 has the smallest minimum and the largest maximum.

b. $\|f_1 - f\| = \sqrt{\frac{2}{3}\pi(\pi^2 - 6)} \approx 2.84684$, $\|f_2 - f\| = \sqrt{\frac{1}{3}\pi(2\pi^2 - 15)} \approx 2.22775$, and $\|f_3 - f\| = \frac{1}{3}\sqrt{\pi(6\pi^2 - 49)} \approx 1.88855$.

Review Exercises: Chapter 4

2. **a.** 1. $\langle p, p \rangle = (p(0))^2 + (p'(0))^2 + (p''(0))^2 \geq 0$. If $\langle p, p \rangle = (p(0))^2 + (p'(0))^2 + (p''(0))^2 = 0$, then $p(0) = p'(0) = p''(0) = 0$. This implies that the three coefficients of p are all zero. Thus, $p = 0$.

2. $\langle p, q \rangle = p(0)q(0) + p'(0)q'(0) + p''(0)q''(0) = q(0)p(0) + q'(0)p'(0) + q''(0)p''(0) = \langle q, p \rangle$

3. $\langle rp, q \rangle = (rp)(0)q(0) + (rp)'(0)q'(0) + (rp)''(0)q(0) = r(p(0)q(0) + p'(0)q'(0) + p''(0)q''(0)) = r\langle p, q \rangle$

4. $\langle p_1 + p_2, q \rangle = (p_1 + p_2)(0)q(0) + (p_1 + p_2)'(0)q'(0) + (p_1 + p_2)''(0)q''(0) = (p_1(0) + p_2(0))q(0) + (p_1'(0) + p_2'(0))q'(0) + (p_1''(0) + p_2''(0))q''(0) = p_1(0)q(0) + p_2(0)q(0) + p_1'(0)q'(0) + p_2'(0)q'(0) + p_1''(0)q''(0) + p_2''(0)q''(0) = p_1(0)q(0) + p_1'(0)q'(0) + p_1''(0)q''(0) + p_2(0)q(0) + p_2'(0)q'(0) + p_2''(0)q''(0) = \langle p_1, q \rangle + \langle p_2, q \rangle$

b. $\sqrt{17} \approx 4.12311$ **c.** $\frac{16\sqrt{15}}{15} \approx 4.13118$

d. $\frac{\sqrt{930}}{30} \approx 1.01653$ **e.** $\sqrt{40} \approx 6.32456$

7. **a.** $\left(\text{proj}_g(f)\right)(x) = -\frac{1}{12}g(x) = -\frac{1}{4}x + \frac{1}{12}$

b. $f(x) = \left(-\frac{1}{4}x + \frac{1}{12}\right) + \left(x^2 + \frac{1}{4}x - \frac{1}{12}\right)$

c. $\left(\text{proj}_f(g)\right)(x) = -\frac{5}{3}f(x) = -\frac{5}{3}x^2$

d. $g(x) = \left(-\frac{5}{3}x^2\right) + \left(\frac{5}{3}x^2 + 3x - 1\right)$

9. **a.** $\langle v, e_2 \rangle = \langle 4e_1 + 3e_2 + e_3, e_2 \rangle = 4\langle e_1, e_2 \rangle + 3\langle e_2, e_2 \rangle + \langle e_3, e_2 \rangle = 4 \cdot 0 + 3 \cdot 1 + 0 = 3$

b. $\langle v, w \rangle = \langle 4e_1 + 3e_2 + e_3, e_1 - 2e_2 \rangle = 4\langle e_1, e_1 \rangle + 3\langle e_2, e_1 \rangle + \langle e_3, e_1 \rangle + 4(-2)\langle e_1, e_2 \rangle + 3(-2)\langle e_2, e_2 \rangle + (-2)\langle e_3, e_2 \rangle = 4 - 6 = -2$

12. **a.** $\left(\frac{1}{\sqrt{2}}, 0, \frac{1}{\sqrt{2}}, 0\right) \cdot \left(-\frac{1}{\sqrt{6}}, \frac{2}{\sqrt{6}}, \frac{1}{\sqrt{6}}, 0\right) = 0, \left(\frac{1}{\sqrt{2}}, 0, \frac{1}{\sqrt{2}}, 0\right) \cdot \left(\frac{1}{\sqrt{2}}, 0, \frac{1}{\sqrt{2}}, 0\right) = 1,$ and $\left(-\frac{1}{\sqrt{6}}, \frac{2}{\sqrt{6}}, \frac{1}{\sqrt{6}}, 0\right) \cdot \left(-\frac{1}{\sqrt{6}}, \frac{2}{\sqrt{6}}, \frac{1}{\sqrt{6}}, 0\right) = 1.$

b. $\dim S = 2$

c. $\text{proj}_S(0, 1, 2, 3) = \sqrt{2}\left(\frac{1}{\sqrt{2}}, 0, \frac{1}{\sqrt{2}}, 0\right) + \frac{4}{\sqrt{6}}\left(-\frac{1}{\sqrt{6}}, \frac{2}{\sqrt{6}}, \frac{1}{\sqrt{6}}, 0\right) = (1, 0, 1, 0) + \left(-\frac{2}{3}, \frac{4}{3}, \frac{2}{3}, 0\right) = \left(\frac{1}{3}, \frac{4}{3}, \frac{5}{3}, 0\right)$

d. $\|(0, 1, 2, 3) - (\frac{1}{3}, \frac{4}{3}, \frac{5}{3}, 0)\| = \|(-\frac{1}{3}, -\frac{1}{3}, \frac{1}{3}, 3)\| = \sqrt{\frac{1}{9} + \frac{1}{9} + \frac{1}{9} + 9} = \frac{2\sqrt{21}}{3} \approx 3.05505$

e. $v_3 = (-\frac{1}{3}, -\frac{1}{3}, \frac{1}{3}, 3)$, so $e_3 = \left(-\frac{\sqrt{21}}{42}, -\frac{\sqrt{21}}{42}, \frac{\sqrt{21}}{42}, \frac{3\sqrt{21}}{14}\right).$

14. **a.** $\langle p_1, p_2 \rangle = \int_0^1 (2\sqrt{3}x - \sqrt{3})\, dx = 0$, $\langle p_1, p_1 \rangle = \int_0^1 1^2\, dx = 1$, and $\langle p_2, p_2 \rangle = \int_0^1 (2\sqrt{3}x - \sqrt{3})^2\, dx = 1.$

b. $\left(\text{proj}_S(q)\right)(x)$
$= \langle q, p_1 \rangle p_1(x) + \langle q, p_2 \rangle p_2(x)$
$= \frac{1}{3}p_1(x) + \frac{\sqrt{3}}{6}p_2(x) = x - \frac{1}{6}$

c. $q_3(x) = q(x) - \left(\text{proj}_S(q)\right)(x) = x^2 - x + \frac{1}{6}$ is orthogonal to both p_1 and p_2.

d. $p_3 = \frac{1}{\|q_3\|}q_3$, so $p_3(x) = 6\sqrt{5}(x^2 - x + \frac{1}{6}) = 6\sqrt{5}x^2 - 6\sqrt{5}x + \sqrt{5}.$

Section 5.1

1. a. $A = \begin{bmatrix} 0 & 0 & 1 & 0 \\ 1 & 0 & 0 & 0 \\ 0 & 1 & 0 & 1 \\ 1 & 0 & 0 & 0 \end{bmatrix}$

b. $A^2 = \begin{bmatrix} 0 & 1 & 0 & 1 \\ 0 & 0 & 1 & 0 \\ 2 & 0 & 0 & 0 \\ 0 & 0 & 1 & 0 \end{bmatrix}$. The *ij*-entry of this matrix is the number of routes

for sending a message for computer *i* to computer *j* in two steps.

c. $A^3 = \begin{bmatrix} 2 & 0 & 0 & 0 \\ 0 & 1 & 0 & 1 \\ 0 & 0 & 2 & 0 \\ 0 & 1 & 0 & 1 \end{bmatrix}$

d. $A + A^2 + A^3 = \begin{bmatrix} 2 & 1 & 1 & 1 \\ 1 & 1 & 1 & 1 \\ 2 & 1 & 2 & 1 \\ 1 & 1 & 1 & 1 \end{bmatrix}$. So it is possible for each computer to

send a message to any computer on the network (including itself) in one, two, or three direct steps.

6. a. $\begin{bmatrix} 1 & 0 \\ 0 & 1 \end{bmatrix}, \begin{bmatrix} 1 & 0 \\ 0 & -1 \end{bmatrix}, \begin{bmatrix} -1 & 0 \\ 0 & 1 \end{bmatrix}, \begin{bmatrix} -1 & 0 \\ 0 & -1 \end{bmatrix}$

b. Let $A = \begin{bmatrix} a & b \\ c & d \end{bmatrix}$. Then $A^2 = \begin{bmatrix} a^2 + bc & ab + bd \\ ac + cd & bc + d^2 \end{bmatrix} = \begin{bmatrix} 1 & 0 \\ 0 & 1 \end{bmatrix}$ gives the

four equations: $a^2 + bc = 1, b(a+d) = 0, c(a+d) = 0$, and $bc + d^2 = 1$. If $a + d \neq 0$, then $b = 0, c = 0, a^2 = 1$, and $d^2 = 1$. This leads to solutions listed in part a. If $a + d = 0$, then we can let c and d be arbitrary real numbers and solve for a and b in terms of these variables: $a = -d$, and $b = \frac{1-d^2}{c}$ if $c \neq 0$. If $c = 0$, then b is arbitrary, but $a = -d = \pm 1$.

10. The *ik*-entry of AB is $\sum_{j=1}^{n} a_{ij} b_{jk}$, so the *il*-entry of $(AB)C$ is

$$\sum_{k=1}^{p} \left(\sum_{j=1}^{n} a_{ij} b_{jk} \right) c_{kl} = \sum_{k=1}^{p} \sum_{j=1}^{n} a_{ij} b_{jk} c_{kl} = \sum_{j=1}^{n} \sum_{k=1}^{p} a_{ij} b_{jk} c_{kl} = \sum_{j=1}^{n} a_{ij} \left(\sum_{k=1}^{p} b_{jk} c_{kl} \right)$$

Since the *jl*-entry of BC is $\sum_{k=1}^{p} b_{jk} c_{kl}$, the *il*-entry of $(AB)C$ is the *il*-entry of $A(BC)$.

17. a. $A(\mathbf{v}_0 + \mathbf{v}) = A\mathbf{v}_0 + A\mathbf{v} = \mathbf{b} + \mathbf{0} = \mathbf{b}$

b. $A(\mathbf{v}_0 - \mathbf{v}_1) = A\mathbf{v}_0 - A\mathbf{v}_1 = \mathbf{b} - \mathbf{b} = \mathbf{0}$

c. By part a, any element of $\mathbf{v}_0 + S$ is a solution to $A\mathbf{x} = \mathbf{b}$. Let $\mathbf{x}$ be any solution of $A\mathbf{x} = \mathbf{b}$. Then by part b, we have that $\mathbf{x} - \mathbf{v}_0 \in S$. Thus, we can write $\mathbf{x} = \mathbf{v}_0 + (\mathbf{x} - \mathbf{v}_0)$ and see that $\mathbf{x}$ is an element of $\mathbf{v}_0 + S$. It follows that the set $\mathbf{v}_0 + S$ is equal to the solution set of $A\mathbf{x} = \mathbf{b}$.

Section 5.2

4. a. $\begin{bmatrix} -9 & 4 \\ 7 & -3 \end{bmatrix}$

    b. $\begin{bmatrix} -\frac{7}{25} & -\frac{3}{25} & \frac{8}{25} \\ \frac{12}{25} & -\frac{2}{25} & -\frac{3}{25} \\ \frac{8}{25} & \frac{7}{25} & -\frac{2}{25} \end{bmatrix}$

    c. This matrix is not invertible.

    d. $\begin{bmatrix} -1 & -6 & 3 \\ \frac{1}{3} & 3 & -\frac{4}{3} \\ -\frac{1}{3} & -1 & \frac{1}{3} \end{bmatrix}$

    e. $\begin{bmatrix} \frac{3}{2} & 1 & \frac{1}{2} \\ 1 & 2 & 1 \\ \frac{1}{2} & 1 & \frac{3}{2} \end{bmatrix}$

    f. $\begin{bmatrix} -5 & 18 & 14 \\ -1 & 4 & 3 \\ 7 & -25 & -19 \end{bmatrix}$

8. a. $\begin{bmatrix} 1 & 1 \end{bmatrix}\begin{bmatrix} 2 \\ -1 \end{bmatrix} = [1 \cdot 2 + 1 \cdot (-1)] = [1]$, the 1×1 identity matrix.

    b. $\begin{bmatrix} 2 \\ -1 \end{bmatrix}\begin{bmatrix} 1 & 1 \end{bmatrix} = \begin{bmatrix} 2 & 2 \\ -1 & -1 \end{bmatrix}$, which is not the 2×2 identity matrix.

    c. A left inverse of a 1×2 matrix must be a 2×1 matrix in order for the product to be the 2×2 identity matrix. Now $\begin{bmatrix} a \\ b \end{bmatrix}\begin{bmatrix} 1 & 1 \end{bmatrix} = \begin{bmatrix} a & a \\ b & b \end{bmatrix}$, which is not the identity matrix for any values of a and b.

9. a. The right inverse C of the $m \times n$ matrix A must be an $n \times m$ matrix. To show that $\{A\mathbf{e}_1, \ldots, A\mathbf{e}_n\}$ spans $\mathbb{R}^m$, let $\mathbf{v}$ be an arbitrary element of $\mathbb{R}^m$. Then $C\mathbf{v} \in \mathbb{R}^n$, so we can write $C\mathbf{v} = r_1\mathbf{e}_1 + \cdots + r_n\mathbf{e}_n$ for some scalars $r_1, \ldots, r_n$. Now $\mathbf{v} = I\mathbf{v} = (AC)\mathbf{v} = A(C\mathbf{v}) = A(r_1\mathbf{e}_1 + \cdots + r_n\mathbf{e}_n) = A(r_1\mathbf{e}_1) + \cdots + A(r_n\mathbf{e}_n) = r_1(A\mathbf{e}_1) + \cdots + r_n(A\mathbf{e}_n)$.

    b. By the Comparison Theorem, the number of elements in this spanning set is greater than or equal to the number of elements in a basis for $\mathbb{R}^m$. Thus, $m \le n$.

    c. The left inverse C' must be an $n \times m$ matrix. To show that $\{A\mathbf{e}_1, \ldots, A\mathbf{e}_n\}$ is linearly independent, suppose $r_1(A\mathbf{e}_1) + \cdots + r_n(A\mathbf{e}_n) = \mathbf{0}$ for some scalars $r_1, \ldots, r_n$. Now $\mathbf{0} = C'\mathbf{0} = C'r_1(A\mathbf{e}_1) + \cdots + r_n(A\mathbf{e}_n)) = C'r_1(A\mathbf{e}_1)) + \cdots + C'r_n(A\mathbf{e}_n)) = r_1(C'A\mathbf{e}_1)) + \cdots + r_n(C'A\mathbf{e}_n)) = r_1((C'A)\mathbf{e}_1) + \cdots + r_n((C'A)\mathbf{e}_n) = r_1(I\mathbf{e}_1) + \cdots + r_n(I\mathbf{e}_n) = r_1\mathbf{e}_1 + \cdots + r_n\mathbf{e}_n$. Since $\{\mathbf{e}_1, \ldots, \mathbf{e}_n\}$ is linearly independent, $r_1 = \cdots = r_n = 0$.

    d. By the Comparison Theorem, the number of elements in this linearly independent set is less than or equal to the number of elements in a basis for $\mathbb{R}^m$. Thus, $m \ge n$.

    e. The inequalities in parts b and d give that $m = n$. That is, A is a square matrix.

12. First consider the case where $a = 0$. Then $ad - bc \ne 0$ if and only if $b \ne 0$ and $c \ne 0$. We can reduce A to a matrix with two leading 1s by interchanging the two rows of A, multiplying the new top row by $\frac{1}{c}$, and multiplying the new bottom row by $\frac{1}{b}$. This is possible if and only if $b \neq 0$ and $c \neq 0$. Thus, rank $A = 2$ (and hence A is invertible) if and only if $ad - bc \neq 0$.

Now consider the case where $a \neq 0$. Multiply the top row of A by $\frac{1}{a}$. Add $-c$ times the new top row to the bottom row to give 0 as the 21-entry. The 22-entry of the resulting matrix will be $-\frac{bc}{a} + d = \frac{ad-bc}{a}$. Thus, in this case also, rank $A = 2$ (and hence A is invertible) if and only if $ad - bc \neq 0$.

Section 5.3

3. **a.** $P^2 = \begin{bmatrix} \frac{4}{9} & \frac{1}{3} & \frac{1}{6} & \frac{1}{6} \\ \frac{1}{3} & \frac{4}{9} & \frac{1}{6} & \frac{1}{6} \\ \frac{1}{9} & \frac{1}{9} & \frac{1}{3} & \frac{1}{3} \\ \frac{1}{9} & \frac{1}{9} & \frac{1}{3} & \frac{1}{3} \end{bmatrix}$

   b. The 11-entry of P^2 gives this probability as $\frac{4}{9}$.

   c. The 43-entry of P^2 gives this probability as $\frac{1}{3}$.

   d. Since all entries of P^2 are positive, this is a regular Markov chain.

   e. The equilibrium vector is $\mathbf{s} = \begin{bmatrix} .3 \\ .3 \\ .2 \\ .2 \end{bmatrix}$.

   f. The clothing store and the hardware store both have ads played 30% of the times, whereas the two grocery stores have ads played only 20% of the times.

4. The probability of change is the 21-entry or the 12-entry of P^n. We want $\frac{1-(1-2p)^{1000}}{2} < .00001$. This is equivalent to $p < \frac{1}{2}(1 - .99998^{0.001}) \approx 10^{-8}$.

7. **a.** The five states are the five categories. The transition matrix is $P =$

   $\begin{bmatrix} \frac{1}{2} & \frac{1}{3} & 0 & 0 & 0 \\ \frac{1}{2} & \frac{1}{3} & \frac{1}{3} & 0 & 0 \\ 0 & \frac{1}{3} & \frac{1}{3} & \frac{1}{3} & 0 \\ 0 & 0 & \frac{1}{3} & \frac{1}{3} & \frac{1}{2} \\ 0 & 0 & 0 & \frac{1}{3} & \frac{1}{2} \end{bmatrix}$. The initial distribution vector might be $\mathbf{v}_0 = \mathbf{e}_k$

   if the market is in state k at the beginning of the observation period.

   b. $\mathbf{s} = \begin{bmatrix} \frac{2}{13} \\ \frac{3}{13} \\ \frac{3}{13} \\ \frac{3}{13} \\ \frac{2}{13} \end{bmatrix}$ **c.** $\frac{2}{13} \approx 15\%$

Section 5.4

1. If the particle starts in position 3, the expected number of steps until it is absorbed is $1 + 2 + 1 = 4$. If the particle starts in position 4, the expected number of steps until it is absorbed is $\frac{1}{2} + 1 + \frac{3}{2} = 3$. The symmetry of the model suggests that the expected absorption times from positions 2 and 4 should be equal. Since a step from position 3 leads either to position 2 or to position 4, it is reasonable that the expected absorption time from position 3 is one more than that from position 2 or 4.

3. The transition matrix is $P = \begin{bmatrix} 1 & 0 & \frac{1}{3} & 0 & 0 \\ 0 & 1 & 0 & 0 & \frac{2}{3} \\ 0 & 0 & 0 & \frac{1}{3} & 0 \\ 0 & 0 & \frac{2}{3} & 0 & \frac{1}{3} \\ 0 & 0 & 0 & \frac{2}{3} & 0 \end{bmatrix}$. The fundamental ma-

trix is $(I - Q)^{-1} = \begin{bmatrix} \frac{7}{5} & \frac{3}{5} & \frac{1}{5} \\ \frac{6}{5} & \frac{9}{5} & \frac{3}{5} \\ \frac{4}{5} & \frac{6}{5} & \frac{7}{5} \end{bmatrix}$. The ij-entry is the expected number of

visits to the ith nonabsorbing state for a random walk that starts in the jth nonabsorbing state.

5. Arrange the list of states so the two absorbing states are first: (TT, TT), (tt, tt), (TT, Tt), (TT, tt), (Tt, Tt), (Tt, tt). The transition matrix is

$$P = \begin{bmatrix} 1 & 0 & \frac{1}{4} & 0 & \frac{1}{16} & 0 \\ 0 & 1 & 0 & 0 & \frac{1}{16} & \frac{1}{4} \\ 0 & 0 & \frac{1}{2} & 0 & \frac{1}{4} & 0 \\ 0 & 0 & 0 & 0 & \frac{1}{8} & 0 \\ 0 & 0 & \frac{1}{4} & 1 & \frac{1}{4} & \frac{1}{4} \\ 0 & 0 & 0 & 0 & \frac{1}{4} & \frac{1}{2} \end{bmatrix}.$$

b. The two sequences of steps $(TT, tt) \rightarrow (Tt, Tt) \rightarrow (TT, Tt) \rightarrow (TT, TT)$ and $(Tt, tt) \rightarrow (tt, tt)$ show that it is possible to reach an absorbing state from any state.

c. The fundamental matrix is $(I - Q)^{-1} = \begin{bmatrix} \frac{8}{3} & \frac{4}{3} & \frac{4}{3} & \frac{2}{3} \\ \frac{1}{6} & \frac{4}{3} & \frac{1}{3} & \frac{1}{6} \\ \frac{4}{3} & \frac{8}{3} & \frac{4}{3} & \frac{4}{3} \\ \frac{2}{3} & \frac{4}{3} & \frac{4}{3} & \frac{8}{3} \end{bmatrix}$. The ij-entry is

the expected number of visits to the ith nonabsorbing state given that the system started in the jth nonabsorbing state.

Review Exercises: Chapter 5

2. **a.** $(A+B)^2 = (A+B)(A+B) = A(A+B)+B(A+B) = A^2+AB+BA+B^2$

b. $(A+B)^2 = A^2+AB+BA+B^2 = A^2+AB+AB+B^2 = A^2+2AB+B^2$

c. $(A+B)^2 = A^2+2AB+B^2 \Longrightarrow A^2+AB+BA+B^2 = A^2+2AB+B^2 \Longrightarrow$
$BA = AB$

5. $\begin{bmatrix} \cos\theta & -\sin\theta \\ \sin\theta & \cos\theta \end{bmatrix}\begin{bmatrix} \cos\theta & \sin\theta \\ -\sin\theta & \cos\theta \end{bmatrix} = \begin{bmatrix} \cos^2\theta + \sin^2\theta & 0 \\ 0 & \cos^2\theta + \sin^2\theta \end{bmatrix} =$
I. By Theorem 5.13, this matrix is also the right inverse.

7. **a.** $\begin{bmatrix} 1 & 0 & 2 \\ 0 & 0 & -1 \\ -3 & 1 & 1 \end{bmatrix}$ **b.** $\begin{bmatrix} 3 & -\frac{5}{2} & 2 \\ -2 & \frac{5}{2} & -2 \\ 1 & -1 & 1 \end{bmatrix}$

10. **a.** $\text{rank } A = \text{rank } B = n \Longrightarrow A$ and B are invertible $\Longrightarrow AB$ is invertible $\Longrightarrow \text{rank } AB = n$

b. Suppose $\text{rank } A < n$. Then there is a sequence of elementary row operations that can be applied to A to result in a matrix with a row of zeros. By Theorem 5.5, applying this sequence of row operations to AB will result in the product of a matrix with a row of zeros times B. The product will have a row of zeros. This contradicts the hypothesis that $\text{rank } AB = n$. Thus, $\text{rank } A = n$. By Theorem 5.12, A is invertible. By Theorem 5.8, A^{-1} is invertible. Hence, $\text{rank } A^{-1} = n$. By part a of this exercise, $B = A^{-1}(AB)$ has rank n.

13. **a.** $P = \begin{bmatrix} 1 & \frac{1}{4} & 0 & \frac{3}{4} \\ 0 & 0 & \frac{1}{4} & 0 \\ 0 & \frac{3}{4} & 0 & \frac{1}{4} \\ 0 & 0 & \frac{3}{4} & 0 \end{bmatrix}$

b. $\frac{1}{4} \cdot \frac{1}{4} + \frac{3}{4} \cdot \frac{3}{4} = \frac{5}{8}$

c. $(I-Q)^{-1} = \begin{bmatrix} 1.3 & 0.4 & 0.1 \\ 1.7 & 1.6 & 0.4 \\ 0.9 & 1.2 & 1.3 \end{bmatrix}$

d. The expected number of visits to the second quadrant is 1.3. Since the particle starts in this quadrant, the expected number of returns is 0.3.

e. $1.3 + 1.2 + 0.9 = 3.4$

Section 6.1

4. Let $f, g \in \mathbb{F}(\mathbb{R})$ and $r \in \mathbb{R}$. Then $T(f + g) = (f + g)(x_0) = f(x_0) + g(x_0) = T(f) + T(g)$, and $T(rf) = (rf)(x_0) = r(f(x_0)) = rT(f)$.

9. Let $\mathbf{v}, \mathbf{w} \in V$ and let $r \in \mathbb{R}$. Then $\mathrm{id}_V(\mathbf{v} + \mathbf{w}) = \mathbf{v} + \mathbf{w} = \mathrm{id}_V(\mathbf{v}) + \mathrm{id}_V(\mathbf{w})$, and $\mathrm{id}_V(r\mathbf{v}) = r\mathbf{v} = r\mathrm{id}_V(\mathbf{v})$.

13. **a.** Since the constant zero function is bounded, $S \neq \varnothing$. Since the sum of bounded functions is bounded, and since a scalar multiple of a bounded function is bounded, S is closed under addition and scalar multiplication. By the Subspace Theorem, S is a subspace of $\mathbb{C}([0, \infty))$.

   b. Let B be a bound for f. That is, $|f(x)| \leq B$ for all $x \geq 0$. For any $x > 0$, we have $\int_0^\infty |e^{-xt}f(t)| \, dt = \lim_{A \to \infty} \int_0^A |e^{-xt}f(t)| \, dt \leq \lim_{A \to \infty} \int_0^A e^{-xt}B \, dt = \lim_{A \to \infty} \left(-\frac{B}{x}(e^{-Ax} - 1) \right) = \frac{B}{x}$. Since $e^{-xt}f(t)$ is continuous, it follows that $\int_0^\infty e^{-xt}f(t) \, dt$ converges.

   c. Let $f, g \in S$ and $r \in \mathbb{R}$. Then $(\mathcal{L}(f + g))(x) = \int_0^\infty e^{-xt}(f(t) + g(t)) \, dt = \int_0^\infty e^{-xt}f(t) \, dt + \int_0^\infty e^{-xt}g(t) \, dt = (\mathcal{L}(f))(x) + (\mathcal{L}(g))(x) = (\mathcal{L}(f) + \mathcal{L}(g))(x)$, and $(\mathcal{L}(rf))(x) = \int_0^\infty e^{-xt}rf(t) \, dt = r\int_0^\infty e^{-xt}f(t) \, dt = r((\mathcal{L}(f))(x)) = (r\mathcal{L}(f))(x)$. Thus, $\mathcal{L}(f + g) = \mathcal{L}(f) + \mathcal{L}(g)$ and $\mathcal{L}(rf) = r\mathcal{L}(f)$.

15. **a.** Since $T(\mathbf{0}) = \mathbf{0}$, we know that $\mathbf{0}$ is in the set. Hence, the set is nonempty. Suppose $\mathbf{v}$ and $\mathbf{v}'$ are in the set and $r \in \mathbb{R}$. That is, $T(\mathbf{v}) = \mathbf{0}$ and $T(\mathbf{v}') = \mathbf{0}$. Hence, $T(\mathbf{v} + \mathbf{v}') = T(\mathbf{v}) + T(\mathbf{v}') = \mathbf{0} + \mathbf{0} = \mathbf{0}$, and $T(r\mathbf{v}) = rT(\mathbf{v}) = r\mathbf{0} = \mathbf{0}$. Therefore, the set is closed under addition and scalar multiplication. By the Subspace Theorem, the set is a subspace of V.

   b. Since $T(\mathbf{0}) = \mathbf{0}$, we know that $\mathbf{0}$ is in the set. Hence, the set is nonempty. Suppose $\mathbf{w}$ and $\mathbf{w}'$ are in the set and $r \in \mathbb{R}$. That is, $T(\mathbf{v}) = \mathbf{w}$ and $T(\mathbf{v}') = \mathbf{w}'$ for some vectors $\mathbf{v}, \mathbf{v}'$ in V. Hence, $T(\mathbf{v} + \mathbf{v}') = T(\mathbf{v}) + T(\mathbf{v}') = \mathbf{w} + \mathbf{w}'$ and $T(r\mathbf{v}) = rT(\mathbf{v}) = r\mathbf{w}$. Therefore, $\mathbf{w} + \mathbf{w}'$ and $r\mathbf{w}$ are in the set. That is, the set is closed under addition and scalar multiplication. By the Subspace Theorem, the set is a subspace of W.

21. $T\left(\begin{bmatrix} 1 \\ 0 \end{bmatrix}\right) + T\left(\begin{bmatrix} 0 \\ 1 \end{bmatrix}\right) = \begin{bmatrix} -2 \\ 3 \end{bmatrix} + \begin{bmatrix} 5 \\ 1 \end{bmatrix} = \begin{bmatrix} 3 \\ 4 \end{bmatrix}$, which is not equal to $T\left(\begin{bmatrix} 1 \\ 0 \end{bmatrix} + \begin{bmatrix} 0 \\ 1 \end{bmatrix}\right) = T\left(\begin{bmatrix} 1 \\ 1 \end{bmatrix}\right) = \begin{bmatrix} 3 \\ 2 \end{bmatrix}$.

23. **a.** $T(2\mathbf{v}) = T(\mathbf{v} + \mathbf{v}) = T(\mathbf{v}) + T(\mathbf{v}) = 2T(\mathbf{v})$.

   b. $T(r\mathbf{v}) = T(\mathbf{v} + \cdots + \mathbf{v}) = T(\mathbf{v}) + \cdots + T(\mathbf{v}) = rT(\mathbf{v})$.

   c. $T(0\mathbf{v}) = T(0\mathbf{v} + 0\mathbf{v}) = T(0\mathbf{v}) + T(0\mathbf{v})$, so $T(0\mathbf{v}) = \mathbf{0} = 0\mathbf{v}$.

   d. If r is a negative integer, then part c and part b applied to the positive integer $-r$ give $\mathbf{0} = T(0\mathbf{v}) = T(r\mathbf{v} + (-r\mathbf{v})) = T(r\mathbf{v}) + T(-r\mathbf{v}) = T(r\mathbf{v}) + (-rT(\mathbf{v}))$, so $T(r\mathbf{v}) = rT(\mathbf{v})$.

   e. Suppose $r = \frac{p}{q}$ for some integers p and q. Then $qT(r\mathbf{v}) = T(qr\mathbf{v}) = T(p\mathbf{v}) = pT(\mathbf{v})$. So $T(r\mathbf{v}) = \frac{p}{q}T(\mathbf{v}) = rT(\mathbf{v})$.

   f. T will need to be continuous in order to extend the homogeneous property from rational multiples to all real multiples.

Section 6.2

1. **a.** $f(x_1) = f(x_2) \Longrightarrow \frac{1}{3}x_1 - 2 = \frac{1}{3}x_2 - 2 \Longrightarrow \frac{1}{3}x_1 = \frac{1}{3}x_2 \Longrightarrow x_1 = x_2$. Thus, f is one-to-one.

   b. $p(1) = 0 = p(2)$. Thus, p is not one-to-one.

   c. $(e^{x_1} - e^{-x_1})/2 = (e^{x_2} - e^{-x_2})/2 \Longrightarrow e^{x_1} - e^{-x_1} = e^{x_2} - e^{-x_2} \Longrightarrow (e^{x_1} - e^{-x_1})^2 = (e^{x_2} - e^{-x_2})^2 \Longrightarrow e^{2x_1} - 2 + e^{-2x_1} = e^{2x_2} - 2 + e^{-2x_2} \Longrightarrow e^{2x_1} + 2 + e^{-2x_1} = e^{2x_2} + 2 + e^{-2x_2} \Longrightarrow (e^{x_1} + e^{-x_1})^2 = (e^{x_2} + e^{-x_2})^2 \Longrightarrow e^{x_1} + e^{-x_1} = e^{x_2} + e^{-x_2}$. Thus, $(e^{x_1} + e^{-x_1}) + (e^{x_1} - e^{-x_1}) = (e^{x_2} + e^{-x_2}) + (e^{x_2} - e^{-x_2}) \Longrightarrow 2e^{x_1} = 2e^{x_2} \Longrightarrow e^{x_1} = e^{x_2} \Longrightarrow x_1 = x_2$. Therefore, s is one-to-one.

   d. $W(0) = (\cos 0, \sin 0) = (1, 0) = (\cos 2\pi, \sin 2\pi) = W(2\pi)$. Thus, W is not one-to-one.

   e. $L\left(\begin{bmatrix} 2 \\ -3 \\ 1 \end{bmatrix}\right) = \begin{bmatrix} 0 \\ 0 \\ 0 \end{bmatrix} = L\left(\begin{bmatrix} 0 \\ 0 \\ 0 \end{bmatrix}\right)$. Thus, L is not one-to-one.

5. **a.** Yes, no   **b.** Yes, no

   c. $(D \circ A)(f) = \frac{d}{dx} \int_0^x f(t)\, dt = f(x)$ by the Fundamental Theorem of Calculus. $(A \circ D)(1) = A(0) = 0 \neq 1$.

10. For any $w \in W$ we have $(h \circ (g \circ f))(w) = h((g \circ f)(w)) = h(g(f(w))) = (h \circ g)(f(w)) = ((h \circ g) \circ f)(w)$.

15. Suppose $v, w \in U$ and $r \in \mathbb{R}$. Then $(T \circ S)(v + w) = T(S(v + w)) = T(S(v) + S(w)) = T(S(v)) + T(S(w)) = (T \circ S)(v) + (T \circ S)(w)$, and $(T \circ S)(rv) = T(S(rv)) = T(rS(v)) = rT(S(v)) = r(T \circ S)(v)$.

20. Let $f(x) = \left(\frac{2x}{x^2+1}, \frac{x^2-1}{x^2+1}\right)$. Then for any $x \in \mathbb{R}$, the point $f(x)$ is on the unit circle and is a point other than $(0, 1)$ on the line through $(0, 1)$ and $(x, 0)$. Suppose (a, b) is a given point on the circle with $b < 1$ and $f(x) = \left(\frac{2x}{x^2+1}, \frac{x^2-1}{x^2+1}\right) = (a, b)$. Compare the second coordinates and check the sign of the first coordinates to conclude that $x = \sqrt{\frac{1+b}{1-b}}$ if $a \geq 0$ and that $x = -\sqrt{\frac{1+b}{1-b}}$ if $a < 0$. In particular, there is only one value of x for which $f(x) = (a, b)$. Thus, f is one-to-one. Also, the image of f is the set of all points on the circle except $(0, 1)$.

Section 6.3

1. **a.** $T\left(\begin{bmatrix} 4 \\ 7 \end{bmatrix}\right) = T\left(4\begin{bmatrix} 1 \\ 0 \end{bmatrix} + 7\begin{bmatrix} 0 \\ 1 \end{bmatrix}\right) = 4\begin{bmatrix} 2 \\ 3 \\ -1 \end{bmatrix} + 7\begin{bmatrix} -5 \\ 1 \\ 1 \end{bmatrix} = \begin{bmatrix} -27 \\ 19 \\ 3 \end{bmatrix}$

   b. $T\left(\begin{bmatrix} a \\ b \end{bmatrix}\right) = T\left(a\begin{bmatrix} 1 \\ 0 \end{bmatrix} + b\begin{bmatrix} 0 \\ 1 \end{bmatrix}\right) = a\begin{bmatrix} 2 \\ 3 \\ -1 \end{bmatrix} + b\begin{bmatrix} -5 \\ 1 \\ 1 \end{bmatrix} = \begin{bmatrix} 2a - 5b \\ 3a + b \\ -a + b \end{bmatrix}$

$$\mathbf{c.} \quad \begin{bmatrix} 2 & -5 \\ 3 & 1 \\ -1 & 1 \end{bmatrix}$$

7. a. Consider the parallelogram whose sides are arrows that represent two vectors $\mathbf{v}, \mathbf{w} \in \mathbb{R}^3$. The image of this parallelogram is the parallelogram whose sides are the arrows $R_\varphi(\mathbf{v})$ and $R_\varphi(\mathbf{w})$. Since the diagonal of the first parallelogram is mapped to the diagonal of the second, we see that the image of $\mathbf{v} + \mathbf{w}$ is $R_\varphi(\mathbf{v}) + R_\varphi(\mathbf{w})$. Similarly, for any $r \in \mathbb{R}$, the arrow that is r times as long as the arrow for $\mathbf{v}$ is mapped to the arrow that is r times as long as the arrow for $R_\varphi(\mathbf{v})$.

$$\mathbf{b.} \quad A_\phi = \begin{bmatrix} 1 & 0 & 0 \\ 0 & \cos\varphi & -\sin\varphi \\ 0 & \sin\varphi & \cos\varphi \end{bmatrix}$$

c. The matrix B_φ for the linear map $S_\varphi : \mathbb{R}^3 \to \mathbb{R}^3$ that rotates each point about the y-axis through an angle of φ is $B_\varphi = \begin{bmatrix} \cos\varphi & 0 & -\sin\varphi \\ 0 & 1 & 0 \\ \sin\phi & 0 & \cos\varphi \end{bmatrix}$.

The matrix C_φ for the linear map $T_\varphi : \mathbb{R}^3 \to \mathbb{R}^3$ that rotates each point about the z-axis through an angle of φ is $C_\varphi = \begin{bmatrix} \cos\varphi & -\sin\varphi & 0 \\ \sin\varphi & \cos\varphi & 0 \\ 0 & 0 & 1 \end{bmatrix}$.

11. The columns of the matrix A are $A\mathbf{e}_1 = A[1]_B = [T(1)]_B = [1]_B = \begin{bmatrix} 1 \\ 0 \\ 0 \\ 0 \end{bmatrix}$,

$A\mathbf{e}_2 = A[x]_B = [T(x)]_B = [2x + 1]_B = \begin{bmatrix} 1 \\ 2 \\ 0 \\ 0 \end{bmatrix}$, $A\mathbf{e}_3 = A[x^2]_B = [T(x^2)]_B =$

$[(2x+1)^2]_B = \begin{bmatrix} 1 \\ 4 \\ 4 \\ 0 \end{bmatrix}$, and $A\mathbf{e}_4 = A[x^3]_B = [T(x^3)]_B = [(2x+1)^3]_B = \begin{bmatrix} 1 \\ 6 \\ 12 \\ 8 \end{bmatrix}$.

Thus, $A = \begin{bmatrix} 1 & 1 & 1 & 1 \\ 0 & 2 & 4 & 6 \\ 0 & 0 & 4 & 12 \\ 0 & 0 & 0 & 8 \end{bmatrix}$.

Section 6.4

2. a. $A = \begin{bmatrix} 2 & 2 & 2 \\ 0 & 2 & 4 \\ 0 & 0 & 2 \end{bmatrix}$

b. $L \circ L^{-1}(p(x)) = L\left(\frac{1}{2}p(x - 1)\right) = 2\left(\frac{1}{2}p((x + 1) - 1)\right) = p(x)$, and $L^{-1} \circ L(p(x)) = L^{-1}(2p(x + 1)) = \frac{1}{2}(2p((x - 1) + 1)) = p(x)$.

c. $A' = \begin{bmatrix} \frac{1}{2} & -\frac{1}{2} & \frac{1}{2} \\ 0 & \frac{1}{2} & -1 \\ 0 & 0 & \frac{1}{2} \end{bmatrix}$

d. $AA' = \begin{bmatrix} 2 & 2 & 2 \\ 0 & 2 & 4 \\ 0 & 0 & 2 \end{bmatrix} \begin{bmatrix} \frac{1}{2} & -\frac{1}{2} & \frac{1}{2} \\ 0 & \frac{1}{2} & -1 \\ 0 & 0 & \frac{1}{2} \end{bmatrix} = \begin{bmatrix} 1 & 0 & 0 \\ 0 & 1 & 0 \\ 0 & 0 & 1 \end{bmatrix}$. By Theorem 5.13,

A' is the inverse of A.

6. We can write any $\mathbf{v} \in V$ as $\mathbf{v} = r_1\mathbf{v}_1 + \cdots + r_n\mathbf{v}_n$ for $r_1, \ldots, r_n \in \mathbb{R}$. Now

$$L_B([\mathbf{v}]_B) = L_B\left(\begin{bmatrix} r_1 \\ \vdots \\ r_n \end{bmatrix}\right) = r_1\mathbf{v}_1 + \cdots + r_n\mathbf{v}_n = \mathbf{v}, \text{ and } \left[L_B\left(\begin{bmatrix} r_1 \\ \vdots \\ r_n \end{bmatrix}\right)\right]_B = [r_1\mathbf{v}_1 +$$

$$\cdots + r_n\mathbf{v}_n]_B = \begin{bmatrix} r_1 \\ \vdots \\ r_n \end{bmatrix}.$$

9. a. $(D^2 + 3D + 2)(x^2 \sin x) = D^2(x^2 \sin x) + (3D)(x^2 \sin x) + 2(x^2 \sin x) = (4x \cos x + (2 - x^2) \sin x) + 3(x^2 \cos x + 2x \sin x) + (2x^2 \sin x) = (3x^2 + 4x) \cos x + (x^2 + 6x + 2) \sin x$

   b. $(D + 1)(D + 2)(x^2 \sin x) = (D + 1)(x^2 \cos x + (2x^2 + 2x) \sin x) = (3x^2 + 4x) \cos x + (x^2 + 6x + 2) \sin x$

   c. Let $f \in \mathbb{D}^{(2)}(\mathbb{R})$. Then $((D + r)(D + s))(f) = (D + s)((D + r)(f)) = (D + s)(f' + rf) = (f' + rf)' + s(f' + rf) = f'' + rf' + sf' + srf = f'' + (r + s)f' + (rs)f = (D^2 + (r + s)D + rs)(f)$.

   d. Let $\mathbf{v} \in V$. Then $(S(R + R'))(\mathbf{v}) = S((R + R')(\mathbf{v})) = S(R(\mathbf{v}) + R'(\mathbf{v})) = S(R(\mathbf{v})) + S(R'(\mathbf{v})) = (SR)(\mathbf{v}) + (SR')(\mathbf{v}) = (SR + SR')(\mathbf{v})$. Also, $((S + S')R)(\mathbf{v}) = (S + S')(R(\mathbf{v})) = S(R(\mathbf{v})) + S'(R(\mathbf{v})) = (SR)(\mathbf{v}) + (S'R)(\mathbf{v}) = (SR + S'R)(\mathbf{v})$.

Section 6.5

1. $P\mathbf{e}_1 = P\begin{bmatrix} 1 \\ 1 \end{bmatrix}_{B'} = \begin{bmatrix} 1 \\ 1 \end{bmatrix}_B = \begin{bmatrix} \frac{1}{3} \\ \frac{1}{3} \end{bmatrix}$, and $P\mathbf{e}_2 = P\begin{bmatrix} -1 \\ 1 \end{bmatrix}_{B'} = \begin{bmatrix} -1 \\ 1 \end{bmatrix}_B = \begin{bmatrix} 1 \\ -1 \end{bmatrix}$.

   Thus, $P = \begin{bmatrix} \frac{1}{3} & 1 \\ \frac{1}{3} & -1 \end{bmatrix}$.

5. The matrix P for changing from B' to B has columns $P\mathbf{e}_1 = P[1 + x^2]_{B'} = [1 + x^2]_B = \begin{bmatrix} 2 \\ 1 \\ 1 \end{bmatrix}$, $P\mathbf{e}_2 = P[1 - x^2]_{B'} = [1 - x^2]_B = \begin{bmatrix} 0 \\ -1 \\ -1 \end{bmatrix}$, and $P\mathbf{e}_3 = $

   $P[x]_{B'} = [x]_B = \begin{bmatrix} 1 \\ 1 \\ 0 \end{bmatrix}$. Hence, $P = \begin{bmatrix} 2 & 0 & 1 \\ 1 & -1 & 1 \\ 1 & -1 & 0 \end{bmatrix}$. By Theorem 6.18, the matrix of T relative to B' is $\begin{bmatrix} 2 & 0 & 1 \\ 1 & -1 & 1 \\ 1 & -1 & 0 \end{bmatrix}^{-1} \begin{bmatrix} 1 & -1 & 2 \\ 2 & 1 & 0 \\ -1 & 0 & 3 \end{bmatrix} \begin{bmatrix} 2 & 0 & 1 \\ 1 & -1 & 1 \\ 1 & -1 & 0 \end{bmatrix} =$

$\begin{bmatrix} -\frac{1}{2} & -\frac{3}{2} & -2 \\ -\frac{3}{2} & \frac{3}{2} & -1 \\ 4 & 2 & 4 \end{bmatrix}$.

6.   **a.** For all $v \in V$ we have $I[v]_B = [v]_B$. Therefore, I is the matrix for changing from B to B.

   b. For all $v \in V$ we have $P[v]_B = [v]'_B$ and $Q[v]_{B'} = [v]_{B''}$. Thus, $(QP)[v]_B = Q(P[v]_B) = Q[v]_{B'} = [v]_{B''}$.

   c. Since QP and I are both matrices for changing from B to B, we conclude from the uniqueness of such a matrix that $QP = I$. Since PQ and I are both matrices for changing from B' to B', we conclude from the uniqueness of such a matrix that $PQ = I$. Therefore, Q is the inverse of P. In particular, P is invertible.

Section 6.6

1.   **a.** The matrix reduces to $\begin{bmatrix} 1 & 0 & -\frac{2}{5} \\ 0 & 1 & -\frac{1}{5} \\ 0 & 0 & 0 \end{bmatrix}$. Hence, $\left\{ \begin{bmatrix} 1 \\ -1 \\ 1 \end{bmatrix}, \begin{bmatrix} 3 \\ 2 \\ 8 \end{bmatrix} \right\}$ is a basis for

the column space, and $\{(5, 0, -2), (0, 5, -1)\}$ is a basis for the row space.

   b. The matrix reduces to $\begin{bmatrix} 1 & 0 & 7 & 0 & 34 \\ 0 & 1 & -2 & 0 & -11 \\ 0 & 0 & 0 & 1 & 2 \\ 0 & 0 & 0 & 0 & 0 \end{bmatrix}$.

Hence, $\left\{ \begin{bmatrix} 1 \\ 2 \\ -1 \\ -2 \end{bmatrix}, \begin{bmatrix} 3 \\ 7 \\ -2 \\ -5 \end{bmatrix}, \begin{bmatrix} 0 \\ 5 \\ 6 \\ 8 \end{bmatrix} \right\}$ is a basis for the column space, and

$\{(1, 0, 7, 0, 34), (0, 1, -2, 0, -11), (0, 0, 0, 1, 2)\}$ is a basis for the row space.

5.   The kernel is all points that project orthogonally onto the origin. This is the line through the origin that is orthogonal to the chosen line. The image is the chosen line.

7.   The kernel is the set of all solutions to the differential equation $y' - y = 0$. Since all solutions are multiples of the exponential function, $\{\exp\}$ is a basis for the solution set.

Section 6.7

1.   **a.** Yes   **b.** Yes   **c.** No   **d.** No

5.   Let w be an arbitrary element of W. Since T is onto, there is $v \in V$ with $T(v) = w$. Since $\{v_1, \ldots, v_n\}$ spans V, we can write $v = r_1 v_1 + \cdots + r_n v_n$. Thus,

$\mathbf{w} = T(\mathbf{v}) = T(r_1\mathbf{v}_1 + \cdots + r_n\mathbf{v}_n) = r_1T(\mathbf{v}_1) + \cdots + r_nT(\mathbf{v}_n)$. Therefore, $\{T(\mathbf{v}_1), \ldots, T(\mathbf{v}_n)\}$ spans W.

9. By Theorem 5.12, an $n \times n$ matrix A is invertible if and only if its reduced row-echelon form has n rows with leading 1s. This is true if and only if the n rows of A form a basis for the row space. In this case, the n-dimensional row space must be all of $\mathbb{R}^n$.

Section 6.8

2. $\mathbb{R}^3$ is the Euclidean space isomorphic to this three-dimensional subspace of $\mathbb{R}^4$.

4. **a.** Since id_V is its own inverse, it is one-to-one and onto. By Exercise 9 of Section 6.1, id_V is linear. Hence, id_V is an isomorphism.

    b. Since T is one-to-one and onto, it has an inverse $T^{-1} : V' \to V$. Since T is linear, T^{-1} is also linear. Since T^{-1} is invertible, it is one-to-one and onto. It follows that T^{-1} is an isomorphism from V' to V.

    c. Since T and T' are one-to-one and onto, Exercises 12 and 13 of Section 6.2 yields that the composition $T' \circ T$ is one-to-one and onto. Since T and T' are linear, Theorem 6.7 yields that $T' \circ T$ is linear. Thus, $T' \circ T$ is an isomorphism.

8. In the magic square $\begin{bmatrix} 8 & 1 & 6 \\ 3 & 5 & 7 \\ 4 & 9 & 2 \end{bmatrix}$, three numbers add up to 15 if and only if they lie in the same row, the same column, or on one of the diagonals. Thus, the fifteen game is isomorphic to tic-tac-toe. In this context, an isomorphism is a function from the nine numbers to the nine squares on a tic-tac-toe board. The function is one-to-one and onto. Also, it preserves winning and losing configurations.

Review Exercises: Chapter 6

3. Let $A, B \in \mathbb{M}(m, n)$ and let $r \in \mathbb{R}$. Then $L(A + B) = P(A + B)Q = P(AQ + BQ) = PAQ + PBQ = L(A) + L(B)$, and $L(rA) = P(rA)Q = rPAQ = rL(A)$.

5. **a.**
$$\begin{array}{ccc} \mathbb{P}_2 & \xrightarrow{\;\;T\;\;} & \mathbb{P}_3 \\ \big[\,\big]_B \downarrow & & \downarrow \big[\,\big]_{B'} \\ \mathbb{R}^3 & \xrightarrow{\;\;\mu_A\;\;} & \mathbb{R}^4 \end{array}$$

    b. $[T(p)]_{B'} = A[p]_B$ for all $p \in \mathbb{P}_3$.

c. $[T(x^3 + 3x - 2)]_{B'} = A[x^3 + 3x - 2]_B = \begin{bmatrix} 1 & 1 & 3 & 1 \\ 0 & 1 & 2 & 1 \\ -1 & 0 & 4 & 0 \end{bmatrix} \begin{bmatrix} 5 \\ -2 \\ 1 \\ 0 \end{bmatrix} =$

$\begin{bmatrix} 6 \\ 0 \\ -1 \end{bmatrix}$. Thus, $T(x^3 + 3x - 2) = 6(x + 1) - 1(x^2 + x) = -x^2 + 5x + 6$.

10. a. If T is onto, then rank $T = \dim \mathbb{M}(2, 3) = 6$. Thus, nullity $T = \dim \mathbb{R}^6 -$ rank $T = 6 - 6 = 0$. By Theorem 6.28, T is one-to-one.

b. If T is one-to-one, then nullity $T = 0$. Thus, $\dim(\text{im } T) = \text{rank } T = \dim \mathbb{R}^6 - \text{nullity } T = 6 - 0 = 6$. By Exercise 21 of Section 3.5, im $T = \mathbb{R}^6$. That is, T is onto.

15. a. Let $\{e_1, e_2\}$ be an orthonormal basis for S. For example, let $e_1 = \left(\frac{1}{\sqrt{2}}, \frac{1}{\sqrt{2}}, 0\right)$ and $e_2 = \left(\frac{1}{\sqrt{3}}, -\frac{1}{\sqrt{3}}, \frac{1}{\sqrt{3}}\right)$. Let $v, w \in \mathbb{R}^3$ and let $r \in \mathbb{R}$. Then $\text{proj}_S(v + w) = ((v + w) \cdot e_1)e_1 + ((v + w) \cdot e_2)e_2 = (v \cdot e_1 + w \cdot e_1)e_1 + (v \cdot e_2 + w \cdot e_2)e_2 = (v \cdot e_1)e_1 + (w \cdot e_1)e_1 + (v \cdot e_2)e_2 + (w \cdot e_2)e_2 = (v \cdot e_1)e_1 + (v \cdot e_2)e_2 + (w \cdot e_1)e_1 + (w \cdot e_2)e_2 = \text{proj}_S(v) + \text{proj}_S(w)$, and $\text{proj}_S(rv) = ((rv) \cdot e_1)e_1 + ((rv) \cdot e_2)e_2 = (r(v \cdot e_1))e_1 + (r(v \cdot e_2))e_2 = r((v \cdot e_1)e_1) + r((v \cdot e_2)e_2) = r((v \cdot e_1)e_1 + (v \cdot e_2)e_2) = r\text{proj}_S(v)$.

b. im $P = S$, so $\{(1, 1, 0), (1, -1, 1)\}$ is a basis for im P.

c. The condition $(a, b, c) \in \ker P$ leads to a system of two equations in three unknowns. A basis $\{(-1, 1, 2)\}$ for the solution set is a basis for $\ker P$.

d. rank P + nullity $P = 2 + 1 = 3 = \dim \mathbb{R}^3$

17. a. Suppose V is infinite-dimensional. We can start with the empty set and apply the Expansion Lemma to produce a linearly independent subset of V with more than dim V' elements. By Exercise 4 of Section 6.7, the images of the vectors in this set form a linearly independent subset of V'. This contradiction of the Comparison Theorem shows that V must be finite-dimensional.

b. Suppose V'' is infinite-dimensional. We can start with the empty set and apply the Expansion Lemma to produce a linearly independent subset of V'' with more than dim V' elements. Since T' is onto, there are elements of V' that T maps to the elements of this set. By Exercise 6 of Section 6.7, these vectors form a linearly independent subset of V'. This contradiction to the Comparison Theorem shows that V'' must be finite-dimensional.

c. By Exercise 4 of Section 6.7, the images under T of the vectors in a basis for V form a linearly independent subset of V'. Since this set spans im T, it is a basis for im T. Thus, $\dim(\text{im } T) = \dim V$. Now dim $V' = \text{rank } T' + \text{nullity } T' = \dim(\text{im } T') + \dim(\ker T') = \dim V'' + \dim(\text{im } T) = \dim V'' + \dim V$.

Section 7.1

3. For $n = 1$, we have $\sum_{k=1}^{1} 1^2 = 1^2 = 1 = \frac{1(1+1)(2 \cdot 1+1)}{6}$. For some fixed $n \geq 1$, assume that $\sum_{k=1}^{n} k^2 = \frac{n(n+1)(2n+1)}{6}$. For this value of n, we have

$$\sum_{k=1}^{n+1} k^2 = \sum_{k=1}^{n} k^2 + (n+1)^2 = \frac{n(n+1)(2n+1)}{6} + (n+1)^2 =$$

$\frac{(n+1)(n(2n+1)+6(n+1))}{6} = \frac{(n+1)(n+2)(2n+3)}{6} = \frac{(n+1)(n+2)(2(n+1)+1)}{6}$. By PMI, the result holds for all natural numbers n.

6. The inequality is false for $n = 2$ and $n = 3$. Let us prove the result for all natural numbers $n \geq 4$. For $n = 4$, we have $4! = 24 \geq 16 = 4^2$. For some fixed $n \geq 4$, assume that $n! \geq n^2$. For this value of n, we have $(n+1)! = (n+1)n! \geq (n+1)n^2 \geq (n+1)(2 \cdot n) = (n+1)(n+n) \geq (n+1)(n+1) = (n+1)^2$. By a modification of PMI, the result holds for all natural numbers $n \geq 4$.

11. Let $n, n+1, n+2$ denote the three consecutive natural numbers. For $n = 1$, we have $1^3 + (1+1)^3 + (1+2)^3 = 36 = 9 \cdot 4$ is divisible by 9. For some fixed $n \geq 1$, assume that $n^3 + (n+1)^3 + (n+2)^3$ is divisible by 9. That is, $n^3 + (n+1)^3 + (n+2)^3 = 9k$ for some integer k. For this value of n, we have $(n+1)^3 + (n+2)^3 + (n+3)^3 = (n+1)^3 + (n+2)^3 + n^3 + 9n^2 + 27n + 27 = 9k + 9n^2 + 27n + 27 = 9(k + n^2 + 3n + 3)$, which is divisible by 9. By PMI, the result holds for all natural numbers n.

14. This result is false; $n = 41$ provides an obvious counterexample.

19. For $n = 1$, we have $\begin{bmatrix} 1-p & p \\ p & 1-p \end{bmatrix}^1 = \begin{bmatrix} \frac{1+(1-2p)^1}{2} & \frac{1-(1-2p)^1}{2} \\ \frac{1-(1-2p)^1}{2} & \frac{1+(1-2p)^1}{2} \end{bmatrix}$.

For some fixed $n \geq 1$, assume that $\begin{bmatrix} 1-p & p \\ p & 1-p \end{bmatrix}^n = \begin{bmatrix} \frac{1+(1-2p)^n}{2} & \frac{1-(1-2p)^n}{2} \\ \frac{1-(1-2p)^n}{2} & \frac{1+(1-2p)^n}{2} \end{bmatrix}$. For this value of n, we have

$\begin{bmatrix} 1-p & p \\ p & 1-p \end{bmatrix}^{n+1} = \begin{bmatrix} 1-p & p \\ p & 1-p \end{bmatrix}^n \begin{bmatrix} 1-p & p \\ p & 1-p \end{bmatrix} =$

$\begin{bmatrix} \frac{1+(1-2p)^n}{2} & \frac{1-(1-2p)^n}{2} \\ \frac{1-(1-2p)^n}{2} & \frac{1+(1-2p)^n}{2} \end{bmatrix} \begin{bmatrix} 1-p & p \\ p & 1-p \end{bmatrix}$. The 11-entry and the

22-entry of this product are both equal to $(1-p)(\frac{1+(1-2p)^n}{2})$

$+p(\frac{1-(1-2p)^n}{2}) = \frac{1}{2}(1-p+p) + \frac{(1-2p)^n}{2}(1-p-p) = \frac{1}{2} + \frac{(1-2p)^n}{2}$

$(1-2p) = \frac{1+(1-2p)^{n+1}}{2}$. The 12-entry and the 21-entry of this product

are both equal to $p(\frac{1+(1-2p)^n}{2}) + (1-p)(\frac{1-(1-2p)^n}{2}) = \frac{1}{2}(p+1-$

$p) + \frac{(1-2p)^n}{2}(p-1+p) = \frac{1}{2} - \frac{(1-2p)^n}{2}(1-2p) = \frac{1-(1-2p)^{n+1}}{2}$.

By PMI, the result holds for all natural numbers n.

21. At $365 \cdot 24 \cdot 60 \cdot 60$ moves per year, $2^{64} - 1 \approx 1.84 \times 10^{19}$ moves will take approximately 5.85×10^{11} years, more than 100 times the current estimates of the age of the earth.

27. a. $x_2 = \frac{1}{2}$, so $\frac{1}{2} \leq x_2 \leq \frac{2}{3}$. For some fixed $n \geq 2$, assume that $\frac{1}{2} \leq x_n \leq \frac{2}{3}$. For this value of n, we have $\frac{3}{2} \leq 1 + x_n \leq \frac{5}{3}$. Hence, $\frac{1}{2} < \frac{3}{5} \leq \frac{1}{1+x_n} \leq \frac{2}{3}$.

Therefore, $\frac{1}{2} \le x_{n+1} \le \frac{2}{3}$. By PMI, the result holds for all natural numbers $n \ge 2$.

b. $|x_3 - x_2| = \frac{2}{3} - \frac{1}{2} = \frac{1}{6} < \frac{16}{81} = \left(\frac{4}{9}\right)^2$. For some fixed $n \ge 2$, assume that $|x_{n+1} - x_n| \le \left(\frac{4}{9}\right)^n$. For this value of n, we have $|x_{n+2} - x_{n+1}| = \left| \dfrac{1}{1 + x_{n+1}} - \right.$

$\left. \dfrac{1}{1 + x_n} \right| = \left| \dfrac{(1 + x_n) - (1 + x_{n+1})}{(1 + x_{n+1})(1 + x_n)} \right| = \left| \dfrac{x_n - x_{n+1}}{(1 + x_{n+1})(1 + x_n)} \right| \le \dfrac{\left(\frac{4}{9}\right)^n}{\frac{3}{2} \cdot \frac{3}{2}} =$

$\left(\frac{4}{9}\right)^{n+1}$. By PMI, the result holds for all natural numbers $n \ge 2$.

c. For natural numbers m and n with $2 \le n \le m$, we have $|x_m - x_n| = |x_m - x_{m-1} + x_{m-1} - \cdots - x_{n+1} + x_{n+1} - x_n| \le |x_m - x_{m-1}| + |x_{m-1} - x_{m-2}| + \cdots + |x_{n+1} - x_n| \le \left(\frac{4}{9}\right)^{m-1} + \left(\frac{4}{9}\right)^{m-2} + \cdots + \left(\frac{4}{9}\right)^n < \dfrac{\left(\frac{4}{9}\right)^n}{1 - \frac{4}{9}} = \frac{9}{5}\left(\frac{4}{9}\right)^n$. Hence, the sequence satisfies the Cauchy criterion for convergence.

d. Let s denote the limit of this sequence. By the continuity of the functions involved, $s = \lim_{n \to \infty} x_{n+1} = \lim_{n \to \infty} \frac{1}{1 + x_n} = \frac{1}{1 + \lim_{n \to \infty} x_n} = \frac{1}{1 + s}$. Since $s > 0$, the quadratic formula gives $s = \frac{-1 + \sqrt{5}}{2}$.

Section 7.2

1. a. 101   **b.** -60 **c.** $\lambda^2 - 3\lambda + 14$ **d.** -12

9. The determinant of a 1×1 matrix consists of one term, which is the single entry from the single row and the single column of the matrix. For some fixed $n \ge 1$, suppose that the result holds for all $(n - 1) \times (n - 1)$ matrices. Let $A = [a_{ij}]$ be an $n \times n$ matrix. Then $\det A = \sum_{j=1}^{n} (-1)^{1+j} a_{1j} \det A_{1j}$. Expand each of the n terms $(-1)^{1+j} a_{1j} \det A_{1j}$. By the induction hypothesis, each of the $(n - 1)!$ terms of $\det A_{1j}$ contains exactly one factor from each row and one factor from each column of A_{1j}. None of these factors comes from the first row or the jth column of A, although all other rows and columns are represented. Thus, when we multiply by $(-1)^{1+j} a_{1j}$, the new terms will contain exactly one factor from each row and one factor from each column of A. By PMI, the result holds for all natural numbers n.

12. a. Let θ be the angle between the vectors (a, b) and (c, d). Then $\cos \theta = \frac{(a,b) \cdot (c,d)}{\|(a,b)\| \|(c,d)\|}$. Multiply through by the denominator and square both sides to obtain $\left((a, b) \cdot (c, d)\right)^2 = \|(a, b)\|^2 \|(c, d)\|^2 (1 - \sin^2 \theta)$. Thus, the square of the area of the parallelogram is

$$\|(a, b)\|^2 \|(c, d)\|^2 \sin^2 \theta = \|(a, b)\|^2 \|(c, d)\|^2 - \left((a, b) \cdot (c, d)\right)^2$$
$$= (a^2 + b^2)(c^2 + d^2) - (ac + bd)^2$$
$$= a^2 c^2 + a^2 d^2 + b^2 c^2 + b^2 d^2 - a^2 c^2$$
$$\quad - 2abcd - b^2 d^2$$
$$= a^2 d^2 - 2abcd + b^2 c^2$$

$$= (ad - bc)^2$$

$$= \left(\det \begin{bmatrix} a & b \\ c & d \end{bmatrix}\right)^2.$$

Therefore, the area is equal to $\left|\det \begin{bmatrix} a & b \\ c & d \end{bmatrix}\right|$.

b. If $ad - bc = 0$, then one or both of the vectors are zero, or (a, b) is $\frac{a}{c}$ or $\frac{b}{d}$ times (c, d), or (c, d) is $\frac{c}{a}$ or $\frac{d}{b}$ times (a, b). In any case, the figure is a degenerate parallelogram, and the area is zero.

c. If the determinant is positive, the angle from (a, b) to (c, d) will be in the counterclockwise direction. If the determinant is negative, the angle will be in the clockwise direction.

d. The absolute value of the determinant of a 3×3 matrix is the volume of the parallelepiped determined by arrows representing the three rows of the matrix. The sign corresponds to the orientation of the three vectors.

Section 7.3

2. $\det \begin{bmatrix} a & b \\ c & d \end{bmatrix} = ad - bc = -(cb - da) = -\det \begin{bmatrix} c & d \\ a & b \end{bmatrix}$. For a fixed $n > 2$,
assume the result holds for all matrices of size $(n-1) \times (n-1)$. Let A be an $n \times n$ matrix. Let B denote the result of interchanging rows p and q of A. Let i be the index of a row other than p and q. Interchanging the rows of the submatrix A_{ij} corresponding to rows p and q of A gives the submatrix B_{ij}. By the induction hypothesis, $\det B_{ij} = -\det A_{ij}$. Therefore, $\det B = \sum_{j=1}^{n} (-1)^{i+j} a_{ij} \det B_{ij} = \sum_{j=1}^{n} (-1)^{i+j} a_{ij}(-\det A_{ij}) = -\sum_{j=1}^{n} (-1)^{i+j} a_{ij} \det A_{ij} = -\det A$. By PMI, the result holds for square matrices of all sizes.

6. Expansion along the first column of a 2×2 matrix $A = [a_{ij}]$ gives $a_{11}a_{22} - a_{21}a_{12} = \det A$. Expansion along the second column gives $-a_{12}a_{21} + a_{22}a_{11} = \det A$. For a fixed $n > 2$, assume the result holds for all $(n - 1) \times (n - 1)$ matrices. The following chain of equalities starts off with the expansion formula for the determinant of an $n \times n$ matrix $A = [a_{ij}]$ along column c. This is followed by the expansion of the determinants of the submatrices along their first rows. The terms are combined and then rearranged to exhibit the expansions of $\det A_{1j}$ along the columns corresponding to column c of A (column $c - 1$ if $j < c$ and column c if $j > c$). The result is the expansion of $\det A$ along the first row.

$$\sum_{i=1}^{n} (-1)^{i+c} a_{ic} \det A_{ic}$$

$$= (-1)^{1+c} a_{1c} \det A_{1c} + \sum_{i=2}^{n} (-1)^{i+c} a_{ic} \left(\sum_{j=1}^{c-1} (-1)^{1+j} a_{1j} \det A_{ic,1j} \right.$$

$$\left. + \sum_{j=c+1}^{n} (-1)^{1+j-1} a_{1j} \det A_{ic,1j} \right)$$

$$= (-1)^{1+c}a_{1c} \det A_{1c} + \sum_{i=2}^{n}\sum_{j=1}^{c-1}(-1)^{i+j+c+1}a_{ic}a_{1j} \det A_{ic,1j}$$

$$+ \sum_{i=2}^{n}\sum_{j=c+1}^{n}(-1)^{i+j+c}a_{ic}a_{1j} \det A_{ic,1j}$$

$$= (-1)^{1+c}a_{1c} \det A_{1c} + \sum_{j=1}^{c-1}(-1)^{1+j}a_{1j}\sum_{i=2}^{n}(-1)^{(i-1)+(c-1)}a_{ic} \det A_{1j,ic}$$

$$+ \sum_{j=c+1}^{n}(-1)^{1+j}a_{1j}\sum_{i=2}^{n}(-1)^{(i-1)+c}a_{ic} \det A_{1j,ic}$$

$$= (-1)^{1+c}a_{1c} \det A_{1c} + \sum_{j=1}^{c-1}(-1)^{1+j}a_{1j} \det A_{1j} + \sum_{j=c+1}^{n}(-1)^{1+j}a_{1j} \det A_{1j}$$

$$= \det A.$$

By PMI, the result holds for square matrices of any size.

8. Since a 1×1 matrix is equal to its transpose, the result holds for $n = 1$. For a fixed $n > 1$, suppose the result holds for all $(n - 1) \times (n - 1)$ matrices. Let $A = [a_{ij}]$ be an $n \times n$ matrix. Then $\det A = \sum_{j=1}^{n}(-1)^{1+j}a_{1j} \det A_{1j} = \sum_{j=1}^{n}(-1)^{1+j}a_{1j} \det(A_{1j})^t = \sum_{j=1}^{n}(-1)^{j+1}A^t_{[j1]} \det(A^t)_{j1} = \det A^t$.

11. $(\det A)(\det A^{-1}) = \det(AA^{-1}) = \det I = 1$. Therefore, $\det A^{-1} = (\det A)^{-1}$.

Section 7.4

1. The probability is 0 that the determinant of a random matrix will exactly equal 0. Therefore, the probability is 1 that the matrix is invertible.

6. **a.** $\det \begin{bmatrix} c & -2 & 2 \\ 1 & 0 & 4 \\ 2 & 1 & -1 \end{bmatrix} = -4c - 16$, and $\det \begin{bmatrix} c & 1 & 2 \\ 1 & 2 & 4 \\ 2 & 0 & -1 \end{bmatrix} = -2c + 1$. Thus,

$f(c) = \dfrac{2c - 1}{4c + 16}$, which is differentiable in its domain $(-\infty, -4)\cup(-4, \infty)$.

b. $\Delta y \approx \dfrac{dy}{dc}\Delta c = \dfrac{9}{4(4.1 + 4)^2}0.002 \approx 0.0000686$. Since $f(4.1) = \frac{2}{9}$, we

have $\dfrac{\Delta y}{y} \approx \dfrac{0.0000686}{0.222} \approx 0.000309$.

c. $\Delta y \approx \dfrac{dy}{dc}\Delta c = \dfrac{9}{4(-4.1 + 4)^2}0.002 = 0.45$. Since $f(-4.1) = 23$, we

have $\dfrac{\Delta y}{y} \approx \dfrac{0.45}{23} \approx 0.0196$.

8. **a.** The columns of X_j are $e_1, \ldots, e_{j-1}, x, e_{j+1}, \ldots, e_n$. Thus, the columns of AX_j are $Ae_1, \ldots, Ae_{j-1}, Ax, Ae_{j+1}, \ldots, Ae_n$. Since Ae_k is the kth column of A, and since $Ax = b$, we have $AX_j = A_j$.

b. When we delete the ith row and the jth column from X_j we obtain either a matrix with a column of 0s (if $i \neq j$) or the identity matrix (if $i = j$). Thus, in the expansion of $\det X_j$ along the jth column, all but the jth term involves a matrix with determinant 0. The jth term is $(-1)^{j+j} x_j \det I_{n-1} = x_j$. Hence, $\det X_j = x_j$.

c. $\det A_j = \det AX_j = (\det A)(\det X_j) = (\det A) x_j$. Provided that A is invertible (so that $\det A \neq 0$), we conclude that $x_j = \dfrac{\det A_j}{\det A}$.

Section 7.5

1. $(2, 5, 1) \times (-1, 2, -1) = \det \begin{bmatrix} \mathbf{i} & \mathbf{j} & \mathbf{k} \\ 2 & 5 & 1 \\ -1 & 2 & -1 \end{bmatrix} = (-5 - 2)\mathbf{i} - (-2 + 1)\mathbf{j} + (4 + 5)\mathbf{k} = (-7, 1, 9)$

5. $\mathbf{v} \times (\mathbf{w} + \mathbf{x}) = \det \begin{bmatrix} \mathbf{i} & \mathbf{j} & \mathbf{k} \\ v_1 & v_2 & v_3 \\ w_1 + x_1 & w_2 + x_2 & w_3 + x_3 \end{bmatrix} = (v_2(w_3 + x_3) - v_3(w_2 + x_2))\mathbf{i} - (v_1(w_3 + x_3) - v_3(w_1 + x_1))\mathbf{j} + (v_1(w_2 + x_2) - v_2(w_1 + x_1))\mathbf{k} = (v_2 w_3 - v_3 w_2)\mathbf{i} - (v_1 w_3 - v_3 w_1)\mathbf{j} + (v_1 w_2 - v_2 w_1)\mathbf{k} + (v_2 x_3 - v_3 x_2)\mathbf{i} - (v_1 x_3 - v_3 x_1)\mathbf{j} + (v_1 x_2 - v_2 x_1)\mathbf{k} = \det \begin{bmatrix} \mathbf{i} & \mathbf{j} & \mathbf{k} \\ v_1 & v_2 & v_3 \\ w_1 & w_2 & w_3 \end{bmatrix} + \det \begin{bmatrix} \mathbf{i} & \mathbf{j} & \mathbf{k} \\ v_1 & v_2 & v_3 \\ x_1 & x_2 & x_3 \end{bmatrix} = (\mathbf{v} \times \mathbf{w}) + (\mathbf{v} \times \mathbf{x})$.

10. $\mathbf{v} \cdot (\mathbf{w} \times \mathbf{x}) = (v_1, v_2, v_3) \cdot \det \begin{bmatrix} \mathbf{i} & \mathbf{j} & \mathbf{k} \\ w_1 & w_2 & w_3 \\ x_1 & x_2 & x_3 \end{bmatrix} = (v_1, v_2, v_3) \cdot ((w_2 x_3 - w_3 x_2)\mathbf{i} - (w_1 x_3 - w_3 x_1)\mathbf{j} + (w_1 x_2 - w_2 x_1)\mathbf{k}) = v_1(w_2 x_3 - w_3 x_2) - v_2(w_1 x_3 - w_3 x_1) + v_3(w_1 x_2 - w_2 x_1) = \det \begin{bmatrix} v_1 & v_2 & v_3 \\ w_1 & w_2 & w_3 \\ x_1 & x_2 & x_3 \end{bmatrix}$. Also, $(\mathbf{v} \times \mathbf{w}) \cdot \mathbf{x} = \mathbf{x} \cdot (\mathbf{v} \times \mathbf{w}) = \det \begin{bmatrix} x_1 & x_2 & x_3 \\ v_1 & v_2 & v_3 \\ w_1 & w_2 & w_3 \end{bmatrix} = -\det \begin{bmatrix} v_1 & v_2 & v_3 \\ x_1 & x_2 & x_3 \\ w_1 & w_2 & w_3 \end{bmatrix} = \det \begin{bmatrix} v_1 & v_2 & v_3 \\ w_1 & w_2 & w_3 \\ x_1 & x_2 & x_3 \end{bmatrix}$.

13. By Theorem 7.16, part g, $\mathbf{v} \cdot (\mathbf{v} \times \mathbf{w}) = \det \begin{bmatrix} v_1 & v_2 & v_3 \\ v_1 & v_2 & v_3 \\ w_1 & w_2 & w_3 \end{bmatrix} = 0$ since the first two rows are identical.

16. If $\mathbf{v} = \mathbf{0}$ or $\mathbf{w} = \mathbf{0}$, then both sides are equal to zero, so the equation holds. Suppose $\mathbf{v} \neq \mathbf{0}$ and $\mathbf{w} \neq \mathbf{0}$. Let $\theta \in [0, \pi]$ be the angle between $\mathbf{v}$ and $\mathbf{w}$. Then $\sin \theta \geq 0$ and $\mathbf{v} \cdot \mathbf{w} = \|\mathbf{v}\| \|\mathbf{w}\| \cos \theta$. Thus, $\|\mathbf{v} \times \mathbf{w}\|^2 = \|\mathbf{v}\|^2 \|\mathbf{w}\|^2 - (\mathbf{v} \cdot \mathbf{w})^2 = \|\mathbf{v}\|^2 \|\mathbf{w}\|^2 - \|\mathbf{v}\|^2 \|\mathbf{w}\|^2 \cos^2 \theta = \|\mathbf{v}\|^2 \|\mathbf{w}\|^2 (1 - \cos^2 \theta) = \|\mathbf{v}\|^2 \|\mathbf{w}\|^2 \sin^2 \theta$. Since none of the factors involved is negative, we can take the square root of both sides of this equation to obtain the desired result, $\|\mathbf{v} \times \mathbf{w}\| = \|\mathbf{v}\| \|\mathbf{w}\| \sin \theta$.

Section 7.6

3. Let A and B be the matrices of the two linear maps relative to the standard basis for $\mathbb{R}^3$. We know that the signs of $\det A$ and $\det B$ correspond to whether the linear maps preserve or reverse orientation. By Theorem 6.13, the matrix of the composition of the linear maps is the product of the two matrices. By Theorem 7.7, $\det AB = (\det A)(\det B) = \det BA$. Thus, the composition is orientation-preserving if both maps preserve orientation or if both maps reverse orientation. The composition is orientation-reversing if one map preserves orientation and the other one reverses orientation.

7.

$$R_x(\mathbf{v}) \times R_x(\mathbf{w}) = \begin{bmatrix} 1 & 0 & 0 \\ 0 & \cos\alpha & -\sin\alpha \\ 0 & \sin\alpha & \cos\alpha \end{bmatrix} \begin{bmatrix} v_1 \\ v_2 \\ v_3 \end{bmatrix}$$

$$\times \begin{bmatrix} 1 & 0 & 0 \\ 0 & \cos\alpha & -\sin\alpha \\ 0 & \sin\alpha & \cos\alpha \end{bmatrix} \begin{bmatrix} w_1 \\ w_2 \\ w_3 \end{bmatrix}$$

$$= \begin{bmatrix} v_1 \\ v_2\cos\alpha - v_3\sin\alpha \\ v_2\sin\alpha + v_3\cos\alpha \end{bmatrix} \times \begin{bmatrix} w_1 \\ w_2\cos\alpha - w_3\sin\alpha \\ w_2\sin\alpha + w_3\cos\alpha \end{bmatrix}$$

$$= \det \begin{bmatrix} \mathbf{i} & \mathbf{j} & \mathbf{k} \\ v_1 & v_2\cos\alpha - v_3\sin\alpha & v_2\sin\alpha + v_3\cos\alpha \\ w_1 & w_2\cos\alpha - w_3\sin\alpha & w_2\sin\alpha + w_3\cos\alpha \end{bmatrix}$$

$$= \big((v_2\cos\alpha - v_3\sin\alpha)(w_2\sin\alpha + w_3\cos\alpha)$$
$$\quad - (v_2\sin\alpha + v_3\cos\alpha)(w_2\cos\alpha - w_3\sin\alpha)\big)\mathbf{i}$$
$$\quad - \big(v_1(w_2\sin\alpha + w_3\cos\alpha) - w_1(v_2\sin\alpha + v_3\cos\alpha)\big)\mathbf{j}$$
$$\quad + \big(v_1(w_2\cos\alpha - w_3\sin\alpha) - w_1(v_2\cos\alpha - v_3\sin\alpha)\big)\mathbf{k}$$

$$= (v_2w_3 - v_3w_2)\mathbf{i}$$
$$\quad - \big((v_1w_2 - v_2w_1)\sin\alpha + (v_1w_3 - v_3w_1)\cos\alpha\big)\mathbf{j}$$
$$\quad + \big((v_1w_2 - v_2w_1)\cos\alpha - (v_1w_3 - v_3w_1)\sin\alpha\big)\mathbf{k}$$

$$= \begin{bmatrix} 1 & 0 & 0 \\ 0 & \cos\alpha & -\sin\alpha \\ 0 & \sin\alpha & \cos\alpha \end{bmatrix} \begin{bmatrix} v_2w_3 - v_3w_2 \\ -v_1w_3 + v_3w_1 \\ v_1w_2 - v_2w_1 \end{bmatrix}$$

$$= R_x(\mathbf{v} \times \mathbf{w})$$

Rotations R_y and R_z work similarly.

8. If you experience difficulty shaking hands with yourself, you can feel confident that your thumbs point in opposite directions.

Review Exercises: Chapter 7

3. **a.** $\det \begin{bmatrix} 0 & 3 & 2 \\ 1 & 5 & 1 \\ -4 & 2 & -1 \end{bmatrix} = 0\det \begin{bmatrix} 5 & 1 \\ 2 & -1 \end{bmatrix} - 3\det \begin{bmatrix} 1 & 1 \\ -4 & -1 \end{bmatrix} +$

$$2\det\begin{bmatrix} 1 & 5 \\ -4 & 2 \end{bmatrix} = 0\cdot(-7) - 3\cdot 3 + 2\cdot 22 = 35$$

b. This matrix is the result of multiplying row 2 of the matrix in part a by 10.

Thus, $\det\begin{bmatrix} 0 & 3 & 2 \\ 10 & 50 & 10 \\ -4 & 2 & -1 \end{bmatrix} = 10\det\begin{bmatrix} 0 & 3 & 2 \\ 1 & 5 & 1 \\ -4 & 2 & -1 \end{bmatrix} = 10\cdot 35 = 350.$

c. This matrix is the result of adding 2 times row 1 to row 2 of the matrix in part b. Thus, $\det\begin{bmatrix} 0 & 3 & 2 \\ 10 & 56 & 14 \\ -4 & 2 & -1 \end{bmatrix} = \det\begin{bmatrix} 0 & 3 & 2 \\ 10 & 50 & 10 \\ -4 & 2 & -1 \end{bmatrix} = 350.$

7. $\det(A + A') = \det\begin{bmatrix} a+a' & b+b' \\ c+c' & d+d' \end{bmatrix} = (a+a')(d+d') - (b+b')(c+c') =$

$(ad - bc) + (ad' - b'c) + (a'd - bc') + (a'd' - b'c') = \det\begin{bmatrix} a & b \\ c & d \end{bmatrix} +$

$\det\begin{bmatrix} a & b' \\ c & d' \end{bmatrix} + \det\begin{bmatrix} a' & b \\ c' & d \end{bmatrix} + \det\begin{bmatrix} a' & b' \\ c' & d' \end{bmatrix}$

11. **a.** $\det\begin{bmatrix} 1 & a & 1 \\ a & -1 & -1 \\ 1 & 1 & 1 \end{bmatrix} = (1-a)(1+a)$, and $\det\begin{bmatrix} 1 & a & 3 \\ a & -1 & -1 \\ 1 & 1 & 1 \end{bmatrix} = (3-a)(1+a).$

Thus, the coefficient matrix is not invertible for $a = 1$ and $a = -1$. In fact, the solution set is empty for $a = 1$ and is infinite for $a = -1$. In particular, f is a well-defined function on $(-\infty, -1) \cup (-1, 1) \cup (1, \infty)$.

b. $f(a) = \frac{3-a}{1-a}$ is differentiable at all points of the domain of f.

c. $f'(a) = \frac{2}{(1-a)^2}$, and $f'(2.15) \approx 1.51$. Thus, $\Delta y \approx f'(2.15)\Delta x \approx 1.51(0.004) = 0.00605.$

12. **a.** $(\mathbf{i} \times \mathbf{i}) \times \mathbf{j} = \mathbf{0} \times \mathbf{j} = \mathbf{0}$ is not equal to $\mathbf{i} \times (\mathbf{i} \times \mathbf{j}) = \mathbf{i} \times \mathbf{k} = -\mathbf{j}.$

b. $(\mathbf{u} \times \mathbf{v}) \times \mathbf{w} = (u_2v_3 - u_3v_2, -u_1v_3 + u_3v_1, u_1v_2 - u_2v_1) \times (w_1, w_2, w_3) =$

$\det\begin{bmatrix} \mathbf{i} & \mathbf{j} & \mathbf{k} \\ u_2v_3 - u_3v_2 & -u_1v_3 + u_3v_1 & u_1v_2 - u_2v_1 \\ w_1 & w_2 & w_3 \end{bmatrix} = (-u_1v_3w_3 + u_3v_1w_3 -$

$u_1v_2w_2 + u_2v_1w_2)\mathbf{i} - (u_2v_3w_3 - u_3v_2w_3 - u_1v_2w_1 + u_2v_1w_1)\mathbf{j} + (u_2v_3w_2 - u_3v_2w_2 + u_1v_3w_1 - u_3v_1w_1)\mathbf{k} = ((u_1w_1 + u_2w_2 + u_3w_3)v_1 - (v_1w_1 + v_2w_2 + v_3w_3)u_1)\mathbf{i} + ((u_1w_1 + u_2w_2 + u_3w_3)v_2 - (v_1w_1 + v_2w_2 + v_3w_3)u_2)\mathbf{j} + ((u_1w_1 + u_2w_2 + u_3w_3)v_3 - (v_1w_1 + v_2w_2 + v_3w_3)u_3)\mathbf{k} = (\mathbf{u}\cdot\mathbf{w})\mathbf{v} - (\mathbf{v}\cdot\mathbf{w})\mathbf{u}.$

c. $\mathbf{u} \times (\mathbf{v} \times \mathbf{w}) = -(\mathbf{v} \times \mathbf{w}) \times \mathbf{u} = -(\mathbf{v}\cdot\mathbf{u})\mathbf{w} + (\mathbf{w}\cdot\mathbf{u})\mathbf{v} = (\mathbf{u}\cdot\mathbf{w})\mathbf{v} - (\mathbf{u}\cdot\mathbf{v})\mathbf{w}.$

Section 8.1

2. **a.** $f(x) = e^{2x}$

b. For any real number r, the function defined by e^{rx} is an eigenvector of D associated with the eigenvalue r.

9. **a.** $\begin{bmatrix} 2 & 0 \\ 0 & 5 \end{bmatrix}$

b. Use the numbers $\lambda_1, \ldots, \lambda_r$ as entries on the diagonal with repetitions if necessary to fill out the diagonal. Use zeros for all the other entries.

11. **a.** $\det(\lambda I - A) = \det \begin{bmatrix} \lambda - a & -b \\ -b & \lambda - a \end{bmatrix} = (\lambda - a)^2 - b^2 = (\lambda - a - b)(\lambda - a + b)$. Thus, the eigenvalues of A are $a + b$ and $a - b$.

b. The eigenvalues of A are distinct if and only if $b \neq 0$.

15. **a.** We are given that $A(\mathbf{v}) = \lambda \mathbf{v}$. Thus, $A^2 \mathbf{v} = A(A\mathbf{v}) = A(\lambda \mathbf{v}) = \lambda(A\mathbf{v}) = \lambda(\lambda \mathbf{v}) = \lambda^2 \mathbf{v}$. That is, $\mathbf{v}$ is an eigenvector of A associated with the eigenvalue λ^2.

b. For any natural number n, we have $A^n \mathbf{v} = \lambda^n \mathbf{v}$. For $n = 1$, we have $A^1 \mathbf{v} = A\mathbf{v} = \lambda \mathbf{v} = \lambda^1 \mathbf{v}$. For some fixed value of $n \geq 1$, assume that $A^n \mathbf{v} = \lambda^n \mathbf{v}$. For this value on n, we have $A^{n+1} \mathbf{v} = (AA^n)\mathbf{v} = A(A^n \mathbf{v}) = A(\lambda^n \mathbf{v}) = \lambda^n(A\mathbf{v}) = \lambda^n(\lambda \mathbf{v}) = \lambda^{n+1} \mathbf{v}$. By PMI, the result holds for all natural numbers n.

c. By Exercise 10, A^{-1} exists if and only if 0 is not an eigenvalue of A. In particular, if A is invertible, then $\lambda \neq 0$. Multiply both sides of the equation $A\mathbf{v} = \lambda \mathbf{v}$ by A^{-1} to obtain $\mathbf{v} = I\mathbf{v} = (A^{-1}A)\mathbf{v} = A^{-1}(A\mathbf{v}) = A^{-1}(\lambda \mathbf{v}) = \lambda(A^{-1}\mathbf{v})$. Hence, $A^{-1}\mathbf{v} = \lambda^{-1}\mathbf{v}$. That is, $\mathbf{v}$ is an eigenvector of A associated with the eigenvalue λ^{-1}. For some fixed value of $n \geq 1$, assume that $A^{-n}\mathbf{v} = \lambda^{-n}\mathbf{v}$. For this value of n, we have $A^{-(n+1)}\mathbf{v} = (A^{-1}A^{-n})\mathbf{v} = A^{-1}(A^{-n}\mathbf{v}) = A^{-1}(\lambda^{-n}\mathbf{v}) = \lambda^{-n}(A^{-1}\mathbf{v}) = \lambda^{-n}(\lambda^{-1}\mathbf{v}) = \lambda^{-(n+1)}\mathbf{v}$. By PMI, for any natural numbers n, $\mathbf{v}$ is an eigenvector of A^{-n} associated with the eigenvalue λ^{-n}.

Section 8.2

1. **a.** $A = I^{-1}AI$

b. $A \sim B$ means $A = P^{-1}BP$ for some nonsingular matrix P. Multiply both sides on the left by P and on the right by P^{-1} to obtain $PAP^{-1} = P(P^{-1}BP)P^{-1} = (PP^{-1})B(PP^{-1}) = IBI = B$. Thus, $B = (P^{-1})^{-1}AP^{-1}$.

c. We are given that $A = P^{-1}BP$ and $B = Q^{-1}CQ$ for some nonsingular matrices P and Q. Substituting the expression for B into the expression for A gives $A = P^{-1}(Q^{-1}CQ)P = (P^{-1}Q^{-1})C(QP) = (QP)^{-1}C(QP)$.

3. If A is similar to I, then for some nonsingular matrix P we have $A = P^{-1}IP = I$.

9. $\begin{bmatrix} 1 & 0 \\ 0 & 1 \end{bmatrix}$ and $\begin{bmatrix} 1 & 1 \\ 0 & 1 \end{bmatrix}$ have the same characteristic polynomial $(\lambda - 1)^2$. However, by Exercise 3, only the identity matrix is similar to I. In particular, the second matrix is not similar to I.

12. Let $A = \begin{bmatrix} a & b \\ c & d \end{bmatrix}$.

a. $\det(\lambda I - A) = \det \begin{bmatrix} \lambda - a & -b \\ -c & \lambda - d \end{bmatrix} = (\lambda - a)(\lambda - d) - bc = \lambda^2 - (a + d)\lambda + (ad - bc) = \lambda^2 - (\operatorname{tr} A)\lambda + \det A$

b. $A^2 - (\text{tr } A)A + (\det A)I = \begin{bmatrix} a & b \\ c & d \end{bmatrix}\begin{bmatrix} a & b \\ c & d \end{bmatrix} - (\text{tr } A)\begin{bmatrix} a & b \\ c & d \end{bmatrix} +$

$(\det A)\begin{bmatrix} 1 & 0 \\ 0 & 1 \end{bmatrix} = \begin{bmatrix} a^2 + bc & ab + bd \\ ac + cd & bc + d^2 \end{bmatrix} - \begin{bmatrix} a^2 + ad & ab + bd \\ ac + cd & ad + d^2 \end{bmatrix} +$

$\begin{bmatrix} ad - bc & 0 \\ 0 & ad - bc \end{bmatrix} = \begin{bmatrix} 0 & 0 \\ 0 & 0 \end{bmatrix}$

15. **a.** We are given that $T(\mathbf{v}) = \lambda\mathbf{v}$ for a nonzero vector $\mathbf{v} \in \mathbb{R}^n$. Now, $A[\mathbf{v}]_B = [T(\mathbf{v})]_B = [\lambda\mathbf{v}]_B = \lambda[\mathbf{v}]_B$. Since the coordinate vector function is one-to-one, $[\mathbf{v}]_B$ is nonzero. Thus, $[\mathbf{v}]_B$ is an eigenvector of A associated with the eigenvalue λ.

b. We are given that $A\mathbf{v} = \lambda\mathbf{v}$ for a nonzero vector $\mathbf{v} \in \mathbb{R}^n$. Now, $[T(L_B(\mathbf{v}))]_B = A[L_B(\mathbf{v})]_B = A\mathbf{v} = \lambda\mathbf{v} = \lambda[L_B(\mathbf{v})]_B = [\lambda L_B(\mathbf{v})]_B$. Since the coordinate vector function is one-to-one, $T(L_B(\mathbf{v})) = \lambda L_B(\mathbf{v})$. Since L_B is one-to-one, $L_B(\mathbf{v})$ is nonzero. Thus, $L_B(\mathbf{v})$ is an eigenvector of T associated with the eigenvalue λ.

Section 8.3

2. **a.** $P = \begin{bmatrix} 1 & 2 & 4 \\ 0 & 1 & 2 \\ 1 & -2 & -3 \end{bmatrix}, P^{-1} = \begin{bmatrix} 1 & -2 & 0 \\ 2 & -7 & -2 \\ -1 & 4 & 1 \end{bmatrix}$, and $P^{-1}AP = \begin{bmatrix} 2 & 0 & 0 \\ 0 & -2 & 0 \\ 0 & 0 & 3 \end{bmatrix}$.

b. $P = \begin{bmatrix} 1 & 0 & 3 \\ 0 & 1 & 1 \\ 1 & -4 & 0 \end{bmatrix}, P^{-1} = \begin{bmatrix} 4 & -12 & -3 \\ 1 & -3 & -1 \\ -1 & 4 & 1 \end{bmatrix}$, and $P^{-1}AP = \begin{bmatrix} 1 & 0 & 0 \\ 0 & 1 & 0 \\ 0 & 0 & -1 \end{bmatrix}$.

c. The only eigenvalue has a one-dimensional eigenspace. Since $\mathbb{R}^3$ does not have a basis of eigenvectors, the matrix is not diagonalizable.

d. The only eigenvalue has a two-dimensional eigenspace. Since $\mathbb{R}^3$ does not have a basis of eigenvectors, the matrix is not diagonalizable.

e. The two eigenvalues each have one-dimensional eigenspaces. Since $\mathbb{R}^3$ does not have a basis of eigenvectors, the matrix is not diagonalizable.

f. The only eigenvalue has a one-dimensional eigenspace. Since $\mathbb{R}^3$ does not have a basis of eigenvectors, the matrix is not diagonalizable.

5. **a.** $\det(\lambda I - A) = (\lambda - (1-p))^2 - p^2 = (\lambda - (1-p) - p)(\lambda - (1-p) + p) = (\lambda - 1)(\lambda - 1 + 2p)$. So the eigenvalues are 1 and $1 - 2p$.

b. $\left\{\begin{bmatrix} 1 \\ 1 \end{bmatrix}\right\}$ is a basis for the eigenspace associated with $\lambda = 1$, and $\left\{\begin{bmatrix} -1 \\ 1 \end{bmatrix}\right\}$ is a basis for the eigenspace associated with $\lambda = 1 - 2p$.

c. The change of basis matrix is $P = \begin{bmatrix} 1 & -1 \\ 1 & 1 \end{bmatrix}$. So $P^{-1}AP = \begin{bmatrix} \frac{1}{2} & \frac{1}{2} \\ -\frac{1}{2} & \frac{1}{2} \end{bmatrix}$

$\begin{bmatrix} 1-p & p \\ p & 1-p \end{bmatrix}\begin{bmatrix} 1 & -1 \\ 1 & 1 \end{bmatrix} = \begin{bmatrix} \frac{1}{2} & \frac{1}{2} \\ -\frac{1}{2} & \frac{1}{2} \end{bmatrix}\begin{bmatrix} 1 & -1+2p \\ 1 & 1-2p \end{bmatrix} = \begin{bmatrix} 1 & 0 \\ 0 & 1-2p \end{bmatrix}$.

d. $\begin{bmatrix} 1-p & 1 \\ p & 1-p \end{bmatrix}^n = \begin{bmatrix} 1 & -1 \\ 1 & 1 \end{bmatrix}\begin{bmatrix} 1 & 0 \\ 0 & 1-2p \end{bmatrix}^n\begin{bmatrix} \frac{1}{2} & \frac{1}{2} \\ -\frac{1}{2} & \frac{1}{2} \end{bmatrix}$

$\qquad = \begin{bmatrix} 1 & -1 \\ 1 & 1 \end{bmatrix}\begin{bmatrix} 1 & 0 \\ 0 & (1-2p)^n \end{bmatrix}\begin{bmatrix} \frac{1}{2} & \frac{1}{2} \\ -\frac{1}{2} & \frac{1}{2} \end{bmatrix}$

$\qquad = \begin{bmatrix} 1 & -1 \\ 1 & 1 \end{bmatrix}\begin{bmatrix} \frac{1}{2} & \frac{1}{2} \\ \frac{-(1-2p)^n}{2} & \frac{(1-2p)^n}{2} \end{bmatrix}$

$\qquad = \begin{bmatrix} \frac{1+(1-2p)^n}{2} & \frac{1-(1-2p)^n}{2} \\ \frac{1-(1-2p)^n}{2} & \frac{1+(1-2p)^n}{2} \end{bmatrix}$

7. a. Let **v** be an eigenvalue associated with λ. Then λ**v** = A**v** = A^2**v** = λ^2**v**. Since **v** is nonzero, $\lambda = \lambda^2$ by Exercise 12 of Section 1.3. Thus, $\lambda = 1$ or $\lambda = 0$.

b. For any **v** $\in \mathbb{R}^n$, let **w** = A**v** and **x** = **v** − A**v**. Then **v** = **w** + **x**. Also, A**w** = $A(A$**v**$) = A^2$**v** = A**v** = **w** = 1**w**, so **w** $\in E_A(1)$; and A**x** = $A($**v** − A**v**$) = A$**v** − A^2**v** = A**v** − A**v** = **0** = 0**x**, so **x** $\in E_A(0)$.

c. It follows that the union of a basis for $E_A(1)$ and a basis for $E_A(0)$ spans $\mathbb{R}^n$. By Theorem 8.14, the union is linearly independent. By Theorem 8.13, A is diagonalizable.

Section 8.4

1. a. $(P + Q)^t = P^t + Q^t = P + Q$

b. $(rP)^t = rP^t = rP$.

c. $(P^{-1})^t P = (P^{-1})^t P^t = (PP^{-1})^t = I^t = I$. By Theorem 5.13, $(P^{-1})^t = P^{-1}$.

d. $\begin{bmatrix} 1 & 2 \\ 2 & 0 \end{bmatrix}\begin{bmatrix} 1 & 1 \\ 1 & 0 \end{bmatrix} = \begin{bmatrix} 3 & 1 \\ 2 & 2 \end{bmatrix}$ is not symmetric.

e. $(P^t)^t = P^t$

5. If A is orthogonal, then $(\det A)^2 = (\det A^t)(\det A) = \det(A^t A) = \det I = 1$. Thus, $\det A = 1$ or $\det A = -1$.

7. a. $\begin{bmatrix} \frac{1}{\sqrt{2}} & -\frac{1}{\sqrt{2}} \\ \frac{1}{\sqrt{2}} & \frac{1}{\sqrt{2}} \end{bmatrix}\begin{bmatrix} 2 & 3 \\ 3 & 2 \end{bmatrix}\begin{bmatrix} \frac{1}{\sqrt{2}} & \frac{1}{\sqrt{2}} \\ -\frac{1}{\sqrt{2}} & \frac{1}{\sqrt{2}} \end{bmatrix} = \begin{bmatrix} -1 & 0 \\ 0 & 5 \end{bmatrix}$

b. $\begin{bmatrix} \frac{1}{\sqrt{2}} & \frac{1}{\sqrt{2}} \\ \frac{1}{\sqrt{2}} & -\frac{1}{\sqrt{2}} \end{bmatrix}\begin{bmatrix} -1 & 4 \\ 4 & -1 \end{bmatrix}\begin{bmatrix} \frac{1}{\sqrt{2}} & \frac{1}{\sqrt{2}} \\ \frac{1}{\sqrt{2}} & -\frac{1}{\sqrt{2}} \end{bmatrix} = \begin{bmatrix} 3 & 0 \\ 0 & -5 \end{bmatrix}$

c. $\begin{bmatrix} \frac{2}{\sqrt{5}} & 0 & -\frac{1}{\sqrt{5}} \\ \frac{2\sqrt{5}}{15} & \frac{\sqrt{5}}{3} & \frac{4\sqrt{5}}{15} \\ \frac{1}{3} & -\frac{2}{3} & \frac{2}{3} \end{bmatrix}\begin{bmatrix} -2 & -2 & 2 \\ -2 & 1 & -4 \\ 2 & -4 & 1 \end{bmatrix}\begin{bmatrix} \frac{2}{\sqrt{5}} & \frac{2\sqrt{5}}{15} & \frac{1}{3} \\ 0 & \frac{\sqrt{5}}{3} & -\frac{2}{3} \\ -\frac{1}{\sqrt{5}} & \frac{4\sqrt{5}}{15} & \frac{2}{3} \end{bmatrix} = \begin{bmatrix} -3 & 0 & 0 \\ 0 & -3 & 0 \\ 0 & 0 & 6 \end{bmatrix}$

d.
$$\begin{bmatrix} \frac{1}{\sqrt{2}} & 0 & -\frac{1}{\sqrt{2}} \\ -\frac{1}{\sqrt{6}} & \frac{2}{\sqrt{6}} & -\frac{1}{\sqrt{6}} \\ \frac{1}{\sqrt{3}} & \frac{1}{\sqrt{3}} & \frac{1}{\sqrt{3}} \end{bmatrix} \begin{bmatrix} 1 & 1 & 1 \\ 1 & 1 & 1 \\ 1 & 1 & 1 \end{bmatrix} \begin{bmatrix} \frac{1}{\sqrt{2}} & -\frac{1}{\sqrt{6}} & \frac{1}{\sqrt{3}} \\ 0 & \frac{2}{\sqrt{6}} & \frac{1}{\sqrt{3}} \\ -\frac{1}{\sqrt{2}} & -\frac{1}{\sqrt{6}} & \frac{1}{\sqrt{3}} \end{bmatrix}$$

$$= \begin{bmatrix} 0 & 0 & 0 \\ 0 & 0 & 0 \\ 0 & 0 & 3 \end{bmatrix}$$

e.
$$\begin{bmatrix} \frac{1}{\sqrt{2}} & -\frac{1}{\sqrt{2}} & 0 & 0 \\ \frac{1}{\sqrt{2}} & \frac{1}{\sqrt{2}} & 0 & 0 \\ 0 & 0 & \frac{1}{\sqrt{2}} & \frac{1}{\sqrt{2}} \\ 0 & 0 & \frac{1}{\sqrt{2}} & -\frac{1}{\sqrt{2}} \end{bmatrix} \begin{bmatrix} 2 & 3 & 0 & 0 \\ 3 & 2 & 0 & 0 \\ 0 & 0 & -1 & 4 \\ 0 & 0 & 4 & -1 \end{bmatrix} \begin{bmatrix} \frac{1}{\sqrt{2}} & \frac{1}{\sqrt{2}} & 0 & 0 \\ -\frac{1}{\sqrt{2}} & \frac{1}{\sqrt{2}} & 0 & 0 \\ 0 & 0 & \frac{1}{\sqrt{2}} & \frac{1}{\sqrt{2}} \\ 0 & 0 & \frac{1}{\sqrt{2}} & -\frac{1}{\sqrt{2}} \end{bmatrix}$$

$$= \begin{bmatrix} -1 & 0 & 0 & 0 \\ 0 & 5 & 0 & 0 \\ 0 & 0 & 3 & 0 \\ 0 & 0 & 0 & -5 \end{bmatrix}$$

f. The matrix is similar to itself, a diagonal matrix.

Section 8.5

2. **a.** By the Fundamental Theorem of Calculus, $\frac{d}{x}\int_c^x a(t)\,dt = a(x)$. Thus, $y'(x) = a(x)\exp\left(\int_c^x a(t)\,dt\right) = a(x)y$.

b. Suppose y_1 satisfies the differential equation. Then

$$\left(\frac{y_1}{y}\right)' = \frac{y_1'y - y_1 y'}{y^2} = \frac{a(x)y_1 y - y_1 a(x)y}{y^2} = 0.$$

Thus, $\frac{y_1}{y} = k$, a constant. That is, $y_1 = ky$.

c. If y_1 also satisfies the initial condition, then $y_0 = y_1(c) = k\exp\left(\int_c^c a(t)\,dt\right) = k\exp 0 = k$.

6. Since $\det A$ is the sum and difference of products of entries of A, the composition of the determinant function with the matrix function A is continuous. Since $\det A(t_0) \neq 0$, there is an open interval containing t_0 such that for any t in the interval, $\det A(t) \neq 0$. For t in this interval then, $A(t)$ is invertible.

9. $(A^n)' = \sum_{k=1}^n A^{k-1}A'A^{n-k}$. For $n = 1$, we have $(A^1)' = A' = IA'I = A^0A'A^0 = \sum_{k=1}^1 A^{k-1}A'A^{1-k}$. For some fixed $n \geq 1$, suppose the result holds. For this value of n, we have $(A^{(n+1)})' = (A^n)A + A^nA' = \left(\sum_{k=1}^n A^{k-1}A'A^{n-k}\right)A + A^nA' = \sum_{k=1}^n A^{k-1}A'A^{n+1-k} + A^nA'A^0 = \sum_{k=1}^{n+1} A^{k-1}A'A^{n+1-k}$. By PMI, the result holds for all natural numbers n.

11. For $t < 0$, we have $W(t) = \det\begin{bmatrix} 1 & 0 \\ 0 & 0 \end{bmatrix} = 0$, and for $t \geq 0$, we have $W(t) = \det\begin{bmatrix} 0 & 0 \\ 0 & 1 \end{bmatrix} = 0$. Suppose $a\varphi_1 + b\varphi_2 = 0$. Evaluate at $t = -1$ to obtain

$$\mathbf{0} = a\boldsymbol{\varphi}_1(-1) + b\boldsymbol{\varphi}_2(-1) = a\begin{bmatrix} 1 \\ 0 \end{bmatrix} + b\begin{bmatrix} 0 \\ 0 \end{bmatrix} = \begin{bmatrix} a \\ 0 \end{bmatrix}. \text{ Thus, } a = 0. \text{ Evaluate}$$

at $t = 1$ to obtain $\mathbf{0} = a\boldsymbol{\varphi}_1(1) + b\boldsymbol{\varphi}_2(1) = a\begin{bmatrix} 0 \\ 0 \end{bmatrix} + b\begin{bmatrix} 0 \\ 1 \end{bmatrix} = \begin{bmatrix} 0 \\ b \end{bmatrix}.$ Thus,

$b = 0.$

15. **a.** $e^{-15}\boldsymbol{\varphi}_1(t) + 2e^{12}\boldsymbol{\varphi}_2(t) = \begin{bmatrix} e^{5(t-3)} \\ 2e^{-4(t-3)} \end{bmatrix}$

b. $\frac{3}{5}\boldsymbol{\varphi}_1(t) + \frac{4}{5}\boldsymbol{\varphi}_2(t) = \begin{bmatrix} \frac{6}{5}e^{-t} + \frac{4}{5}e^{4t} \\ -\frac{3}{5}e^{-t} - \frac{2}{5}e^{4t} \end{bmatrix}$

c. $-\boldsymbol{\varphi}_1(t) + 2\boldsymbol{\varphi}_2(t) - \boldsymbol{\varphi}_3(t) = \begin{bmatrix} 2e^{-t} - e^{-2t} \\ e^{-2t} \\ -e^{t} + 2e^{-2t} \end{bmatrix}$

d. $\frac{1}{2}e^{2}\boldsymbol{\varphi}_1(t) + \frac{3}{2}e^{2}\boldsymbol{\varphi}_2(t) + 3e^{4}\boldsymbol{\varphi}_3(t) = \begin{bmatrix} e^{2(t+1)} + 3e^{4(t+1)} \\ 3e^{2(t+1)} - 3e^{4(t+1)} \\ 4e^{2(t+1)} + 3e^{4(t+1)} \end{bmatrix}$

Review Exercises: Chapter 8

3. **a.** $16 + 3 + 2 + 13 = 34, 5 + 10 + 11 + 8 = 34, 9 + 6 + 7 + 12 = 34$, and
$4 + 15 + 14 + 1 = 34.$

b. $A\begin{bmatrix} 1 \\ 1 \\ 1 \\ 1 \end{bmatrix} = \begin{bmatrix} 34 \\ 34 \\ 34 \\ 34 \end{bmatrix}$, so $\begin{bmatrix} 1 \\ 1 \\ 1 \\ 1 \end{bmatrix}$ is an eigenvector associated with the eigenvalue 34.

c. The other eigenvalues are 0, 8, and -8.

d. $\begin{bmatrix} 1 \\ -3 \\ 3 \\ -1 \end{bmatrix}$ is an eigenvector associated with 0, $\begin{bmatrix} 2 \\ -1 \\ 0 \\ -1 \end{bmatrix}$ is an eigenvector associ-

ated with 8, and $\begin{bmatrix} 1 \\ 0 \\ 1 \\ -2 \end{bmatrix}$ is an eigenvector associated with -8.

e. Let $P = \begin{bmatrix} 1 & 2 & 1 & 1 \\ -3 & -1 & 0 & 1 \\ 3 & 0 & 1 & 1 \\ -1 & -1 & -2 & 1 \end{bmatrix}$. Then $P^{-1} = \begin{bmatrix} -\frac{1}{16} & -\frac{3}{16} & \frac{3}{16} & \frac{1}{16} \\ \frac{7}{16} & -\frac{3}{16} & -\frac{5}{16} & \frac{1}{16} \\ -\frac{1}{16} & \frac{5}{16} & \frac{3}{16} & -\frac{7}{16} \\ \frac{1}{4} & \frac{1}{4} & \frac{1}{4} & \frac{1}{4} \end{bmatrix}$, and

$P^{-1}AP = \begin{bmatrix} 0 & 0 & 0 & 0 \\ 0 & 8 & 0 & 0 \\ 0 & 0 & -8 & 0 \\ 0 & 0 & 0 & 34 \end{bmatrix}.$

10. **a.** $T(e^x) = e^{x-1} = e^{-1}e^x$

b. $T(e^{2x}) = e^{2(x-1)} = e^{-2}e^{2x}$. Thus, e^{2x} is an eigenvector associated with the eigenvalue e^{-2}.

c. For any real number $r > 0$, we have $T(e^{(-\ln r)x}) = e^{(-\ln r)(x-1)} = e^{(\ln r)}e^{(-\ln r)x} = re^{(-\ln r)x}$. Thus, r is an eigenvalue of T.

d. $T(\sin(\pi x)) = \sin(\pi(x-1)) = (-1)\sin(\pi x)$. Thus, $\sin(\pi x)$ is an eigenvector of T associated with the eigenvalue -1.

e. Write an arbitrary negative real number as $-r$ for $r > 0$. Then
$T(e^{(-\ln r)x}\sin(\pi x)) = e^{(-\ln r)(x-1)}\sin(\pi(x-1)) = e^{\ln r}e^{(-\ln r)x}(-1)\sin(\pi x)$
$= (-r)e^{(-\ln r)x}\sin(\pi x)$. Thus, $-r$ is an eigenvalue of T.

f. If $T(f) = 0f$, then $f(x-1) = 0$ for all $x \in \mathbb{R}$. Thus, f is the zero function.

14. **a.** The eigenvalues of A are 0 and 49.

b. $\left\{ \begin{bmatrix} 2 \\ 0 \\ -1 \end{bmatrix}, \begin{bmatrix} 0 \\ 3 \\ 1 \end{bmatrix} \right\}$ is a basis for $E_A(0)$, and $\left\{ \begin{bmatrix} 3 \\ -2 \\ 6 \end{bmatrix} \right\}$ is a basis for $E_A(49)$.

c. $\left\{ \begin{bmatrix} \frac{2}{\sqrt{5}} \\ 0 \\ -\frac{1}{\sqrt{5}} \end{bmatrix}, \begin{bmatrix} \frac{2}{\sqrt{245}} \\ \frac{15}{\sqrt{245}} \\ \frac{4}{\sqrt{245}} \end{bmatrix}, \begin{bmatrix} \frac{3}{7} \\ -\frac{2}{7} \\ \frac{6}{7} \end{bmatrix} \right\}$

d. $P = \begin{bmatrix} \frac{2}{\sqrt{5}} & \frac{2}{\sqrt{245}} & \frac{3}{7} \\ 0 & \frac{15}{\sqrt{245}} & -\frac{2}{7} \\ -\frac{1}{\sqrt{5}} & \frac{4}{\sqrt{245}} & \frac{6}{7} \end{bmatrix}$, and $P^{-1}AP = \begin{bmatrix} 0 & 0 & 0 \\ 0 & 0 & 0 \\ 0 & 0 & 49 \end{bmatrix}$.

16. **a.** $\begin{bmatrix} 2 \\ 1 \end{bmatrix}$ is an eigenvector associated with the eigenvalue 6, and $\begin{bmatrix} -3 \\ 2 \end{bmatrix}$ is an eigenvector associated with the eigenvalue -1.

b. $\varphi_1(t) = e^{6t}\begin{bmatrix} 2 \\ 1 \end{bmatrix} = \begin{bmatrix} 2e^{6t} \\ e^{6t} \end{bmatrix}$ and $\varphi_2(t) = e^{-t}\begin{bmatrix} -3 \\ 2 \end{bmatrix} = \begin{bmatrix} -3e^{-t} \\ 2e^{-t} \end{bmatrix}$ define functions that form a basis for the solution space.

c. $-\frac{1}{7}\varphi_1(t) + \frac{4}{7}\varphi_2(t) = \begin{bmatrix} -\frac{2}{7}e^{6t} - \frac{12}{7}e^{-t} \\ -\frac{1}{7}e^{6t} + \frac{8}{7}e^{-t} \end{bmatrix}$

Index